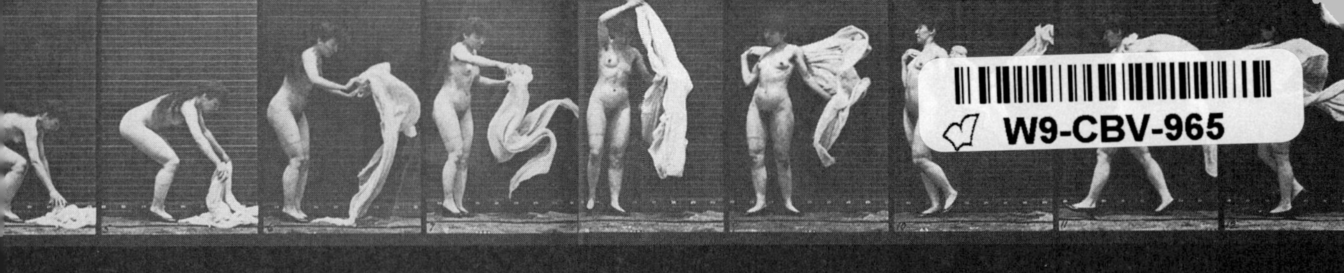

NEW DIGITAL EDITION

The Animation Book

A complete guide to animated filmmaking—from flip-books to sound cartoons to 3-D animation

Kit Laybourne

Preface by *George Griffin* Introduction by *John Canemaker*

Three Rivers Press • New York

Published by Three Rivers Press, 201 East 50th Street, New York, New York 10022. Member of the Crown Publishing Group.

Original edition published by Crown Publishers, Inc., in 1979. First revised paperback edition printed in 1988.

Random House, Inc. New York, Toronto, London, Sydney, Auckland
www.randomhouse.com

THREE RIVERS PRESS and colophon are trademarks of Crown Publishers, Inc.

Printed in the United States of America

Design by June Bennett-Tantillo

Library of Congress Cataloging-in-Publication Data

Laybourne, Kit.

The animation book : a complete guide to animated filmmaking—from flip-books to sound cartoons to 3-D animation / by Kit Laybourne.—Rev. ed.

Includes index.

1. Animation (Cinematography). I. Title.

TR897.5.L39 1998

778.5'347—dc21 97–32774

ISBN 0-517-88602-2

10 9 8 7 6 5 4

DEDICATION

THIS BOOK IS FOR
GERALDINE
MY ONE & ONLY

CONTENTS

ACKNOWLEDGMENTS

During the 1995–1996 academic year, I sat in for my friend John Canemaker, teaching his Advanced Animation seminar at New York University's Tisch School of the Arts. It would be a pathetic understatement to say that I learned more than I taught that year. There was no way I could stay ahead of that particular undergraduate/ graduate class. Everyone seemed to have some sort of computer competency, and we collectively pooled our knowledge in what became a joint exploration of how the new digital tools and techniques served the traditional ones of film animation. No fewer than five of my students that year have contributed to this volume. We are colleagues.

Others have played large roles in the seat-of-the-pants education that has issued this digital edition of *The Animation Book.* I am particularly grateful to a number of visionaries at NYU's Center for Advanced Technology who worked with me on a CD-ROM about animation. This volume owes a large debt to Jack Schwartz, Cynthia Allen, Ken Perlin, Daniel Russo, Clilly Casteglia, Dan Moss, and Dan Schrecker.

I have always been a sucker for new technologies. For just over a year, starting in early 1996, I took a job as head of the Digital Production Studio at Tele-TV Media, a company set up by three of the Bell systems to explore the fusion of TV, telephones, gaming, and the Internet. I am particularly indebted to Japhet Asher, George Escobar, Jane Buckwalter, Mike Lasky, and the studio's team of interactive designers, artists, and software engineers.

A very special thanks to Michael Dougherty, who was one of my students at NYU and then joined my staff at Tele-TV. Michael is the embodiment of the kind of reader whom I hope this book will nourish—a multimedia animator par excellence, equally at home with a down-shooter or a hard drive.

It was at Tele-TV that I started working with Trista Gladden. She and I retreated into the Hollywood Hills for the six months in which text and pictures came together. I could not have completed this project without Trista's intelligence, tenacity, and good company.

The twenty years between the first edition and the current one I spent as a producer, writer, and director of animation and live-action productions. My treasured partner was Eli Noyes, who has indelibly shaped my creative understanding of animation. Together we built Noyes and Laybourne Enterprises, an independent animation studio in a loft that occupied two floors of a TriBeCa building in downtown New York City. Eli and I found opportunities to explore virtually every tool and form of animation, thanks to great gigs in broadcast graphics, advertising, and TV programming. Some of our best projects are sampled in these pages, and we herewith salute the staff members who created them with us: Kathy Minton, ChiChi Pierce, Brian O'Connell, Stuart Dworeck, Diane Fazio, Shawn Cuddy, Jeff Schon, Adam Bernstine, and others. As I look back, I am proud at how we pushed our collective envelopes —along with those of many independent animator friends—through *Liquid Television,* a TV series developed in the early 1990s for MTV. Thanks to Judy McGrath, Abby Terkuhle, and John Payson at MTV and to our soul mates at San Francisco's (Colossal) Pictures, including Japhet Asher, Prudence Fenton, Drew Takahashi, and Lawrence Wilkinson .

Rounding up production stills and film-grabs is a huge task in a book like this. Here is a long list of folks who went out of their way to locate visual references or to help in other ways: Joe Ahlbum, Kim Arnold, Doug Aberle, Valeska Bailey, Sarah Baisley, Dave Bastian, Alison Brown, Carlos Casso, Brad deGraf, Patric DuGuette, Tony Eastman, George Evelyn, Jeff Fino, Prudence Fenton, Steve Gold, John Hays, Traci Johnson, Todd Kessler, George Lacroix, Dave Masters, Brian O'Connell, George O'Dwyer, Darwin Peachey, Jeff Schon, Robert Scull, Henry Selick, Sandy Serling, Carl Stavens, Cricket Stettinius, Drew Takahashi, Abby Terkuhle, Terry Thoren, Susan Trembley, Mike Turoff, Peter Wallach, Chris Wedge, Mark Welch, and Gunnar Wille. All were generous in wanting to excite and inform the next generation of animators.

Thanks to Patrick Sheehan, my editor at Crown Publishers, who provided as much help as Jake Goldberg did for the original edition. Thanks also to June Bennett-Tantillo for her design and production efforts for this book.

On the following pages, I have packed together short bios of all those who collaborated with me in writing chapters and case studies. I am hugely indebted to each of them. As you work through the volume, return here to see who did what. You'll find they are an amazingly diverse group.

While I delight in the good company of those who are acknowledged here (and in all those independent animators whose work remains from the first edition), I'm afraid I alone must take the heat for any mistakes, inaccuracies, omissions, and other grievous faults.

Art Bell was a cofounder of Alias Research, a 3-D computer graphics software firm, where he was responsible for product design and development of 3-D software used in numerous motion pictures, including *Terminator 2, The Abyss, The Lion King, Jurassic Park,* and *Beauty and The Beast,* as well as for use in automotive design at most car companies, including Honda, Porsche, General Motors, Ford, Volvo, and BMW. Currently, Art lives in Vermont, mixing hockey with furniture restoration and software design at American Happware, a start-up focused on making insanely easy, cool software.

John Canemaker became a professional animator in 1973 after acquiring a Master of Fine Arts degree in film from New York University, and has been the head of his own production company since 1981. He has designed and directed a variety of award-winning animated films and commercials. His writing has included six critically acclaimed books on animation and over a hundred articles for periodicals such as *The New York Times, Time,* and *Film Comment*. Currently, Canemaker is a tenured Associate Professor and Chair of the Animation Program at New York University Tisch School of the Arts.

Jan Cox is Director of International Animation at Manga Entertainment, Inc., and compiles several annual animated short film festivals that tour the world. Her early animation influences are *Beany and Cecil*, *Rocky and Bullwinkle*, and *Mighty Mouse*. Jan worked in advertising for nine years before becoming solely devoted to animation in 1992, when she joined Mellow Manor Productions in San Diego. She also pursues an interest and lifelong love of fine art, which she feels animation to be a part of.

Michael Dougherty created two animated short films, *Crayons* and *Season's Greetings,* while a student at NYU's Animation Program. These works have gone on to tour the festival circuit and win awards from the Academy of Television Arts and Sciences, the Hamptons International Film Festival, the Chicago Underground Film Festival, the Annecy Animation Festival, and the Academy of Motion Picture Arts and Sciences. In addition to his personal work, he has also created animation for MTV, the Cartoon Network, and Tele-TV. Michael currently works as an animator for Nickelodeon's *Blue's Clues*.

Dsquared, Inc., is a New York City–based multimedia studio specializing in animation and other "time-based design." The history of the company's founders, Dan Schrecker and Daniel Moss, makes Dsquared perfectly suited to write this book's chapter on 3-D computer animation. Schrecker's predigital animation experience, which included the use of clay, cels, time lapse, and line drawings, combined with Moss's background in architecture, sculpture, and 3-D design, allows Dsquared to approach computer animation with a firm grasp on classic technique as well as a solid technical understanding, partly gained from NYU's Interactive Telecommunications Program.

George D. Escobar pursues a dual career in technology and entertainment. He has developed interactive TV and Internet products as vice president of Product Technologies for Tele-TV and Bell Atlantic. Escobar has also worked in Hollywood as a creative executive, an assistant to the director, story analyst, and producing intern for several networks and production companies. Currently, George is producing an independent animated feature and developing an Intranet animation support system. He has written five feature screenplays and is a graduate of the American Film Institute's Center for Advanced Film and Television Studies.

Trista Gladden graduated from Cornell University with a B.A. in English and traveled abroad to assist in developing a junior high school English language curriculum for the Department of Education in Nagano, Japan. Following this, she moved to L.A. and began working at Direct Cinema Ltd., a documentary and short film distributor, where she first became interested in animation after screening Pixar's *Tin Toy, Red's Dream,* and *Luxo, Jr*. Trista then moved to Tele-TV Media, where she was wrangled into working on *The Animation Book* with Kit Laybourne. Currently, Trista is assistant producer for *Hank the Cowdog*, a project in development for Nickelodeon.

Athomas Goldberg is a research scientist at New York University's Media Research Laboratory, where, along with Professor Ken Perlin and others, he is developing IMPROV, a system for authoring real-time behavior-based 3-D environments. He has presented this work in a number of published papers and demonstrations at numerous conferences and symposia, including ACM SIGGRAPH and the American Association for Artificial Intelligence. Before coming to the Media Research Lab, Athomas spent several years as an illustrator, theatrical lighting and set designer, and performance artist, after studying film production at NYU's Tisch School of the Arts.

George Griffin studied political science at Dartmouth and has been doing animation in New York City since the late 1960s. His diverse work includes abstract musical studies, reflective essays on process, cartoon narratives, television commercials, and flip-books. His latest film, *A Little Routine*, illustrating an extended dialogue with his six-year-old daughter, won the Unicef Award at the 1995 Ottawa Animation Festival.

Derek Lamb, born in the U.K., made his first films with the National Film Board of Canada during the early 1960s. He has worked extensively in commercial, educational, and experimental production in the United States, Canada, and Europe. For several years he taught animation and screenwriting at Harvard and McGill Universities. He was Executive Director of Animation, National Film Board of Canada, from 1976 to 1981. His work has received numerous international prizes, including a 1980 Academy Award for *Every Child*, as writer and producer. Recent productions include codirector and writer on *Meena*, a television series on the rights of girls, through UNICEF South Asia, and writer-producer for *PEEP*, a television series for preschool, in collaboration with Kai Pindal and Jeffrey Schon.

Randy Lowenstein is an independent animator and filmmaker. He received his MFA in film at New York University, where he specialized in traditional as well as 2-D computer animation. In 1996, he joined Jennifer Taylor to form a commercial production company that specializes in combining traditional and computer animation.

Eli Noyes has been making moving images of one sort or another since he was in his teens, many decades ago. He has worked as an animator, designer, documentary filmmaker, cameraman, art director, director of commercials, principal in several production companies, executive producer, and creative director. He has always experimented with new kinds of production: animation of all sorts, blends of live action and animation, bluescreen, video, and computer graphics. Eli has studied architecture, acting, directing, drawing, screenwriting, and dance. He is married with two children and is currently developing programs for Disney/ABC Cable Networks that anticipate the eventual fusion of the television with the computer.

Greg Pair works at AMPnyc Deluxe Animation, the creative triumvirate of the dashing Michael Adams, street-smart Ted Minoff, and techno-wiz Greg Pair. This new production studio on the New York City scene has received praise and won awards for its animated work since its first year in business. Together with their fearless consortium of freelancing friends, the members of AMPnyc are prepared to run the gauntlet toward the goal of creating outstanding commercial animation as well as originating fresh concepts for animated series.

Dave Palmer, the animation director of Nickelodeon's *Blue's Clues*, co-designed Blue and developed the show's method of animation combining traditional forms, Photoshop, and After Effects. He received his master's degree in film animation from NYU and got his undergraduate degree, a B.F.A. in film, photography, and visual arts, from Ithaca College. Dave began his relationship with *Blue's Clues* as a freelance animator after getting an opportunity to work on the pilot episode.

John Payson drew cartoons for the *Harvard Lampoon* in college. After graduating, he worked at MTV: Music Television for over twelve years. John produced many of MTV's animated IDs and wrote and directed many promos and short films for the channel, including the original short *Joe's Apartment*. He was also supervising producer for the first two seasons of MTV's animated series *Liquid Television*. In 1996, John wrote and directed a full-length version of *Joe's Apartment*, his first feature film. John lives in a cockroach-infested apartment in New York City.

Jason Porter does not have a computer science or art degree. It has even been rumored that he has only fifteen units of college. What he always has had is an interest in visual aesthetics and an aptitude for technology. That and the willingness to work 100 hour weeks has vaulted him to the lofty position of head of digital media at Wild Brain, Inc., where he oversees digital production for television, feature films, commercials, and new media projects. Jason was born and raised in San Francisco and thinks L.A. is a quaint little town.

Jennifer Taylor is a twenty-six-year-old independent filmmaker who employs experimental animation and motion graphic techniques in her films. She produces a large portion of her personal and commercial work on a high-end Macintosh workstation, using Adobe After Effects, Photoshop, and Premiere. Jennifer received a B.A. in film from Bard College, and an M.F.A. in film from New York University. She currently lives and works in New York City.

Jane White joined Protozoa in November of 1996 as the director of development and an executive producer. She has been working in the area of new media and entertainment since 1987 when she was a cofounder of ABC News Interactive. After leaving ABC in 1992, she was executive director of children's products for Paramount Interactive. Before returning to the Bay Area in 1996 she was director of development at Viacom New Media. In this capacity she identified appropriate CD-ROM development for Nickelodeon.

Muybridge meets Wild Brain: Back when photography was in its infancy, an American scientist named Eadweard Muybridge harnessed the new medium's ability to "freeze" motion. Rigging multiple still cameras to shoot a sequence of photos at precise intervals, Muybridge conducted a multiyear investigation into the nature of movement itself. His seminal volumes, *Animal Locomotion* (1887), *The Human Figure in Motion* (1899), and *Animals in Motion* (1901) have been treasured resources since the birth of animation and have since been republished by Dover Books.

In the cover for the digital revision of *The Animation Book*, animators and designers at Wild Brain, a top San Francisco animation studio, took successive frames from a few of Muybridge's studies and rendered them using the range of animation techniques discussed in this volume. President and cofounder John Hays directed the design efforts with coordination by Jeff Fino, Sarah Shen, and Diane Tateishi. The photography was done by Carter Tomassi, and all the stylistic variations sampled in the cover art were created by Wild Brain staffers, including Chris Carter, Roger Dondis, Lee Hong, Amber MacLean, JT O'Neal, Cindy Ng, Jason Porter, Vaughn Ross, Robin Steele, and Dave Thomas.

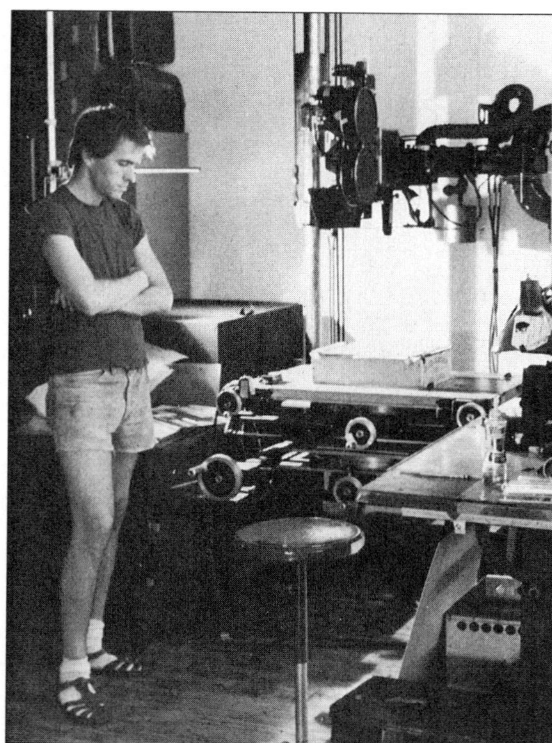

A

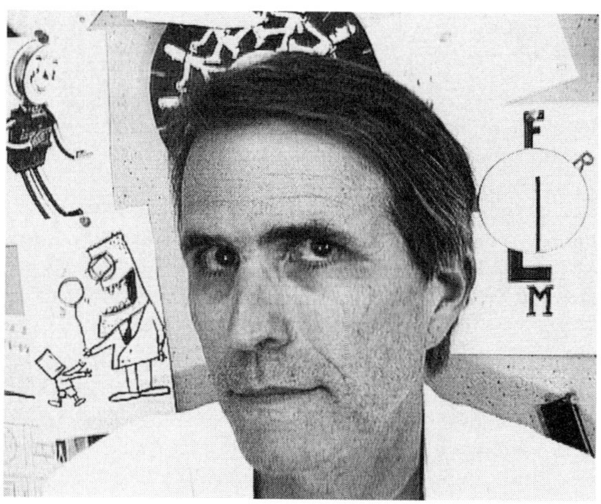

B

George Griffin stands out among a handful of people who have built and sustained a robust tradition within American Independent Animation. Portrait *(A)* shows George circa 1979, about the time he wrote the original Preface for this book. Portrait *(B),* circa 1997, was taken for this new edition, and was submitted with a wry note that displays the wonderful way George can "read" an image. Here's a snippet: "What to say about that first photo (with the distance and alleged wisdom afforded by time)? I think it captures the myth of the independent filmmaker who has wrested an artifact of the industrial age from the assembly-line studio and tamed it into producing quirky, personal films. It also emphasizes the technical, artisanal nature of animation. And maybe it suggests a certain art-worker heroism, somewhat like that Eakins painting of steelworkers outside their factory. Who was it said freedom of the press belongs to those who own the press?

"But like all myths, that image also distills and filters out certain important truths, e.g., you don't really need that monstrous iron contraption to make a film (a Bolex on a copy stand works fine) and, in fact, it may inhibit your experimental energies by making everything too smooth or predictable. This is probably the main reason I resisted a 'George in 1997' in front of my Mac monitor, preferring instead to reference the sketches on the bulletin board."

Three frames from Griffin films: *(C), Flying Fur,* 1981; *(D), New Fangled,* 1991; and *(E), A Little Routine,* 1994. A videotape collection of George Griffin's films and digital projects is available under the title "Griffiti," available for $40 (prepaid) from Metropolis Graphics at 349 West 4th Street, New York, NY 10014.

C

D

E

PREFACE

A new method of making animated films has surfaced in the last decade and with it a new generation of artists who use the medium primarily for self-expression. The new animators assume direct responsibility for nearly every aspect of the filmmaking process: concept, drawing, shooting, even camera stand construction. This reclamation of creative authority contrasts sharply with the impersonal assembly-line production system of the studio cartoon industry and returns animation to its original experimental impulse as embodied in the work of Windsor McCay, Emile Cohl, Hans Richter, and Oskar Fischinger.

GEORGE GRIFFIN
NEW YORK CITY, 1978

A lot has happened in the twenty years since the appearance of Kit Laybourne's *The Animation Book*. The cartoon feature boom, witty and satirical TV series, edgy television graphics and advertising, even the Internet's current jerky-mation—all indicate a major resurgence of graphic expression. For those of us genetically disposed to creating alternate worlds of unreal moving pictures, these are interesting times.

As with any artistic explosion, there are inevitable gains, losses, and realignments. Consider the expansion of the animator's task from marking and scoring on paper to directly manipulating a wide variety of material (clay, sand, puppets) to complex digital layering, impossible before computer technology. The dream of melding the spontaneous twitch of design with the methodical flow of choreography now seems a reality: Digital charcoal will smudge only when and where you want. Animators can manipulate time and space before choosing a delivery medium (film, video, CD-ROM, flip-book), thereby freeing animation from a particular medium's limitations. But this brave new world isn't without troubling side effects (eyestrain, repetitive stress syndrome, problems associated with being sedentary, expensive-obsessive-upgraditis), and you may miss those charcoal-smudged hands.

Despite the enormous upheavals in technology, the independent animator's artistic role remains essentially the same: to draw time, to construct a model of ideas and emotions, using any means available. Whether you draw a sequence of abstract doodles or construct a cartoon narrative, your chief goal should be to resist the conventional formulas by seeking your own voice and style.

Animators are drawn to this art by the wide range of pleasures it offers. For some it's the complete sense of control, down to the very elemental interval of frame or pixel. Others love the ability to mime and act through a character, make a story, put on a grandiose show. Yet others are attracted to the meditative processes, akin to those involved in cooking and pottery making, that will ultimately yield a performance of kinetic art.

The book you are now holding is a versatile tool. It can act as compass, road map, and manual; if you are lucky, you might actually get lost, and invent or discover a new route. Use it to learn how to make things move. Then whatever movies now dancing in your head can find expression for the rest of us.

GEORGE GRIFFIN
NEW YORK CITY, 1998

A

B

C

D

John Canemaker writes with great insight and warmth about the art of animation and the artistry of great animators. Here are covers from four of his books: *(A), Felix—The Twisted Tale of the World's Most Famous Cat (*Da Capo, 1991); *(B), Before the Animation Begins (*Hyperion, 1996); *(C), Windsor McCay—His Life and Art (*Abbeville, 1987); and *(D), Tex Avery* (Turner, 1996). *Images courtesy John Canemaker.*

INTRODUCTION

For close on twenty years, Kit Laybourne's *The Animation Book* has been *the* animation book: the most engaging, fun, and informative guide to self-expression through moving-image making ever published.

Two generations of students, as well as countless teachers, professionals, and fans of animation—the twentieth century's most significant new art form—have found Kit's book an essential source of encouragement, stimulation, and inspiration. Now comes the updated edition, which is welcome indeed, and timely, to say the least.

As ever, the book bursts with essential information regarding traditional methods of ''making 'em move": cels, stop-motion puppets, clay, pixilation, flip-books, paint-on-glass, sand, cutouts, and drawing directly onto film, among other techniques. The multitude of processes and theories are again profusely illustrated (a distinctive feature of the original book) using production stills, original art, frame blowups, and now screen-grabs off computer monitors. The new material—about a third of the book—covers the hows and whys of emerging computer-based tools and techniques.

Whether discussing the new, old, or hybrid worlds of animation, *The Animation Book* remains a vehicle to broaden knowledge, sharpen perceptions, and stimulate the imagination. Its text continues to be playful, user-friendly, and accessible: a guiding voice rather than a dictating one. As such, it aptly reflects the personality and educational values of the book's author.

It has been my pleasure to call Kit Laybourne a friend and colleague for over twenty years. In observing his impressive career as teacher, producer, director, and author, I have come to think of him primarily as a teacher on a grand scale. In his television productions, CD-ROM projects, writings, workshops, lectures, and the classes he has conducted from New York to Singapore (and points in between), Kit always seeks to inform, guide, and raise artistic consciousness. His enthusiasm for learning and encouraging the human spirit to fulfill its artistic potential shines through, whatever the medium.

The heart and soul of *The Animation Book* remains Kit's passion for animation as a form of personal expression. It is an ideal that he and I share, and one that I try to instill in my students as chair of New York University's Tisch School of the Arts Animation Program.

It was, therefore, lucky for both me and my students when Kit agreed to teach my Advanced Animation production class for one year while I was on a recent sabbatical. I was more than a little relieved; Kit's presence and our mutual belief that animation—whether you use pencils or pixels—is a great and unique medium through which artists can express something personal and meaningful ensured that there would be a continuity of philosophy in the classroom.

My students were delighted. They found in Kit personally the qualities he has put in his book: accessibility, friendliness, inspiration. In turn, Kit's year at NYU Tisch influenced his thinking about animation and the new technology, and he noted he was "wonderfully pushed by our students, who've grown up with computers in the way we grew up with TV." I was pleased and proud to learn that several of the Advanced Animation students contributed to the updated sections of *The Animation Book*.

The book arrives at exactly the right moment to introduce and/or explore with readers old and new the brave new world of digital animation and how it fits with traditional animation. In the last decade, the merging of old and new methods of creating moving imagery has profoundly affected the medium as both industry and art. Thus, happily, the life of *The Animation Book* has been extended, and its valuable information, wonderful can-do spirit, and love of all kinds of animation will survive into the new millennium to challenge and influence animators yet to be born.

JOHN CANEMAKER
ANIMATOR, AUTHOR, TEACHER
NEW YORK, 1998

E

F

G

John Canemaker is also an independent animator whose work includes *(E)*, PBS's *What Do Children Think of When They Think of the Bomb?* © The Icarus Company, 1983; *(F)*, feature animation in *The World According to Garp* © Warner Bros., 1981; and *(G)*, a personal film entitled *Bottom's Dream* © John Canemaker, 1983. *Images courtesy John Canemaker.*

WELCOME TO A SPANKING NEW EDITION

You've got your hands on the first book that covers two generations of animation. There is *animated filmmaking*—a venerable art form that can trace its origins back to the birth of movies. And now there is *digital animation*—an energizing hybrid that applies new computer technology to the same fundamental process of creating movement from a series of motionless images.

This edition of *The Animation Book* represents a major makeover. There are over eighty pages of new text, with hundreds of illustrations and frame-grabs. Seven chapters are wholly new and cover topics that didn't even exist when the book was first published in 1979. Only three of the technique chapters remain untouched by computers. Everywhere you'll find images that represent new sets of tools and new creative domains we've barely discovered.

Yet—as you shall see—the old and the new worlds of animation exist quite comfortably together. This is because the prodigious strength of computers has been successfully engineered to support the traditional aesthetic and production tasks that reside in making any piece of animation. Thus you will find comments about digital approaches side by side with traditional topics such as storyboarding, pixilation, and ink and paint.

Wherever one looks in the world of animation, the lines between old and new are blurry. The data that make up movies and TV increasingly come to us encoded as a bunch of zeros and ones. Computers can read and rearrange these signals with blistering speed. Everything meets and mixes in a digital pot. Drawings that have been shot onto a piece of celluloid motion picture film are seamlessly composited with drawings that have been fashioned on a computer, pixel by pixel. And the stew gets richer yet with the digital mixing of images from a number of sources including new-generation videocams, old movies, and electronic character generators.

This volume will help you sort out what parts of animation are best done on film and what parts are best done on computers. As often as not, either choice will work—although the artistic vision of the piece is inevitably affected by that choice. The overlap of traditional and digital is less confusing than you might expect. Many animators switch easily from sheets of paper and pencils to computers and digital drawing tablets. So integrated are the new forms of animation with the old that the phrases used to describe them are often used interchangeably. By learning about the venerable film technology, you won't be wasting time. In fact, the animator always gains

A

B

C

D

Kit Laybourne *(A),* critiquing a storyboard in a class at NYU's Tisch School of the Arts. Although Kit taught for almost ten years early in his career, the recent year at NYU was a one-time gig. He took over the Advanced Animation Seminar during the 1995–1996 school year when John Canemaker was on sabbatical. Kit is an independent producer with a special interest in animation. He is currently working on a project, *Hank the Cowdog,* that combines animated characters within a live-action world that has been digitally manipulated. *Photo by Karl Staven.*

In *(B), (C), and (D),* students in the NYU animation program promoted their annual screening of undergraduate and graduate animation with posters.

in understanding and control when he or she explores both traditional and digital alternatives. And this dual exploration is exactly what you will find throughout this new edition of *The Animation Book.*

SAME OLD HIDDEN AGENDAS

Like the first and second editions, this book aims to do more than provide you with a comprehensive, up-to-date introduction to the world of animation. Be warned that it will try to coax you into becoming an animator. Sometimes you will be asked to grab a pencil and work right within these pages. At various places, you'll find "projects"—short assignments for animated pieces you can undertake on your own. Everywhere, you'll find plenty of examples and case studies.

In preparing materials to explain about computer animation, I made the conscious choice to stick with a focus on animation with an independent spirit—by which I mean works that are made by artists. These works range from personal shorts to creator-driven films and television series like *Toy Story* and *Rugrats.* Each of these were made not strictly for commerce, but in order to convey someone's vision. It's my deep conviction that when learning about an art form, one should keep as close as possible to the vitality, vision, and passion of the individual artist. You get good taste by tasting good things.

HOW TO WORK THE BOOK

The volume is divided into three sections. It begins with Part I: Fundamentals. The purpose here is to give you a warm-up. You'll find activities in drawing, generating ideas, and playing with various animation toys and principles. These early pages also provide a quick preview of the equipment you will be using when you produce your own films and computer-generated animations.

Part II: Techniques is about production. Here you will find an amazing spectrum of distinct categories of animation. Each technique is unique. Each is discussed thoroughly. Each teaches something about the categories that follow it. This section of the book also contains information about basic production steps that are common to all the different techniques: working with sound, storyboarding, and production planning.

Part III: Tools is a catalogue. It describes and gives rough costs for all the equipment you will need to undertake your own work—be it film- or computer-based. In this new edition of *The Animation Book,* emphasis is given to the "low end" of computer technology. This is in keeping with my not-so-secret agenda of encouraging readers to make their own animations. For those who already know how to animate in film and video formats, this section should expand creative choices using computer tools. You'll probably want to consult the hardware and software chapters while you are working through Part II. For those who are familiar with digital animation but know squat about traditional film tools, Part III will yield a quick and indispensable overview. The last chapter carries annotated listings of many places where you can get additional help—from books, periodicals, and Web sites to professional associations, festivals, and training programs.

Within all three parts, examples and case studies attempt to break down the overall material into manageable pieces. I hope the extensive use of cross-references will increase your tendency to browse and to use the book in a circular, self-programmed way. Please be encouraged to move through these pages freely, even impulsively.

KIT'S CHICKEN SOUP

Before you launch into this new edition, I want to offer some psychological nourishment. Hear me out as I suggest a single destination and some strategies to get you there.

Style is the ultimate achievement. I don't think there can be a more difficult goal or a more rewarding one than developing a unique vision through an art form like animation. We're not talking here about a quick trip to the corner store. Finding your style is a cross-country trip. It's a protracted journey into the self. Some survival tips:

- *Start from Strength.* The best place to begin is where your interest is strongest. Read along until you come to something you find yourself eager to try, and then do it. But when you begin to lose interest, move on. Don't agonize when you want to quit a particular project. Move from strength to strength, cranking up your momentum and self-confidence.
- *Build a "Studio."* Claim a space that can become the symbolic home for your work as an animator. There is no need to be fancy. When I started my own work I allocated a single drawer to the new enterprise. It was a modest studio, to be sure. But it gave me an important "psychological space." For many today, a computer hard disk is a good place to build your own studio. Commit to it a specific amount of energy, time, and money.
- *Develop a Project Mentality.* Imagine you were setting out to be an oil painter. No one could expect a masterpiece on your first canvas. Making film or digital animation is every bit as difficult to master as oil painting. So give yourself a break. Just as a painter begins with a sketch pad, allow yourself the psychic freedom to develop your skills at a reasonable pace. This book suggests different "projects" that have been carefully formulated. Try as many of them as you can. But don't be overly self-critical. Don't judge yourself by first efforts. Remember, it's not the product of your work that counts as much as the process of your learning.
- *Keep an Animator's Diary.* I strongly recommend that you consistently keep track of what you are learning and how you are feeling. Any system will do. An easy one is to designate a special notebook as the place to keep a running record of your projects and ideas. There are lots of technical notes and lots of sketches that you will want to preserve.
- *Take the Long View.* Gauging your abilities as an independent animator ought to be a very long-term goal. This book will identify all the individual skills that comprise the art and craft of animation. There are lots of things to master and no one has ever been able to do all of them. Walt Disney didn't. None of the independent animators mentioned in this book would say that they have total facility and mastery. I certainly don't have them. You won't either.

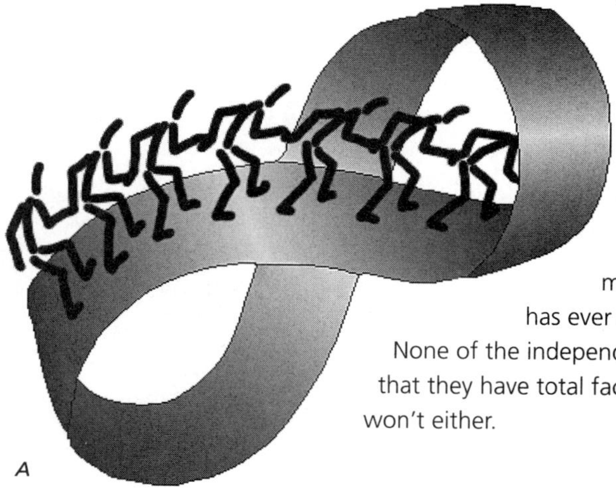

A

And it's not important that you do. What is important, however, is for you to make ongoing assessments of areas where you are strong and areas where your skills are less developed. This is critical because it will allow you to guide your own development, to explore widely, and to innovate. You must be able to gauge your own skills, interests, goals, and talents if you are to realize the ultimate goal of finding your own style.

KIT LAYBOURNE
JANUARY 1998

B

Möbius interface: Kit Laybourne is a sucker for new technology. In collaboration with the Voyager Company and then the Center for Digital Multimedia at NYU, he completed a full working prototype for a game-based CD-ROM about animation. Titled *The Animation Kit,* this project had four goals: (1) to develop a new genre of digital publishing that grabs the attention of computer-gaming young people and moves them toward material of substance in the arts and humanities; (2) to explore the history, as well as the aesthetic and creative processes, of animation; (3) to provide access to the works and personalities of outstanding independent animators; and (4) to challenge interactive and multimedia publishers by setting new standards for the pedagogy of computer-based learning.

Like lots of visionary projects, this one has never quite gotten off the ground. Yet.

The Möbius strip pictured here is the main interface for *The Animation Kit.*

Figure *(A)* shows the "Gulag" of marching stick figures, who tread slowly on both the inside and the outside of this infinite Möbius surface.

(B) shows the home page for the CD-ROM. By clicking on one of the stick figures, the user is taken to sections on flip books, time-lapse animation, motion graphics, rotoscoping (the drawing was to be replaced by a photo of Kit in his shorts), cutout animation, cameraless techniques, and character animation. Each section would provide a simple animation engine to let users create their own works and send them to friends via the Web. (When we undertook the demo on flip-books, the "simple" engine proved very expensive to create, because there was no standard application we could plug into.)

In any event, the tribe of stick figures created for the CD-ROM project have found life in these pages. Although extremely simple, the drawings display the structural elements of human anatomy and they also show how just a few expressive lines can suggest the broad range of human activities and emotions.

Viewmaster: The cover of the original edition of *The Animation Book* was adapted from a work titled *Viewmaster,* created by George Griffin. The original film is composed of eight disks, including the one seen here. With the camera framed tight on one of the images, the cycle of runners was shot two frames at a time on the animation stand, with each disk carefully repositioned around the same center point in successive frames. The resulting illusion became that of watching as a parade of eclectic runners—of different scales and different styles—moved across the frame. Only at the end of the film was there a pull-back that revealed that what one had been watching was, in fact, a form of the familiar toy, the Viewmaster disk that fascinated an earlier generation of children.

The new cover for this digital edition echoes the same subject matter—people in motion and, in a different way, samples the range of expressive techniques available for today's generation of animators. *Courtesy George Griffin and Metropolis Graphics.*

I · FUNDAMENTALS

The goal of this book is to help you discover your own unique style as an independent animator. That's a lofty objective but an achievable one. It requires work. It requires experimentation. In the long run, finding a style means understanding a lot about yourself as well as about the thing you've set out to accomplish. This makes it worth doing well. And it makes it fun.

Animation requires a few innate skills and talents. Without these you might as well close the book right now. As it turns out, however, the essential skills are *not* the ones you'd expect.

An animator must be fascinated with the way things move. He or she must be a keen observer of the world. An animator must be something of an actor. The ability to give a cartoon figure character depends on the ability to feel a character within yourself. If a film is to work, what happens must happen with purpose. Actions must be motivated. So every animator has to be an actor, even if it's just a closet actor.

Like designers, fine artists, and inventors, animators need to enjoy the process of identifying a problem and then working out its solution. Sloppy thinking and lazy execution guarantee failure. An animator has got to know how to think.

Finally, everyone who has ever done it will agree that the art of animation requires liberal measures of patience, precision, perseverance, and pride. Please note, however, that these are not innate skills as much as they are personality traits, or elements of character.

It's all in the head. If you want to animate, if you've always been fascinated by cartoons and cameras and computers, if you like making things move, then you've got all the requisite talents. Don't just take my word for it. Make up your own mind about what it takes by exploring the notions and the problems that follow.

1 ▪ Basic Skills

The biggest single misconception about animation is that you need to be an artist to do it, that you need to know how to draw. To disprove this, just skim through the pages of this book and you'll find frame enlargements from many films that don't have a single drawn line in them. You can be a prizewinning animator and never touch a pencil or paintbrush.

For some people, the ability to express themselves in visual terms, in images, has been blocked. Most usually this is expressed as "But I can't draw." It's an all-too-familiar phrase. It's also a self-defeating phrase, for it shuts down an individual's innate impulse to draw. Worse, a fear of drawing separates a person from that special kind of thinking that involves making marks—a kind of thinking that is commonly acknowledged as among the most creative known to man. I'm referring to visual thinking—a form of problem solving and communication that is quite different from verbal thinking, the mental process most valued within formal education.

Please stop reading and find a pencil. Locate the illustrations in this chapter that are referenced as Figures 1.1 through 1.4. Read their captions and do as they say, making pencil marks right on these pages.

1.1 Improvise: Add a quick line or two to each of the circles as fast as you can so that they become different things. If you can get through this set in two minutes, try twenty more.

VISUAL THINKING

Whether or not you feel that your artistic impulse has atrophied or even that it has never existed, these exercises are

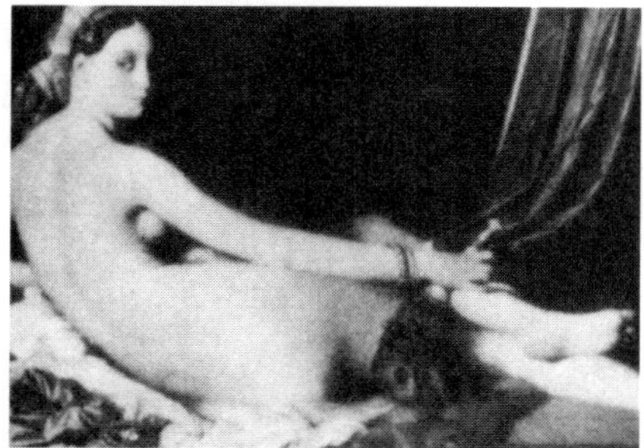

1.2 Recycle: With just a few quick pencil strokes, try to copy each of these items in corresponding frames on the opposite page. It is important to work very fast in this exercise. When you've finished, study the visual style of the resulting sketches. How would you characterize the kind of markings you've made (light, exact, cluttered, bold, loose, free, strong, delicate)?

intended to pump new energy and awareness into your use of eye and hand and imagination. As their caption titles will tell you, each of the four exercises asks you to perform a specific mental process—but to do it entirely through images, not words. As you tackle these problems (and others like them that appear later in this chapter), I hope you will notice that what you are actually doing is not "drawing" per se, but rather a form of thinking that takes place through the dialogue between hand and eye, between pencil and paper. In visual thinking it is the *process* of working graphically that counts, rather than the product of that thinking. In fact, it's very important not to confuse drawing to extend your thinking with drawing to communicate a well-formed idea. As you'll experience, visual thinking also has to do with the working out of ideas, with idea generation.

DRAWING AND ART

So much for calisthenics. I do hope you plunged right into the activities associated with Figures 1.1 to 1.4. If not, I recommend you do so now. They'll warm up your eye and limber up your fingers. And by doing them you'll show yourself how visual thinking—like thinking through words, writing, or other symbolic code systems—is a skill that one acquires, a skill that requires constant practice if you want to stay in peak form.

As we now move into a discussion on drawing technique, it may be important to step back for a moment and note that the most distinctive thing about our species is that we are symbol makers. Just as each human being is born with the impulse to speak, so each of us is also wired to create images. While it is true, of course, that some people are endowed with unique gifts or abilities for verbal or graphic expression, it is also true that a facility in any expressive mode is something that must be developed. In other words, you can learn to draw. Artistic ability is not some mystical God-given gift. As with any type of expression, there are some skills that must be acquired through practice and by studying the basic principles and conventions of each form.

I grant that even the greatest self-discipline and tireless instruction won't equip you to sketch like Michelangelo or Leonardo, but a few exercises and tips can definitely improve the look and the impact of your drawings. Just as important, self-instruction may also give you more self-confidence.

Creative Clichés. A basic trick in drawing is to select and use just those details most often associated with what you are drawing. A simple box, for example, is immediately recognized as a house if it has a chimney belching smoke. But with a rounded rectangle drawn on one of its sides, the box becomes a television set. Similarly, suppose you want a character to look old. What clichés convey old age? A cane? A stooped walk? What else? White hair and old-fashioned spectacles? Symbols or attributes of old age are, by definition, all clichés, but used effectively in drawings, they can communicate quickly, directly, and clearly. As you try the problem in Figure 1.4, keep in mind that the special art of caricature depends entirely on the artist's ability to determine which details make a particular individual recognizable.

Deep Structure. Probably the best way to make a quantum jump in your drawing abilities is to look past or beneath the visual detail of any image to those basic geometric forms that constitute its structure. Because simple geometric shapes are easy to create, they become ideal building blocks in the act of drawing. Figure 1.5 features some examples taken from Preston Blair's book *Animation—How to Draw Animated Cartoons.* What holds for Pluto can easily be integrated into your own character design and drawing.

After you study the "deep structure" of the examples, try to construct your own character using just spheres. Note that spheres are the easiest of all shapes to draw because they don't change shape when viewed from different angles.

"Less Is More." What the architect Mies van der Rohe said about buildings is true also about drawing for animation. Simplify. Cut out all nonessential details. Simplify still more. As you watch animated films on television or as you study the individual frames reproduced in these pages, note how much is left out. See how simple and lean the drawings are, how uncluttered and unfussy the graphic techniques.

Graphic Styling. Just as there is more than one way of seeking knowledge and skinning a cat, so there is more than one way of representing the same thing in graphic terms. As an animator, you will prefer some drawing styles and you will dislike others. Some graphic modes will be beyond your reach and you'll have no feeling for others. Be that as it may, you

1.3 Explore: What objects can you find inside these scribbles? Use a marker to add details and emphasize form. When you've worked awhile with this one, explore your own random scribbling.

should consciously explore new approaches. You should try to extend your repertoire for creating images. In fact, much of your eventual success and satisfaction will come from being able to exercise different options in selecting a graphic idiom. The case study in Figure 1.7 takes a familiar household object and shows just a few of the different ways it can be styled.

Line and Body Languages. The quality of a line—its thickness, its precision, its tension, its relationship to other lines—should be examined, evaluated, and controlled in your drawing. Here is a sampling of "rules" that cartoonists and animators often employ: heavier lines attract our attention; angular lines suggest a feeling of tension; smooth and free lines create a peaceful, happy feeling. As with lines, so with bodies. A closed figure is "tight" or "angry." An open gesture or position is "friendly" or "relaxed." Frame enlargements throughout the book will suggest the range of lines that can be used and their various impacts.

Framing and Composition. Animation, like most graphic art forms, takes place within a frame. The changing composition within this rectangle is a factor continually affecting the impact of one's drawing. The most important tip about good composition is this: Avoid the center. It is a *dead* center. Positioning something in the middle of the frame creates a feeling of total stasis. Similarly, stay away from an even division of the frame or a composition that is too balanced. Sometimes, of course, you'll *want* your images to be centered and stable.

Importance of Background. Man is a pattern-seeking animal. We all search for meaning in all things. Context (background and surroundings) can be used to relay much information about the "subject" of a drawing. Almost any frame from a good animated film will make this clear, and this book is filled with them. While it's true that backgrounds should be simple and never detract the viewer's attention from the main component of the drawing, be cagey in the selection of objects and details you place in the background in order to convey a strong feeling of place and atmosphere.

Color Psychology. Try in your drawing to exploit those standard conventions and cultural assumptions that your audience will share with you. For example, warm colors (reds and oranges) are perceived as active and they tend to come forward in a drawing or painting (at least in our culture they do). Cool colors, on the other hand, tend to recede into the background and are perceived as inactive. Everyone knows that a red face is angry, a blue face sad, and a green face sick. Incorporate such shorthand symbols into your drawing and coloring. But don't neglect to consider colors that do *not* conform to our cultural expectations. The best animated films use color as an equally powerful graphic element to line, shape, and even movement itself. The use of color is very, very important.

Depth Cues. Here's a quick but by no means complete list of visual cues that will create an impression of spatial depth within a drawing: heavier lines come forward and stand out; foot placement indicates the ground; place the horizon line high in the frame; use more detail in the foreground; employ perspective lines to indicate the relationship of the subject to the background; place shadows underneath people. Figure 1.6 provides a simple drawing that incorporates some of these cues. Can you find each of them?

1.4 Visual clichés: Add whatever details you wish to transform the same basic body form into these different characters: a pirate, a hippie, an old woman, a creature from outer space, a monster. A few extra outlines are provided so you can try different solutions and create new characters.

DIGITAL DRAWING

Drawing on a computer is hard. It feels awkward. It takes longer. It requires hard thinking. The learning comes slowly.

Okay. Let's deal at the outset with these unhappy facts. Let's start with a step back. Consider how difficult it is to attain real mastery of the traditional elements of drawing—the ones we've been considering thus far in this chapter. Hard enough, right? First there were years of scribbling and coloring with blunt crayons. Later came writing. Year after year of schooling drilled us in how to hold a pencil and master penmanship well enough so that others could decipher our markings and read what we had written. Seen from this perspective, even though we've all been making marks on paper since we were toddlers, and even though few of us claim to be great artists, you can still bet that every single person who has the skills to read this book possesses the mastery required to put a line onto paper exactly where he or she wants it.

You can forget any such degree of instant facility when it comes to drawing on a computer screen. It's almost like being a toddler again. The very meaning of the words *drawing* and *paint* must be relearned because they take on special nuances when you begin working on a computer.

1.5 Deep structure: A valuable approach to drawing cartoon characters is to begin with circular and rounded forms. These examples were created by one of the masters of character animation, and are reproduced from *Animation—How to Draw Animated Cartoons* by Preston Blair, published by Walter Foster. This modest book is extremely useful for analyzing animated drawings. Mr. Blair provides a step-by-step approach to designing and drawing cartoons.

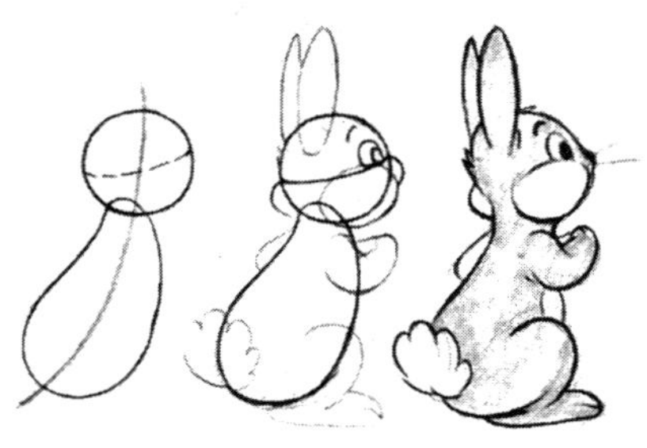

So where do you begin with digital drawing? For openers, you need to understand that there are three fundamental modes of computer image making or computer graphics.

Bit-map Images. You probably know that the computer screen is actually composed of tiny dots called *pixels.* There are exactly seventy-two pixels in a square inch of computer screen resolution. One way to create an image is by delineating a patch of pixels—maybe a thin line or maybe a solid shape of colors—that is known as a *bit map.* If you want to move a bit-mapped image, you have to encircle all the dots that make it up and select them as a group. Bit-mapped images are rendered (and saved) by a group of computer software applications that are called *paint programs.* Although the act of creating a bit-map image is much like painting on paper, problems come when you try to move or resize the image: In the former case you can leave a gaping "white hole" on the computer screen and in the latter case you can find your enlarged bit map has the dreaded "jaggies." The benefit of paint programs is that you can edit and draw pixel by pixel. The animator has great control and flexibility.

Vector Images. Another way to create a computer graphic is by treating it as a discrete object that the computer generates from data stored as a mathematical equation. Vector images (also known as object-oriented graphics) are created with what are called *draw programs.* The benefits of working with vector images is that crisp appearance is retained if one changes the object's size. Also, draw programs make it easy to move images around independently. The downside is that you can't fine-tune an object, adding a pixel here or changing the color of a pixel there.

Three-dimensional Images. The most radically innovative form of computer graphics is 3-D. Here the animator uses "wire frame" constructions to simulate a three-dimensional image on the two-dimensional computer screen, making an animation more realistic. After the basic shape is constructed, 3-D software programs allow you to fill in the surfaces with colors and textures. Once the complicated basic image has been laboriously entered into computer memory, it serves as an object that can be manipulated with a flexibility that matches the most positive aspects of both bit-map and vector graphics. During frame-by-frame creation of an animated sequence, the 3-D image can be turned, rotated, repositioned, and even given subtle changes in reflection and lighting effects. It is this addition of lighting and shadows that gives 3-D animation its true-to-life look.

I lied when I said that there are only three types of graphics programs. In fact, there are another two—although I think of them as hybrids of the basic kinds described above. *Paint/draw programs* do as their title suggests—they create both bit-mapped and object-oriented images, placing them on different levels that work like the clear acetate sheets or "cels" of traditional animation, letting the artist see and manipulate digital drawings on each level. *PostScript-based illustration programs* create a special kind of digital file that scales perfectly to all sizes. This type of graphics application was created to help computers work with letter shapes and type fonts. I suspect that readers who have worked only with computer word-processing programs have marveled at how easy it is to swap type fonts. You can also enlarge or reduce something that has been typed onto the screen. A simple click of the mouse can blow up a letter form to huge scale. Another click and a word or phrase is reduced to miniature size, perfectly proportioned. You'll be pleased to know PostScript graphics programs can also be used to create animation.

1.6 Suggesting depth: This very simple drawing was done using Painter, a software program that mimics many of the traditional techniques used by artists. The image incorporates a number of the conventions commonly used to convey the dimension of depth.

But how do you actually make a digital drawing?

Regardless of the type of graphics program, computer images share a common pool of *input devices* and *imaging tools.* You can certainly use a *mouse* to scribe a digital graphic in bit-map, vector, or 3-D form. But for most people it is a lot easier to use a *digital drawing tablet* that has a pen you hold the same way you hold a pencil or paintbrush.

Figure 1.9 provides a helpful catalogue of the imaging tools that have become standard in computer graphics. There is not space in this volume to describe and provide an example of how each of these tools is operated with either mouse or tablet. Nor, fortunately, is there such a need. All graphics applications come with computer-based tutorials and program manuals that will give you a comprehensive, item-by-item introduction to each tool. Digital drawing is hard but it is not counterintuitive. The hard part comes in gaining the same level of control and dexterity that we are all accustomed to when we take pencil to paper. It just takes time, that's all. Just about any computer you can get your hands on will have a graphics program of one kind or another. And just about anyone who

1.7 CASE STUDY: The Range of Rendering

Here is a jumbo project that deserves as much time as you can give it. The assignment is to choose a single object and explore the range of ways you can render it using "old" media and "new media." The goal is *not* to create a series of masterpieces. Instead, your goal is to make many, many different images of the same thing.

On the left page are five examples of traditional approaches. *(A)* shows a photo of a normal telephone. Variations shown include a pencil drawing, *(B),* a still life made by ripping (not cutting) basic shapes from a sheet of construction paper, *(C),* a rendering formed by building up density with a rubber stamp, and *(D),* a bas-relief version created by building up layers of masking tape, *(E).* Here are some other approaches for you to try: Draw the shadows cast by your object, but not the object itself; draw a huge enlargement of one detail; make a pointillist version by tapping dots with a Magic Marker; do a high-contrast version that eliminates all gray tones and emphasizes zones that are either black or white (no single lines); experiment with renderings of the object done with colored pencils, crayons, watercolors, and ink.

On the right page are digital ways of making images. The source photo, *(F),* was made with a Sony digital still camera (DSC-F1) that can store up to fifty-eight images when working in its "standard" mode of image quality that yields a 640 x 480 image compressed in JPEG at 64 KB (if this is all "geek" to you, the hardware and software chapters should provide assistance).

C

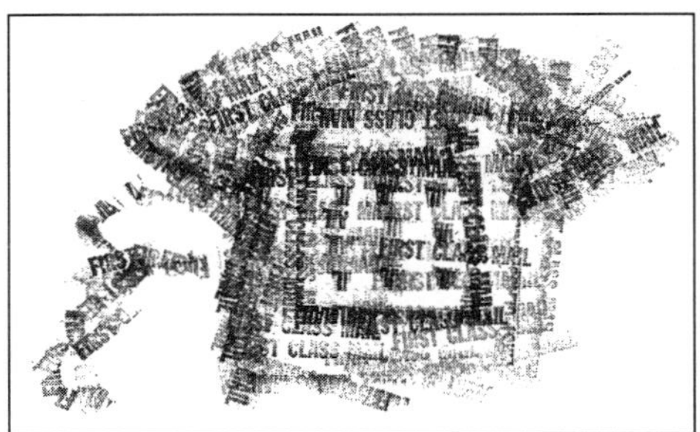

D

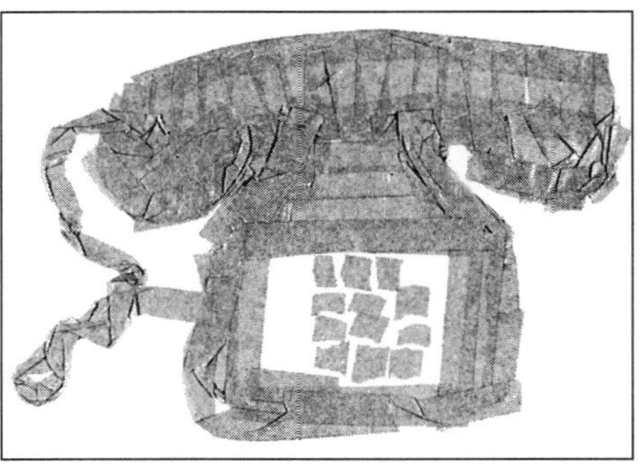

E

A

B

F

H

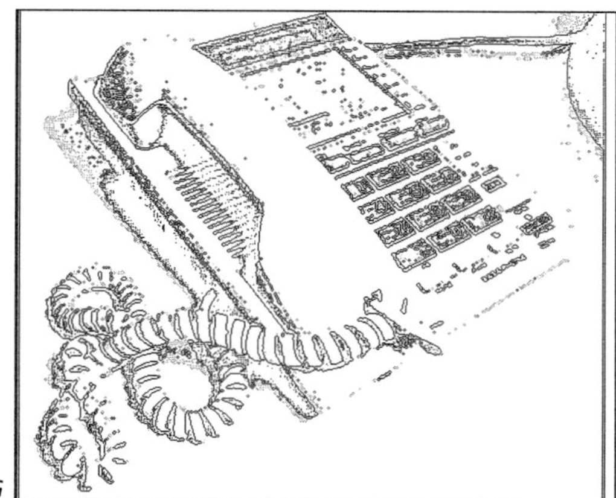
G

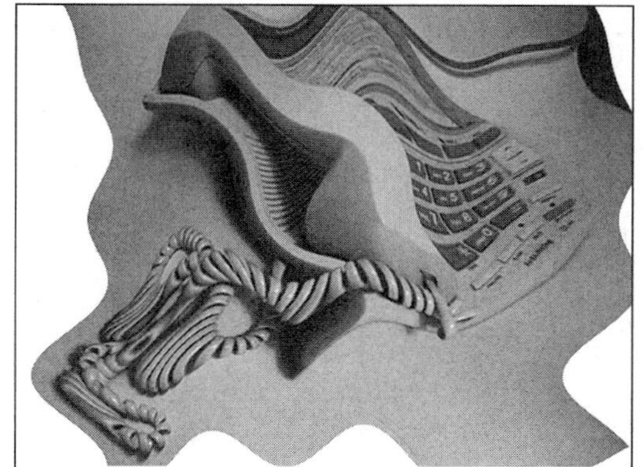
I

The four variations here were all created using the standard image processing software called Photoshop. Image *(G)* shows the image with a "Trace Contour." Image *(H)* samples "Crystallize." Image *(I)* is the "Wave" effect. And image *(J)* uses Photoshop's multiple layers of alpha channels to combine "Blur" and "Hi-Contrast" effects. All of these digital treatments come from a standard roster that appears when one pulls down the "Filters" menu within Photoshop. Not once was a mouse used (much less a graphics tablet) in creating the variations sampled above.

There are plenty of other filters and effects one can explore with a program like Photoshop. Try "Gaussian Blur," "Spherize," "Mezzotint," "Lens Flare," "Emboss," "Twirl," "Add Noise," "Solarize," or "Find Edges" — among many, many more. The possibilities will become mind-numbing — yet creatively stimulating — as you experiment by pushing the same basic image through some of the different drawing, painting, and 3-D software packages discussed throughout the book.

J

1.8 Three modes of digital drawing: This simple little bug is shown as rendered by wholly different types of graphics programs. The Bit-Mapped Bug, *(A)*, was done in a program called Painter. Pixel-by-pixel manipulation gives the artist a hand-drawn sensibility, complete with lines that vary in thickness and different tonal ranges, corresponding to the pressure put upon the drawing stylus. The Vector rendering, *(B)*, was done in Adobe Illustrator. Here the Benzier handles are shown. These are the control points that the artist manipulates in shaping (or reshaping) the mechanical-looking lines of consistent weight. Finally, there is a full 3-D rendering done in a program called Ray Dream Studio, *(C)*, complete with texture-mapped surfaces and lighting. Chapter 23: Computer Software describes in more detail the different approaches to digital drawing. *Courtesy George Escobar.*

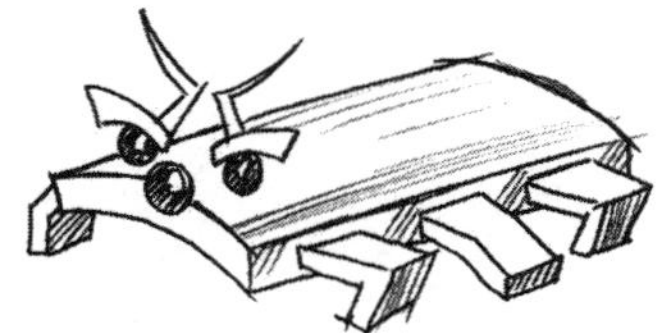

B

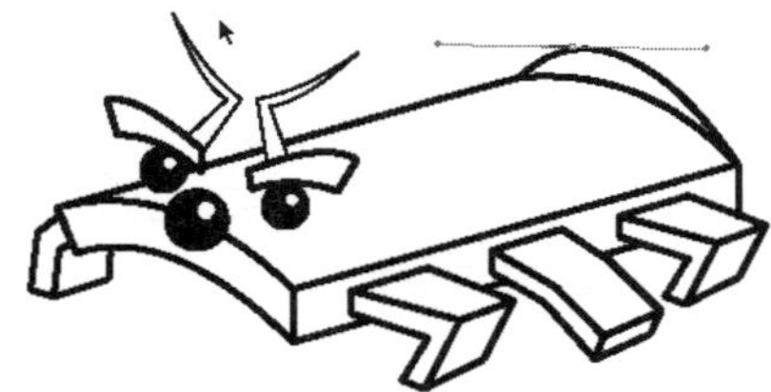

C

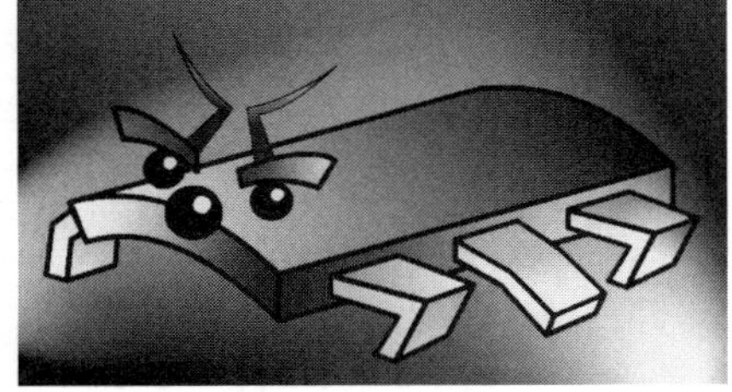

has worked on a computer will be able to give you a quick introduction to the basic operational techniques that allow one to start making images.

Doing your own computer graphics gets easier and easier with each new generation of computer and each fresh version of computer applications. Over the past few years, the designers of the major graphics programs—regardless of type—have come to adopt a uniform language of shortcuts and procedures for making and manipulating computer graphics. This is great for the animator and digital artist because it means that once you have put in the time and self-schooling to function reasonably well in one program (be that Illustrator, Painter, FreeHand, or any other major application), you are not going back to square one when you need to work in another kind of program.

Before moving on to the next topic under Basic Skills, let me reassure those readers who know zero about computers that Chapter 23 provides a ground-up introduction to computer software and that Chapter 24 does the same thing for computer hardware. If you scoot ahead to these chapters now, you will find text and illustrations that provide greater detail about the three basic modes of computer graphics. As for the varied and different techniques of bringing such images to life, almost every chapter between here and there will chart different areas of the digital domain.

ART IN MOVEMENT

Animation is art in movement. More, it is the art of movement. In an animated film, drawings are not static, as they must be in these pages. Whether it's on a movie screen, TV set, or computer monitor, the drawings come alive, and it is the quality of that life that matters, not the quality of a particular image or frame of film. Whether it is a drawing or a lump of clay or a puppet or a collage or whatever, the animator places life and meaning into his or her material by making it move.

Movement within animation is something that simultaneously exists on many levels and speaks in many ways. Movement conveys story, character, and theme. It creates tension through the development of expectation and its release, through the arousal of curiosity and its resolution. Movement creates a structure for the passage of time. It is also intimately related to music, dialogue, and other elements of the audio portion of the film or video project. Movement is the essential magic of cartoons. Why else did the first people to see motion pictures spontaneously label the experience "movies"? Movement is the beginning and the end.

If you've tried the various problems contained in this chapter's illustrations, I'm hoping you'll have experienced the following three truths: that *visual thinking is different from other kinds of thinking,* that *drawing skills are acquired as well as inborn,* and that *animation is art in movement.*

These truisms form a security blanket. You should never again feel terror when asked to pick up a pencil. But don't worry if you're still not quite convinced that you, a klutzy drawer, can still create magnificent and powerful animated films. Against a lifetime of being told that "art" is based on the ability to draw with representational precision, these few pages of encouragement and exercise can only begin to make you comfortable with whatever abilities you possess and can develop further. But keep at it. Here are four commonsense strategies that can help you solidify and extend your acquisition of greater drawing skills:

Planned Doodling. For a week or two carry around with you a cheap pad of paper and a pencil. Whenever you can, make doodles. It doesn't matter what you draw. It matters only that you draw. So do as much as you can and don't worry about saving the sketches. There will be time enough for that.

Sketch File. Find an old box, envelope, drawer, or file folder. Every time you make a drawing that interests you in some way, slip it into the file. If you are a computer jock, make a digital file of drawings you made, clip art, and images that you can scan into the sketch file. Note that a drawing doesn't have to please you. It should just be interesting to you in some way. And don't worry about using the file. Just keep it.

Self-inventory. Systematically study how you go about drawing. Keep track of the way you seem to prefer working. Here are some questions you can ask yourself: Do you like to work with large or small formats? Which drawing implements do you prefer? Do you draw quickly or slowly? Do you like to try copying what others have drawn? Do you like to draw people or objects? What kinds? I encourage you to think up other questions and generally search for patterns among your attitudes, preferences, beliefs, and of course, your artwork itself.

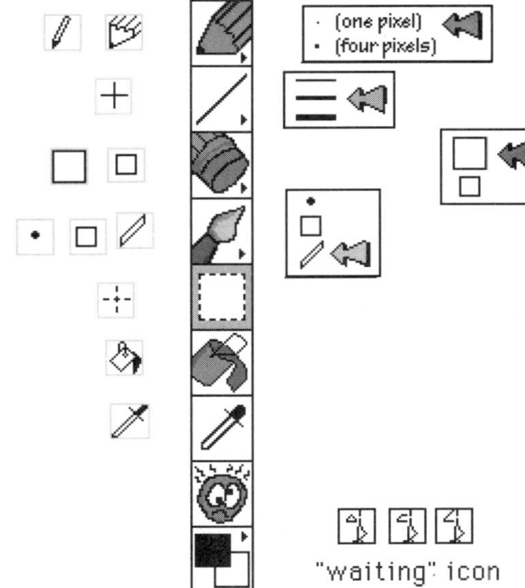

1.9 Generic graphics tools: Although today's sophisticated graphics packages display a bewildering range of tools, protocols, and custom controls, the basic tool set is represented in the menu of on-screen icons shown here, which Kit Laybourne designed for use in a CD-ROM project. From top to bottom in the center column: a pencil to signify drawing freehand; a straight line maker; an eraser; a paintbrush; a box to select a screen area (for moving or deletion); a color "fill" tool; a color "picker"; an undo button; and a pair of boxes that show foreground and background colors.

Each tool can be modified in various ways. To the right of the central column of icons are "pop-out" submenus that the user employs to modify the corresponding tool. To the left are cursor icons that show how the mouse's screen marker will appear, depending on which variation of the tool has been selected. The "waiting icon" suggests a three-image cycle that would show a stick figure scratching its head. This would appear on-screen while the computer is processing information. Like the spinning wheel or clock icons you are familiar with, the purpose here is to assure the user that although nothing may be happening on the screen, the computer has received its orders and is busy "thinking."

1.10 The atoms of computer graphics: So, you wanna get up close and personal with a *pixel*? Here's a huge magnification of two icons shown in the preceding illustration. The word *pixel* comes from "picture element." You should be able to make out individual pixels (the smallest square unit) and you can probably even get a sense of different colors — shown by the various gray scales on the paintbrush tip, for example. All desktop computers share the same resolution of 72 dots per inch (dpi) — which is also the number of pixels per inch. If your screen has a measurement of 8" by 10", you'll be looking at exactly 472,320 pixels. It's good to know that any graphics program will let you zoom in to the "atomic" structure of pixels. Go ahead, cruise in as close as you can. Maybe, like the author, you'll find it is totally engrossing to design images, working at this most minute of levels.

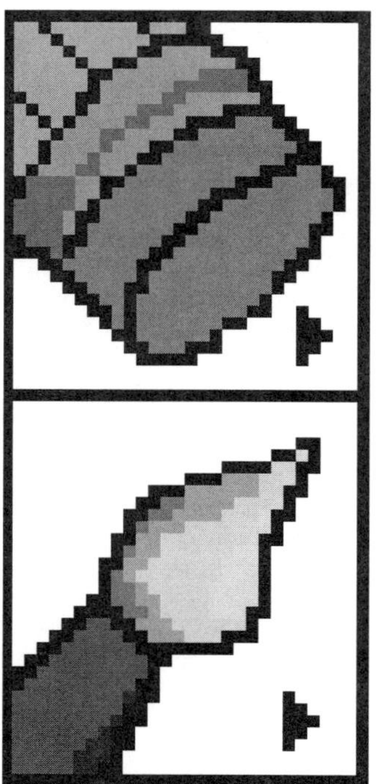

Favorite Things. Don't waste time trying to draw things that don't interest you. Work only with stuff that is exciting to you. Take notes (literally) of what you are drawn to. Then draw it. Such a "passion list" can be immensely helpful in locating themes, characters, settings, effects, and other elements you can incorporate into layer productions.

GETTING IDEAS

Is the creative process really a mystery? Many would say so. They view inspiration as a divine gift; invention as a form of genius; insight as something you luck into; and original thinking in general as a mystical experience. Creativity, most people would agree, is something you are born with. Either you've got it or you don't.

Dr. Abraham Maslow, a pioneer in the field of humanistic psychology, spent much of his life studying how creative individuals manage to function as they do. According to Maslow, primary creativity "is very probably a heritage of every human being and is found in all healthy children." So how come there's so little real creativity surrounding us? Thomas Edison offered this quantification of the creative process he so regularly seemed to experience: "Invention," he said, "is 1 percent inspiration and 99 percent perspiration." This entire volume is based upon the axiom that everyone is creative and that with hard work and encouragement, everyone can come up with good ideas.

Insight and creativity happen when one is playing around. This doesn't mean goofing off, however. Getting ideas requires a delicate and determined playfulness—one that encompasses a variety of approaches, demands all the discipline you can manage, and usually ends up requiring plenty of sheer faith.

Because visual design is so essential to animation, it is likely that if you're already interested in animation; you will already have a built-in mechanism for finding new ideas. It is your own taste for graphics combined with your own abilities in drawing and image making that will lead you in idea hunting. You'll probably get a lot of help in tracking down good ideas from a set of latent tendencies most animators have toward being gag writers, mimes, shrinks, junk collectors, long-distance swimmers, jewelers, magicians, and all-around nonconformists. Here are ten strategies that can help you locate and stockpile good ideas.

Idea Book. Purchase a large-format (at least 11 by 14 inches) spiral-bound sketchbook with sheets of quality white drawing paper. This book will become a home for your ideas. Use it freely. Jot down anything that strikes you as even remotely relevant as a film or part of a film you might one day make. Plunk into your book images or quotations or recollections of any kind.

The large size of the pages is important because it will promote a comparison of different ideas and visual elements. The absence of ruled lines forces you to write out ideas in unfamiliar ways. The wire binding allows you to lay the book flat for drawing. This kind of book also encourages the addition of odd materials—clippings, postcards, doodles, collages, and so on.

Weird Combinations. Search for comparisons and analogies in all things. You might carry around with you each day some sort of provocative idea and try to combine it with experiences and observations you make as your day proceeds. Pick just one object for, say, five minutes and see how, if, why, and with what effect it does or does not combine with any other object or environment or whatever. How does a toothbrush relate to the morning newspaper? What is a toothache like?

Dream Diary. Maslow said that creativeness emerges from our unconscious. "In our dreams," he wrote, "we can be . . . more clever, and wittier, and bolder and more original. . . . With the lid taken off, with the controls taken off, the repressions and defenses taken off, we find generally more creativeness than appears to the naked eye."

Write down your dreams just after you've awakened. You will find that your own unconscious communication system is a terrific source of good ideas. With practice you'll have no trouble remembering dreams in rich and fruitful detail. Try it. You'll like it. And it works.

Recycling Ideas. There isn't a new idea under the sun. Sometime before—someplace—someone must have come up with every possible combination of ideas. But this fact should only serve as an encouragement for copying ideas, borrowing them, refining them, making them your own. Go about this purposefully. When you come across an idea that is really attractive to you, jot it down. As you later work over these ideas you will select further and invariably infuse and alter what you like in your own way. The process of editing and transforming makes a found idea your idea—to the limited extent that any idea can ever belong to anyone.

You might even begin with this book. Draw a big arrow to notions or images or techniques that catch your interest. Better, write them on the inside of the covers, where other ideas can join them. Better still, put them into your idea book.

Brainstorming. Here is a creative problem-solving technique that requires a group of people. The object of brainstorming is for the group to come up with as many alternative solutions to a problem as they can. Quantity is valued but quality is not judged. To brainstorm well, everyone participating must spontaneously call out every idea that enters into their heads. Allow no self-censorship. You don't say to yourself or to others, "Gee, I'm not sure if this is off-the-wall but . . ." Instead, you immediately spit out the idea. You free-associate.

Brainstorming works on the theory that the spontaneous generation of ideas will produce novel and valuable solutions and that a group will come up with a large number of better ideas than will the same number of people working individually. The first part of a brainstorming session is given over to generating ideas. This is usually given a time limit—perhaps five minutes. Then there is a second part in which the ideas are evaluated. This can be a creative step too. Ideas can combine in interesting ways; discussion can lead to refinements or new directions that the shout-out speed of the first part of the session does not encourage.

The Back Burner. Sometimes the best way to solve a problem is to forget it. The mind's subconscious will keep working at a problem even after the conscious portion of the mind has gone on to other concerns. This incubation period is essential to creativity and it is the major reason why working under too tight a deadline will be unproductive if original thinking is required. So as a conscious strategy, allow yourself time for ideas to simmer and perk on the back burners of your mind.

Mixing Media. As media theorist Marshall McLuhan pointed out, a tremendous release of raw energy occurs when information that is packaged in one medium is translated into another medium. A good place to look for ideas is within the other arts. This process of cross-fertilization has been long valued and used by artists of all forms: dance, theater, writing, mime, music, sculpture, painting, and so on.

The Composite Medium. All films, videos, and computer-based motion graphics are made up of other expressive forms that combine the performing arts and the plastic arts, that unite the world of science and technology with the world of arts and letters.

Just one ingredient of these media's composite forms can fire your imagination. For example, "sound" is made up of music, dialogue, narration, and sound effects. Any one of these can provide the source of a new idea on which to base your own project. Each element is important. And all parts work together, even as they are bound together physically on the screen's surface and bound together in the intellects and emotions of those who perceive them.

Materials. For animators, one of the richest sources of ideas comes from the material of animation itself. This will become evident as you study and try your hand at the techniques discussed in this book. You will discover, for example, that the character of sand being moved across a sheet of white Plexiglas will produce its own ideas. And so it will be as you animate other objects, work in clay, draw on registered paper sheets, combine collage imagery, or design a cutout character. In animation, every material must be used differently. Every material has its unique characteristics and potentials and these must be exploited fully by the artist.

Hanging Tough. When nothing else seems to work, try brute force. Muscle your way to a good idea. Let inspiration come out of perspiration. The very discipline of work—of spending time at it even when it feels unproductive—will often yield a sudden breakthrough. If you spend twenty minutes every day, seven days a week, working in your idea book, then I promise that the discipline of doing this alone will get you through. Something worth pursuing will emerge. And eventually, a pursuit of one kind or another will get you somewhere.

You must trust yourself. All great artists have counted on such faith to get them through the "dry spells." Virtue is its own reward. Keep working.

2 ■ Cameraless Animation

The ancestors of animated films were a series of nineteenth-century mechanical toys that created the illusion of movement. None of these gadgets had any direct application within the worlds of commerce and science. Their existence was based solely on their ability to delight. They were toys in the real and best sense.

Over a hundred-year period, inventor after inventor further refined these optical devices. Along the way, machines combined with the emerging technology of photography to create yet another entertainment, the movies. Mechanical toys were wedded to celluloid and gave birth to cartoons.

It is both interesting and instructive to retrace the genealogy of animation and to study some of those crazy gadgets that led to contemporary animation techniques. If you try your hand at making some of these machines, not only will your appreciation of them be enhanced, but you'll pick up some valuable information and experience concerning the perceptual and mechanical foundations of animation, the techniques of designing pictures that move, and the process of spontaneous invention itself. As a bonus, you'll be creating materials that can be used later in making real cartoons.

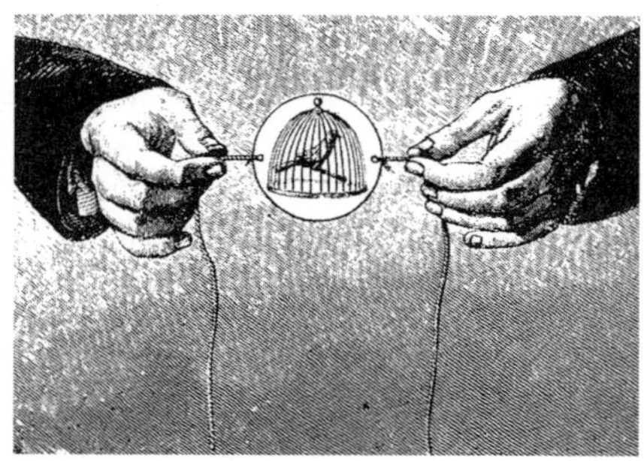

2.1 A thaumatrope: This 1826 French engraving shows how to spin a thaumatrope and it also suggests the device's effect. In this case, one side of the disc bears the image of a bird and the other side carries that of an empty cage. The superimposed image is created, of course, only when the device is being spun between fingertips. *Courtesy Stanford University Art Museum.*

THE THAUMATROPE

Animation prehistory begins with a simple device named the *thaumatrope*. This optical toy was in wide circulation in the early nineteenth century, and it may have been known far earlier than that.

The toy is simplicity itself: a disc that is attached to two pieces of string. When the disc is twirled by the operator's hands, images placed on either side of the disc are perceived together as a single image (Figure 2.1). Twirling the disc superimposes images upon each other by means of a perceptual phenomenon known as the *persistence of vision.* Our eyes hold on to images for a split second longer than they are actually projected, so that a series of quick flashes is perceived as one continuous picture.

Using a piece of heavy cardboard and some string, you can re-create the bird-and-cage effect or try out something more personal.

Project: Plastic Surgery. Locate a black-and-white photograph of yourself, not more than a couple of inches in length and width, in which your head is fairly large within the frame. Center your image and then cut this out and mount it on a piece of round cardboard. Punch two small holes at the opposite edges of the disc and attach a string to each. Design a number of alternative images to be attached to the reverse side of your thaumatrope: a beard, a scar, a hat, a mask, a missing tooth, or whatever else you'd like to superimpose on your own face. Try different effects using colors. See what happens when you place the photograph of a movie star on the opposite side. Twirl up a storm.

A

B

2.2 Discs for Plateau's phenakistoscope: A simple wooden handle was used to hold these discs as the viewer spun the wheel while facing a mirror and sighting through the slits in the disc's surface. The old etching, *(A),* shows a variation of the phenakistoscope. To operate this device correctly, the viewer positions an eye close to the surface with slits. As the disc is spun, the animated movement is perceived by sighting through the series of slits to the series of drawings beyond. Plateau's fascination with mechanical movement is apparent in that moving machines provide the "content" within two of his discs, *(B). Courtesy Stanford University Art Museum.*

THE PHENAKISTOSCOPE

In 1832 a native of Belgium named Joseph Plateau invented the first machine that really created the illusion of sustained movement. His invention, the *phenakistoscope,* is a spinning wheel that bears a series of drawn images and viewing gates that frame the viewer's vision of the drawings (Figure 2.2).

Project: Mirror Movies. In Figure 2.3 you'll find a one-quarter-size pattern for creating your own phenakistoscope. Reproduce a full-size version, 9 inches in diameter, and attach it to a backing surface. After cutting viewing slits at the indicated places, draw something in the twelve frames outlined on the surface of the stylus. Note that you can actually try two different movements on a single sheet of paper by using the outside zone—*x*—for one set of drawings and the inside zone—*y*—for another set.

THE ZOETROPE AND PRAXINOSCOPE

It wasn't long before a new generation of inventors refined and extended Mr. Plateau's device. Among the most ingenious of the new toys were the *zoetrope* and the *praxinoscope.* Both machines provided more convenient projection devices for their drawings. Both extended the number of drawings that could be used—and hence the duration of movement itself.

The zoetrope is a revolving drum that has slits in the sides, spaced equally. By looking through these slits as the

2.3 Phenakistoscope stylus: This particular configuration of twelve wedges seems to work well, but you are encouraged to experiment with other formats. Use the sections marked *x* for one set of drawings and the sections marked *y* for the second set. Beside the stylus (which has been reduced in scale) are three possible drawings you might want to try: a sports car that zooms through the frame; an airplane that flies directly at you; and an exotic plant that you can make grow. The finished phenakistoscope (cut from cardboard or another heavy paper stock) can be mounted onto a pencil by securing its center to the pencil's eraser with a pushpin. Go to the mirror and watch the drawings come alive and move at various speeds.

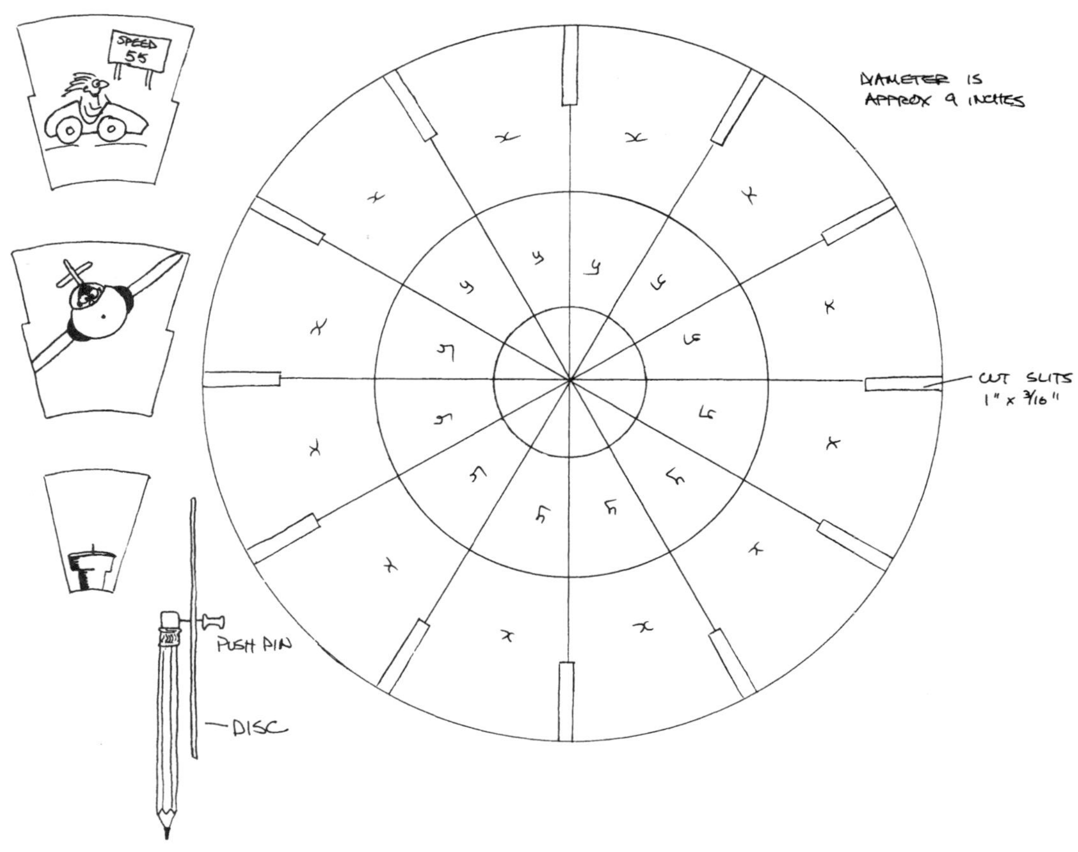

drum is spun, the viewer is able to catch glimpses of a series of drawings that have been created on a strip of paper and then placed inside the drum (Figure 2.4). The larger the drum's diameter, the longer the "movie," and, of course, the same drum could present different strips of drawings. Incidentally, the machine's name means "wheel of life." It was so titled by Pierre Devignes in the year 1860, although earlier versions of the same basic device had been developed in England by William Horner around 1834.

2.4 A zoetrope: The photograph of early motion equipment, *(A)*, shows the circular metal drum and wooden base of a zoetrope. To the left of the zoetrope is a thaumatrope disc and behind that a device that was cranked rapidly to animate a series of drawings on cards. The long strip in the foreground is the artwork that is placed inside the zoetrope's drum. Photograph *(B)* shows four nineteenth-century British zoetrope strips. *(A), courtesy Museum of Modern Art/Film Stills Archives; (B), courtesy Stanford University Art Museum.*

The praxinoscope represents a refinement on the zoetrope. The slits are replaced by a set of mirrors that spin in the center of the drum. You can try to make your own version of the prism device by using a shiny plastic material, which is available in art stores. The finished mirror structure is placed over the spindle of a record player. When the machine is turned on, the outside band of images is animated as one looks into the revolving mirrors. The same drawings you have made for the zoetrope can be modified for use in a model praxinoscope.

The inventor of the praxinoscope was Emile Reynaud, and in 1892 he opened the world's first movie theater in Paris. Reynaud's Théâtre Optique projected a "movie" that was

A

B

made of individual drawings on a long piece of paper. The show lasted just a few minutes. In many ways, Reynaud's invention parallels the modern film projector.

FLIP-BOOKS

Remember those animated drawings you did as a kid on the dog-eared edges of a textbook? By flipping through the pages, you could make the characters or the design move. Sometimes you could buy a small flip-book at the local novelty store, or you'd discover one already created in a comic book, or get one as a Cracker Jack prize.

Flip-books also invite comparison with the technique of cel animation, one of the most sophisticated of all animation techniques. Each page in the flip-book corresponds to an individual piece of artwork that, along with all the other drawings, makes up a movie when it's filmed by an animation camera. The binding of the pad or flip-book acts as a *registration* system —a way of keeping things precisely sequenced and lined up. The act of thumbing through the pages of the flip-book is the act performed first by the camera and then by the projector, if working in film, or by the scanner and then the computer, if working digitally.

2.5 Reynaud's praxinoscope theater: By the turn of the century, this relatively sophisticated device was being commercially distributed in Europe and North America. Emile Reynaud combined this invention with a magic lantern to create his famous Théâter Optique. Located in Paris and operating between 1892 and 1900, the Théâter Optique entertained sizable audiences by projecting an extended series of animated drawings that had been made on long strips of translucent paper. *Courtesy Stanford University Art Museum.*

The quickest and easiest way to make your own flip-book is to purchase a small, unruled pad of white paper. A convenient size is 5 by 7 inches, although smaller 3- by 5-inch books also work. You'll find these pads in any good stationery store. With one of these pads and a pencil, you're ready to start.

The first drawing is made on the last page of the pad. When the next page is permitted to fall forward and cover the page you've just drawn, you will be able to see through the new sheet well enough to make out the preceding drawing. You may now redraw or trace the first drawing, but not exactly. In order to create movement, you must alter each successive drawing in some minor way. These minute changes accumulate or build up to produce the illusion of movement.

The process of completing a drawing, covering it with a new sheet, redrawing, recovering, and so forth is continued until you work your way to the first page of the pad. To see the results of your labor, hold the book in one hand so that you can flip through the pages, back to front, with the other hand.

Standard index cards provide a superior alternative to the bound pages of a small notepad. The drawing technique is similar. With index cards a registration system is achieved by lining up the cards as you draw, one on top of the other. Index cards are thicker than regular paper. This makes them flip more efficiently, but it also makes them more difficult to see through as you draw one image on top of the preceding one. To remedy this you may want to make yourself a *light table.* This consists, very simply, of a piece of transparent or translucent glass or Plexiglas on which you draw, with a bright light projecting upward from beneath this surface, making it easy to see through a number of index cards or pages of regular paper. Chapter 20 provides more information on making or purchasing a light table.

Index cards provide greater flexibility than notepads. For one thing, you can easily throw away a particular drawing without weakening the binding. Similarly, you can insert one or more cards should you determine that you need more drawings to smooth out a particular movement. Most important, index cards allow you to reorder the sequence of a finished flip-book. This means that you can rearrange a finished flip-book so that it is viewed from front to back. Between showings a strong rubber band will easily hold the cards together.

With either pads or index cards, you have some options on just where to draw whatever it is you want to put on the paper. Figure 2.7 shows two common placement systems. Technically and aesthetically, there seems to be very little difference between the two. However, if you think you may eventually want to film one of your flip-books, you should work in an area with a width-to-height ratio of 4 to 3. This is the standard ratio for most movie, television, and computer screens.

Project: Circle Boogie. Using either a pad or index cards, try to solve the following problem. Your first drawing is to be a circle, roughly 1½ inches in diameter. In the following twelve pages, transform the circle into another object—for example, the circle could evolve into a set of lips. And in the following twelve sheets, try to get back to the original circle. But as you

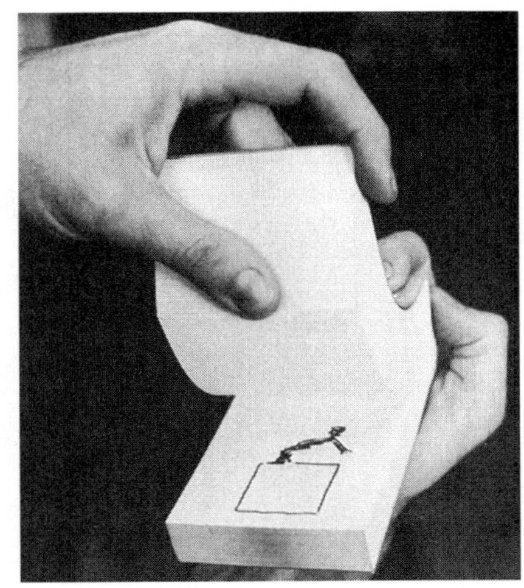

2.6 Previewing a flip-book: A finished flip-book with many pages is being flipped by its creator, George Griffin. This way of holding and fanning the pages will work with almost any kind of flip-book.

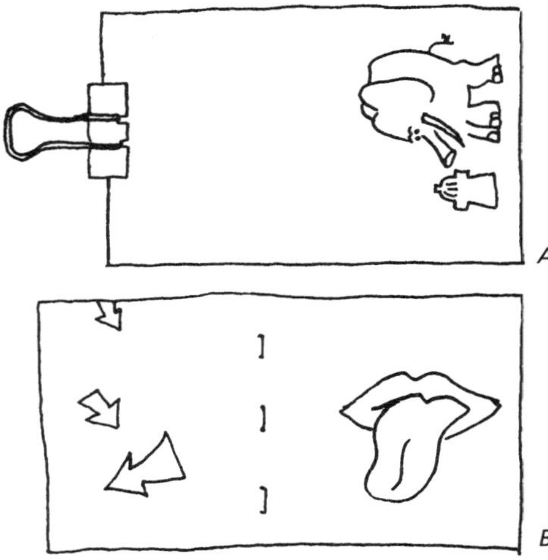

2.7 Two flip-book formats: The scale and dimensions here are highly arbitrary. Flip-book *(A)* uses a #20 (small) binder clip to hold its pages firmly. One's hand can perform the same function. Flip-book *(B)* is stapled in the center. This arrangement creates two drawable, flippable surfaces on the same set of stapled pages. Select a paper stock and a size of drawing that will suit your preferences and resources.

do this, follow a different route from the one you took in the first twelve drawings. You might, for example, have the lips blow out a bubble-gum bubble that grows until it hides the lips and becomes the same size as the original circle.

If you follow these directions, you will have a movement in twenty-four drawings, from a circle to something else to a circle again. Now repeat this process a second time, going to a different shape or object before returning on sheet 48 to the original circle. When you flip the completed book, you should see a movie that, while it lasts just a few seconds, creates a clear visual "beat." Do other variations within this basic structure and add these to your flip-book. Vary the amounts of movement between drawings and the degree of complexity of the transformations. Add color. Try having a number of things happening at once.

Sneak Preview. Save these first flip-books. Later, with an animation camera or with a computer and scanner, you will be able to turn them into movies. Instead of appearing as

2.8 Flip-book as film and publication: These drawings represent part of the actual layout prepared by independent animator George Griffin for formal publication of one of his flip-books.

cramped drawings on small pages that are flipped with unavoidably irregular speed, these flip-books can be produced into real movies, huge in scope, gracious in form, unfettered in presentation.

CAMERALESS FILMS

You can make your own animated films without a camera and without photographically developing the film itself. This is the technique of cameraless animation often called "scratch the doodle" filmmaking. It's a good place to begin your exploration of animation. It is cheap and fast. Few tools are required and you can see the results of your work immediately.

But there are more important reasons why cameraless animation is the best place to start working with actual film. First, the technique allows you to get to know the size and characteristics of the celluloid strips that comprise the physical material known as film. Second, drawing on film gives you a good area for experimentation with the perceptual phenomena that allow the movies to "move." Finally, cameraless animation is relatively simple. You don't need any previous experience or knowledge to create your very first world premiere.

A friendly warning: Although cameraless animation is direct and simple, the technique is deceptively difficult to master. If you want to fashion a film that really works, you will need to do a lot of experimentation and try a number of different ways of working.

Clear leader is the term given to a strip of celluloid film that has no photographic image on it. Lay the clear leader out on a flat surface, draw directly on it and, presto, you've made an animated film. Most often, 16mm film is used for cameraless animation. It is easily available, inexpensive, and the surface is wide enough to permit control.

Sixteen-millimeter clear leader comes with either one or two sets of sprocket holes running along its outside edges. These are termed *single-perforated* and *double-perforated* clear leader respectively. The single-perforated type is recommended because it makes it easier to determine on which side of the film to work. If you draw on the "wrong" side, your image will be projected in reverse.

There are two important facts that you need to know

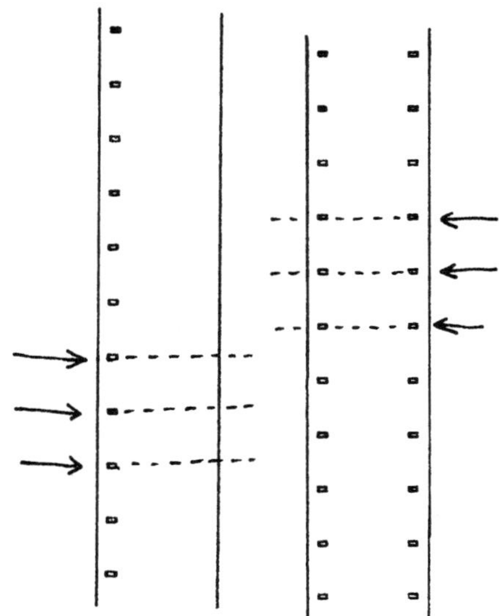

2.9 Frame lines: The frame lines are indicated by the arrows and dotted lines. They go straight across the film. Sometimes you may be working with leader that has only one set of sprocket holes. Location of the frame remains the same.

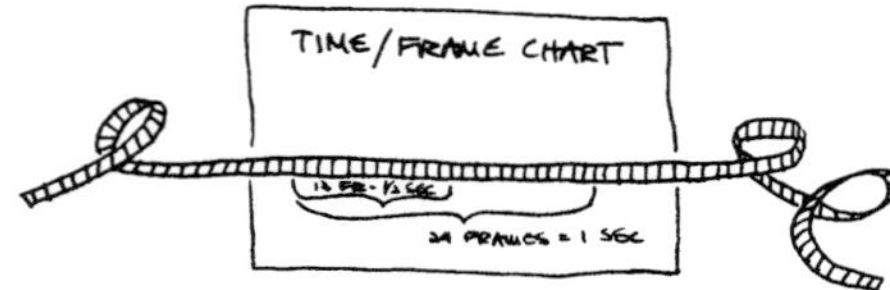

2.10 Time/frame chart: A sheet of white paper or any flat surface can be used to make a chart that indicates what number of frames will equal what duration of time when the leader is projected at the standard sound speed of 24 frames per second. You may, of course, modify the chart to indicate those time intervals that you'll be using frequently in your movie. Remember that a recruiting visual "beat" will make almost any film more exciting to watch. Leave a few feet of blank film at the head of your strip. This is required in threading the projector.

before attempting your first film. The first has to do with *projection speed.* At normal sound projection speed, 16mm film is projected at a rate of *24 frames per second* (fps). The standard projection speed for super 8mm film is *18 frames per second.* In the discussions that follow, I'll be talking about 16mm leader and the normal 24-fps speed.

Frame lines are another important concept. You will need to know where a single image or frame is placed on a strip of clear 16mm leader. The frame lines that mark off the area of the film that is actually projected are to be found crossing the width of the film opposite each sprocket hole.

Among the drawing materials that work well for marking on the acetate surface are *felt-tipped pens.* Make sure your markers adhere to the acetate base of 16mm clear leader. Many of those that promise to write on "anything" don't. *Grease pencils* work too. Because they are not completely translucent, grease-pencil colors are muted when projected. There are some kinds of *paints* that will work well on acetate. But you will have to do a test or two to determine how these paints hold up. On drying, some paints crack and flake off the film. This can clog the gate of the film projector and necessitate careful cleaning after every use. There are also special *inks* that can be applied to the leader with either brush or pen. Check with your local art supply store.

PERSISTENCE OF VISION

One of the first things you will need to know as an animator is, quite simply, what the human eye is capable of seeing. Quite obviously, everything in animation depends upon the viewer's recognition of an image and his ability to follow its movement.

The moving pictures of film don't actually move. All you have to do is look at a piece of film and you'll be reminded that, in fact, the medium is made up of a series of still images. It is the human eye and brain that make movies move. More accurately, the illusion of movement on film is created by a physiological phenomenon called the *persistence of vision,* as mentioned earlier. When a single image is flashed at the eye, the brain retains that image longer than it is actually registered on the retina. So when a series of images is flashed in rapid order, as a movie projector does, and when the images themselves are only slightly changed, one to the next, the effect is that of continuous motion. This very remarkable illusion is the perceptual foundation of film and television.

Project: Charting Visual Thresholds. How little can the eye actually see? If twenty-four individual images are projected during one second of film, can the eye see just one of these? Do different people have different perceptual thresholds? Can the viewer's eye be trained to see with new perceptual sensitivity?

You can conduct some experiments to answer such questions. Holding a length of clear leader against a white sheet of paper, create a time/frame chart like that shown in Figure 2.10.

FRAMELESS STYLE

A distinctive kind of cameraless animation is achieved by marking on the film's surface *without* reference to the individual frames that comprise a length of film. Markers, paints, or other media are applied to the clear leader in broad pat-

2.11 Work on 16mm leader: *From left to right,* the samples show: two abstract ways of drawing on clear leader without reference to frame lines; an abstract technique that uses a stamp cut into a pencil eraser; an abstract treatment in which the entire surface is colored and then decorated; a carefully registered sequence in which a star grows larger and smaller; a representational narrative (a speedboat pulling a water skier); a registered series of abstract graphic forms that are scratched from black leader; two more examples of black leader with abstract patterns scratched onto the emulsion; and, finally, a piece of clear leader with the word *end.*

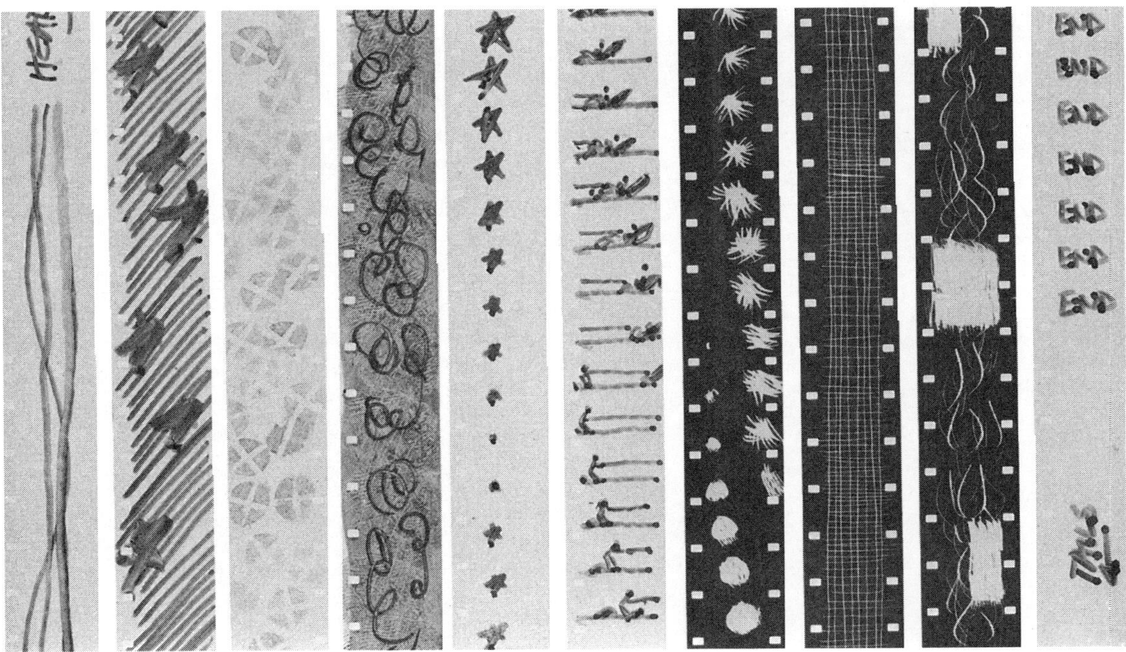

2.12 Drawn-on-plus-some: This photograph shows 4½ frames from *Uncle Sugar's Flying Circus,* a 2½-minute film produced without a camera by independent filmmaker Warren Bass. In addition to drawing with markers directly on clear leader, the techniques employed include hole punches, transfer-type printing black-and-white images onto color film stock using color filters, alternating black and clear frames, and punching images out of a 35mm slide and taping them with Mylar into holes punched in the 16mm leader. *Courtesy Warren Bass.*

terns that cross frame-line boundaries at will. The results are striking, unlike anything that a camera can record. And they are impossible to describe in words. Mixtures of color, movement, and shape bounce off the movie screen with psychedelic effect. This is not to say, however, that the frameless style of drawing on the leader always achieves a pleasing and provocative viewing experience. Quite the contrary. As an animator you will have to purposefully create a changing set of carefully fashioned markings if you want to come up with a film that is interesting to watch. A series of tests is the only way you can identify styles or effects that work for you; take the best of these and integrate them into a unified expression. Include in your experimentation the study of what effects are produced by different sorts of drawing and coloring styles and by various drawing materials. Experiment with different printing techniques—for example, the use of a sponge and ink, or an eraser used as a stamp, or your fingertips as printing tools. When you've completed some experiments on your own, I recommend that you see some cameraless films by Norman McLaren and other animators.

FRAME-BY-FRAME STYLE

The second general style of marking on clear leader is centered upon using the frame lines. By modifying the shapes or the positions of images that are repeated on subsequent frames, the illusion of motion is created when the film is projected. Some samples are provided in Figure 2.11.

An important point to remember when you are drawing in this manner is that the individual frames are very small. Your working surface is much smaller than the smallest postage stamp. Its actual dimensions are six sixteenths of an inch wide and five sixteenths of an inch high. Remember these four elements as you try to create a recognizable image and then repeat it, with slight variations, frame after frame:

Simplicity. Reduce whatever it is you are drawing to its absolute minimum of details.

Tools. If you're working representationally you will need drawing implements with fine points. A pen that leaves ink drops, for instance, is useless. A good magnifying glass can help control your marking tool.

Motor Skills. To a large extent, success in working representationally depends upon your own personal abilities as a draftsman. Physical control must be very exact.

Registration. In order to have the drawings appear to move consistently and somewhat smoothly, it is important to devise a system that gives you the ability to place each individual drawing in just the right place on each frame.

REGISTRATION DEVICES

This last item may very well be the most important. Fortunately, you can create a device that will help you achieve some measure of accuracy in placing one drawing in the proper relationship to the drawings preceding and following it. The need for registration is basic to all kinds of animation. Different techniques require different ways of registering the positions of camera and object. For cameraless animation, there are a variety of registration systems that can be used.

Graph Paper. Place your clear leader on a piece of graph paper and use the existing grid lines to help match specific places on one frame with those on the following frame. The sprocket holes are used to establish standard reference points to the graph paper beneath the film.

Discarded Film. Place the clear leader on top of a piece of discarded 16mm film that has been photographically exposed and developed in the normal way—that is, "used" film. Line up the sprocket holes as you begin working. You'll discover that the "old" image gives adequate reference between one of your doodles and those preceding and following it.

Hand-Drawn Registration Chart. Before you start working, place a short length of 16mm clear leader on a piece of white paper. With a pencil or pen, trace the outline of the film's edges and sprocket holes. Remove the leader. Draw a series of horizontal lines to show the frame lines crossing the film's surface at each sprocket hole. Draw one or two vertical lines down the length of the film tracing parallel to both edges of film. Finally, you can make two diagonal lines that intersect each frame. The result of all this measuring should be a series of frames that are exactly like each other. You place your leader over this chart as you draw, successively moving the film up onto the chart as the work continues.

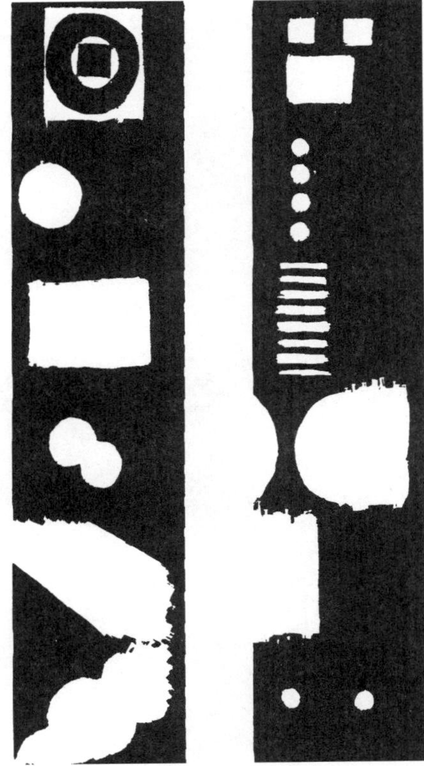

2.13 Work on black leader: Two film clips from Pierre Herbert's *Op Hop* suggests the strong shapes and sharp edges that can be explored through scraping off the emulsion on black leader. This film was done in 35mm format, allowing the artist a larger working surface and therefore greater control. *Courtesy National Film Board of Canada.*

SCRATCHING ON BLACK LEADER

Using the same general techniques and knowing the same basic facts, you can create a different kind of cameraless animation by using *black leader* instead of clear leader. Black leader (often called camera leader or opaque leader) is readily available from film laboratories or equipment sales/rental outlets.

It is easy to get a sharp, clean, thin line by scraping off the black emulsion (Figure 2.13). The resulting white lines (when projected) can be easily colored with felt-tipped pens and many people find that the results are the most pleasing form of cameraless film. In "scratch" films, the screen is black except for the images that have been etched onto the surface of the leader. The problem, of course, is that it's far more difficult to get accurate registration with black leader than with clear leader.

Scissors, straight pins, or any other sharp, pointed object is good for scraping off the emulsion. The first time you try this technique, be sure to test both sides of the leader to determine which has the emulsion coating. Otherwise you can scrape and scratch all day without producing any clear space through which the projector's light will pass. The most effective tool for scratching on black leader is a *silk-screen line cutter,* a sharp metal loop attached to the stem of a paintbrush.

PROJECTING CAMERALESS ANIMATION

Cameraless animation requires a lot of working time and yields relatively little viewing time. Here are some hints on how to stretch out the screening of your films and, in the process, extend their impact upon an audience.

Loops. If your piece of finished film is not too long (between 5 and 15 feet), you can thread it through the projec-

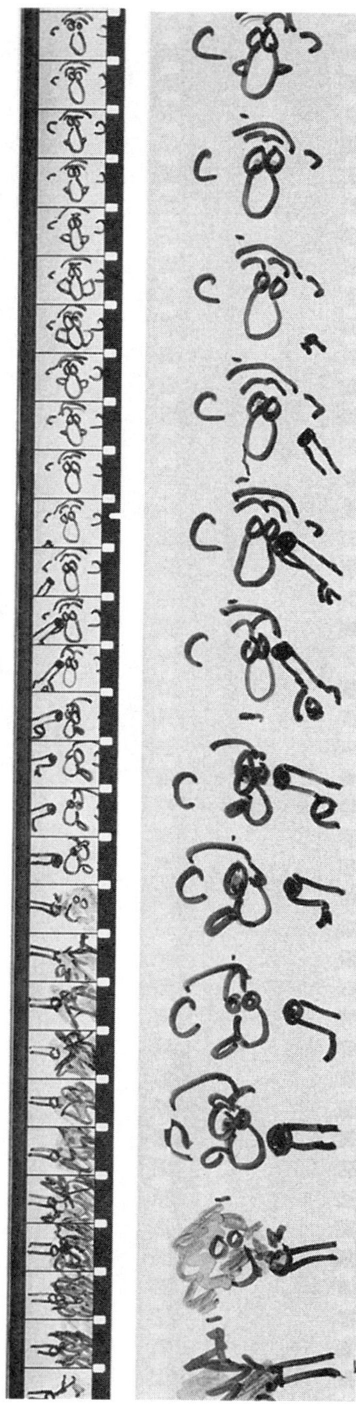

2.14 35mm cameraless animation: The size of a 35mm frame allows one to work with more detail than could be undertaken in 16mm format. Here are some samples of experiments by the author. The clips include a 16mm reduction print taken from the same series of 35mm hand-drawn films; a cartoon sequence drawn onto clear leader; a series of tracings from small photographs; drawings added to a "found" piece of film from a 35mm television commercial; abstract series of circular shapes on clear leader; and designs scraped into the emulsion on black leader.

SAFETY FILM

tor in a way that allows it to repeat itself continuously without rethreading. First thread the film normally and allow the front end to run out of the machine for 3 or 4 feet. Stop the projector. Take the front end (the "head") and splice it to the rear end of your movie (the "tail"). You will have to manually feed the film out the rear of the 16mm projector and into the front so that the film doesn't snarl up or touch the floor.

Silent Speed. The universal speed for 16mm sound films is 24 frames per second. This is an international standard. It is the slowest speed at which a film's sound track can move and still generate reproducible sound through the projector's amplification system. But many 16mm projectors are equipped with a switch that allows them to operate at a *silent speed* of 18 frames per second. At this slower speed our persistence of vision still works, so that visual continuity is maintained. When you project your cameraless film at silent speed, it takes longer going through the machine and hence it takes longer to see.

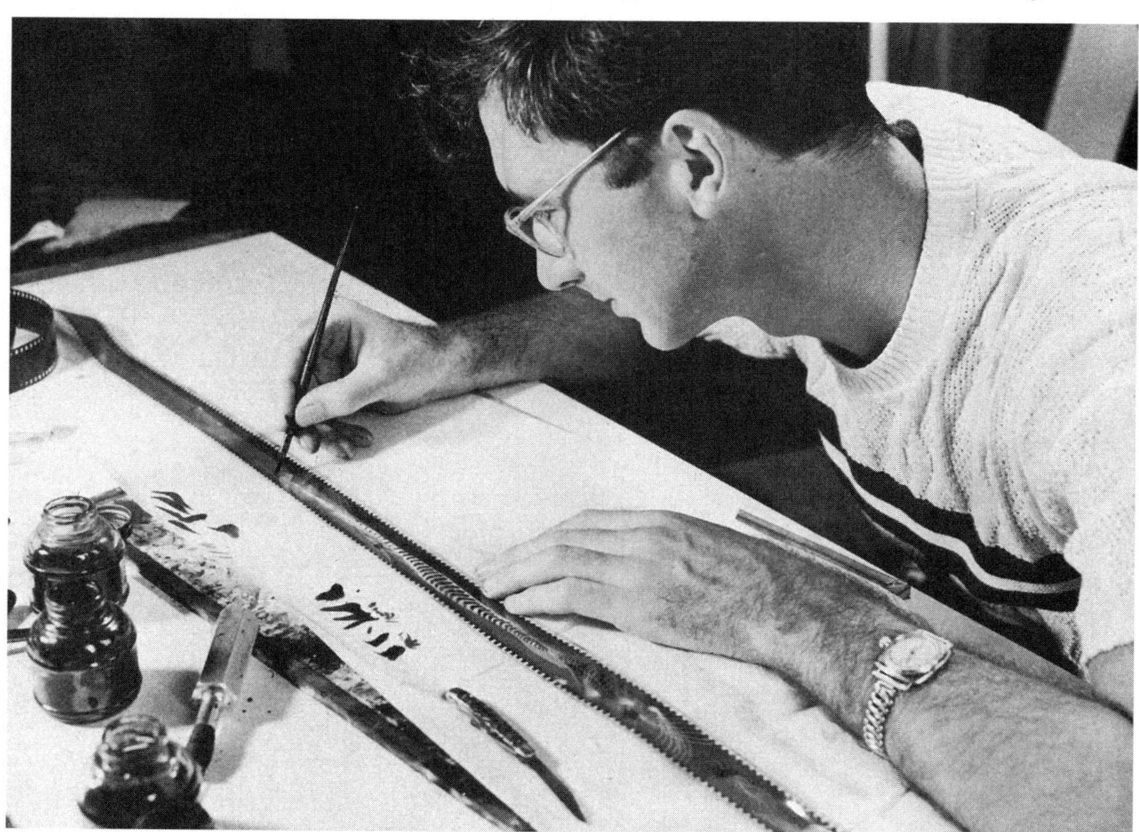

A

The difference between 18 and 24 frames per second may not seem significant, but it is. This reduction in projection speed will add one fourth more time to a film's running time. A 15-second film at sound speed will run 20 seconds at silent speed. The difference is significant, at least to the artist.

Forward/Backward. Another way to extend the viewing experience is by projecting film in reverse. Most 16mm projectors have this capability. Simply thread and run the projector normally and then, when the last frame of your work has been

B

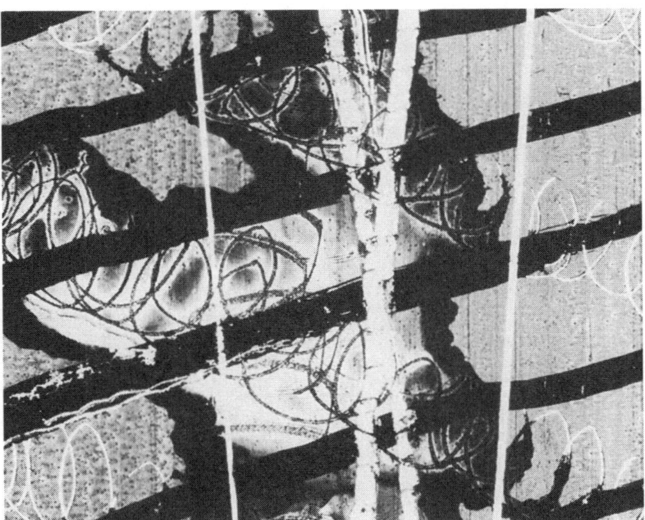

C

2.15 Norman McLaren's work: Although other artists had worked directly on film before him, Norman McLaren is recognized as the primary explorer, refiner, and popularizer of cameraless animation techniques. This series of stills shows his work and that of Evelyn Lambert, a colleague at the National Film Board of Canada. *(A)* shows McLaren painting directly onto a strip of 35mm clear leader; *(B)* shows Lambert applying patterns to the film's surface with ink and roller; *(C)* and *(D)* show frame enlargements from *Begon Dull Care,* a clear and opaque frame respectively. *Courtesy National Film Board of Canada.*

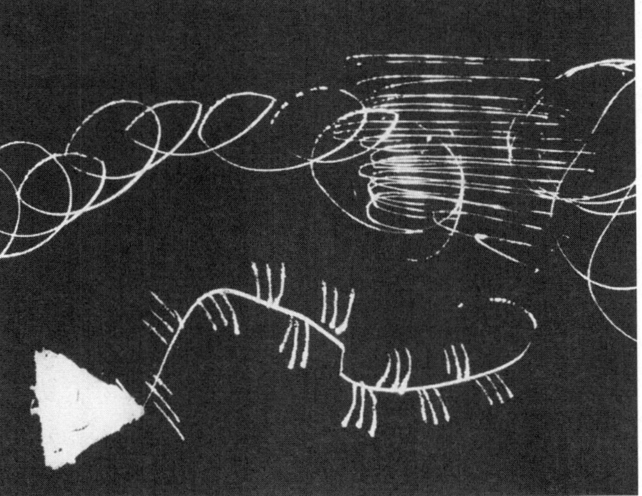

D

THIS PAGE CAN BE MADE INTO A
FLIP BOOK!
CUT ON LINES WITH A RULER AND BLADE. ASSEMBLE IN ORDER, AND STAPLE TOGETHER.
READY TO FLIP
SLIPPIN' AND A SLIDIN'

exposed, shut off the projector, put it into reverse mode and see the entire film again, back to front. You will need at least a 3-foot tail on your film so that it doesn't come undone or pass through the projection gate before you reverse the projector.

Musical Accompaniment. Whenever you can, play music as you screen your cameraless animation. It's weird to discover that no matter what kind or what tempo of music you select, it always somehow seems to work with the visual segment. And if you experiment with enough different music tracks, you'll come upon one that will appear to have been made just for your film. Prerecorded audio may be the easiest way to provide musical accompaniment to your movies. But it is also valuable to try creating your own original tracks to go with these movies.

2.16 A Tony Eastman workout: On the facing page is a do-it-yourself project designed by New York character animator Tony Eastman. Here Tony helps us see what a mistake it is to think that a progression of animated images must evolve gradually. Although the thirty poses in *Slippin' and a Slidin'* may seem pretty disconnected, the human eye has a powerful ability to merge them into a single swift action. To prove this is so, you've got to follow the directions and turn the page into a flip-book. If you can, try to output this full-page image onto a paper stock that has more than the ordinary thickness (you can use old-fashioned copier technology or you can scan the image and print it out on a computer printer). After you have studied the flipbook forward and backward, you might even want to try adding some "in-betweens." Chapter 14: Line and Cel Animation, will give you a structured introduction to the not-so-mysterious techniques of character animation. *Reproduced with the permission of author Tony Eastman and Metropolis Graphics. Originally appears as one of a series of flip-books published by Metropolis Graphics.*

35MM CAMERALESS FILMMAKING

Take a look at the pieces of 35mm film printed in Figure 2.14. Next to all those 16mm frames you've been studying and working with, it will look pretty big. A gracious landscape indeed. Because it's so gloriously large (to the animator, at least) 35mm clear leader is the most effective and luxurious medium for creating drawn-on-film images and sound tracks. Anything that can be done on 16mm leader can be done better on 35mm leader because the larger size affords greater control, greater detail, and greater ease in drawing.

The primary developer and popularizer of hand-drawn films is the Canadian filmmaker Norman McLaren, who has worked for more than thirty years at the National Film Board of Canada. McLaren's wonderful movies are now commercially available, and prints of them exist in most school and public library film collections.

Because of the size of 35mm film, it is helpful to use an accurate registration system when you want to repeat or vary slightly an image from one frame to the next. Working carefully, you can draw a registration strip by hand, making a series of frames carry the same visual patterns and then placing a piece of fresh leader on top of the grid system as you work. A better system is to get hold of a piece of 35mm film on which a grid has been filmed. You can use any piece of used 35mm film with whatever image exists as a registration guide.

The use of 35mm clear leader makes it far easier to create and control a handmade sound track. Here is Norman McLaren's own description of the technique he invented:

"I draw a lot of little lines on the sound-track area of the 35mm film. Maybe fifty to sixty lines for every musical note. The number of strokes to the inch controls the pitch of the note: the more, the higher the pitch; the fewer, the lower is the pitch. The size of the stroke controls the loudness: a big stroke will go 'boom,' a smaller stroke will give a quieter sound, and the faintest stroke will be just a little 'm-m-m.' A black ink is another way of making a loud sound, a mid-gray ink will make a medium sound, and a very pale ink will make a very quiet sound. The tone quality, which is the most difficult element to control, is made by the shape of the strokes. Well-rounded forms give smooth sounds; sharper or angular forms give harder, harsher sounds. Sometimes I use a brush instead of a pen to get very soft sound. By drawing or exposing two or more patterns on the same bit of film, I can create harmony and textural effects."

In order to project cameraless 35mm film it is necessary to have it *optically reduced* by a laboratory to a 16mm format. And 16mm can even be reduced to a super 8mm format, assuming, of course, that you don't have access to a 35mm projector—the kind of jumbo machine that you see in commercial movie theaters.

Reducing the film is a relatively simple process but requires a laboratory with special facilities. Be certain to check with your lab before you undertake a 35mm project. You may need to send your finished film to a lab that specializes in such work.

3 ▪ Tooling Up

A general introduction to filmmaking and computer tools is all that remains before you're ready to begin making animated movies, videos, and computer creations. This chapter will do just that, and quickly.

To get things under way, I'll first preview the basic film hardware you'll need to be familiar with. Then I'll go over the basic computer hardware. Between the two you will get a useful introduction to the various pieces of gear required for almost all of the techniques described in the chapters to come. The goal here, however, is only to orient the reader. A far more comprehensive and detailed description of equipment will be found in the seven chapters of Part III: Tools. Let me say again that I strongly encourage you to flip forward whenever you have a specific technical question, need further information about a particular tool, or are just plain curious to know more.

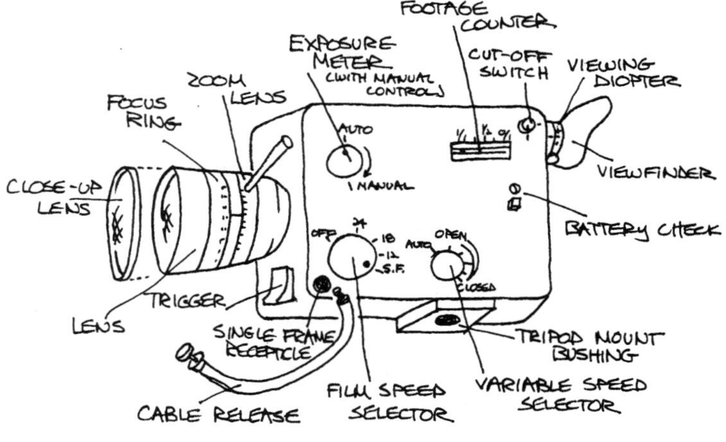

3.1 The basic camera: Whether you film with super 8mm or with 16mm equipment, the features indicated here are all basic to animated filmmaking.

THE BASIC FILMMAKING SETUP

Camera. A motion picture camera must have two features before you can animate with it. First, it must have the capabil-

3.2 Tripod capabilities and standard positions: The only absolute requirement of a tripod is that it be able to hold the camera absolutely motionless during and between exposures. Line drawing *(A)* shows the directions in which a good tripod head should be adjustable. Drawings *(B)* and *(C)* suggest very different but commonly used positions for the camera-mounted tripod during filming.

ity to make single-frame exposures, usually done by means of a cable release, a device that screws into the camera's trigger and allows the operator to release one frame at a time, as opposed to shooting the camera at full, live-action speed. An animation camera must also have a lens that can be focused on a relatively small field; 8½ by 10 inches is a working minimum, although some techniques will not require a field this small. Usually all motion picture camera lenses will take an auxiliary close-up lens or a diopter. These gadgets screw into place in front of the standard lens and act as magnifying agents. Figure 3.1 identifies these and other features of the basic camera.

Whether you work in super 8mm or 16mm format (see the Film Formats section of this chapter), your motion picture camera should have an exposure setting, worked either manually with the assistance of a light meter or by means of an automatic, built-in exposure metering system. Most super 8mm cameras have the automatic system. It's also helpful if the camera you use has a manual override, which allows you to set the lens opening or aperture by yourself. Other common features are a footage counter, a drive mechanism with a spring- or battery-powered motor, and a variable speed setting.

Zoom lenses are almost standard on today's inexpensive movie cameras. Having one built in or being able to mount a zoom on your camera will be helpful for various techniques in animation, although it's not essential. There are other special features and accessories, such as intervalometers, fade and dissolve mechanisms, variable shutters, frame counters, and backwind mechanisms that are nice to have, but they're gravy.

Tripods and Animation Stands. Common to every technique of animation is this inflexible requirement: The camera must be held in exactly the same position throughout the filming of a sequence. Great energy and endless gadgetry have gone into designing ways to hold a camera steady.

The simplest way of securing a camera is with a *tripod.* Generally, the bigger and stronger (and more expensive), the better. A good tripod will get you through every single filming requirement in this book. Beyond its requisite firmness, a tripod's best feature is its flexibility. Figure 3.2 shows three of the many positions in which a tripod holds an animation camera.

Animation stands are actually just sophisticated and specialized variations of tripods. Stands are less flexible but allow more precision and control in what is the most common camera position, pointing down onto a surface. Chapter 20 contains information on animation stands and their various features.

Lighting and Film. Animation almost always takes place indoors and thus requires artificial lighting. The precise kind of lighting depends on the kind of film stock that is being used. Most animators choose to use color film that is formulated for "indoor" or "tungsten" lighting conditions. There are many types of color (and black-and-white) films and each will require different degrees of brightness and different measurements of what is called color temperature. More on all this later.

Other Paraphernalia. To see what you've filmed, a few additional items are required. Most important (and expensive) of these is a motion picture projector. It's nice, of course, if you are able to project the completed film onto a screen, but this isn't really necessary; a white wall will do. It's also nice, but not essential, to record and then play back a sound track that accompanies the animated film. If you want to edit your film or join two rolls of processed film onto a single reel, then a viewer and a splicer are required.

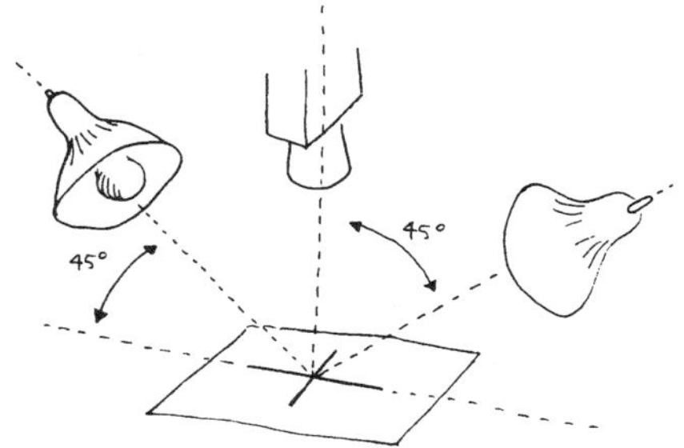

3.3 Basic lighting: For even illumination when shooting indoors, a pair of lights are mounted at a 45-degree angle to the camera's axis and to the surface being filmed. Pictured in this sketch are two photoflood bulbs that have been mounted in standard metal reflectors.

FILM FORMATS

As an independent animator, you have a choice of two production formats: 16mm and super 8mm. The following is a basic analysis of the similarities and differences between them. Incidentally, the word *format* is given a broad meaning here. It includes the film stock, film equipment, and even the process of filmmaking. In other words, format is an entire *system* for filmmaking. Selecting one of these two production systems is one of the most important decisions you will have to make.

Film Gauge. This refers to the actual physical makeup of what goes through the camera—the film stock. Paragraphs of discursive prose won't provide as clear a definition as you'll get by a quick look at Figure 3.4. Study the differences in width, perforations, and image area.

Image Area. The actual projected *image area* on 16mm film has three and a half times as much area within a single frame as that of the super 8mm image. The proportions of the frame's rectangle are the same in both gauges—horizontal to vertical dimensions form a proportion of roughly 4 to 3. This is called the *aspect ratio* and it is usually written 1.33:1.

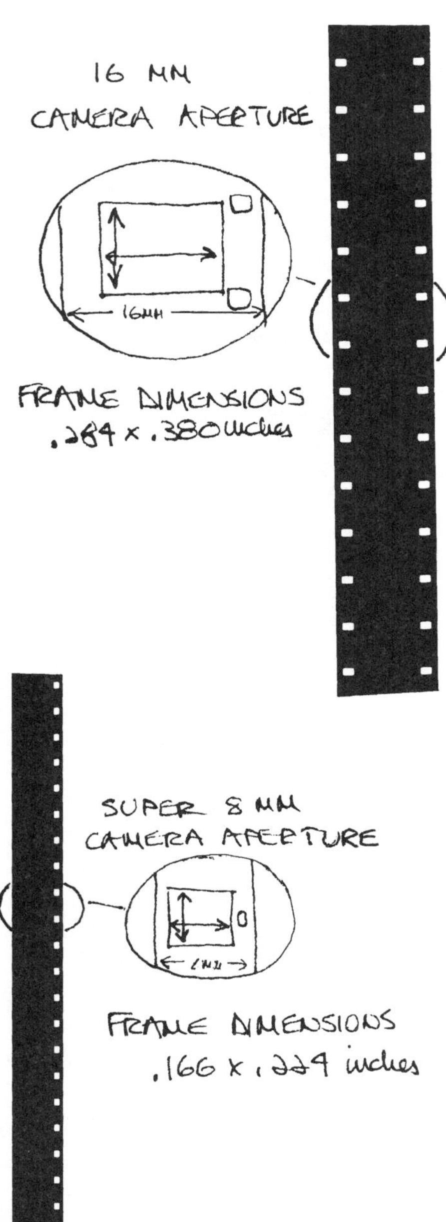

3.4 Film gauge: The relative sizes of 16mm and super 8mm film are easily compared by lining up a strip of each format. The line drawings suggest the detail of the formats and give precise measurements for each.

Cost. Sixteen-millimeter stock is three to four times more expensive than the same screen time of super 8mm stock. In animation, however, the cost of film stock is less significant in overall budget terms than in live-action forms of filmmaking. Even the most prolific animator shoots a lot less film than his or her live-action counterpart. A standard shooting ratio of exposed footage to footage in a final edited film is roughly 15:1 or 20:1 for documentary films and closer to 2:1 for animation.

Quality. It's not easy to compare the quality of super 8mm and 16mm gauges because it's not easy to decide exactly what quality is. According to technical definitions that measure grain, sharpness, and color accuracy, 16mm is always of a better quality than its super 8mm equivalent. But to the naked eye, the quality of gauges is very difficult to distinguish. A well-projected and well-exposed super 8mm sequence is virtually indistinguishable from a 16mm image.

Selection of Films. In general, there is little difference between super 8mm and 16mm formats. The high-resolution and low-speed stocks that are best for animated filmmaking are available in both formats.

Equipment Systems. Up until the mid-1970s, it would have been easy to claim that the 16mm format had a clear advantage over super 8mm in terms of sophistication and the technical quality of production tools. This is no longer true. By the mid-1990s, the design, workmanship, and reliability of the best super 8mm filmmaking systems had become equivalent to that of the 16mm systems. This goes right down the line, from cameras to tripods to editors to sound systems to projectors to laboratory services.

THE BASIC COMPUTER SETUP

While a lot of the animation and special effects you see in feature films are done using high-end computer systems that cost hundreds of thousands of dollars, the average computer user can still create beautiful animation using cheaper hardware and software. It may surprise some people that a large portion of computer graphics you see on TV shows and advertising—and in movies too—is made with the very same computer gear that might be sitting on your desk at home.

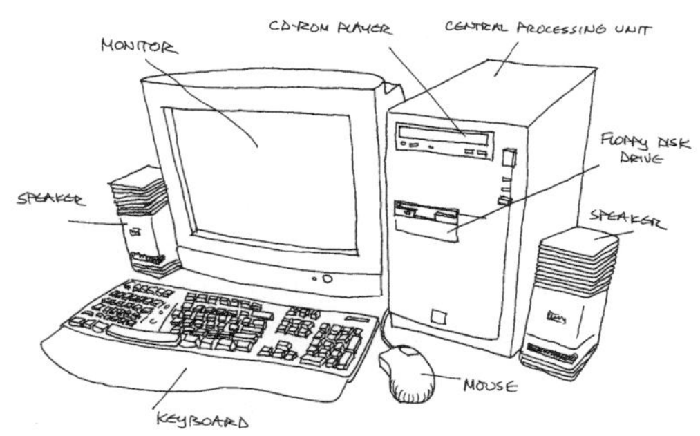

3.5 The basic desktop rig: The components of a generic computer setup are shown here. There are lots of accessories that you will want to consider as you get into digital animation: drawing tablets, cameras, portable storage units, scanners, and more. Chapter 24 will give you the rundown on all the goodies. And Chapter 23 goes over the software you will want to put inside the box.

Choosing a Computer Platform. Picking a computer is a difficult choice, even for the most techno-savvy buyer. Advice from friends and coworkers can help, but every computer is different from the next, and everyone has their personal favorites. The first big choice is deciding on a *platform.* For the average user, the two major platforms in the computer industry are the *Apple Macintosh* (*Mac* for short) and the *PC* (standing for *personal computer*). Both platforms are capable of producing wonderful animation since they often share the very same software packages. Yet each has its own set of advantages and disadvantages, including speed, disk storage space, price, and ease of use.

Macintosh Computers. When Macs were first introduced in 1984, they broke new ground by being extremely user-friendly. The icon-based interface (which revolutionized the industry by using the familiar metaphor of an office and desktop) empowered even the most computer-illiterate. Since then, Macs have managed to lead the pack when it comes to ease of use and they have a brand loyalty rarely seen in the consumer market. They are often described as the "creative person's computer" since it's often a stereotype that creative people are technically handicapped, and because Macs are famous for their powerful graphic design and desktop publishing capabilities. Macs typically come with everything you need to get started, including software, hardware, and especially sound and video capabilities. There are multitudes of models to choose from, each one having its own set of advantages.

PCs or "Windows" Computers. PCs are by far the most common kind of desktop computer. They are sometimes referred to as *DOS* computers—a reference to the *Desktop Operating System* that was created way back when by a fledgling company called Microsoft and was subsequently adopted by IBM for its early home computers. The terms *Windows* and *Windows NT* are sometimes used as well to refer to the PC universe because Windows is used on virtually all PCs. These are, of course, references to the operating systems created by Bill Gates's Microsoft Corporation and the terms have become synonymous with the desktop computer industry.

PCs are actually made by a wide range of companies—Compaq, Hewlett-Packard, IBM, Canon, Toshiba, Epson, Hitachi, and Sharp, just to name a few. Because of their broad adoption in corporate and technological America, PCs have

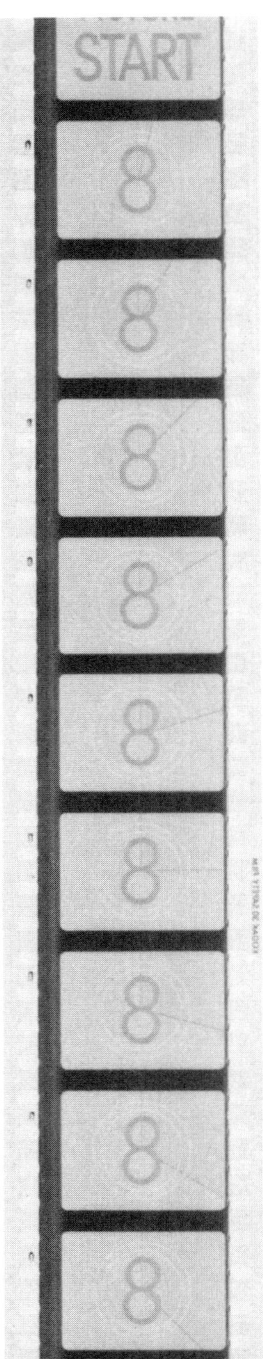

been generally classified as computers for businesspeople since they were relatively difficult to use unless you had some training or an innate skill at working with them. More and more, however, the use of Microsoft products has changed all that, providing PCs with an interface eerily similar to that of Macs, thus making them almost as easy to set up and use.

If you plan on teaming up with other animators who already have computers, it might be a good idea to buy the same platform they currently use. That way files and peripheral devices such as scanners or external hard drives can be easily shared and there won't be any conflicts between software. If you are a student studying animation, you will probably want to invest in the platform used in your school's program.

There are many other things to consider before buying a computer, no matter which platform you decide on. Animation isn't like writing reports or term papers. The software used, the storage space needed, and the file types involved in producing even the simplest animation demand a lot from a computer.

FILM AND DIGITAL DEPENDENCY

The farmers and the cowboys should be friends.

The time of rivalry and mutual skepticism between animators working in film and television and those working with computers is quickly passing, I am glad to report. For a while it seemed that differing attitudes, aesthetics, and working techniques formed a gulf between animators working in the various media formats. But that phase is history. The reason? *Everything has gone digital.* Feature films that are shot in the traditional 35mm format are today routinely transferred into a digital format—the language of computers—where they are edited and scored. It is exclusively within the digital domain that contemporary moviemakers concoct those dazzling special effects. Television has gone digital, too. Cameras record onto digital tape decks operating on international D-2 (the *D* is for "digital") and Beta SP standards. Virtually all film and TV editing—plus compositing of live-action with graphic materials—is migrating toward what are termed *nonlinear,* digital systems. They are so called because changes can be made in the edit list at any place, and opticals such as dissolves and freeze-frames can be instantly previewed. In the future, television broadcast, cablecast, and switched broadband networks will all operate exclusively in digital formats that can speak with each other. And, of course, computer animation is flourishing at both the low end (treated in this book) and with more expensive digital platforms provided by companies such as Silicon Graphics.

Because everything has gone digital, everything can mix. Footage shot on a 16mm animation stand can be seamlessly wedded to a music track lifted off a CD recording. A three-dimensional character generated on the computer screen can be transported onto the television screen or transferred onto film stock and projected onto a movie screen. The future is clearly one in which computer, TV, and film technology are going to operate codependently. In the process, each media form will be freed to do what it does best. And new tools and techniques for image building will lead to new levels of storytelling. All of us who delight in animation are in for some mighty fine times.

But we're getting ahead of ourselves. Let's spend a moment considering the fundamental relationship that has already emerged between film and digital animation.

Computers Are Good at Familiar Things. The next part of this book will prove, beyond a doubt, that digital tools and processes have made themselves indispensable in many of the traditional techniques of film animation. As it turns out, computers are very, very useful in at least three broad tasks that accompany *any* form of animation: storyboarding, making audio tracks, and production management. You will see in following chapters how off-the-shelf desktop animation tools and software packages provide alternatives to many animation techniques that, until a few years ago, seemed like they could only be done in the photographic realm of film. The chapters on cutout and stop motion will show, for example, that both techniques can be done on a computer. Digital methods have made it much easier and much more effective to create stunning animation from still images. The venerable (but really awkward) technique of rotoscoping has been born again through the computer. There are lots of steps within the traditional processes of character and cel animation where digital tools can help and sometimes replace cumbersome filmmaking gear.

Computers Are Creating New Ground. It won't take you long to appreciate how computers are helping traditional animators by taking away some of the drudgery often associated

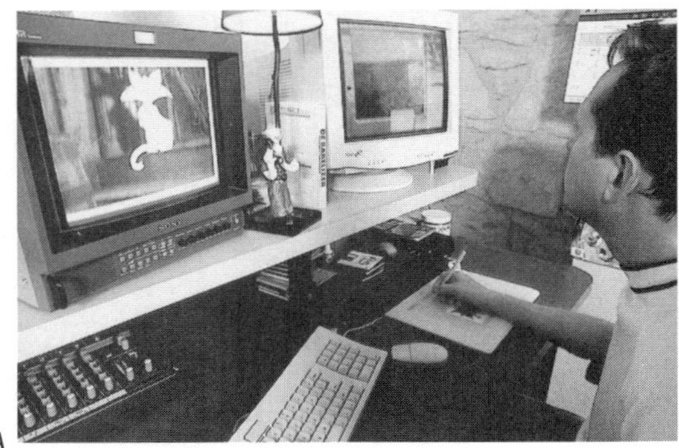

A

B

C

with hand-drawn animation. But computers are also the tools that help create groundbreaking forms of animation, the likes of which couldn't exist without the aid of microchips. A large chunk of this new aesthetic turf can be found in *computer-generated images* (*CGI,* for short) such as the 3-D animated characters in *Toy Story,* and the special effects combined with live action in *Jurassic Park, Space Jam,* and *Terminator 2—Judgment Day.* Computer technology has propelled visual effects to a whole new level, making once-impossible effects commonplace.

This level of animation may seem overwhelming and out of reach to a computer novice at first. Yet the uncharted visual world of three-dimensional images, morphing, and composited effects can be explored on home computers. Many of the people working in Hollywood's visual effects industry got their start in computer animation at home, using average Macs and PCs.

Nor is the revolution over. Chapter 17: Animation Frontiers will introduce developments from the cutting (bleeding!) edge of computer animation, where the processes of technical innovation and creative exploration are just gearing up. As exciting as the new tools are, far more exciting and far more important is their potential power to create groundbreaking genres of animation that can stir the human imagination and speak to us in that deep way in which any genuine piece of art touches us.

It's easy to be rah-rah about digital animation. But this isn't to say that computers will turn everyone into master animators. Nor is it safe to think that animation will suddenly transform the entertainment

3.6 The digital high end: Although this book is about accessible and relatively inexpensive or "low-end" tools for digital animation, it's a good idea to know something about the top-of-the-line tool sets. This seems a good place to do just that. This series of photographs was taken in Austin, Texas, at a top creative facility called 501 Group. Together they give an image of the professional-level tool set that is routinely used by animators who are putting together projects for corporate and television clients.

(A) shows a standard Power Macintosh workstation that is connected via ethernet and by T1 Internet connection to other tools in the 501 facility. In *(B)* you look over the shoulder of a graphics designer working on an SGI (Silicon Graphics) platform that can do 2-D animation, compositing, rotoscoping, and 3-D animation. *(C)* gives you a look into the videotape room where racks of multiformat machines connect and store the various production suites. Final output (and initial input) can accommodate any combination of digital tape formats that are used in professional TV production: Digital Betacam, D5, D2, 1", U-Matic, VHS, S-VHS, HI-8, and Sony DV. *Photos courtesy of 501 Group.*

industry. Computers are still just tools—machines that will only help a talented beginning animator to become more efficient and more creative. Digital animation has a long way to go before it can boast a classic body of work that matches that of traditional film animation, yet those of us who have followed the birth of digital animation can get pretty fired up about all the changes in technology and all the opportunities computers have created!

HOW TO CHOOSE BETWEEN FORMATS

If all beginning animators were forced to select between super 8mm and 16mm production, or between a full-blown Mac and a PC workstation, they'd face a confusing barrage of offsetting factors and conflicting recommendations. For digital animators, arriving at the computer store with a fat bankroll wouldn't help either, because that just increases the bewildering array of choices.

Fortunately, few neophyte animators ever face such decisions. For most of us, it's not *which* format to go with but rather *how* to get going at all. Polemics are replaced with pragmatics. You use what you can afford and what you can get your hands on.

If you are faced with the decision of whether to work in film or digital format, trust your own common sense. Ask yourself the following questions:

What Is Available? If you already have a computer, check out what it will take to purchase the software and hardware required to start animating. If you can get your hands on some filmmaking equipment, you should certainly try to build your exploration of animation around that equipment—regardless of it being 16mm or super 8mm format. Snoop around to see what tools are available. Can you con someone into letting you borrow their gear?

How Much Do You Want to Spend? Money matters. Take a cold, hard look at what you can comfortably invest in your passion to animate. Chapter 24: Computer Hardware should help you develop a rough working budget for whatever kind of film or computer project you have in mind. And there is more instruction on budgeting in Chapter 18: Production Planning.

3.7 Splicing: A long shutter speed suggests the flurry of activity in preparing 16mm camera original and magnetic recording stock for subsequent printing at a film laboratory. Pictured are a portable hot splicer and a four-gang synchronizer with sound head. *Photo by Elissa Tenny.*

What Kind of Animation Will You Be Doing? Sometimes a decision on format can be made by choosing the sort of cartoon you want to make. If it's frame-by-frame drawing techniques that hold your fascination, then filmmaking techniques are the way to go. On the other hand, if you want to use still photographs to weave a montage with rich effects and dense visuals, then you will want to move directly to the computer.

Who Will Be Seeing Your Work? The advent of home video and computer technology has provided exciting new distribution potential. Now almost anyone can produce cartoons and share them with others, either on videotape or as digital files sent between computers. The future seems to hold great possibilities for compressing cartoons and sending them over the Internet. CD-ROMs—and their next DVD generation—will help expand your audience. Unless you are satisfied screening your work for a limited number of people (fellow students, for example), film technology can no longer be looked to as a good distribution medium. The 16mm film distributors who flourished when this book was first written have mostly gone out of business and few institutions or individuals own and operate super 8mm or 16mm film projectors. Certainly the most universal way to circulate your work is by transferring it from either film or computer to VHS video.

STARTING OUT

I am a strong believer in inductive learning. I trust experience. All the questions and decisions that will face you as an animator are best taken on at that precise moment when the questions spontaneously present themselves, and not before. What is important is to get yourself under way, to begin working in animation. After that, matters take care of themselves as one tool or technique leads to another and as there is a comfortable back-and-forth between the film and digital worlds.

Whether your next step is to scrounge or borrow a basic animation setup or you already have all the equipment you need to get started, it's my hope that you are now prepared to actually make some animation, to explore various techniques, to get deeper into this great art form. Your orientation is complete. The following fifteen chapters stand ready to help you on your way.

II · TECHNIQUES

Here come fifteen chapters about the nitty-gritty of creating your own animation.

There are two *megacepts* you should know. What's a megacept? It's an idea or perspective that is so broad it applies to *all* fields of information. Megacepts are so encompassing that they are often hard to see. (Does a fish know it lives in water?) That's why I want to offer two such large perspectives to introduce Part II: Techniques.

DIGITAL IS DIFFERENT

This new volume of *The Animation Book* introduces the convention of case studies. Just about every time the discussion turns to digital animation techniques, you will find an extended example that offers the opportunity to look over someone's shoulder as he or she shows you, step by step, how a particular scene was built using a particular software program.

Such case studies weren't needed when this book focused solely on film technology because filmmaking tools are pretty much the same, regardless of whether one is working in super 8mm, 16mm, or 35mm format. In the digital world, this is not so. Specific software applications have distinct techniques and significantly different on-screen interfaces and tools. To know the nitty-gritty about computer animation you have to get much more specific than was necessary with filmmaking.

But there are hundreds of software applications. Which "app" warrants its own case study? I have chosen ten. In my judgment these are the established leaders in their categories. They come from respected publishers, are widely available off-the-shelf, and operate on the standard configurations of desktop computing hardware.

Learning digital techniques requires a hands-on approach. Learning film-based techniques can be more conceptual. Bouncing between the two will require you to be conscious of switching gears. To help you keep a pragmatic scale of reference, I try to anchor coverage of digital techniques by making the case studies highly personal. You'll meet an eclectic bunch of smart animators working on a broad range of nifty projects.

BOUNDARIES ARE ARBITRARY

Classification can be a pretty dubious undertaking. Most of the techniques that follow are well-established ones that take their categorization from materials such as clay, 3-D, cutouts, and line and cel animation. Other techniques are singled out for the tools and processes they use, such as digital ink and paint and rotoscoping, for example. Three of the fifteen chapters are cross-technique: Working with Sound, Storyboarding and Animatics, and Production Planning each deal with a set of problems encountered in any animation project. The sequence of chapters has been loosely structured so that a beginner will quickly gain animation's basic precepts and concepts (yeah, maybe even a few megacepts too) that will be useful in understanding later, more complicated topics. Under the catch-all title of Animation Frontiers, Chapter 17 highlights a handful of areas that will one day warrant treatment as chapters in their own right.

The point here is that our art form is very much alive and growing. As you study the various animation techniques in the following chapters, I hope you will appreciate that the boundaries between these genres are simply conventions. In moving toward your own work, disregard all such schemata. Plunge in where you want. Experiment freely. It's in the breaking out from and cross-fertilization of established techniques that the best new forms of expression are found.

4 ■ Animating Objects

A

B

4.1 Materials: That just about anything can be animated is illustrated by these two frame enlargements. In *Bags,* by Tadeusz Wilkosz, simple household objects plus a bag create a drama of aggression, *(A).* Simple cookies with drawn heads constitute all the players in Frenc Varsanyi's *Honeymation, (B).* The films were produced in the former Czechoslovakia and Hungary, respectively. Eastern European animators have developed strong traditions in animating objects and puppets. *Courtesy Pyramid Films.*

Above all, animation is the art of movement. The accomplished animator can bring to life just about anything—a series of drawings or a tin can. Unfortunately, there are no prescriptions or formulas for animating an object. You just have to develop a feel for movement—what animators call "touch." It can be learned only through experience, though there are some general guidelines that will help you along. One of the quickest and easiest ways to develop your understanding of the nature of animated movement is to experiment by animating a series of small objects.

MATERIALS

It is an indication of the medium's power that even the most prosaic and ubiquitous of household items can be used to create interesting animation, pieces that move an audience in emotional and conceptual terms. And you can do this with anything—with pipe cleaners, pennies, pop-tops, pajamas, plates, penknives, pansies, peas, pills, and pins. As long as you can lift it and move it, it's possible to animate it. You can make a group of pencils roll themselves across your desk, have some salt and pepper shakers do a jig, or even make a roomful of chairs and tables play at bumper cars. For the beginner, it's a good idea to start off with smaller items to get the hang of things, and when you're a little more experienced maybe then you can try bringing your living room couch to life.

Before discussing how animators treat such objects, I want to offer you a view of a distant summit and point out two trails for your ascent.

If object animation seems awfully simple, let me assure you that it is the very direct first step toward one of the animation art form's loftiest peaks. *Stop-motion animation* is the name of the technique that was used to create such movies as *The Nightmare Before Christmas* and *James and the Giant Peach,* and the wonderful *Wallace and Gromit* feature and series of shorts by British animator Nick Park. Here, simple objects are replaced by fully articulated puppetlike characters or by a series of slightly modified 3-D models. A stationary camera is replaced by a *motion-control camera* rig: a 35mm camera mounted on a computer-controlled arm that plots its way, millimeter by millimeter, through a landscape of stop-motion elements with an accuracy that can be repeated exactly in successive takes.

Object animation is an excellent place to begin one's apprenticeship because it can be so easily undertaken with both filmmaking and digital tools. The two creative pathways start out with equipment configurations that are both relatively fast and cheap.

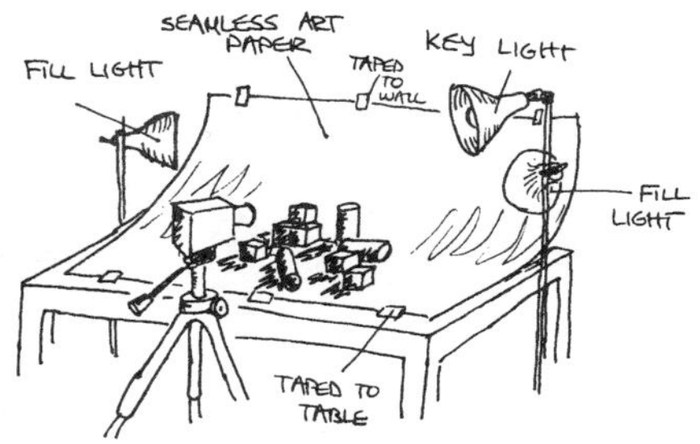

4.2 Diorama setup: The illustration suggests a common setup for animating small objects. Many lighting combinations are possible. Here a *key* light focuses on the objects while two *fill* lights illuminate the background. The use of seamless art paper, available at art stores, creates the illusion of a "limitless" background.

THE SETUP

Figure 4.2 shows a typical arrangement for animating objects. Until recently, the camera was always a film camera—super 8mm or 16mm formats being most common. But, as noted on page 52, today you can shoot by hooking up a computer to video cameras—Hi-8 or VHS formats being most common. Whichever recording device you employ, it is the objects you've chosen to animate that will dictate the setup required in shooting them.

The heart of the animation, as we've been saying, is movement. Anything that detracts the viewer's attention from what is being animated works against the movie. For this reason, backgrounds in all forms of animation tend to be simple, nondistracting, and absolutely rock steady. Certainly this is true in animating objects. If the placement of the camera is to be vertical, that is, pointing straight down at the surface to be photographed, then a solid-colored piece of paper can rest underneath the objects themselves. Naturally, the color of this background should help emphasize the objects on top of it. Often the camera will be mounted in a horizontal position and pointed toward what amounts to a small stage with both floor and backdrop. In this case, seamless art paper forms an effective background material that has no edges or structural details that can detract attention from the animation. Such backgrounds appear to recede into a depthless void. This kind of setup is called a *diorama.*

As for lighting, although different objects will present unique lighting problems, generally a form of "flat" lighting is used that minimizes the existence of strong shadows that overpower the movement of the object being animated. This is achieved by the use of at least two light sources, three if possible. The illumination of one tends to eliminate the shadows caused by the other. However, sometimes lighting must be used more dramatically. For instance, it may be important to create a bright background against which the objects will stand out clearly. A controlled absence of lighting can create a similar effect. In this case the lights are focused only on the objects being filmed, while a black background, unlit, provides contrast for the animation itself. Lighting is a creative universe unto itself. Try out different lighting schemes that experiment with mood. For example, you can animate shadows themselves by moving your lights around while shooting. One of the benefits of shooting your animation digitally is that what you see on your computer screen is what you get. It's impossible to overexpose or underexpose your image, a frequent problem for beginning animators who use film.

A

B

4.3 Overhead setup: Canadian Laurent Coderre is shown working with tiny chips of wood in his film *Zikkaron, (A).* Note the simplicity of the setup: a motionless camera held over a sheet of black matte board, *(B). Courtesy National Film Board of Canada.*

THE VIDEO/COMPUTER TOOL SET

Due to the recent growth of digital technology, there are now many kinds of electronic cameras that make it easy to shoot object animation. Not only can you hook up a standard VHS camcorder to your computer for video input, but there are also dozens of digital cameras made especially for capturing images for the computer, all of which vary in features and price.

The Connectix Quickcam is an extremely cool-looking digital camera. Once hooked up to your computer, it's easy to use the eyeball-shaped gizmo to take still photos of your family and friends, to do videoconferencing over the Internet, or—for our purposes here—to record bits of video. When the signals from the camera are routed through a digital video application such as Adobe Premiere, the Quickcam can create video clips frame by frame. There are currently two models on the market. Both are small enough to fit in the palm of your hand. When placed in their plastic holder, Quickcams are easy to move around, rotate, and tilt up and down. There is also a threaded mount on the bottom of each camera, so that you can mount one on a tripod. The gray-scale model was the first one available and if you're on a slim budget, it's the perfect item to start off with. If color is a must, Connectix also has a full-color Quickcam available, which provides crisper photos and better video.

While a Quickcam is great way to get a taste of object animation, there are multitudes of consumer-level video cameras that perform at a higher grade, and that cost more. A Hi-8 video camera or a standard VHS camcorder can be hooked up to the computer using the proper cables and the video-in jacks. All such cameras can be mounted easily onto a tripod and come with zoom lenses and a tiny viewfinder monitor that is useful for composing shots before the frame-at-a-time recording begins.

4.4 Animated beads: Four frames from Ishu Patel's *Beadgame* suggest a more sophisticated style of animating small objects. Here tiny plastic beads, the kind used in beaded belts and costumes, become the medium creating complex transformations of line and color. The "ray" effect in photos *B* and *C* was created with a computer-controlled movement of the camera while the shutter is opened on successive single-frame exposures. *Beadgame* is the product of a highly sophisticated camera and stand, in the hands of a highly experienced animator. *Courtesy National Film Board of Canada.*

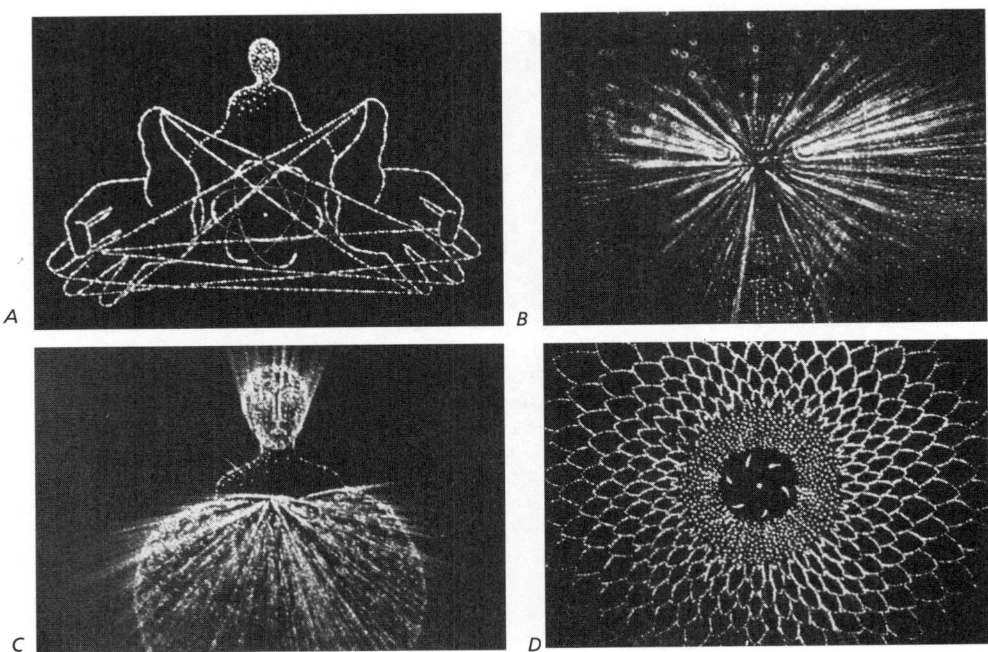

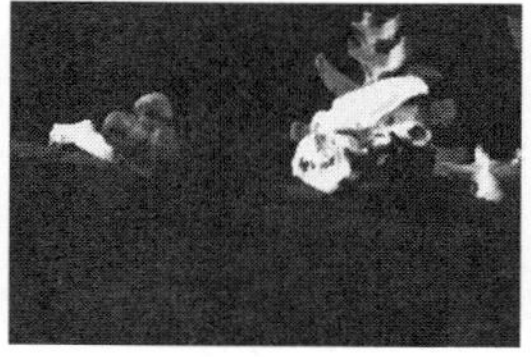

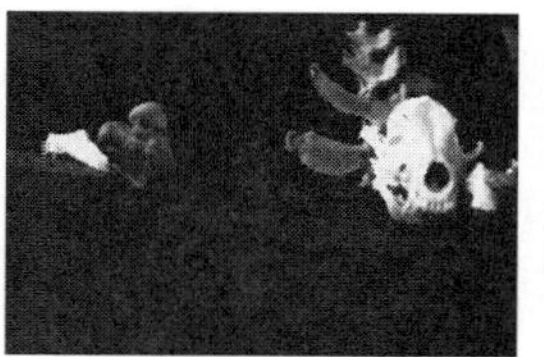

A

Kit Laybourne's office in New York City became an impromptu studio for producing an example of Digital Object Animation. Kit and colleague Michael Dougherty thought a coyote skull and some cattle bones might make an interesting and eerie composition, and one that would be easy to animate. Figure *(A)* shows samples of the finished piece, which was ultimately turned into a QuickTime screensaver. The action consisted of the coyote scull spinning around and then attacking the viewer.

B

Setting Up

The setup was extremely simple, with source materials captured directly onto Kit's Mac 8100. In *(B)* you can see that we created the diorama using a tabletop as the surface and a black sheet as a backdrop to nicely emphasize the stark white of the skull and bones. We used both natural sunlight and household lamps to light the scene. The hardware setup is a multimedia-configured Mac running Premiere video software and a QuickCam from Connectix positioned at the same height as the bones. The QuickCam can create QuickTime movies in either black-and-white or color, depending on what model you choose, and can create movies at 160 x 120, 320 x 240 (half size), or 640 x 480 (full size) screen resolutions. We used the black-and-white QuickCam.

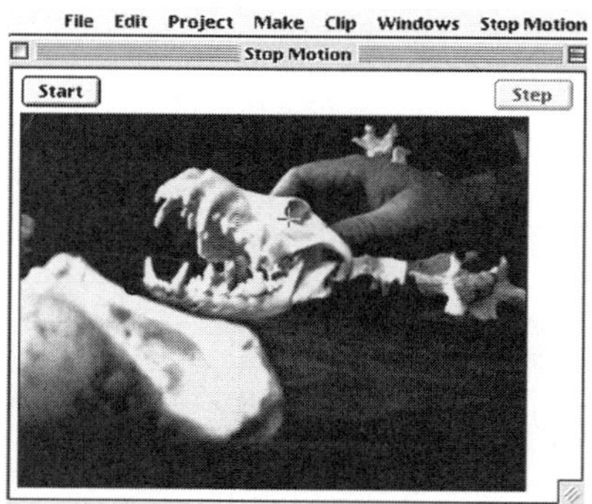

C

Composing the Frame and Animating

Before beginning the animation, the frame was composed and lights were positioned to create the best-looking frame, *(C)*. We rehearsed the movement of the skull, playing out different actions and their corresponding impact. Previewing is easy to do, since Premiere gives you a display of exactly what the QuickCam sees and what will be recorded in the animation.

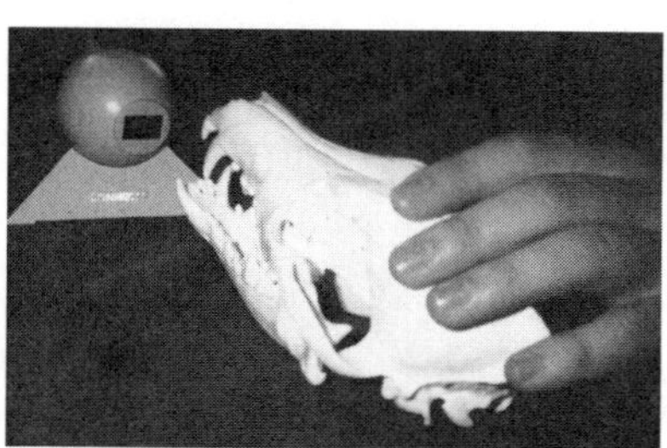

D

With the setup and composition fixed, it was time to bring the dead back to life. The skull was moved ever so slightly and then a new frame was taken using Premiere's stop-motion feature. This move, click-a-picture, move-again process was repeated until the scene was completed. *(D)* is a production still of the final frame in the sequence, which corresponds to the last frame of the piece, as shown in *(A)*.

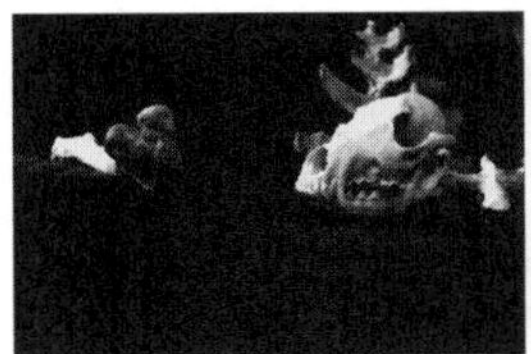

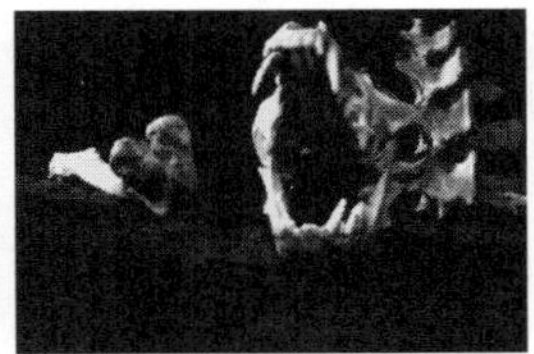

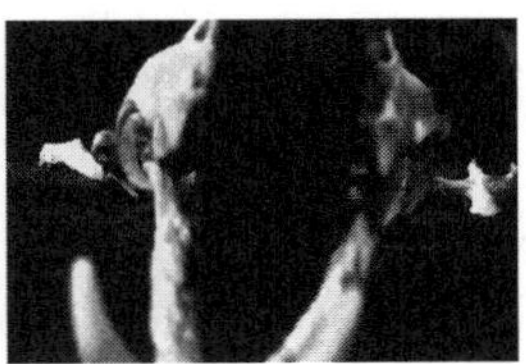

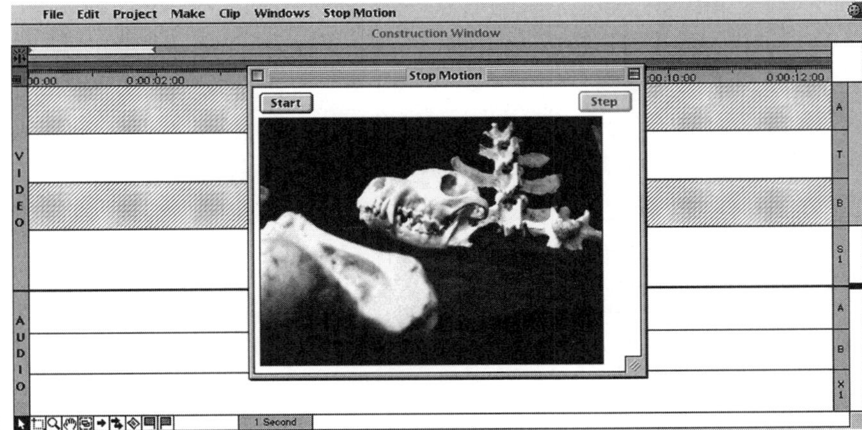

E

Premiere Interface

The stop-motion feature in Premiere, *(E),* allows you to create animation frame-by-frame using the video signal from a QuickCam or other video source hooked up to your computer.

Note that the "Step" button captures the current frame, and even makes a *click* noise similar to traditional film cameras.

Once completed, the animation was saved as a QuickTime movie and labeled "Clip: skull movie," as seen along the top of the file, *(F).* There's no trip to the lab. You can play back the object animation as soon as it's done and look for mistakes. Those are the playback controls at the bottom of *(F).*

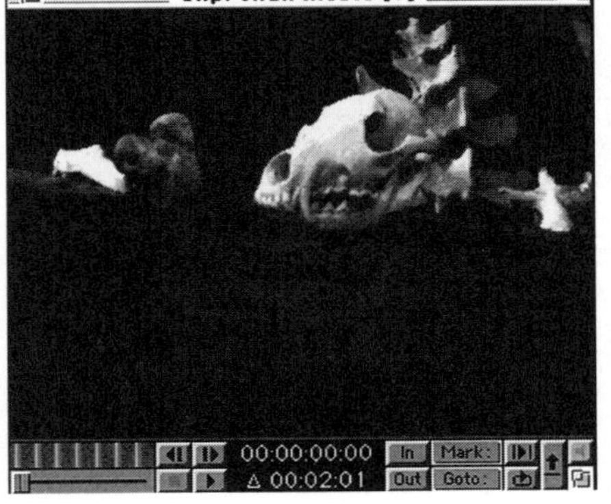

F

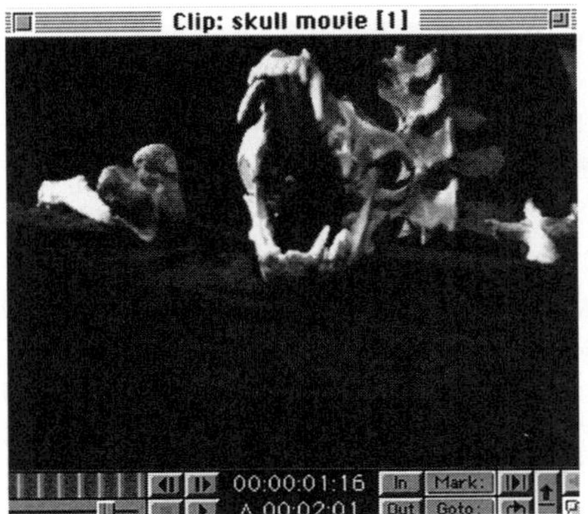

G

This little piece of animation, *(G),* was quite short — about two seconds long — and took about a half hour to complete, including setup and animation. It was created at 320 x 240 (half size) resolution and is small enough to export and fit on a floppy disk. Perhaps we'll send it out next Halloween.

No animals were harmed in the making of this QuickTime clip.

4.6 Distance/speed relationships: This series of photographs tries to suggest two aspects of relative movement upon the screen. Moving the same *real* distance, the pawn appears to move more quickly when it is passing a stationary castle, *(A),* than when it passes a castle that is also moving, although not at the same rate, *(B)*. The proximity of the pawn to the camera in the third series, *(C),* will cause it to appear to be moving faster than in the first series, despite the fact that in both the pawn is covering an identical distance in real space.

Along with a computer and a camera, you need a software program such as Premiere from Adobe. Premiere has an excellent stop-motion recording feature that can use the camera's video signal as the source. When recording in Premiere's stop-motion mode, a window pops up showing you what the camera sees so that you can compose your image, and when you shoot a frame it even makes a nifty *click* noise that mimics an old film camera. The case study on the previous pages gives you a step-by-step introduction to animating objects with a computer setup.

GIVING THE OBJECTS LIFE

Let's say that you've got a camera and a handful of loose change in your pocket. Hook the camera up to your computer, place it on a tripod, surround it with lights, and point it down at the surface of a table.

How can you make the money move? When is something moving too fast for the eye to follow? Is it possible to imbue objects with recognizably human kinds of movement? You'll have to find out the answers to such questions by your own experimentation. Here are some general parameters.

How Many Frames? The most thorough way of animating an object is to shoot one frame of it, alter the object's position slightly, and then shoot another frame. The camera and background must not move during this frame-by-frame shooting process. It is the absolute steadiness of the background and camera angle that permits us to perceive the illusion of movement brought about by incremental changes in the object's position. During the projection of the film or the playback of the video, the tiny incremental changes pile on top of each other and the persistence-of-vision phenomenon lets us experience the illusion of movement on the screen. Changing the position of an object before each exposure requires shooting twenty-four separate movements in order to make one second of film or thirty separate movements per second of video.

Fortunately, "smooth" movement can still be perceived when two exposures are snapped before changing the position of the object. This lets the animator create the same quality of movement in the same amount of screen time by making half the number of changes to the object's position. This process is called *shooting on twos* or simply *shooting twos*. It is a basic element of animation technique. Unless there is a special reason why single exposures must be made, animators will usually photograph two frames before altering an object being animated. After all, this saves half the work, and the visual results of shooting on twos and shooting on ones, as seen in the finished animation, are not significantly different. So unless otherwise stated, *it is always assumed that shooting in either film or digital formats will always be done by shooting on twos.* This holds for nearly every animation technique. The habit of clicking two frames at a time will become second nature.

Sometimes, however, it is possible and desirable to shoot on threes, fours, or even fives. Extending the number of exposures for each position causes a flickering or jumpy quality to the movement. The larger the number of exposures between movements, the greater the "visual stutter." But smoothness is not always appropriate. You may want to have objects moving in a jerky, high-energy fashion. There is only one way to learn the different effects caused by exposing a different number of frames: You must experiment. And if you're shooting your object animation digitally, you'll discover that experimentation is easier and results are instantaneous since you don't have to process any film. As you begin to know applications like Premiere inside and out, you will also find that digital tools allow you to fix a broad spectrum of problems. Whereas the film animator must reshoot an entire scene, the digital animator can manipulate single frames, saving time, energy, and moola.

What Distance? What Speed? Because projection and playback speed is usually a constant (24 fps in film and 30 fps in video), the greater the physical distance you move an object between exposures, the faster it will appear to move, assuming the same lens and camera location. The closer you space its positions, the more slowly and smoothly an object will appear to move in the finished animation. Note that movement is always perceived relative to something else—usually the background or the frame of the image itself. "Relative" or "apparent" speed is worth elaboration. Figure 4.6 illustrates how to make an object move quickly or slowly.

There is no limit to how smoothly or slowly you can animate an object. There is, however, a perceptual limitation on the maximum rate of movement you can use. If the object is placed in a position that does not permit it to at least partially

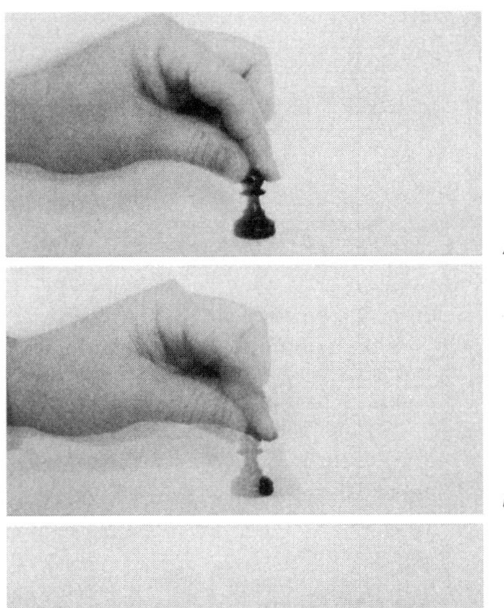

4.7 Flutter: A photograph shot with a long exposure time, *(A)*, suggests a very small change between an initial and a final position. If the first and last positions were shot on an animation camera (one wouldn't see the hand moving the piece), the movement would appear very smooth and very slow. Photograph *(B)* suggests a larger change in position. This would be perceived as a faster movement if the first and second positions were filmed with an animation camera. Note the overlap of position—the dark area. Because of this overlap, the motion on the screen will appear smooth. Two stationary pawns, *(C)*, suggest a degree of change between the first and second positions that would create a "flutter," or "strobe," effect.

overlap or touch its preceding position, its motion may appear to "flutter" when played back. Figure 4.7 shows the problem of visual continuity.

Movement and Style. Here's where the real fun begins. In the final analysis, the effectiveness of animation is not simply a matter of how far a movement goes or what number of frames it commands. What's more important by far is the *way* it moves and the feelings that this movement evokes. Animation is really the art of making things move with style, and the common gift of all great animators is an ability to observe and then re-create movement in ways that resonate deeply within all of us. To the good animator, movement becomes a concrete material that has color, portrays mood and motivation, and is textured in overtones of human passion and human meaning.

4.8 Animating blocks: *Tchou-Tchou* is a fourteen-minute adventure story featuring five characters that are created from children's blocks and inhabit a block world. In the film's climax, the hero and heroine, *(A),* and a friendly bug, *(B),* outsmart a ferocious dragon composed of a long row of blocks, *(C).* Animator Co Hoedeman and a colleague help indicate the scale and complexity of this production's set, *(D).* *Courtesy National Film Board of Canada.*

A

B

C

D

5 ■ Cutout Animation

WRITTEN WITH DAVE PALMER

Animation excels at telling stories, especially the kinds of stories that are impossible to put on film using real actors and real locations. Fantasy landscapes and fantasy characters have dominated the growth of animation from its earliest days. Unfortunately, the full-cel techniques most often used by professional animators and commercial studios in developing such narrative films are time-consuming and difficult for an individual animator to undertake. So independent animators have invented *cutout* animation, a simplified form of cartooning that lets them work effectively with characters and stories.

Something is always lost when a simple technique replaces a complex one. So it is with cutouts. As you will see, the movements of the characters are restricted. Stories tend to be more concrete and simpler with this technique than with its full-animation counterpart. But the benefits outweigh the disadvantages. Cutout animation in film and digital formats makes accessible to the individual artist a broad and rich universe of creative possibilities. Just in case you think of cutouts as being a "simple" technique—like training wheels for the real stuff—flip ahead to the case study at the end of this chapter.

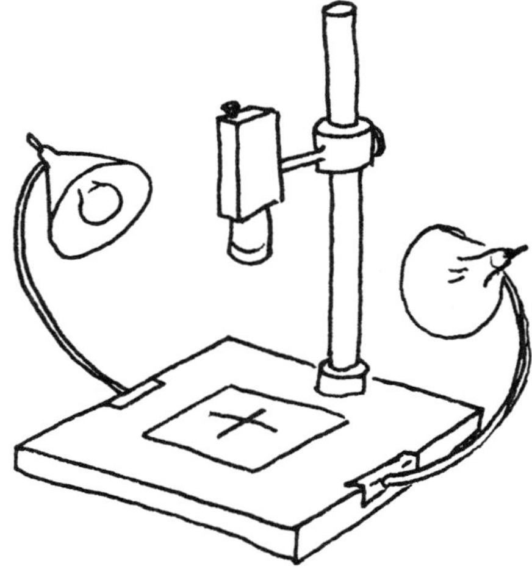

5.1 Standard setup: The camera is positioned directly above the background on which the cutout characters will be animated. A simple copy stand (as illustrated) or a tripod can easily hold the camera, and lights are mounted at both sides to provide even illumination.

STORIES

Cutouts work best with plots featuring lots of physical action that takes place on a broad scale. Conversely, it's difficult with

5.2 Jointing: As this character walks, its arms and legs will move. The composite figure *(A)* is actually composed of four separate cutouts, as indicated in *(B)*. These movable parts can be fastened to the body by means of a piece of thread sewn through the cutout's body. The ends of the thread are then anchored with masking tape at the appropriate places on the reverse sides of both jointed elements, *(C)*. Another joining technique uses metal fasteners that are punched through the cutouts and opened on the back side of the body section, *(D)*.

cutouts to deal effectively with the nuances of story as revealed through the detailed movement of figures and backgrounds. Here's an example that will make the distinction clear. Say you have a story in which a character gets very angry at his parked car when a wheel falls off. In cutouts, the anger would have to be shown, for example, by having the character leap to the roof of the car and start jumping up and down on it. Cutout techniques would *not* be effective, for instance, in showing a slow flush building in the character's face or in catching a subtle kind of rage that builds in the character as he glares at his vehicle.

The natural bias of cutout techniques is toward broad action. Keep this in mind when selecting stories to work with. As with all kinds of animation, the limits and the unique characteristics of a particular technique are most often bound up in the materials that are used. This will become clearer as you begin to prepare the characters and background of your first cutout film.

CHARACTERS

As its name suggests, cutout animation is achieved by moving figures that have been drawn on a piece of paper and then cut out. The force of gravity (and a glass pressure plate, or *platen*) holds these figures flat against the background or "scene." The animator's own hand moves the cutout pieces across the scene. Positioned overhead, the animation camera clicks off two exposures between each movement. A typical cutout animation setup is illustrated in Figure 5.1.

Quite obviously, this technique saves the animator a great deal of work, since he or she can use one drawing again and again. But this also means that the figures themselves cannot be changed. This apparent dilemma is solved by the process of *jointing.* Depending on what a character will be doing in the story (and depending on the animator's patience), each cutout is designed so that it has some movable parts: arms, legs, hands, a head, and even some facial expressions.

There are three common ways of connecting the moving parts of a cutout character or object:

Thread and Tape. A piece of thread is attached with

masking tape to the reverse side of each separate piece of paper. The thread keeps the pieces aligned and still allows them to move. The thread works as a hinge. The shorter it is, the easier you'll find it is to work the characters under the camera.

Metal Fasteners. A small metal fastener can be used to connect various portions of a character. Such clips are very effective and hold up well in use. But the connecting mechanism can be seen by the camera. In some films this matters, in some it does not. The choice is aesthetic.

Gravity. Plain old gravity can be used to keep one part of a figure aligned with another part. This is the easiest way to design characters, but you may find them difficult to manipulate because each part has to be moved individually during shooting and it's all too easy to inadvertently move one piece when you only intended to move its neighbor.

There are times when the gravity method is the only way to go. It makes no sense, for example, to connect an eyeball with the eye socket in which it rests. *Overlay* is a generic term that refers to a series of separate cutouts that are used at different times within the same character or object (see Figure 5.3).

Even in cutout animation you'll encounter situations where you will need to create a different version of the same character. Such moments occur when a character's physical appearance must change drastically for dramatic effect, like when zapped by a bolt of electricity, or when it is impossible for your cutout to accommodate a new body position that a character is forced to assume, like when a character stops walking across the scene and turns to face the camera.

Sometimes, too, you will need to draw and cut out a detail of your character that will subsequently be filmed as a close-up shot: for example, when it is necessary to show an extreme close-up of a bug crawling on someone's nose, or a hand holding a bottle of poison, or a foot slipping on a banana peel.

In designing cutout characters, always keep in mind what your story requires a particular character to do physically. What sort of personality do you wish to create? Your story itself should help in determining where and how to use jointings, overlays, and close-ups.

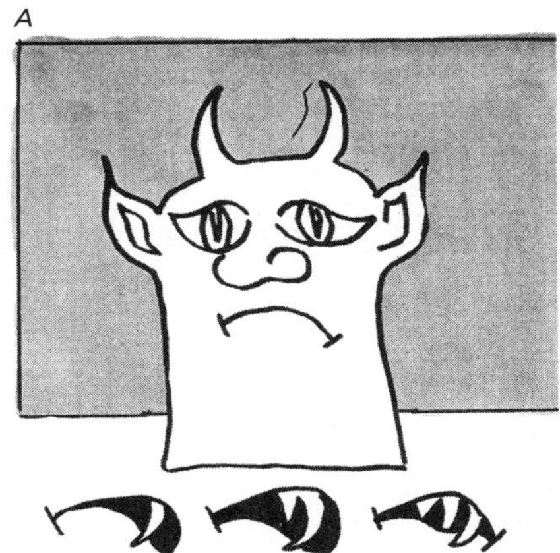

5.3 Overlays: This lopsided and toothy collection of mouths would give the creature in *(A)* a very wicked style of speaking. By placing the "explosion" overlay over the magician's hand in scene *(B)*, taking it out and putting it in again on successive pairs of frames and then overlaying the hand with the rabbit, an effective piece of magic can be achieved.

A

B

C

D

E

5.4 Cutout styles: For every general trend that one can observe within a particular technique, there are always plenty of contrary examples. The six films sampled here make this point quite eloquently. All employ cutout techniques but each has a distinctive graphic style, and the last two don't have the kind of narrative structure that I've associated with cutout animation. The films cited are: *(A), Cecily* by Paula Reznickova—*courtesy Learning Corporation of America; (B), Crocus* by Suzan Pitt Kraning—*courtesy Serious Business Company; (C), My Financial Career* by Grant Munro and Gerald Potterson—*courtesy National Film Board of Canada; (D), Our Lady of the Spheres* by Larry Jordan—*courtesy Serious Business Company;* and *(E), Shout It Out Alphabet Film* by Lynn Smith—*courtesy Phoenix Films.*

FILM TECHNIQUES

Working Scale and Paper Stock. Cutouts should always be made to a relatively large physical scale. By using a 12- by 16-inch field, or larger, you can make characters about 4 to 6 inches tall, with plenty of detail, and their moving parts will be easy to manipulate during filming. Furthermore, if you are filming with a camera that has a zoom lens, large-size artwork allows you to zoom in for close-ups instead of having to create new drawings. Another tip: Stick with the same scale throughout your film. This way you can use the same characters within different scenes. It makes things easier, too, if you draw your characters and overlays on heavyweight paper stock. This keeps the edges from curling, and the movable figures will last longer.

Backgrounds. While the location of a story is often an important element in a cartoon, it is best to wait until you have designed your characters and central props before you fashion the background on which they will move. One reason for this procedure is to ensure that you end up with a background that is big enough for the cutouts you've made. But a more important reason has to do with design. You don't want to create a background so busy that your characters get lost in it. The exaggerated example in Figure 5.6 will help you to remember this point. It is important to balance all the graphic elements of your movie. Make backgrounds simple. Use muted colors. Keep nonessential action to a minimum. It is always better to tend toward understatement in designing a background. Otherwise you risk upstaging your characters and the story they have to tell.

Don't neglect the dramatic potential of the background,

5.5 A cutout production: These three photographs show the working scale, *(A)*, the cutouts, *(B)*, and a sample frame, *(C)*, from Evelyn Lambert's *Fine Feathers.* The animator herself is seen at work. A camera is positioned over her head and is activated by a foot switch. *Courtesy National Film Board of Canada.*

A

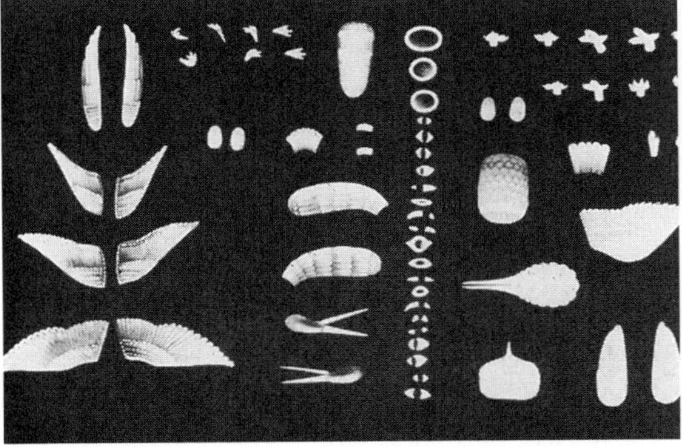

B

C

5.6 Background design: Keep it simple.

however. And don't be afraid to cut into your background, literally. The two examples in Figure 5.7 show that sometimes you can use a cutout within the background to relay important information. A well-designed background can also solve key dramatic problems.

SPECIAL EFFECTS

If necessity is the mother of invention, ingenuity is the father. The successful invention of a new technique is one of the greatest satisfactions an animator can have. So don't give up when your story calls for something to happen that seems impossible to achieve with cutout techniques. Improvise. To give you an idea of the difficulties that can be overcome, here are some ingenious inventions that have been rediscovered many times by independent animators.

The Stormy Day. Rain or snow is painted on a clear sheet of acetate and pulled across the background and characters while the scene is being animated, a frame at a time. Lightning effects can be achieved by varying the positions and brightness of the filming lights. To show an earthquake, either the table holding the cutouts or the camera itself is given a healthy shake while film runs through the camera at live-action speeds.

20,000 Leagues Beneath the Sea. A glass tray filled with water is suspended between the camera and the artwork. The water is stirred with a fingertip before the shutter is released. You can also place a piece of blue acetate over the camera lens. Yet another alternative is to shoot the camera through the bottom of a colored glass jar. This has the effect of keeping part of the scene in focus and part out of focus as the jar is rotated during filming. A final possibility is to apply a series of greenish and bluish washes directly to the film after it has been shot and returned from the laboratory.

Project: A Cutout Title Sequence. Make a 15-second film in which a single character (person, animal, object, or whatever) enters a black field and in some appropriate way produces the letters of your name, one at a time. This exercise will let you experiment in fashioning a character with movable parts, as well as teach you how to move the figure so that it has a characteristic personality. You will also learn to handle a series of nonjointed elements—in this case, the letters of your name.

COMPUTER TECHNIQUES

The principles and techniques of computer-based cutout animation are exactly the same as those of the traditional, under-the-camera, film-based style. Only the tools are different.

But what a difference the digital tools make, starting with the cutout objects themselves. Using the computer, you can assemble a very eclectic range of elements, including shapes cut out of textures and fabrics, drawings done on paper, and bits of clay. Manipulating a mix of object types under the conventional film camera presents many difficulties in lighting and background choices. Such problems vanish in a digital realm, where every image becomes a set of pixels. Issues of scale become simpler as well. Working in real space under a film or video camera, one needs to prepare various cutout elements so that they are sized appropriately to each other. Using a computer, it becomes easy to scale up a dime, for example, so it appears the same size as an apple. Digital files can be easily color-corrected, brightened, distorted, cropped, rotated, or otherwise altered.

Two pieces of computer hardware are required for cutouts: a scanner and a digitizing camera. Chapter 24: Computer Hardware provides more information on these tools.

Once the cutouts are inside the computer, two pieces of software are required: *Photoshop* and *After Effects,* both from the Adobe Corporation. Using Photoshop, the animator can literally cut out the image from its background and place it on a transparent level. Here is where colors and sizes and shapes can be tweaked. Photoshop also has the capability to gather the individual elements that make up a particular cutout character and form these into a composition that can thereafter be operated as one object, even though it maintains all its layered characteristics. The *Blue's Clues* case study that immediately follows shows this very clearly.

Adobe's After Effects is a software application that accomplishes the actual animation of the cutouts and subsequent output of completed scenes as QuickTime digital movies. QuickTime has become a standard format for storing and playing back digital animation files. Because After Effects is compatible with and accepts layered Photoshop documents, the animator maintains minute control over all the different parts of the characters, with all the elements grouped together and properly registered.

5.7 Backgrounds: Pulling down the single cutout (blind and hand) in *(A)* would provide a simple way to end a movie. Note that the windowpanes could be cut out so that the scene outside could be changed by simply sliding a different drawing under the window. The gumball machine background in *(B)* is slotted to allow the gumballs to tumble out of the black hole.

So creatively rich and so powerful is After Effects that this software has given birth to its own animation genre, called *motion graphics*. Chapter 10 provides a focused exploration.

Animating on the computer will liberate you from the danger of spending hours under the camera only to ruin a sequence because a light moved slightly during the shoot or an unexpected sneeze scattered your paper characters all over the floor. Digital techniques permit one to work with finely detailed and complex cutouts that would be way too difficult or time-consuming to animate in traditional film techniques. One wayward snip with the scissors doesn't send you back to the drawing board. Both software applications allow you to hit a keystroke or click on an icon to undo an error or unwanted change and pick up where you left off.

5.8 Dave and Blue: The star of the show peeks over Dave Palmer, the animation director at Nickelodeon's *Blue's Clues*. *Courtesy Dave Palmer and* Blue's Clues, *used with permission of Nickelodeon.*

The tremendous benefit of digital cutout animation is its changeability. If you decide after seeing a shot that you really want some particular action to happen slightly sooner, or later, or not at all, you simply have to slide a few key frames or layers to correct the problem. The traditional film methodology of reshoots, film processing times, and editing causes even the simplest of revisions to take hours or days to complete. Because digital files remain changeable or "liquid" throughout the process (in fact, digital files within a program remain malleable forever), many problems can be changed in a few minutes and for a lot less money than ever before. And this means that your animations aren't totally done until you're totally happy with them.

A

B

C

D

E

5.9 Silhouette animation: The medieval love story of *Aucassin and Nicolette* has been animated with silhouette techniques by German animator Lotte Reiniger. The delicate characters are hand-cut from black construction paper and have intricate moving parts. The design of the film and the construction of the cutout silhouette characters are similar to the techniques developed by Ms. Reiniger in 1923–1926 when she created *The Adventures of Prince Achmed,* the world's first full-length animated film. Two frames from *Aucassin and Nicolette* are shown in *(A)* and *(B).* Produced at the National Film Board of Canada, the film presents characters that are set against backgrounds of colored-tissue paper and gels, lit from below a glass table and filmed from above, *(C).* The great asset of silhouette animation is that any jointing of the characters is fully hidden from the viewer by the filming process. The contrast between the black cutouts and the brightly illuminated background easily camouflages jointings. These simple techniques pioneered by Lotte Reiniger, *(D)* and *(E),* are capable of tremendous detail and grace. *Photos (A) and (B) courtesy of the National Film Board of Canada, and photos (C), (D), and (E) by Lois Siegel.*

5.10 CASE STUDY: Digital Cutouts

A

Blue's Clues is the first television series to be animated with cutouts entirely on desktop computers. The show is designed for two- to five-year-olds and plays daily on the Nickelodeon cable network.

There are twenty half-hour episodes in a season, each clocking in at exactly 24:30. This adds up to over eight hours of animation, all of which is created, produced, designed, animated, and edited by a group of about thirty people. A single show is taken from script to videotape master in about ten weeks, and in this case study, Animation Director Dave Palmer runs through the production cycle.

B

Blue's Clues presents a mix of real objects and cutout characters that are composited on screen with its live-action host, played by actor Steve Burns. The four screen grabs in *(A)* are typical. The show follows the adventures of a cartoon puppy named Blue and her owner, Steve, as they search throughout a storybook world for items with blue pawprints on them. These items are clues that will reveal the answer to that particular episode's question, such as, "What game does Blue want to play?" or "What story does Blue want to read?"

The scene we will be following takes place at a beach, where Steve and Blue meet groups of animals and play a game identifying different rhythms.

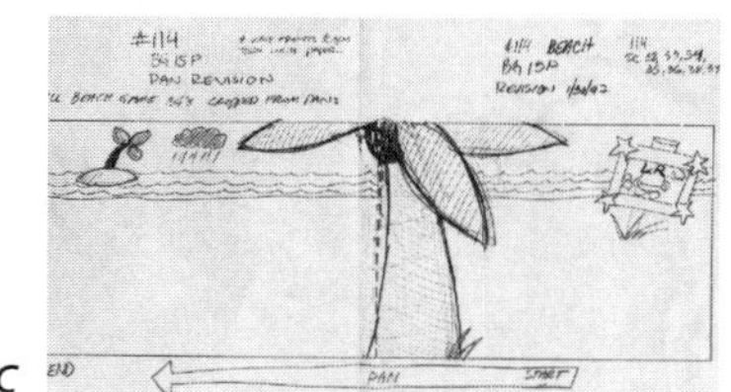

C

Character and Background Design

An episode begins with a script and a storyboard. Our overall goal is to create a participatory environment in which our viewers will literally get out of their chairs to yell out answers to Steve's questions and play along with the games in each show.

At *Blue's Clues,* we push ourselves to surprise the kids with great characters and unique, interesting environments. The design process begins with character model sheets and background sketches. *(B)* and *(C)* provide examples of both. Once a character or background has been finalized on paper, the individual *elements* (or pieces that will comprise the character or background) are scanned on a flat scanner or digitized with a video camera. The photograph in *(D)* shows the range of cutouts and objects we used in this show. *Blue's Clues* is based on traditional cutout animation techniques, yet has its own distinct aesthetic. We use real objects: items made out of clay or plastic, cutout felt pieces, textured papers and different fabrics — anything that is tactile and fun. (E) is a sampling of fabrics and textures that we scan directly into the computer and use to give our characters and scenes a handmade look, like a children's storybook.

D

When a character or item requires a 3-D look, like the eggs in *(F),* the objects are created with clay or Play-doh and digitized with a video camera, as in *(G).* Any video camera you have access to can accomplish this goal if you have the software to con-

E

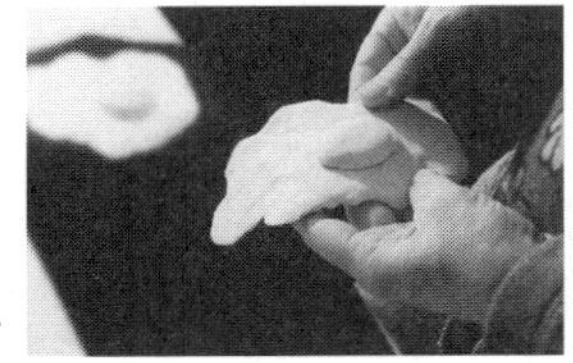

F

nect it to your computer. The parts that constitute the crab character from our scene, *(H),* include a clay body and eyes, and cutout paper for the legs, eye stalks, and mouths.

The background drawings are scanned into the computer, imported into Photoshop, and used as templates for the actual backgrounds used in the show, *(I)*. Everything created by our digital designers is made to fit a video frame, which is 640 pixels by 480 pixels at 72 dots per inch. Sometimes, however, we choose to work larger. Backgrounds, for instance, can stretch to 2,000 pixels wide by 570 high, so they can be moved horizontally behind the footage of Steve to create the illusion of a long pan.

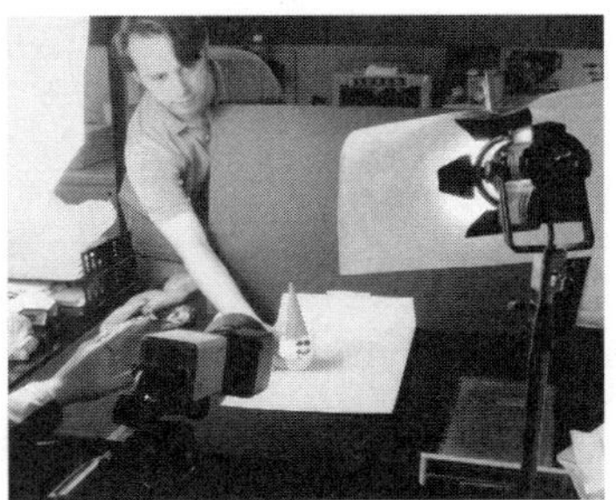

G

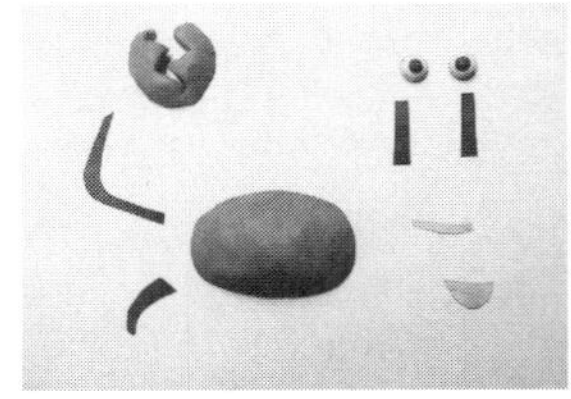

H

Finishing and Layering Elements

All scans and digitized pictures of 3-D objects are imported into Photoshop to be refined and prepped for animation. Photoshop is an expansive and versatile tool that can clean up edges of scans, cut shapes out of large pieces of fabric or texture, add or delete parts of a scan, or simply add highlights and shadows. At this Digital Design stage, we often change the colors of a particular element to brighten it up and make it fit within the aesthetic of the whole show. *(J)* shows one of our designers refining the crab elements seen on page 70 in the screen shot, *(K)*. Notice how a path is being made around the element of the crab's body. The path will be used to remove the unwanted background information and smooth out the shape's edges.

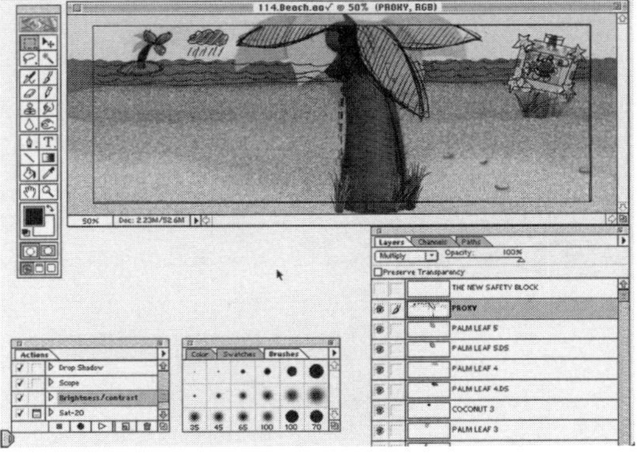

I

Once the arms, legs, bodies, heads, mouths, eyes, and other body parts of our crabs are finished, all the items are placed into one Photoshop document, *(L)*. This "layering" step allows us to keep the different parts registered, and to add drop shadows underneath various parts of the united object. Registering the parts of any character is important because it ensures that every part of the character is in its proper place. This is especially critical if you're using replacement mouths for lip sync or if you need to move one element around a specific point on another — like the crab's claws, *(L),* which we split into two halves and joined at the bottom. We did this so that each claw can be opened and closed independently and yet remain connected to the crab's "wrist."

J

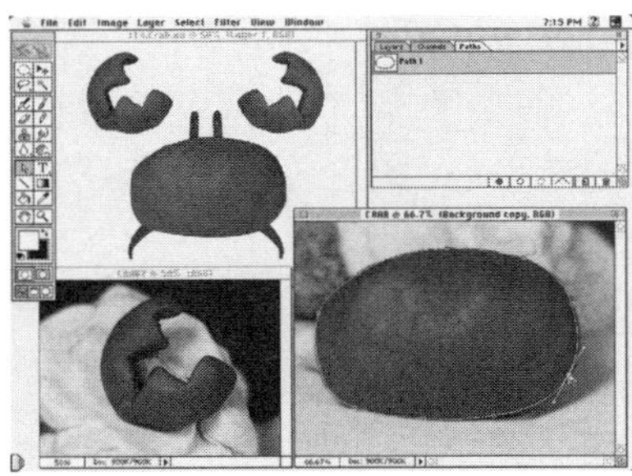

K

Overall, there are about thirty layers to the crab. Can you imagine what a nightmare it would be if you wanted to change the crab's position, rotation, or scale in the shot, and each eye, leg, claw, and replacement mouth had to be moved and repositioned piece by piece? By combining all of these parts into one layered Photoshop document, we can place the crab over the background and move it from point to point as just one piece.

L

N

Animation and Compositing

(M) shows the crab's finished Photoshop document just as it was imported into After Effects as a composition. See how the Photoshop layers remain in the same position and order, and retain the same name. The animator, *(N)*, can bring the creature to life, working with the familiar and universal principles of animation, like anticipation, exaggeration, and squash and stretch. Each of the crab's parts can be moved, flipped, rotated, scaled, or made more or less opaque. Replacement animation, a traditional cutout technique, is easily simulated by making different layers visible or invisible via After Effects' opacity key frames. The lip sync for characters in *Blue's Clues* is often done with replacement mouths. The animator creates the illusion of talking by switching between six or so different mouth shapes, each shape corresponding to a different sound.

When the animator receives the edited studio footage, the bluescreen has been deleted from around Steve so that the live-action can be layered into the scene. You can see all the pieces coming together in *(O)*. The composition of the entire shot is labeled Comp 1. Steve's footage is the layer in Comp 1 labeled 11432 (the number of this particular shot), and the background of the shot is labeled 11415P, which means it is a long background created for a pan in shot 11415. The rest of the layers in Comp 1 are also compositions, either of background or foreground animations, like the rain cloud, or of other characters in the shot, like the familiar crab character. One of the crab compositions, labeled Crab 01, is opened here for the animator to be able to work in both the crab composition and the composition of the entire shot at the same time.

There is a lot going on in this beach scene. The animator has three animated characters to choreograph with each other, all of them having to stay within boundaries set by Steve's footsteps and eyeline. While working in such multiple compositions it is important to keep everything clearly labeled and organized. For instance, the three crabs are labeled Crab 01, Crab 02 and Crab 03, and any one of them can be accessed by clicking on its particular Composition window in the upper right, *(O)*. Synchronizing all of the compositions (via the Preferences menu), allows the animator to move to the same frame within all the

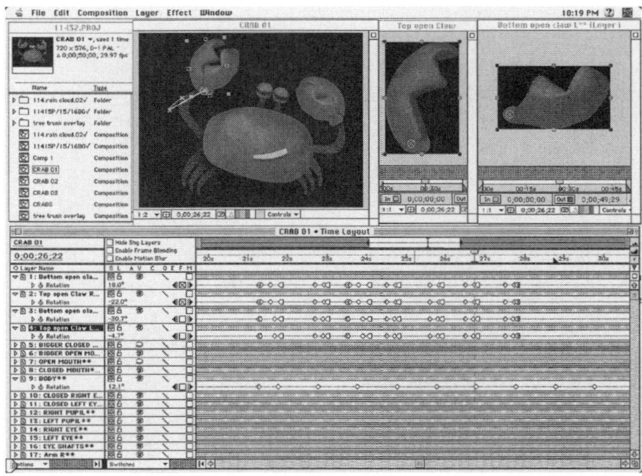

M

individual compositions that are used within the same scene. This makes it easier to choreograph scenes and to properly time reactions so that characters play off each other's dialogue and movements.

In this scene, for example, we want to synchronize the movements of the crabs to a beat track included in Steve's footage. The beats can be located down to the frame by accessing the waveform of the audio, and when the animator places the time marker on a beat, each of the crab compositions will jump to the same corresponding frame. The animator then hops from one crab composition to the next, animating each crab, and then moves to the next beat in the composition of the entire shot to repeat the process.

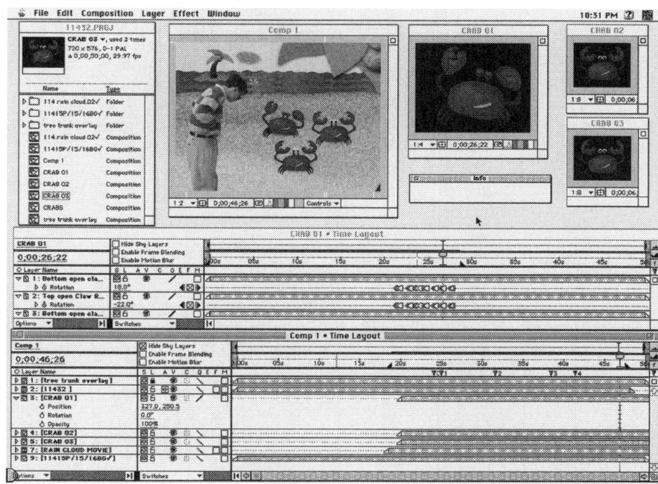

I guess you can see how a *Blue's Clues* scene can get pretty complicated, which is why planning ahead and using as few layers as possible is key. Some simple scenes will take a *Blue's Clues* animator half a day to animate, while more complex and longer scenes can take two weeks or more to complete.

Rendering and Editing

Our final step is to render the finished animation as a QuickTime movie. This is often a long process, and could take ten or twenty minutes for a short scene with a few layers, or ten hours or more for a long scene with lots of layers and digital effects.

Once a scene is rendered, the animator can watch it in "real time," or thirty frames per second, to judge the timing and action of the characters, and to make sure the background elements are the correct size and in the right spot. Often at this first look we will see something in the shot that needs to be improved or changed. Each episode of *Blue's Clues* is tested with children throughout the production process to ensure that the show is working on an entertainment and educational level, and revisions based on these testing sessions are an expected and necessary part of that process. If revisions need to be made, the animator merely opens up the saved Project file for the shot, makes the changes by moving layers or key frames, or substituting one element for another, and re-renders the scene. The beautiful thing about working digitally is that if you need to make changes, they can usually be done in a few minutes with little trouble — hence avoiding the hours of reshooting and added processing costs involved in traditional animation.

When all of the scenes of a particular show are completed and rendered out, the scenes are collected and the soundtrack is mixed. The show, animations and soundtrack, are then transferred from the hard drive to videotape with a Macintosh-based digital editing system, ready for broadcast.

Credits: Executive Producer: Todd Kessler **Producer/Head Writer**: Angela Santomero **Designer/Producer:** Traci Paige Johnson **Animation Director:** Dave Palmer **Animators:** Chris Boyce, Robert M. Charde, Michael Dougherty, Olexa Hewryk, Nancy Keegan, Holly Klein, Scott Klossner, Danial Nord, Jennifer Oxley, Anne Nakasone, Trixy Wattenbarger, and Joe Silver **Supervising Producer:** Jennifer Twomey-Perello **Research Director:** Alice Wilder **Art and Animation Manager**: Soo Kyung Kim **Assistant Art Director:** Christian Hali **Art Production Assistant:** Ian Chernichaw **Digital Designers:** Yo-Lynn Hagood, Jane Howell, Adam Osterfeld **Editor:** L. Mark Sorré **Assistant Editor:** David Burger **Technical Manager:** Boris Beaubian

6 ■ Time Lapse and Pixilation

Our perception of the world and the things in it can be vastly extended through animation. Everyday processes that were once hidden from the naked eye are suddenly revealed. Things that happen over long periods of time become quick events. Patterns of movement that have always been seen at "natural" speeds become startling as film or digital animation collapses time. In this chapter you will meet two of animation's most delightful perception benders: time lapse and pixilation.

TIME-LAPSE ANIMATION

In time-lapse animation, every frame is exposed at a predetermined interval, which may range from a few moments to a few days. A familiar example will provide a handy definition to this technique. Out of a plot of earth we see a small green shoot grow and reach toward the sun. Within a few seconds of viewing time, this shoot becomes a plant with leaves. It continues growing. Then it buds and finally we witness the gentle explosion of a flower's blooming.

Essentially, the time-lapse technique alters our perceptions by collapsing time. This compression of the normal sequence of events will usually reveal either a process of generation or one of destruction. By thinking of this kind of animation in just these terms, and by seeing beyond the growing flower image, you will quickly find new and exciting possibilities. Here are a few.

Nature. The organic patterns clouds make as they sweep across the sky, the movement of shadows on a brick wall, the passing of a thunderstorm, the path of a snail, the arrival of a snowfall, the development of a chick embryo.

Man. Crowds of people moving in and out of buildings, the traffic patterns of a highway or an airport, the demolition of a house, the construction of a skyscraper.

Art. The development of an oil painting, the erosion of a sand sculpture, the decoration of a Christmas tree, the scribbling of graffiti upon a wall.

All computer draw and paint programs will allow you to record (Save As) successive evolutions of an image. These can be formatted with QuickTime movies or other playback engines that string together the series of versions into a single animation.

Many of the phenomena cited above have already become topics of films, and the experience of viewing any of these movies for the first time is pure revelation. Familiar things are discovered in startling new ways.

A

B

C

6.1 Time lapse: In Derek Lamb's *House-moving, (A),* a time-lapse technique records the disassembly, transportation, and relocation of an entire colonial home. Walerian Borowczyk's *Renaissance* shows the reconstitution of a room filled with mementos. The frame enlargement in *(B)* shows a photograph that is reconstructed in one sequence. Oskar Fischinger's *Motion Painting No. 1, (C),* produced in 1949, presents the step-by-step evolution of an abstract painting. *Photo (A) courtesy Phoenix Films, (B) courtesy Pyramid Films, (C) courtesy The Museum of Modern Art Film Stills Archive.*

INTERVALS

The exact amount of time between exposures for time-lapse photography will depend upon the nature of the subject you select. For example, documenting the construction of a skyscraper might require just a few frames per day. Shooting could go on for months. In comparison, it would take less time and you would use smaller intervals between exposures if you wanted to catch the drama of a melting ice-cream cone.

Here's how to determine the interval you'll need. Begin by studying carefully the subject of your film. What is the movement or change you want to explore? Next, determine how much "real time" it takes for the action you've selected to complete itself. Now decide upon the best overall duration for your finished movie. This will be its "screen time." Because the projection rate is always constant (24 or 18 frames per second in film, 30 frames per second in video, and between 8 and 15 frames per second on the less-constant computer), you can now work out the correct interval between exposures. Multiply the *screen time in seconds* by the appropriate *projection speed.* The answer is the total *frame count* for the finished pro-

6.2 Weeks become seconds: This series of key frames from *Organism* shows the demolition of an entire office building. In this film by Hilary Harris, time-lapse and pixilated sequences compare the flow of a city's life with that of the human body as seen through macrocinematography. *Courtesy Phoenix Films.*

ject. Calculate the *real time* of your subject in the smallest units of time that are practical (generally in seconds or in minutes). Divide the *real time* by the *frame count*. The result is the *interval between frames*.

An example will make this calculation process less daunting. Suppose you want to make a movie or tape about office workers streaming out of a building at the end of the workday. You visit the location and study the action (and select the best camera position), and you determine that most of the "action" takes place in a ten-minute period. In order to emphasize the "flushing" effect of the spectacle, you decide that the entire event ought to take place on the screen in just five seconds. You know you'll be playing back at 24 fps. Here are the computations:

Step 1: 5 seconds x 24 fps = 120 frames

Step 2: 10 minutes x 60 seconds = 600 seconds

Step 3: 600 / 120 = 5-second intervals

SETUP

Camera Mount. Like all forms of animation, time-lapse techniques require a stationary camera. A solid tripod is generally used, although it's possible, with ingenuity, to mount a camera just about anywhere—even on the nose of a 747 in order to let the world experience a transcontinental trip in one minute.

Triggering Exposures. The actual exposure of each frame can be done either manually—by a shutter release or the tap of a computer key—or by a specially designed motor called an *intervalometer,* which is built into the software of many digital cameras. With a motor or using a computer, the camera can be set to take a picture at a predetermined interval. More information on traditional and digital cameras can be found in Part III: Tools.

Exposure Level. One of the problems of filming over a long period of time is that the lighting of your subject is apt to change between the individually exposed frames. A built-in, through-the-lens light-metering system can help in keeping a relatively constant exposure value when filming outside. If you're working under lights, it's possible to get a timing device that will turn on your lights a few moments before the picture is taken and then turn them off. If you are dealing with a complicated lighting problem, be certain to do a thorough test before undertaking the entire filming.

Other Variables. Try running your film backward or inverting the captured frames if you are working digitally. By reversing the usual start-to-finish order, you can heighten reality: a flower can "run in terror" from an approaching blight by appearing to pull in its petals and shoot back into the earth. Shoot at night using a long exposure time for each frame. Shoot with the camera mounted in an unusual position. Shoot with decreasing time intervals between frames, so that whatever you play back will appear to be going faster and faster.

Project: Super Doodle. Set up your camera so that it looks over your shoulder at a blank piece of paper. Begin doodling on the paper. Using either a foot release, an intervalometer, or a friend, film for a period of 10 minutes at a time interval of 3 seconds between exposures. Pay no attention to the clicking of the camera as you draw. Don't worry about arriving at a "finished" picture at the end of a given 10-minute period.

With a 3-second interval, your finished film will take about 8 seconds to project. Not only will you witness the evolution of the drawing, but the camera will also have recorded the motion of your hands and shoulders. With luck, you may be able to gain a fresh perspective on the style and energy with which you doodle.

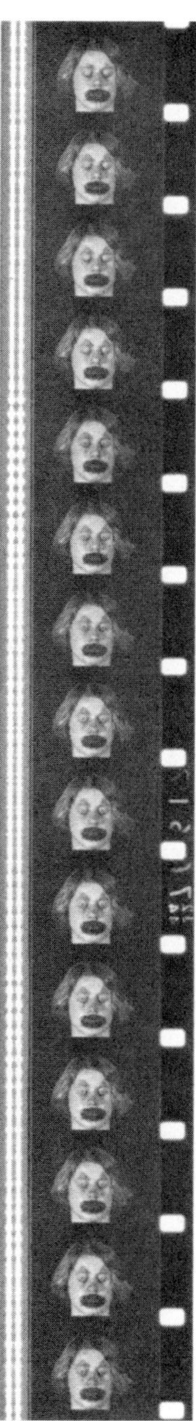

6.3 Animated face: An enlarged strip from a student's 16mm film that includes both time-lapse and pixilation techniques.

PIXILATION

Pixilation is a specialized technique for animating people. The camera records occasional frames of some natural or real-time event, but because of the intermittent filming, the effect in the resulting film is that of an unnatural movement somewhat like an old silent movie. What is impossible in real life becomes commonplace with pixilation.

In Europe the common expression for a cartoon is "trick film." Pixilation is trick film at its trickiest. And one of its best tricks is the way the term itself hides its own roots. I've been unable to locate even an unauthoritative explanation of how the term originated. But whatever that may be, and although the technique admittedly bears a close relationship to both

A

B

C

D

E

F

6.4 Pixilation: Norman McLaren's Oscar-winning film *Neighbors* is credited with introducing the technique of pixilation in 1952. Described as the most eloquent plea for peace ever filmed, *Neighbors* shows how a neighborly misunderstanding escalates into genocide. The enlarged strips of 35mm film—*(A), (B),* and *(C)*—show the positions that the two principal actors had to hold for the single-frame exposures. The 360-degree sliding handshake soon turns into the 360-degree nose-to-nose staring battle. The duel with pieces of a picket fence, *(D),* marks the acceleration of hostilities during which both actors show grimmer faces with each sequence. *(E)* shows the final face worn by animator and actor Grant Munro. Interestingly, the movie's climax (in which wives and children are slaughtered) was cut from prints because the sequence's effect was so shocking to sensibilities of the time. Since then, the National Film Board of Canada has reinserted the original ending, *(F)*. The production still, *(G),* shows the camera setup and the highly simple and stylized set. McLaren is in the white T-shirt. *Courtesy National Film Board of Canada.*

G

6.5 Pixilated heroes: Notwithstanding the serious topic and sobering impact of *Neighbors,* pixilation usually lends itself to humorous, zany animated productions. This is evident in *Blaze Glory,* a parody of the cowboy hero by Chuck Melville and Len Johnson. *Courtesy Pyramid Films.*

time-lapse and small-object animation, the expressive function of pixilation is clear. The technique is almost always used for humor. And the effects one can create with it are absolutely astonishing.

In pixilation, a stationary camera records a stationary, posed subject, shooting on twos. Between exposures, the character moves to a new position. The process of taking exposures is often very slow and requires great patience and concentration. For the performer, great physical agility is often needed. Because the subject in pixilation is usually a living person, complete control of the pose is difficult, and this forces the animator to guesstimate much of the time. Trial and error rules pixilation.

EFFECTS

What the technique lacks in precision, it makes up for in its sheer display. Here's a beginning catalogue of characteristic effects.

Locomotion. Pixilation permits weird ways of moving. A person can appear to fly, which is achieved by having the subject jump into the air precisely as the camera clicks off a single frame, over and over again. Or an individual may move like an automobile. Here the actor moves about on the ground but carefully assumes the same "driving" position during each double frame exposure, with hands holding an imaginary steering wheel and feet on an imaginary brake and an accelerator. There are endless variations: skidding, bouncing, floating, swimming, even flipping end over end.

Entrances and Exits. There are all sorts of tricks for entering or leaving the screen. People can come up out of the ground (a hat, clothing, and other props are required for the first few frames). They can appear through a wall. They can flash in and out of the picture, or they can just "pop" on and off the screen.

Speed. Pixilation gives the effect of speeding up things. If a sidewalk crowd was filmed at 3 frames per second (instead of the usual 24 fps), the crowd would appear to be racing along in a jerky, old-time-movie style. The slower the rate of exposure, the faster the crowd will begin to move.

If you want to see just how fast you can make things go,

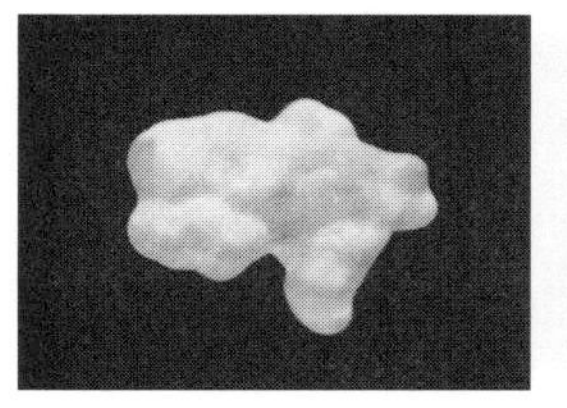

A

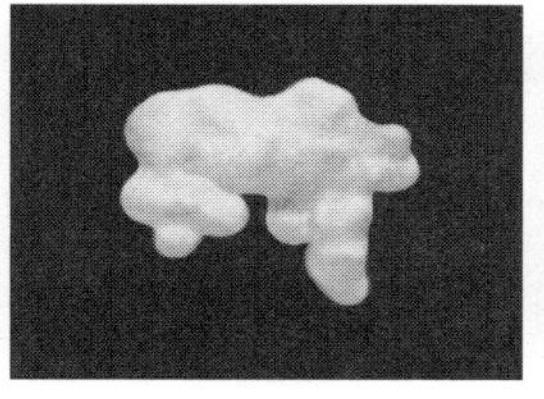

B

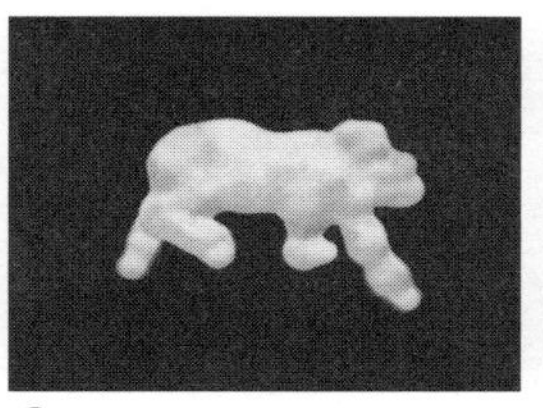

C

D

E

6.6 Time lapse meets cel: In this short animated piece for *Sesame Street,* animator Peter Wallach worked with time-lapse footage he shot of the New York skyline. *(A), (B), (C),* and *(D)* sample some hand-rendered positions of a cloud as it transforms into the shape of a cat. Using digital compositing technology, the time-lapse formation of moving "real" clouds was joined by the morphing "cat clouds" as seen in *(E). Images courtesy Fly Films and Sesame Street Productions.*

point a camera out the front window of a car and expose a frame every couple of seconds as a friend drives you around town. Here is another special effect you might want to try. Have an actor walk with intentional "slow motion" within a crowd that is walking normally. With some experimentation, you can pixilate the scene so that the character will appear to move almost "normally" while others will appear hyperactive.

People with Objects. Coordinate an interplay of pixilation (animating people) and time-lapse or stop-motion techniques (animating objects). The result will be animated bedlam: a vacuum cleaner swallows a housewife; a chair torments its owner; a necktie embarrasses an overly earnest suitor; a refrigerator produces the evening's meal automatically—then serves up the owner as the entrée.

SHOOTING

Here are some pointers for filming pixilated actions.

Continuity. Have someone working with you who does nothing more than make certain that each actor assumes a pose identical to his or her previous pose after relocating between shots. Precise registration is always a problem in pixilation. After all, it's not easy to hold an awkward position again and again as an animator laboriously builds a scene.

Camera Movement. The more frequently you take exposures, the less rigorously you will need to control the movement of the camera itself during shots, tilts, or zooms. As always, the steadier the camera and the more accurately you can calibrate incremental changes, the better everything will work out. It is wise to use camera movement very sparingly.

Exposures. Pixilation is often done outdoors and the process of filming a scene can take long periods of time. This raises the danger that natural lighting will change significantly between the beginning and the end of shooting. The result can

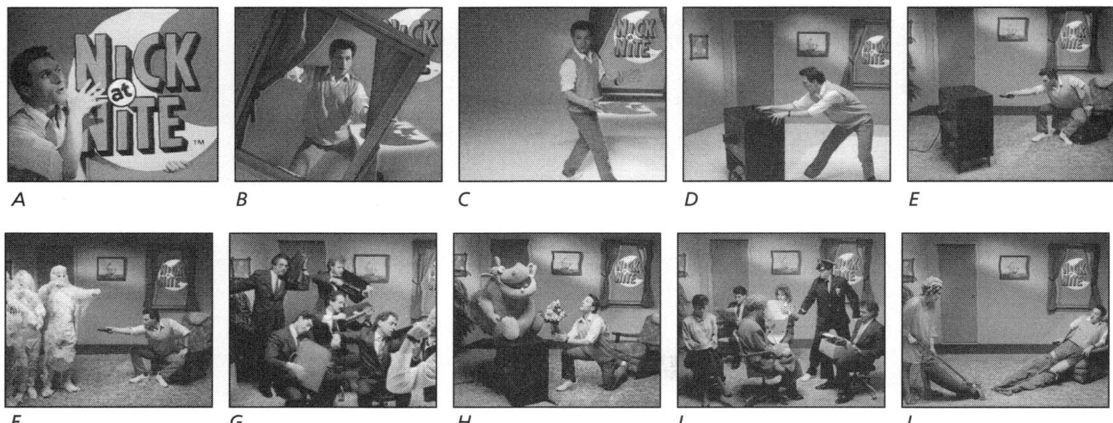

6.7 Nick at Night open: When cable TV service Nick at Night was first launched, its "on-air look" consisted of a series of pixilated ten-second IDs. The first seven seconds were identical for the more than fifty broadcast graphics produced by independent studio Noyes & Laybourne, located in New York's TriBeCa district. *(A), (B), (C), (D),* and *(E),* show key frames from the opening section in which an actor, Alan Mintz, "created" a room with a TV set in it. Alan's action was carefully choreographed starting with his holding the logo, *(A),* and then hanging it near the background of the limbo set, *(B).* Pixilated construction by our actor produced a window frame, *(C),* walls, and a floor and TV set, pulled into the frame, *(D).* The final action consisted of the actor "clicking" toward the TV set with a remote control in his hand, *(E).*

The rest of the frames sample the range of weird surprises that comprised the final three seconds of each ten-second ID: mummies pop onto the screen, *(F)*; a group of MTV Network execs bow their briefcases, like members of a string quartet, *(G)*; a puppet hippo spins on top of the TV, *(H)*; our TV watcher becomes a cop, policing a gang of Nickelodeon staffers, riding their office chairs like bumper cars, *(I)*; and a vacuum cleaner sucking our star's pants off, *(J)*. Production was wonderful lunacy. *Courtesy Nick at Nite. Images used with permission from Nickelodeon.*

be a disturbing effect of arbitrary changes in the picture quality caused by uneven exposures. The best way to avoid this difficulty is to shoot on a clear, sunny day. Still, the shadows will change during the course of the filming.

Background. The camera picks up and magnifies any movement in the background, so if you are planning to shoot outside, pick a day when there is no wind. Check your location to be certain that the background is devoid of cars, cows, or other objects that may suddenly move or disappear during filming.

Project: The Skating, Snaking, and Skidding Film. Set up your camera in an environment that will have little background movement. Find a patient friend to assist in operating the camera. You will be out in front of the camera this time. Begin with a static frame (no camera movement or background movement) through which you "skate," even though you're traveling across the earth and there is no ice. Next, try to coordinate a camera pan that follows you "snaking" past the camera on your stomach. Place obstacles in your way so you can try to smoothly pass over them as a snake might. For a grand finale, you might attempt a short film in which you enter the frame in a live-action walk and then, through pixilation, start skidding as you appear to be trying to stop. Direct your skid toward a tree or similar obstacle. Add more and more "panic" to your body position as you approach the obstacle. Finish yourself off with a terrific crash—one that lifts you high into the air and throws you quite some distance before landing.

7 ■ Working with Sound

WRITTEN WITH MICHAEL DOUGHERTY

Special bonds connect sound with film and video images. But in animation, these bonds are *very* special. The visual elements and the audio track seem to share a more intimate and more creative partnership than exists in any other motion picture form. This point is proved again and again every time you rent and then screen an animated feature. But *why* is this so?

The technology of animation—either film or digital—encourages a higher degree of synchronization than is practical within other forms of moviemaking. In both visual and audio realms, the animator has total control. An image or a sound can be placed with pinpoint accuracy down to tenths of a second. Furthermore, what you see and what you hear share common elements of structure. A musical tune, for example, is often characterized by a simplicity that is not unlike the simplicity of personality we recognize in cartoon characters. In many animated films musical tempo is related to the visual "beats" of the action taking place on the screen.

Music theorists have suggested that the fundamental power of music to move us as it does is a result of various psychological states we experience in listening to the music's structure. Humans are pattern-seeking creatures. It is in our deepest nature to listen for the completion of various patterns fashioned from individual notes, intervals, instrumentations, themes, and rhythms. We automatically anticipate how a familiar melody will play itself out. We wait for that outcome. If

7.1 Great tracks: A selection of frames from cartoons with innovative sound tracks. In *Cockaboody, (D),* John and Faith Hubley informally recorded the conversations of their two small daughters at play and based the entire film on an edited version of that sound recording. Sound effects and dialogue by Eskimos constitute the sound track of Carolyn Leaf's *The Owl Who Married the Goose, (B).* Highly amplified natural sounds are featured in the track for *The Animal Movie, (C),* by Grant Munro and Ron Tunis. In Norman McLaren's *Blinkity Blink, (A),* the animator hand-drew both the images and the sound track directly onto 35mm black leader. *Courtesy of: (A), the Learning Corporation of America; (B), Pyramid Films; (C), the National Film Board of Canada; and (D), McGraw-Hill.*

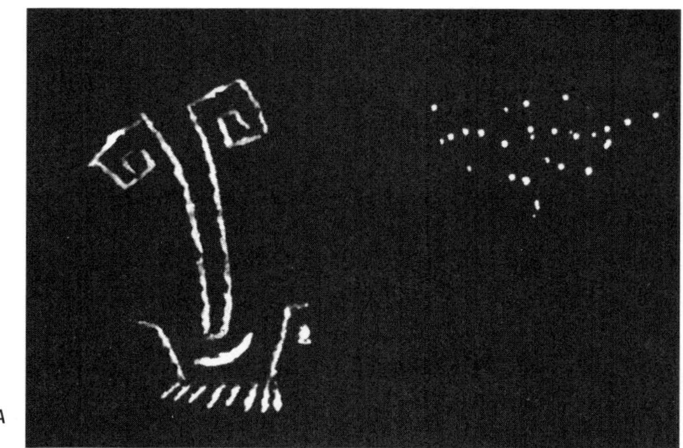

A

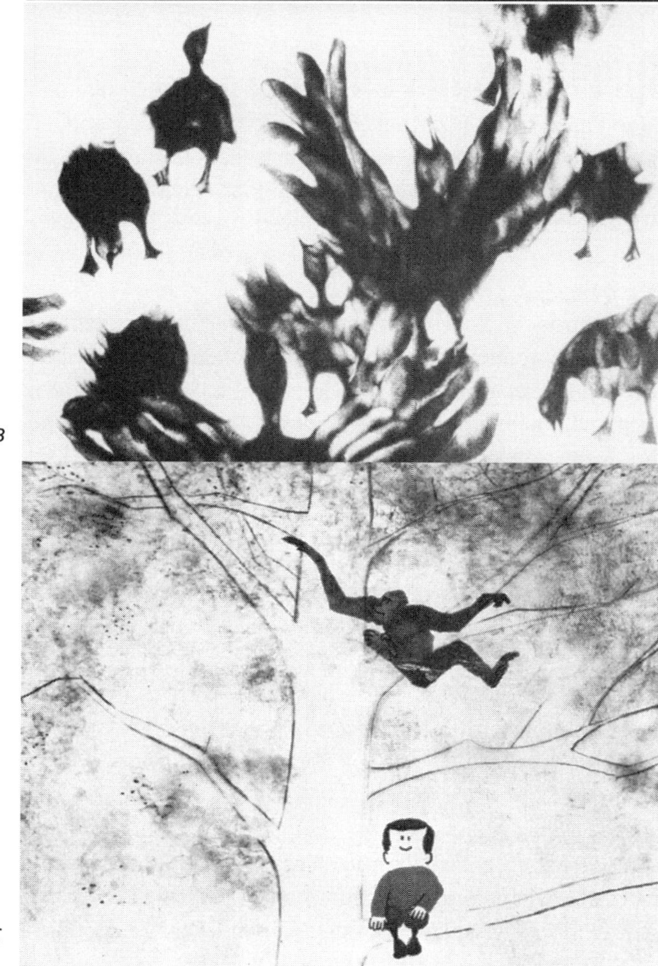

B

C

the composition follows the pattern we are expecting, there is a satisfying sense of closure and "release." If our anticipation is thwarted, we are disturbed and our attention increases, and if we experience a familiar pattern that is embellished in a new and unexpected way, we are delighted. The gifted animator, along with the music director/composer and the editor, understands the psychology of musical design and seeks to trigger the parallel responses through conscious manipulation of various visual patterns: repetition and rhythm; the movement of characters and background; the predictability of a story; and the pacing and editing of the finished work. To the degree that animation can create a flow of patterns, the medium appears to mirror the nature of musical expression.

Theorizing aside, there's no doubt about the terrific power that exists when a close synchronization of animated images and sounds is achieved. No advanced degree is required to appreciate just how rich the relationship is—all you need to do is rent a copy of Walt Disney's *Fantasia* or George Dunning's *Yellow Submarine*. You'll hear it. You'll feel it.

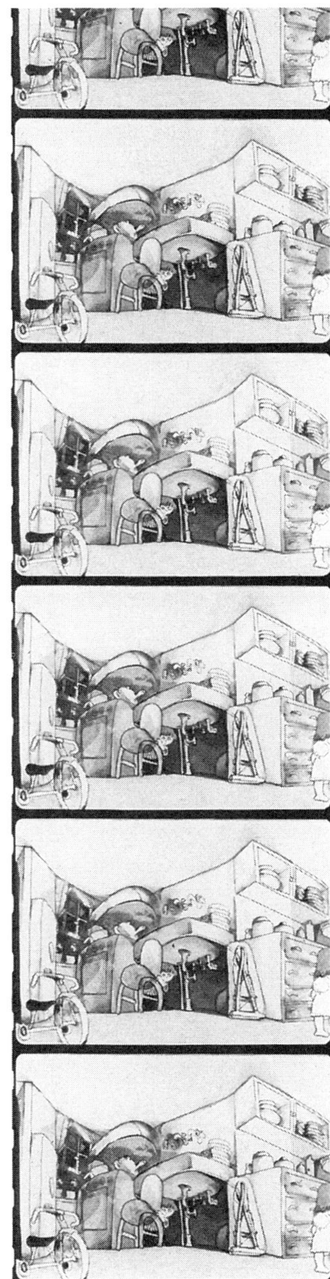

D

THE STUFF OF SOUND TRACKS

It starts out in the quiet of your mind. The animator studies the possibilities for a sound track. The movie's theme, style, length, and technique are all factors in the audiovisual equation. At this initial stage, the goal is to select those audio elements that will best fortify and extend the visual components of the movie. The outcome of this original design phase is a *sound treatment* that identifies in detail all the parts of a film's anticipated final sound track.

There is no standard format for such a treatment. It can be written out or it can be a set of notes jotted onto a storyboard and script. It is important, however, that you consider each of the three basic sound sources.

Music. A particular piece of music can sometimes give birth to ideas for animation. Take yourself off to a music store that lets you sample albums. Check out the music collection at your public library. Allow yourself to listen to selections that may be far outside your normal musical tastes. Choose musical selections that you intuit as "working" with your upcoming project.

Effects. Look up from the book for a moment and listen. Now what'd you hear? Whatever it was, whether it was birds chirping in your backyard, a dog barking in the distance, the hum of your computer fan, or an ambulance screeching down the road, they can all be recorded and used in your animation. Anything that can be recorded, or is already recorded on tape or CD, is easily imported into your computer and used to build a soundtrack. Later in this chapter you'll find more about gathering those Ker-Plumps, Ka-Pows, Zaps, and Zowees.

Voice. In a given film, there can be character dialogue or narration voice-over. There can be speech that is lip-synchronous or non-lip-synchronous. Because lip sync can be tricky when starting out in animation, you might want to look for creative ways of avoiding those scenes that call for precise lip synchronization.

It's important to sit down and think of what overall type of track you want to build. Is your animation similar to the classic cartoons of yesteryear, chock-full of sight gags, exploding dynamite, and falling anvils? Get ready to look for some compact disc featuring stock sound effects. Is your piece more abstract and trying to communicate a mood and feeling instead of a story? Chances are you'll want to experiment with altering everyday sounds or stock music until they fit your needs. Write a list as long as your arm of whatever type of sound comes to mind, no matter how small or insignificant it might seem. Take yourself inside the project you will be making. Imagine every little chirp, rustle, footstep, sigh, or breeze you might hear.

Here's a practical tip that can unleash a wealth of creativity. When thinking about your next 'toon, think about creating a sound track that is entirely original—from voices to sound effects to music. It's not just that there is something nice in the idea of using your own resources. There can be legal complications if you don't. You'll have no problem if you want to use prerecorded sound for a project that will be used only for your own personal enjoyment. But if you anticipate any commercial use of your animation through sale, rental, or admission charges from a public viewing, then you'll eventually have to acquire the rights to any and all prerecorded materials you use. And that costs money!

WHICH COMES FIRST? SOUND OR PIX?

It is often debated whether the sound track or the animation should be created first, and the truth is, there's no real answer.

If you're working in the classic cartoon style where you need accurate lip sync, then you'll definitely want to record a voice track first, analyze it, and create the animation using the voice track as reference. Many other types of projects benefit from following one or more elements from the sound track.

For most people starting out in animation, the picture comes first and accompanying music, sound effects, and voice elements are usually added once the picture itself is completed or "locked." Starting out with a locked-down track can actually be an impediment if your project is experimental or abstract. In such cases it's smart to make the visuals compelling before starting track work. If you can make your piece work without the benefit of any audio, bank on the fact that it will be even more powerful once your sound has been added. A truth whispered among animators is that 70 percent of a show's impact comes from the sound track.

It's usually a good idea to take the time early in a project's creative development to create a very rough sound track, or *scratch track.* Think of this as a loose outline, a first draft. A scratch track is usually built at the same time the storyboard is completed. In fact, it's common practice to fashion a scratch track as the audio portion of an *animatic*—a piece of film, tape, or computer animation that shows storyboard panels in sequence and that runs the same length as the anticipated project. An animatic's track will often have large "plugs" of silence

7.2 Waveforms: These two odd forms are the visual, digital expression of what sound looks like when it is worked with on a computer. *(A)* is a thunderclap and *(B)* is music. There's not a human alive who can look at one of these images and discern what sound it represents. However, anyone can tell from the flattening down of the spikes when a sound ends and silence begins. And that particular piece of intelligence proves hugely useful when you are using digital tools to create sound tracks for your animation.

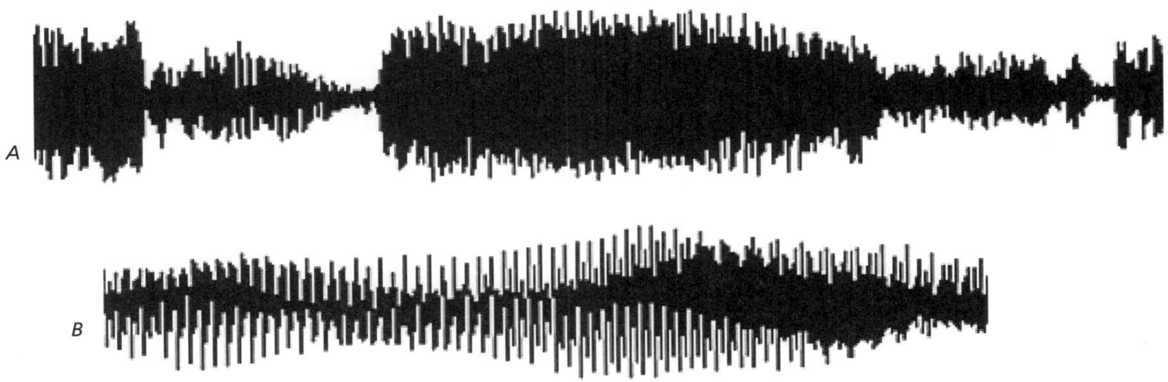

where music or effects will subsequently be placed. But even if there are only a few known elements—perhaps some lines of dialogue, a piece of music you like, a few important sound effects (even if laid in as rough vocals)—whatever elements you can slug into the scratch track will provide valuable guidance as you work on the timing of scenes and go about animating them. You'll find more about animatics and their cousin, the Leica reel, in the following chapter on storyboarding.

TRACK BUILDING

Now let's talk about tools and techniques that allow you to harness all that audio. Toolwise, the major focus here will be on digital technology, for that is the easiest and best way to undertake the process of creating a track for your animated movie—starting with the first scratch track and going all the way to the final mix. At the end of this chapter you'll find some information about the film-based technology that computers now replace.

Here's a breakdown of the track building process.

Separation. Each track is recorded separately and kept as its own entity. Only at the very end of a project's completion will these tracks be mixed. Sometimes two or three individual tracks are recorded, sometimes as many as ten or more. With digital technology, the recording process is easy and it begins by using any audio tape recorder. Even the least sophisticated personal computers have some sort of sound-generating capability—how else would you hear that annoying *quack* or *beep* every time you screw something up? Playing sounds on computers has been around for years, long before playing motion image clips became an everyday feature, and as digital technology has evolved, so have the audio editing and manipulation features. Today it's a common practice to completely create, mix, and output a high-quality sound track using only a desktop computer.

Sound Effects and Music Libraries. When watching your favorite cartoons or TV shows, chances are that what you're listening to is really a lot of the same sound effects repeated over and over again. Think of the classic whistle you hear during Wile E. Coyote's plummet off a cliff, or the metallic smash of an anvil as it lands on a character's head. This repetition of sound effects between one show and another exists because those stock sounds were probably pulled from the same sound effects library. Such commercial libraries have thousands of different selections ready for the animator and/or sound designer to choose and lay onto a single track. Because our ears are in some ways less finely tuned than our eyes, stock audio libraries are standard sources for everyday sounds like footsteps, wind, rain, clocks, birds, or street traffic, and even more for not-so-everyday sounds like explosions, gunfire, spaceships, bloodcurdling screams, or hippo mating calls.

There are also music libraries filled with every type of music imaginable, from classical to hard rock. Few music libraries provide top hits on the pop music charts since that would involve paying the pop artists and composers some serious cash. Instead, you will find musical "beds" and "stings" and "bumpers" that can be purchased for a simple charge per *needle drop.* This term comes from the days when a filmmaker would pay a separate fee for each selection he or she used from a music or sound effects library, each time he or she used it. One didn't pay by the length of the cuts, just by the needle drops. More and more in today's digital world, one can make an outright purchase of an entire music and sound effects library for which no royalties or needle-drop fees are required—you've already paid in the purchase price. Look carefully before you make such a purchase to be sure that the library's music cuts and sound effects are copyright free. One of these collections is a solid investment. You'll have fun listening to the awesome variety of music and sound effects and, by browsing the library, you will probably come up with new ideas for your sound track. If you can, buy the compact disc version (as opposed to audiocassettes) since it will make digitizing your sound a lot easier.

Have Microphone, Will Travel. If you don't want to rely on sound effects libraries, or if you need something so unique it can't be found anywhere, record it yourself! Grab a tape recorder and a microphone and head outside for a day of recording whatever neat sound you come across. If you have access to a laptop computer, it's also easy to hook up a microphone to it and use the computer as your recording device instead of a tape deck. Wandering the streets with your ears doing the looking might seem a bit strange, but it's a normal part of the job for a sound designer. Top guys in the business can often be found creeping through zoos, farms, and con-

7.3 Recording session: Rock star Iggy Pop, seen in a recording session for the *Rugrats* feature film, in production by Nickelodeon Features and Paramount, *(A).* On the other side of the glass that divides the studio from the control booth, composer Mark Mothersbaugh (foreground) works with Engineer Bob Casale, *(B). Photos by Vince Gonzales. Courtesy Klasky/Csupo and Nickelodeon.*

A

B

struction sites, trying to capture that unique grind, grunt, or groan.

If you choose to record your own effects, you'll have entered a time-honored domain of animation called *foley.* At the big feature animation outfits in Hollywood, there are foley studios, where sound designers use whatever means are necessary to create that perfect sound effect needed for a particular animated sequence. This often means taking everyday objects like old pairs of shoes, metal trash cans, rocks, doors, or even fruits and vegetables and manipulating them to create a distinct and effective sound. Foley artists have manufactured stabbing sounds by cutting a melon, punching noises by slapping a side of beef, and the sound of a neck breaking by snapping a stalk of celery. Fun, huh?

Recording Voices. If your animation will involve voices, whether in the form of dialogue or voice-over, you'll need to record these as separate elements from the sound effects and music. The best approach is to use a good-quality microphone and hook it up to a tape recorder or your computer. Try to find a secluded, quiet, soundproof room so that no excess or unwanted noise pops up while recording. Many schools and TV stations have recording booths you can borrow. The rental fees for commercial sound studios are surprisingly modest and these facilities always help the voice talent and director get the best performance.

CAPTURING SOUND IN THE COMPUTER

Once you've gathered all the sound effects, music, and voices for your project, these elements must be "captured" into your computer and stored as sound files. This process is sometimes called *digitizing* and has become extremely easy now that almost all computers come with multimedia features that let you record, play back, and edit.

If you're pulling music or a series of effects off a CD, the process is a breeze. It usually takes nothing more than putting the disc in your CD-ROM drive and dragging over the icons for the desired tracks from the CD to your computer's hard drive. Or you can use special sound-handling software to record the sound effects while you play the CD. Either way, the resulting digital file can be named and saved for subsequent use in your sound track.

The process is easy, too, if your source sounds have been recorded onto audiotape. Most of today's tape decks have either an earphone jack or an audio-out jack in the back. These jacks can be used to connect the tape playback machine to your computer's microphone jack using a standard RCA stereo wire. Once you've hooked your tape player to the computer, any sound played on the tape will be sent to the computer, ready to be digitized. To capture sounds you've recorded yourself, a sound editing or capturing application is a necessity. Both Macs and PCs usually have a built-in sound recording program that is crude but effective.

TRACK LAYOUT AND MIXING

Let's say that every voice, sound, and piece of music needed for your animation has been gathered and captured in your computer. Now comes the task of carefully placing each clip where it belongs and then mixing all those ingredients together to create a beautiful and moving sound track. Thanks to the recent evolution of desktop video and multimedia tools, there are a number of terrific sound editing programs on the market. These range in expense and complexity (see Chapter 23: Computer Software) and have been designed to meet the needs of individuals at different skill levels, from fledgling animators to seasoned pros working in multitrack recording.

Simple Waveform Editors. The most basic audio editing program should let you open up an audio file, look at its *waveform,* and do simple cut-and-paste jobs on it, moving sound effects from one part of the track to another and pasting in other sounds. Even these simple audio editors will usually let you add special effects such as fade, echo, dissolve, reverse, and amplify. The finished audio track can then be saved as a new sound file or, better yet, as a QuickTime movie that can later be laid down with a video track. You'll find that manipulating sounds and music is great fun. Give yourself time to experiment. Everyone gets a kick out of making their voice sound like a chipmunk or hearing how different things sound when played backward.

The drawback to basic audio editing software is that it

7.4 CASE STUDY: Digital Sound

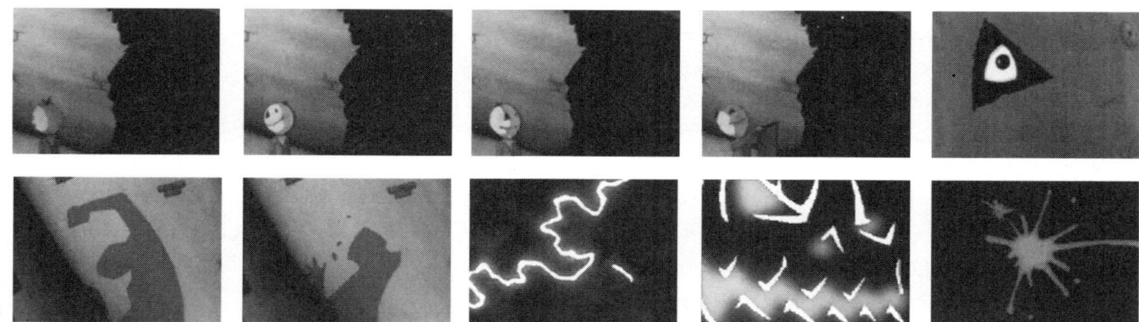

A

Season's Greetings is a 16mm cel animation film created by Michael Dougherty as a student film for the NYU Animation Program. It tells the macabre story of a child trick-or-treating late one Halloween night. The project was completed in just under a year and later went on to win awards from the Student Emmys, the Student Academy Awards, and several other film festivals. The story is told with eye-catching visuals that portray a young innocent who is being followed by a menacing stalker. The movie's track was created digitally once the film itself had been completed in 16mm format. To emphasize the frightening story and help build the terror and suspense that climaxes in the scene studied here, an original musical score and effects tracks were created by Evan Chen, an NYU student studying musical composition. Ten key frames from the sequence are shown, *(A)*. Michael Dougherty's comments follow:

Importing into SoundEdit 16

The audio design process began once the film had been digitized and imported into SoundEdit 16, a sound editing application from Macromedia, *(B)*. The 16mm work print was transferred to 3/4" format video at a postproduction house in New York City, then digitized by playing back the tape into a QuickTime clip on a Macintosh.

The resulting QuickTime clip, *(C)*, was imported into SoundEdit 16, but still lacked audio. Icons representing the animation are displayed at the top, serving as a guide for where to place sounds. You can see the empty tracks, waiting to be filled.

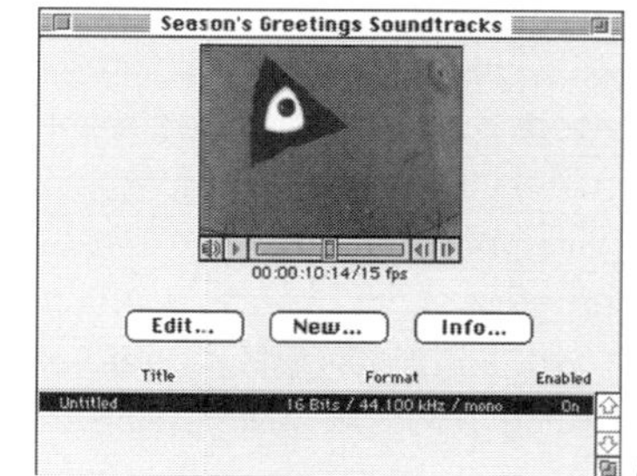

B

Laying Down Tracks and Scoping Waveforms

Two tracks were then added: the musical score and the sound of thunder, which was used as ominous foreshadowing of a (mostly) off-camera fight and murder, *(D)*.

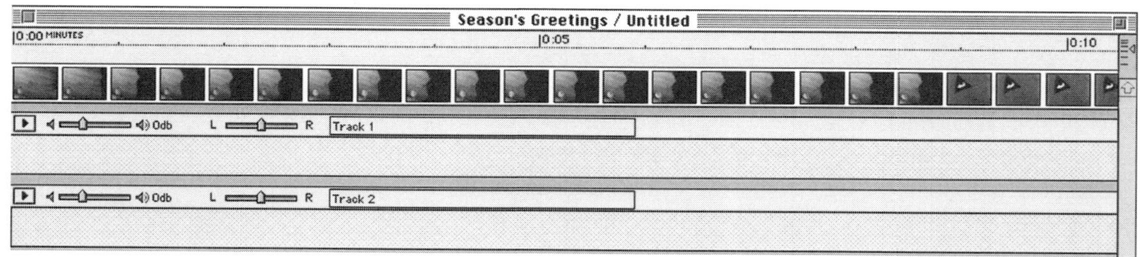

C

More tracks were added, *(E)*, to go with the scoring and thunderclap sound effect. Isolated in its own track, the waveform shows precisely where the scream begins, making it easy to place the sound against the animation.

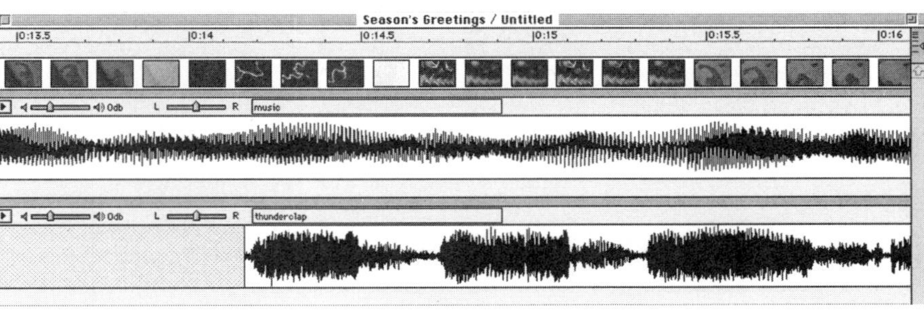

D

The waveform is a visual representation of a sound. The longer and louder the noise, the longer and higher the peaks appear on the waveform. If there is a moment of silence, the waveform will appear as a straight line. One of the great things about working digitally is that it's so easy to slide a sound element one way or another, testing to see where it works best against a picture. Each track can be manipulated independently to allow for the addition of special effects like echoes, fades, and changing the pitch or tone.

E

Mixing Down

Once all the tracks were added, manipulated, and placed at proper times, the multiple tracks (a total of four in this project) could be mixed down to a single sound track that is represented by one waveform, *(F)*.

F

SoundEdit 16 displays the "stats" of the sound track in the space below the movie's sample frame, *(G)*. In this case, we created a 16 bit, 44.1 MHz mono sound track, which is close to CD sound quality and suitable for presentation on video.

Because I wanted to have a traditional 16mm film complete with an optical track, I transferred the mixed track to Digital Audio Tape (DAT), which was taken directly to the lab with the film negatives, turned into an optical master, and married to the negatives to create the answer print. After screening and approving the answer print, several release prints for distribution were made.

Voilà! A finished film with a killer track.

G

Credits: **Directed and Animated by:** Michael Dougherty **Music and Sound by:** Evan Chen **Backgrounds Painted by:** Dan M. Kanemoto **Faculty Supervisors:** Kit Laybourne and John Canemaker rarely lets you mix and manipulate multiple tracks. They usually have only the capability to edit one or two tracks of audio at a time, which can come in handy but doesn't allow the flexibility of mixing larger numbers of sound effects and music together.

Multitrack Audio Editors. For ambitious sound tracks that combine effects with music and dialogue tracks, you'll want to use an audio editing program that can import, edit, and mix multiple tracks and can cut, manipulate, and distort each sound in a million different ways.

The value of control over individual tracks is obvious. If a barking dog effect you've gotten off a sound effects CD is too loud for your taste, you can lower its volume so it's more of a background noise. You could even change the pitch and tone so a giant Rottweiler suddenly becomes a tiny yipping Chihuahua. Using multiple tracks also lets you control the placement of your audio effects, down to a single frame. Don't like the thunder effect starting at the beginning of your animation? Change the in point and it comes in at the end. Any effect you've gathered can also be duplicated an infinite number of times. If you've recorded the chime of a clock that rings three times but you need it to chime for twelve midnight, the sound of a single chime can be copied by selecting its chunk on the waveform and pasted nine more times, all with a few clicks of the mouse.

The case study that begins in Figure 7.4 has been designed to introduce you to the workings of one of the major sound editing applications serving both Windows and Mac computers. *SoundEdit 16* is a simple-to-use, midlevel piece of software that allows you to create and edit multiple tracks, apply a large number of special effects, and output to fourteen different file formats. The simple, customizable interface is a snap to learn. Each audio track can be labeled and is displayed as a waveform that can be edited with the standard cut, copy, and paste functions seen in almost all software packages. Cue points and notes can be marked on the tracks for easy reference, and special effects can be added by just highlighting the part of the waveform you want to change. One of the best features is its ability to open up a QuickTime movie and edit the existing sound track or create a whole new one for it. It has become a standard software package used in the multimedia and digital video industries for CD-ROMs, the Internet, and video productions. The folks at Macromedia who program and publish SoundEdit 16 have begun to employ *plug-ins* (like Adobe Photoshop and After Effects) to give the software room to grow.

DIGITAL AUDIO EFFECTS

Applying a special effect to your audio tracks can give your sound track extra kick, taking a mundane noise and turning it into something new and vibrant. It involves little more than selecting the area on the waveform where you want the effect to appear and then using a pulldown menu where all your options are listed (see Figure 7.5). With such special effects you can turn the voice of a sweet old lady into that of a two-hundred-pound weight lifter or make music sound as if it were emanating from inside a cavern. Think of effects as the audio equivalent to image filters in Photoshop. They're a quick, easy, and painless way of altering preexisting sounds to create an effect that might be way, way too difficult to create on your own.

TRACK ANALYSIS AND LIP-SYNC

Once all the audio elements have been given their own track, you need to analyze them to see how they relate to each other and exactly where they occur. This is not so hard with a single sound effect that appears as a high-spike blip on its track. Using the waveform as a guide, the animator has a visual reference to locate exactly what time a certain sound effect occurs. Many software applications will let you place reference marks and notes right onto the computer screen.

It's more difficult to mark the rhythm of a piece of music. The effort is worth it, of course, when a piece of visual action is matched up to the cadence of music. Animators discovered the power of synchronous sound quite early and the phrase

7.5 Pull-down sound effects: These pieces of user interface from the software program SoundEdit 16 will show you the path to some very cool sound effects which, at a mouse click, you can apply to any sound that you record or get from another source. *(A)* shows the basic control panel for this program. Look familiar? It's just like every sound recorder device you've ever worked. The interface in *(B)* is our destination: a "pull-down" array of choices that you can apply to any chunk of waveform that you have highlighted. *(C)* gives you a chance to study the main tool bar for SoundEdit 16. This is a very versatile program with plenty of creative power. A longer discussion of it is included in Chapter 23.

A

Effects Control

Mix...

Amplify...
Backwards
Bender...
Delay...
Echo...
Emphasize
Envelope...
Equalizer...
Fade In...
Fade Out...
Flanger
Noise Gate...
Normalize...
Pitch Shift...
Reverb...
Smooth
Tempo...

B

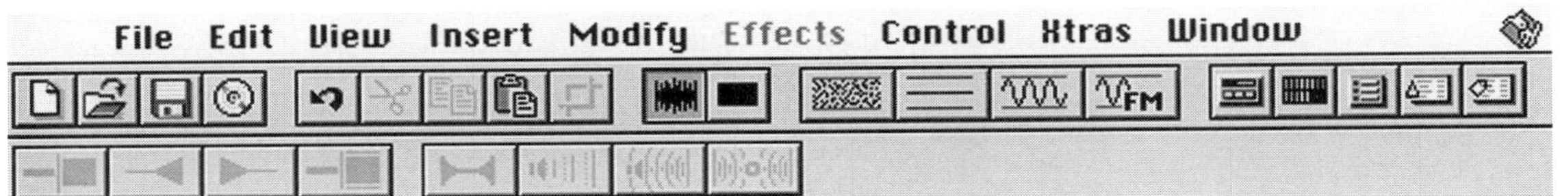

C

"Mickey Mousing" is still used to describe the effect when a character walks along with his footsteps perfectly in sync with a musical beat. It's generally impossible to discern rhythm from a waveform. Therefore, to do your own Mickey Mousing you will need to learn how the computer program you are using allows you to mark visual frames "on the fly," while the sound track is playing. If this capability doesn't exist or you can't figure it out, record your track and then experiment with placing marks on individual frames. By the process of trial and error, you will be able to find exactly which frame matches which beat or other place in the music track where you want to match picture and sound.

The precision with which it is possible to analyze a sound track enables the animator to locate individual words or parts of words. This makes it possible to create cartoon characters who move their lips with uncanny accuracy to the words delivered in the voice track. The effect, of course, is that the character actually talks.

In order to create lip sync, you must analyze each voice track so that you know the location of each vowel and consonant sound in a given piece of dialogue. With this information, you will be able to match each utterance with the appropriate mouth shape. Figure 7.6 shows six different mouth shapes and the distinct sound each represents. Naturally, the way a character is designed will require modifications of these basic shapes. The way Bart Simpson speaks is far different from the lip-sync style used for the Tick. And both of those are different from Buzz Lightyear or the Lion King!

Both track analysis and lip-sync drawing are laborious processes, as you can imagine. Doing it with computers is a lot easier than doing it with film tools. Doing it right requires patience and perseverance. A real mastery of lip-sync animation requires control of head, neck, and body movement and the ability to create gestures appropriate to the meaning of individual words, to the character's distinct personality, and to the dramatic requirements of the scene. If you are interested in pursuing this specialized technique, obtain one of the advanced manuals on the subject, select a software package that deals with lip-syncing, and be prepared for an education. It takes a lot of patience and practice to become proficient in determining what shape of mouth goes with a particular spoken sound. Animators often sit in front of mirrors in order to study the shapes of their own mouths while they try to determine how to match an analyzed dialogue track. Lip sync needs to be accurate. Poor track analysis will result in mouth shapes not matching the voices and that will make your animation look like a bad Japanese monster movie.

7.6 Mouth shapes: This lunatic face, created by animator Michael Dougherty, samples six mouth shapes that correspond to familiar sounds used in human speech. Chapter 14 shows other configurations of mouth shapes that animators use in making their characters lip-sync to spoken words and other guttural utterances.

"c," "d," and other consonants not shown below

"o"

"e"

"u"

"f" and "v"

"w" and "q"

FIX IT IN THE MIX

The creative juice of a sound track is always experienced as a surprising yet welcome jolt right at the end of the animation process, when individual tracks are mixed into a single track that goes with the finished piece. It's essential, of course, that the mix be done "to picture"—that your edited and locked visuals are viewed up against all the individual tracks you have painstakingly digitized and placed in their respective positions.

And the salvation! Those pieces of visual action and character shtick that never really worked as you hoped will suddenly live up to your highest expectations. Sometimes it is only in the final mix that you discover the need for one more effect—or that a piece of sound you've positioned now seems unnecessary and can be dropped.

If you are working in the digital domain and the final animation will live in videotape or on a computer file, the last step is simple. Typically your audio editing software, whether it's a simple waveform editor or a multitrack editor, lets you output your track to any number of sound file formats. The resulting file can then be brought into a video editing or special-effects application like Premiere, After Effects, or an Avid system and laid down onto your digital video version of the animation. If you've done everything right, the picture and sound will match beautifully to create your masterpiece.

7.7 Professional recording and mix: Here is a small — but well equipped — sound recording and mixing facility located in Amarillo, Texas. Because the music, radio, and television businesses all require a steady stream of voice and music recording, you can probably find a professional sound facility close to where you live. The proprietors and technicians of such places are often interested in the work of animators and eager to show off their skills. *Photo courtesy Carlos Casso and The Audio Refinery.*

REMEMBERING MAG STOCK

This chapter would be incomplete if it didn't provide a quick review of the film-based processes for working with sound. Happily, the basic process is pretty similar to the one you've just been reading about.

Source Recording onto Separate Tracks. Music, voice, and sound effects are recorded onto standard ¼-inch audiotape, recorded at a speed of 7.5 inches per second (sometimes 15 ips is used with music to get the highest quality of recording).

Transfer to 16mm Magnetic Recording Stock. The source tapes are then transferred via a Magna-Sync machine onto a form of audiotape that has the exact same physical configuration as 16mm film. (All professional sound studios offer the use of their transfer machines at minimal cost.) A similar format is used intentionally for both movie images and sound recordings; *16 mag stock* (as it is called) allows the filmmaker to locate any given sound according to a specific frame on the visual film. This way, when editing, film and sound can be played simultaneously and it's far easier to keep them in sync than it would be if the film were one size and the audio a different size.

Three-Track Editing. In 16mm film production technology, it is standard operating procedure to use three separate tracks: one for voice, one for music, and one for special effects. If there are many sounds and they overlap, additional tracks can be prepared. Only in the final mix will the filmmaker hear all the tracks being played at the same time.

Track Analysis. There is no film equivalent to the digital waveform to serve as a visual guide for the animator to see what sound is being played at a specific place on the sound track, so in order to mark the locations of specific sounds, a section of 16 mag stock must be analyzed on a frame-by-frame basis. Over and over again the animator listens to the recorded track using special equipment that allows the sound to be played at various speeds. The exact location (found by counting frames from a start mark) is determined for every single significant piece of sound—any words, music notes, recurring rhythms, and sound effects that the animator will want to know the timing of as he or she creates the movie's visual elements.

Analyzing a recorded track with precision requires two tools. The *flatbed editing machine* is a costly piece of specialized equipment with a variable-speed motor that will simultaneously project the film and "read" (play back) the sound track that has been transferred to mag stock. By playing the machine at a slow speed and by going forward and backward over a particular stretch of sound many times, you can identify the exact place where a sound begins or ends.

The other tool used to analyze a sound track is called a *gang synchronizer.* It is equipped with a magnetic soundhead that reads the analog track and then amplifies it with a special speaker, sometimes called a *squawk box.* Figure 7.9 shows how the mag stock is run through the synchronizer. The filmmaker's control over the rewind reels allows him or her to run the track under the soundhead many times until the precise location of a sound is determined.

Bar Sheets and Cue Sheets. The animator needs to keep careful and accurate track of the various elements in a film's sound tracks. A *bar sheet* is the standard notation system by which animators record their analysis of a sound track. The

7.8 A horizontal editing machine: See Chapter 22 for more detail concerning these and other editing systems that can be used for the analysis of recorded soundtracks in super 8mm, 16mm, and 35mm formats.

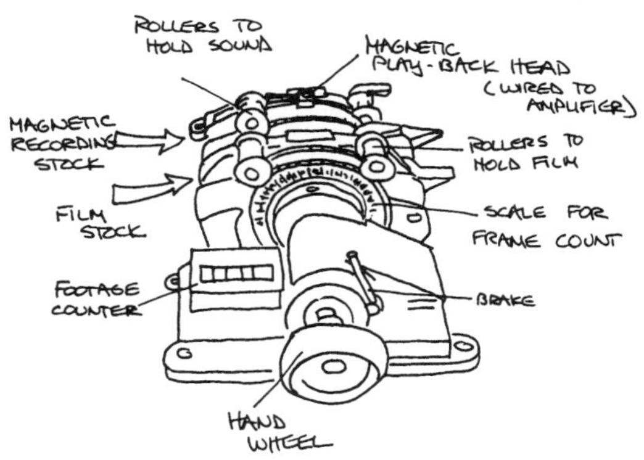

7.9 A two-gang synchronizer: This specialized piece of equipment is available in two-, three-, and four-gang models. Specialized synchronizers are also manufactured that combine formats—for example, 8mm with two 16mm gangs.

sample shown in Figure 7.10 provides places to log words, sound effects, and music. There is nothing sacred about the particular bar sheet shown here. Animators will often note specific frame locations on a cue sheet, the guide that is traditionally used to provide directions to the animation camera operator.

The Multitrack Mix. When the film has been "locked" after final editing, the animator traditionally makes an appointment at the sound recording/mixing studio for a monumentally important session in which all the separate tracks of 16mm mag stock are mixed down to a single track. After lining up the new mixed track with the sync mark at the head of the movie leader, the animator working in film has the experience, for the very first time, of watching the completed film while all the sound tracks play at the same time—a thrilling moment!

Optical Tracks. Figure 7.11 provides a sampling of all the different soundtrack formats that exist within the world of animated filmmaking (as opposed to digital animation). Two of these formats, you will see, have *optical tracks* that are "read" by the movie projector, which then plays the amplified sound track to the film's audience over conventional sound speakers. The process of converting a mixed mag master track to an optical master track (which is used when printing the finished film) is done at the film lab.

Magnetic Tracks. Although all 16mm and 35mm distributed films have optical tracks, there does exist, within both

7.10 A sample bar sheet: A *bar sheet* is the standard notation system by which animators record the analysis of a soundtrack. It has parallel lines to track dialogue (voice), action, sound effects, and musical score. There is nothing sacred about the particular bar sheet format sampled here. You can create a recording system of your own design that will accommodate the special needs you find in a given project.

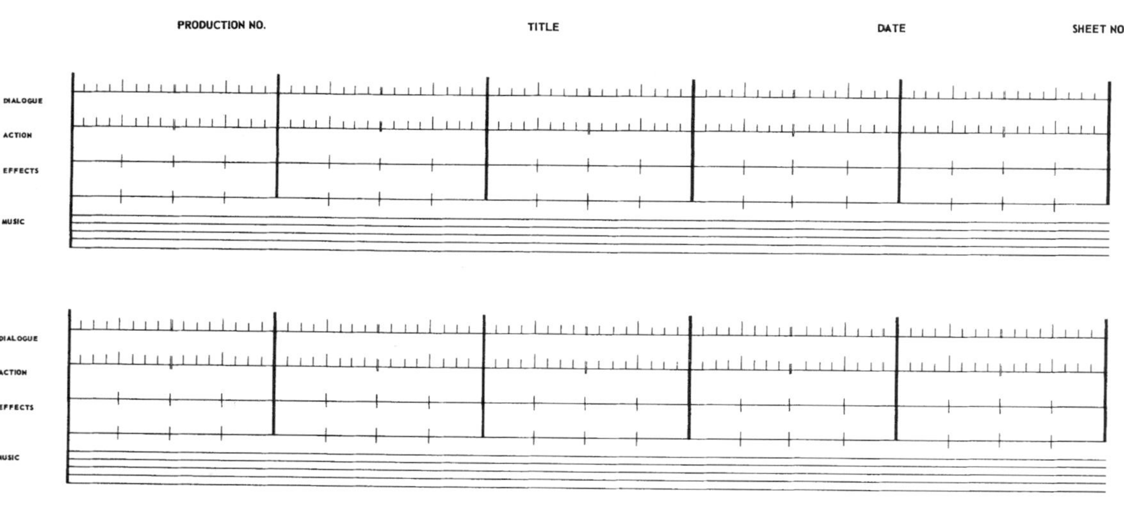

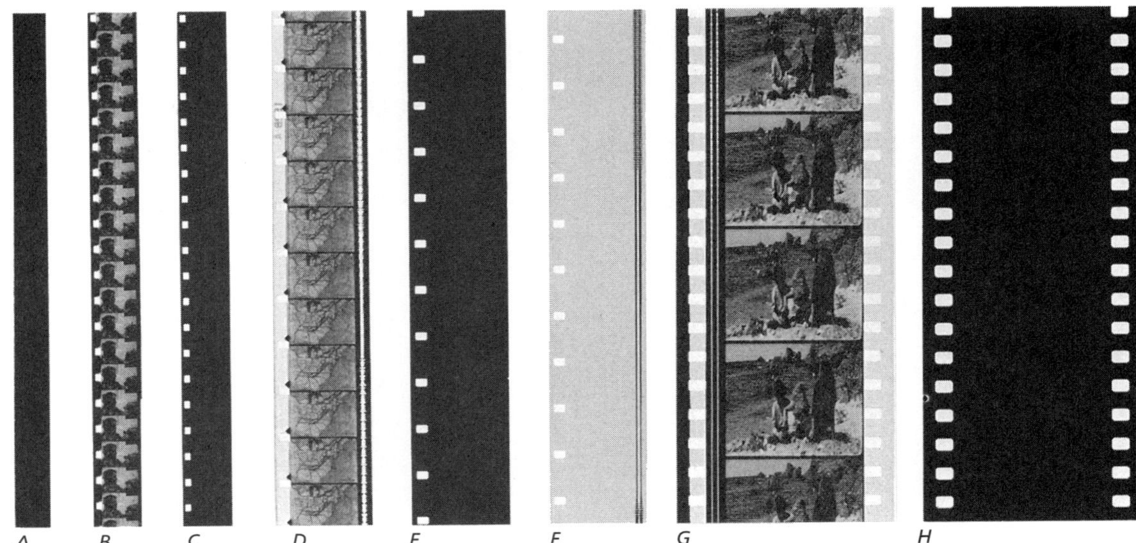

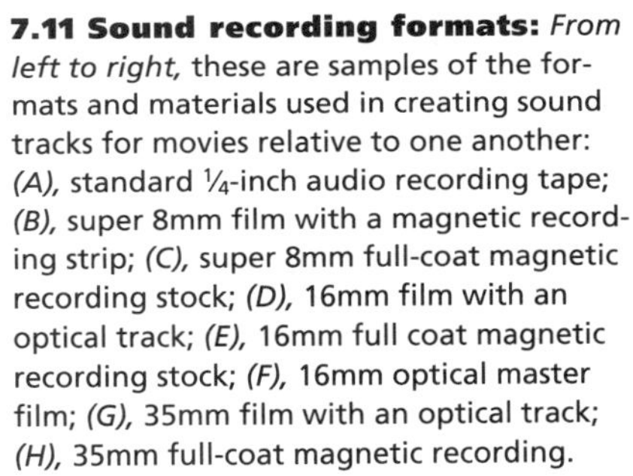
7.11 Sound recording formats: *From left to right,* these are samples of the formats and materials used in creating sound tracks for movies relative to one another: *(A),* standard ¼-inch audio recording tape; *(B),* super 8mm film with a magnetic recording strip; *(C),* super 8mm full-coat magnetic recording stock; *(D),* 16mm film with an optical track; *(E),* 16mm full coat magnetic recording stock; *(F),* 16mm optical master film; *(G),* 35mm film with an optical track; *(H),* 35mm full-coat magnetic recording.

16mm and 8mm film technology, a way of playing back mixed sound tracks via a magnetic strip of audio recording material that is placed on the edge of the celluloid film stock. With a special 8mm projector that has sound recording and sound playback capabilities, it is possible to place a sound track directly on the thin strip of magnetic recording surface. Such projectors accept a number of different sound sources (microphone, line input, etc.).

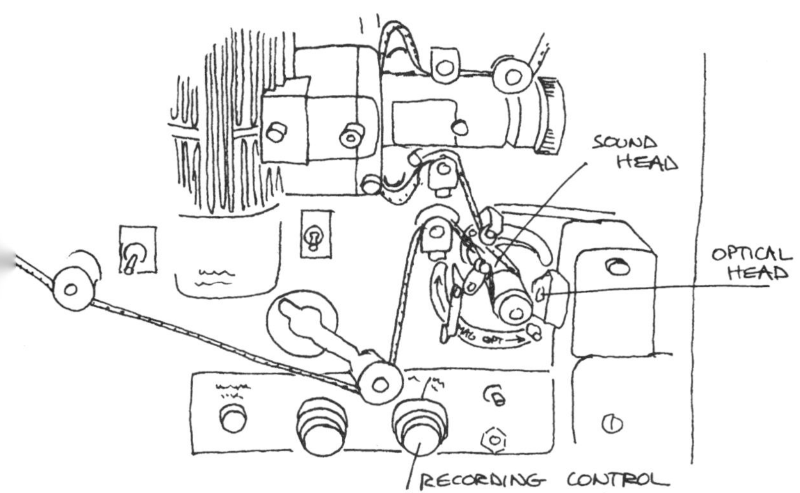

7.12 16mm projector with magnetic sound: Detail of a Bell & Howell model 504 projector. This versatile piece of equipment can both record and play back 16mm magnetic stock. It can also project standard 16mm prints with optical tracks. With simple modifications and using a special threading technique, this model can serve as an interlock projector, a specialized and expensive piece of equipment used to simultaneously project a 16mm print while playing a matched magnetic track in 16mm mag stock format. Occasionally, these classic Bell & Howell magnetic/optical projectors can be purchased used at reasonable rates.

8 ■ Storyboarding and Animatics

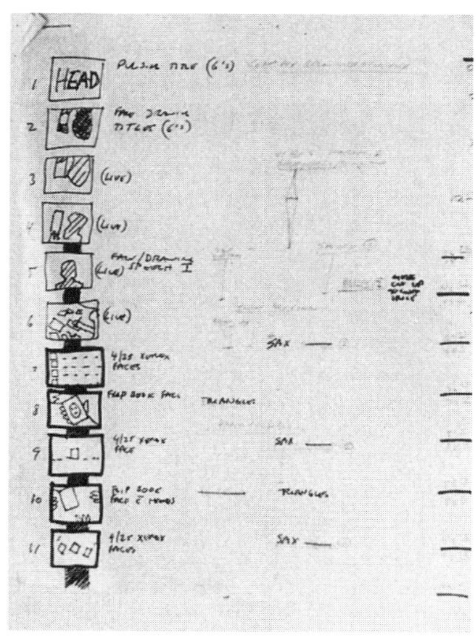

A

8.1 Storyboard styles: Animators approach storyboarding with an endless variety of styles and techniques. *(A)* shows a working storyboard element from George Griffin's *Head,* while the schematic in *(B)* represents the overall structure of that finished ten-minute film. The National Film Board of Canada encourages its animators to use a standard format in presenting finished boards for approval prior to production. In *(C)* the NFBC's template is employed to present the beginning of *Why Me?,* a film by Derek Lamb and Janet Perlman.

Storyboarding is the push that gets a piece of animation going. And if you use a computer, this initial push can take you a long, long way.

In this chapter you'll see that storyboards come in all shapes and sizes, from nonlinear doodles to a wall covered with fancy drawings. But the best part will be when you discover how storyboards absolutely thrive in a digital production world, how a series of simple sketches can become a literal skeleton for your movie. At the chapter's end there is a case study about *Joe's Apartment,* a feature film that stars animated cockroaches.

It's extremely important to start off with a solid concept of what a storyboard is. Think of it as the outline form of a developing film or video. Storyboards aren't just for animation: live-action filmmakers, writers, production designers, and special-effects designers regularly use storyboards to work out their concepts. But the basic form is always the same: a collected series of single pictures, each of which represents a distinct visual sequence or narrative element within the project being developed.

For animators, it would be suicide *not* to storyboard. Unlike live-action filmmaking, where one shoots scenes in a variety of ways and then finds the final form through the editing process, in animation you never want to execute a finished scene that might be discarded, because this process costs too much time and money. The use of small sketches gives the animator a spectacularly clear and inexpensive way to work out his or her creative vision.

Have you noticed how difficult it is to explain the imaginative story lines that cartoons so often follow? It can be downright impossible to describe projects that push into realms of the abstract and the fantastic. A storyboard can help here. It is useful in explaining a concept to people you are working with (animator friends, camerapeople, writers, musicians, etc.), and becomes absolutely essential in explaining a proposed project to producers and clients.

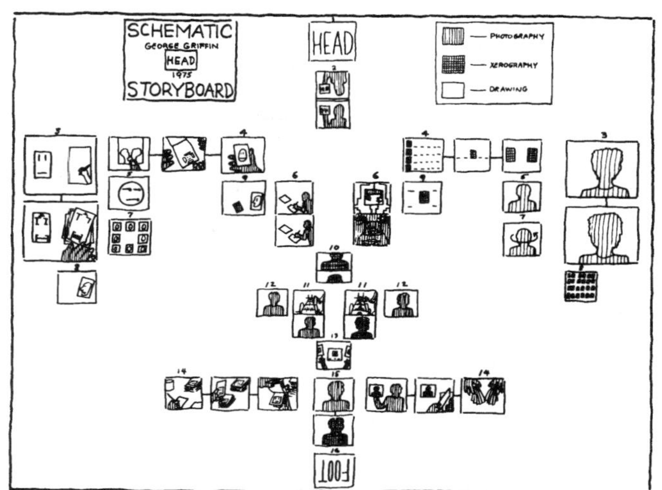

B

STORYBOARD STYLES

Look through the sample storyboards that accompany this chapter. What should be immediately evident is that the act of storyboarding can be undertaken in many different ways. This makes sense: Each movie has its own requirements and its own set of elements that the animator is working out. In addition to the pictorial elements, boards usually include important information about the script and dialogue, camera moves, special effects, and sound effects.

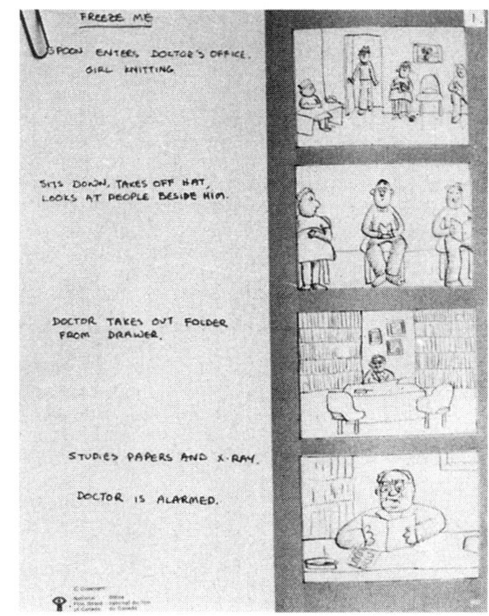

C

Commercial animation production houses and advertising agencies have made the storyboard into a high art form. Their renderings are done by well-paid specialists who place each color drawing on fancy, precut mounting boards. A typical "show" storyboard will also display verbal notations (script and descriptive information) that have been typed and mounted beneath appropriate panels. Figure 8.3 is a photograph of one such highly finished presentation storyboard.

For most independent animators, however, there is nothing precious about a storyboard. It is a tool and an aid, not a piece of art. Sketches are made quickly and revised often. A standard storyboarding procedure is to create each sketch on a different card or sheet of paper. All drawings are then tacked up on a bulletin board. This technique makes it possible to study the overall film-to-be. And it makes it easy to change the order of scenes and to insert or delete an entire set of drawings. At the Disney studios in Burbank, California, there are

8.2 Storyboarding as you go: In many "direct animation" techniques that involve image manipulation under the camera, it is difficult to storyboard onto paper. The clever designer knows, however, that it is important to follow a plan and to work this out in advance with sketches or photographs. Here is Caroline Leaf's animation setup, *(A),* with camera pointing down to the rear-lit area where she works (see Chapter 11: Sand and Paint-on-Glass Animation). Note the smaller drawings on either side of the filming area which Caroline uses to guide her technique of painting directly onto glass. Some of these are shown in the enlargements, *(B).*

A

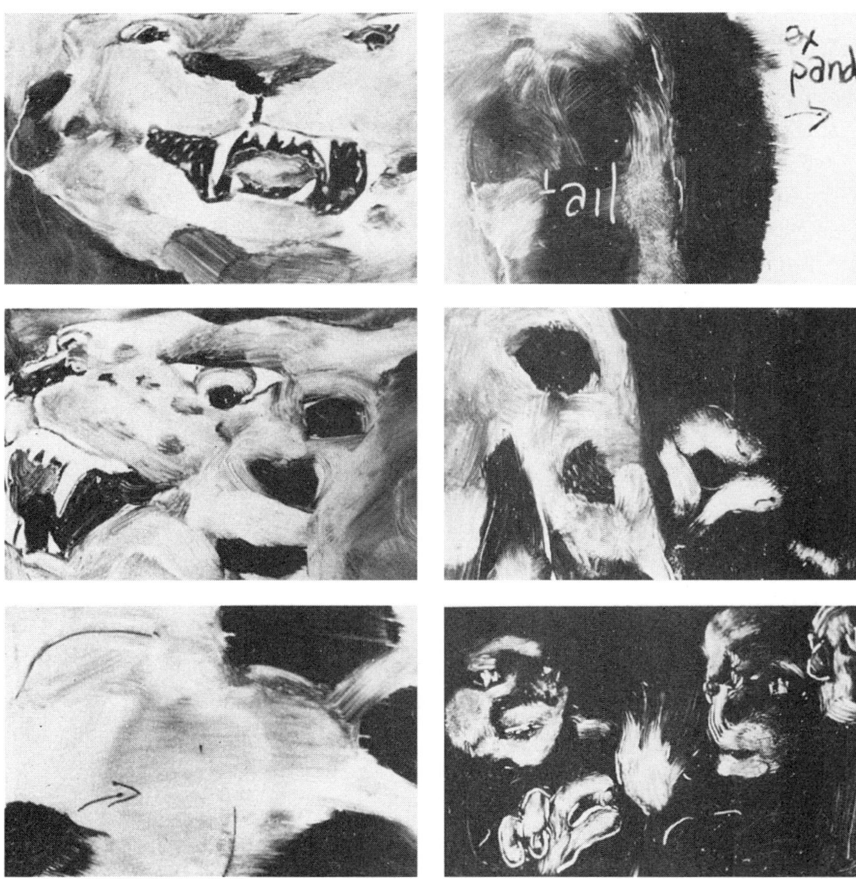

B

entire rooms with cork-lined walls where the principal animators and directors of feature films mount and study thousands of individual sketches used in planning an animation feature.

STORYBOARD FUNCTIONS

A pair of related questions almost invariably presents itself as you prepare a storyboard: "How many pictures should I use?" and "How much detail should I show?" To answer these questions, it is necessary to consider the different functions that a storyboard serves.

Conceptualization. The storyboard is a tool for working out a project's core idea and structure. Even when the notion you've got in your mind seems absolutely detailed and complete, getting it down on paper is always a creative step. The process of visual thinking releases new energy, new insights, and new ideas. Best of all, developing a storyboard lets you see the problems and challenges of the project ahead. In fact, many would say that the real creativity starts only after that inner notion has had its first concrete expression on a storyboard. No matter how informal, that initial set of storyboard panels makes the concept accessible in new ways. You are able to step back from the idea and study it with more objective eyes. Various component elements can now be seen independently of the central concept.

Key Moments. The storyboard should represent all key moments of the film, whether they be a story or an abstract piece. The number of individual drawings needed to do this can vary widely. If you think of a *scene* as screen action taking place during one stretch of time and in one location, then each scene should be represented by at least one drawing. More often, each scene is made up of a number of different *shots*: wide-angle "establishing" shots and various medium shots and close-ups. There should be a separate drawing for each of these.

Flow and Transitions. The process of storyboarding should direct your attention to two particularly tricky structural elements. The first of these is the order or *sequence* of individual shots and scenes. Will the viewer be able to follow the flow of scenes without getting confused? Is emphasis placed in the right place? Is the pacing varied enough to stimulate interest?

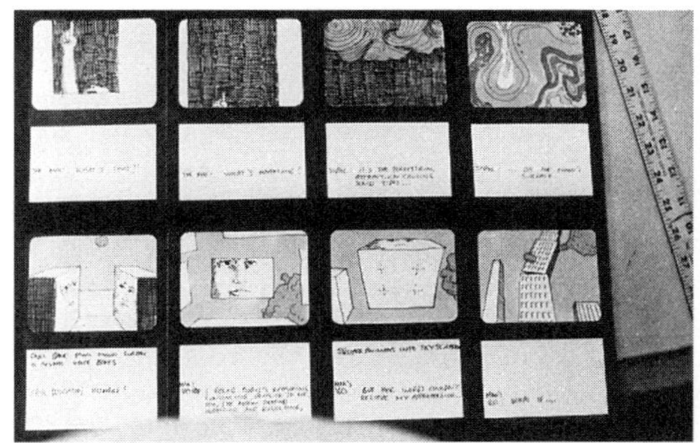

8.3 A "show" storyboard: Drawings and dialogue are formally mounted on a precut storyboard matte. This sample is from an unproduced project by George Griffin.

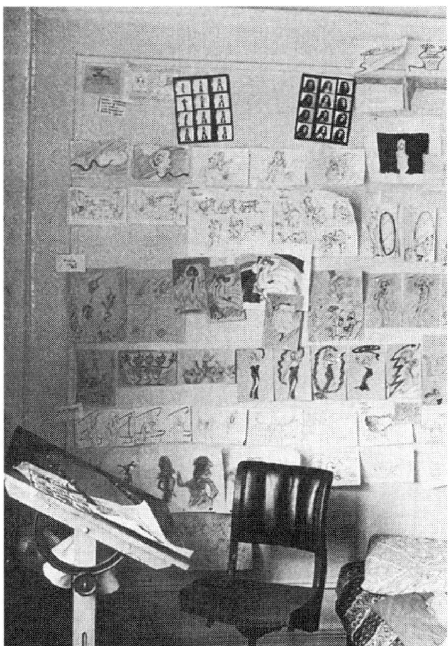

8.4 A wall as tool: The wall beside animator John Canemaker's drawing table is used as in impromptu (but productive) planning/storyboarding tool. *Courtesy John Canemaker.*

The study of your project's visual development and sequencing—of those building blocks of scenes and shots—will lead to a study of the *transitions* between them. Here is one of the most challenging areas of animation design.

Detail, Composition, and Aesthetics. How much detail does there need to be in an individual drawing? The right amount is the amount that works for you and that allows others to understand your plans. A few very rough sketches can do the trick. Or you may find it essential to execute renderings that are every bit as finished as you can make them—formal drawings with full color and complete detail. All drawings in the same storyboard need not be drawn at the same level. However, it's a good idea to have at least one well-detailed image for each new scene that shows the full background, the color palette you are using, the styling of characters, and so on. But once that is done, action sketches can be very schematic.

When you have a full storyboard spread out, you will discover patterns in your visualization process that you didn't know existed. Are frames composed in too predictable a fashion? Is there a rich enough mix of angles and points of view? What balance is there among colors? Is the important information presented with emphasis? Are characters sharply portrayed both by how they look and the detailed action you've given them? Are there enough close-ups? Are there enough establishing shots?

Logistics. Look at the storyboard with a practical eye, not just an artistic one. Does the project call for tools that you don't have? What new techniques will you need to learn? Can you reasonably expect to finish the project in the time you will be able to put aside for it? What is the cost of this production? The list can go on and on and on. Check out Chapter 18 for more about production planning.

DIGITAL STORYBOARDING

When pencil lines on paper become pixel forms on a computer screen, the storyboard starts to become a very different animal. The process begins with an easy step: Drawings done on paper are scanned into the computer or, using a digital drawing tablet, you can draw directly into the computer. Any paint/

draw software application will let you make a set of image files that become the building blocks of the storyboard.

The next task is to bring these images into a single storyboard—most often in the form of a document that is printed out on paper. Today, just about every motion graphics application provides templates specifically designed for storyboarding. You'll find it's a snap to scale the same image file bigger or smaller, depending on whether you want to post individual panels onto a board or squeeze them down to standard 8½-by 11-inch pages. If you can get your hands on the right ink-jet printer, you can wow your colleagues, teachers, and friends with full-color boards.

Remember that digital storyboards are working animals, not just creatures of beauty! There are many different kinds of hard information that should accompany the image files, including dialogue, notes on direction, sound effects, and questions about technique.

BOARDS TO ANIMATICS TO LEICA REELS

Storyboard still images are adequate for discerning the visual look and sequence of a developing project, but once they reside in the digital domain, it becomes possible to add something else: *duration.* This is the great switchover that turns a storyboard into an animatic and, in due time, into a full Leica reel and eventually into the completed animation itself.

Let's talk nomenclature. The term *animatic* is generally used to describe a film, video, or computer-based presentation of the still drawings that comprise a storyboard. Sometimes an animatic will represent movement within a stationary piece of art. For example, if the board calls for a pan across a scene, the animatic emulates the move. Working in film or video, either the camera pans across the single storyboard panel or, alternatively, the camera is fixed and the artwork is moved in front of it. On the computer, an animatic displays the sequence of drawings according to timing instructions given to the clock that hums inside every single computer. Many software programs create animations in this way. The best one of these is After Effects, because it is so adept at adding camera moves and transitional effects to storyboard drawings.

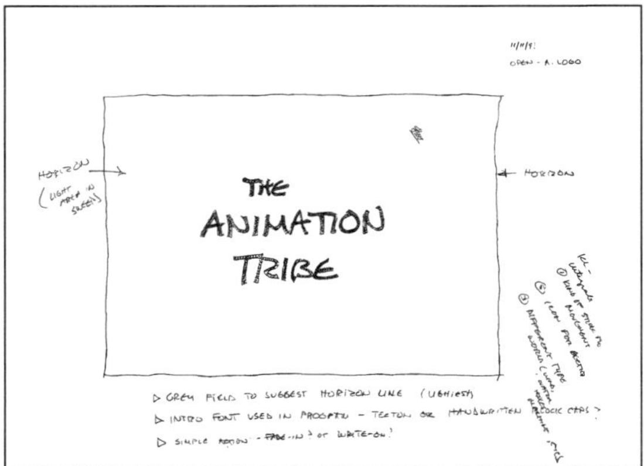

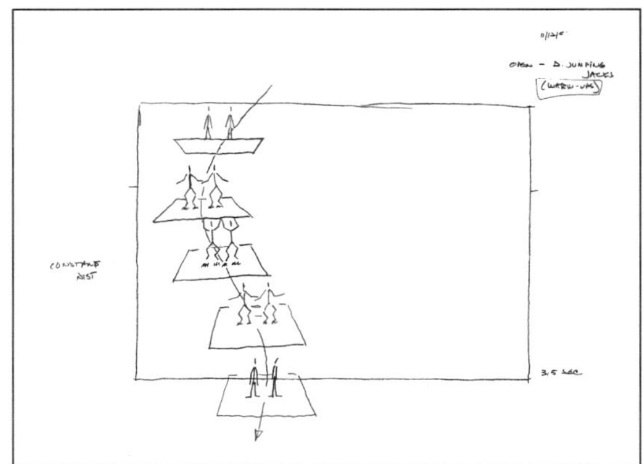

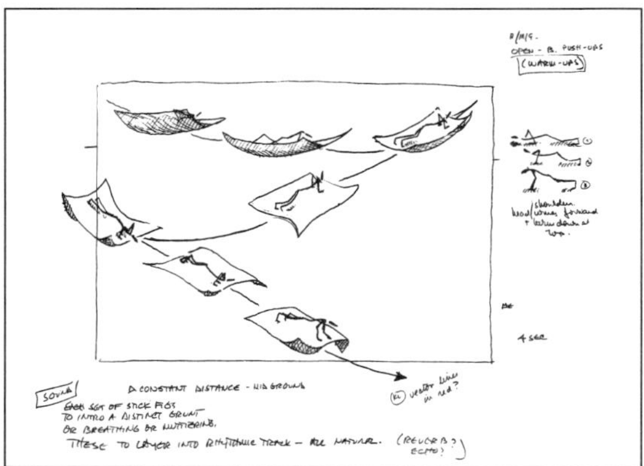

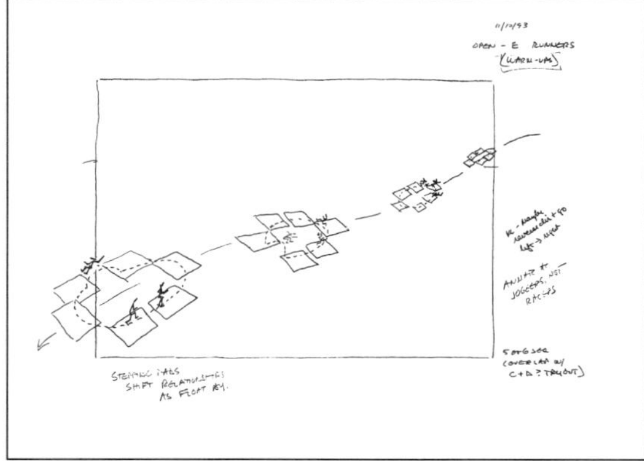

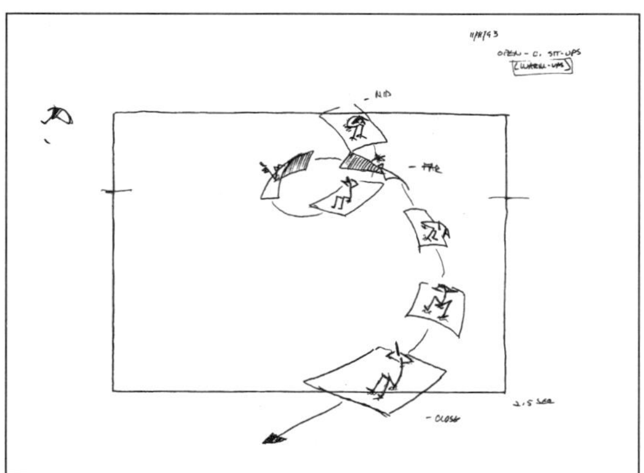

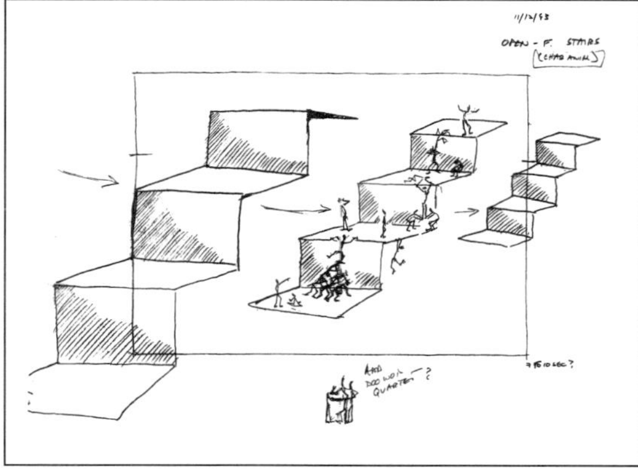

8.5 The animation tribe: These six panels constitute the storyboard for the opening sequence of a CD-ROM about animation. Each drawing represents the action for a short scene. Timings are suggested: Each scene would overlap so that the final open remained about twenty seconds long. Note that the "frame" resides well inside each sheet, leaving room to work out the transition. The writing consists mainly of working notes that Kit Laybourne wrote to himself. You'll also see how ideas are worked out by quick sketches done in the margins.

The great switchover continues its march toward true animation with something called a *Leica reel.* This unusual name comes from an early stage in the development of character animation. In the first studios, a cheaper camera was used to film the registered sheets of pencil drawings. The results were called pencil tests. But when the various scenes were edited together, the edited version became the Leica reel—named after the German camera that filmed the pencil tests. The process of shooting the final art (inked and opaqued cels over watercolor backgrounds) was done by more expensive, specialized cameras with brand names like Oxberry or Acme. Anyway, the name Leica stuck to the aggregate of test footage that had been edited together.

In the Leica reel rendition of the storyboard, two additional elements are added. The first is the *sound track.* Leica reels are always built upon at least one of the audio tracks—usually a voice track. But there are also enriched visual materials. The Leica reel is the repository for test footage: either *pencil tests* in 2-D cel animation, *motion tests* in stop-motion techniques, or *wire frame tests* in 3-D. The Leica reel starts out as an animatic but over the course of the animation testing evolves into a full-motion version of the completed project, albeit a visually impoverished version composed of test footage, not final footage showing fully produced art.

The Leica reels are handled by the editor, who "drops" the full-motion tests into the work print for film or rough video for video versions. Eventually the editor replaces test footage with the finished footage.

Let's summarize the incredible metamorphosis that is possible if you are working digitally. One begins by scanning or drawing individual images that stand for an entire scene. The addition of rough timings makes the board into an animatic. The addition of a sound track and test footage (pencil tests, wire frame tests, motion capture tests) evolve the animatic into a Leica reel. Ongoing art production and rendering transform the Leica into the real McCoy.

At every step in this process, the animator gains a fresh look at the work in progress. The evolution is easy, fast, inexpensive, seamless, and way cool.

8.6 CASE STUDY: Digital Storyboarding a Feature

A

B

Roach makings spread out at the workshop of Fly Films, in New York City.

It's almost a miracle when the spirit of Independent Animation — provocative, original, risk-taking — makes its way undiluted into the high-stakes world of feature films. But this is exactly what happened when John Payson, an MTV producer/director, got a chance to expand a short film he made for TV into a full-blown theatrical release.

In this case study, John Payson describes the ideation and design process through which he guided the miraculous production of *Joe's Apartment.*

A few years ago, I was sitting in my apartment in New York City, hating life and watching cockroaches crawl over the walls. "Well, things could be worse," I said to myself. "At least the roaches don't talk." And then I thought — wait, what if they did? That's how the idea for *Joe's Apartment* was born.

It started out as a three-minute film that aired on MTV, and eventually became a full-length feature film for Geffen Pictures — and all that from one lousy day hanging out in my grungy East Village apartment!

Once I had written the feature script, it was time to start working on the storyboard. Drawing a picture is a great means of conveying an idea, because if you can make somebody *see* your idea, you don't have to *explain* it.

We had a big room in the *Joe's* offices where we pasted up the storyboards as fast as we could draw them, so we could start figuring out the best ways to turn these drawings into reality. If we thought we could make a scene work with live bugs, we tried that first. We actually got one of the live critters to land on someone's nose!

But when the story called for shots that the bugs just couldn't do on their own, like swarm in very specific, controlled patterns, or act in a recognizably human manner, we turned to animation.

Joe's Apartment is mainly a live-action film, but within its 83 minutes, there are 14 minutes of animation: about 2 minutes of stop-motion and 12 minutes of computer-generated (CG) animation on live-action backgrounds.

C

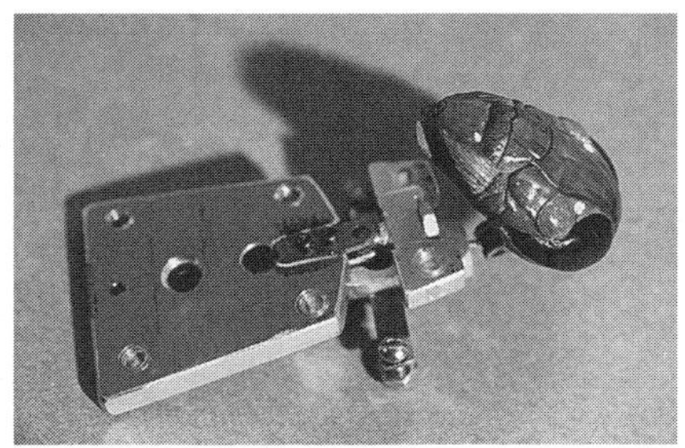

D

E

These close-ups of roach construction show the sophisticated metal armatures around which the superrealistic characters are built. The armatures allow animators to reshape the stop-motion actor during the filming process.

F
Here is an example of "flocking" animation. This shot shows two roaches breaking into Joe's apartment, with swarms of their fellows appearing on all sides.

G

H

I

Roaches on the set and as they appeared in the feature film.

J

Stop motion is a great technique for animating insects because bugs move in odd, quick, jerky ways, and the crisp, no-blur movement of stop motion can capture that really well. Since insects are too small and delicate to manipulate effectively, we built realistic plastic and aluminum models of roaches, six times normal scale. This upscaling enabled us to get real character into the roaches, but it also made it unfeasible to place them in normal-sized environments or interact with real humans. For shots that had lots of roaches, or where the bugs interacted with Joe or real objects, we placed computer-generated roaches on real-life backgrounds.

For the computer-generated animation, we created a "virtual" roach model in the computer. A roach is an ideal kind of object to model in a computer because computers are able to render something hard and shiny, like an insect's shell, much more easily than something stretchy like skin, or soft like fur.

We worked really hard to make the computer model realistic, recreating the slightly translucent quality of their legs, and the greasy gunk they secrete all over their foul little bodies. One of the great advantages of CG animation is that when you build one model, you can replicate it over and over again to create scenes with hundreds of models. With stop motion, you have to build each individual model, which becomes extremely costly and time-consuming. It's lucky that all common household roaches (*Periplaneta americanus*) really do look very similar. We were able to use just one virtual roach model for all of our roach characters, who were individualized only by their voices and the unique way each character was animated.

All the CG animation was composited onto real live-action backgrounds. When we shot these background "plates," we had to make sure there was no unwanted movement in the camera, and that the depth of field was great enough to minimize the movement of the animated bugs in and out of focus. Once we had shot a plate, we always took a frame of "references" — a model roach on a stick (which we termed an R. O. S.), along with various cubes, balls, and grids that

K

This shot incorporates both character animation and flocking animation. The roaches on Joe's chest are hand animated via stop-motion techniques and are representative of character animation, while the roaches approaching him from all sides is an example of flocking animation.

L

This is an example of anamorphic behavior: One of the two roaches standing on the spoon is scratching its head in a very humanlike manner. This is referred to as "character animation."

M

This photo illustrates the production scale of the stop-motion animation.

showed the animators the volume of space visible in the frame and the direction and intensity of the light. Armed with this information, we were able to exactly replicate these characteristics on the virtual characters. It got to be a ritual after a while: After we got the background shot, somebody would bellow out "R. O. S.!" or "Balls and cubes!"

The CG animation in *Joe's Apartment* splits primarily into two types: *character animation*, where individual roaches are animated frame by frame, and *flocking* or *herding animation*, where the computer is allowed to program a degree of random movement to a whole bunch of objects at once. As a rule of thumb, very human-looking action, like when a roach scratches his head in puzzlement, is probably character animation. Swarms, like a shot with hundreds to thousands of roaches menacing the bad guys, tend to be rendered with flocking animation. Some kinds of CG animation are so complex that they mix both types of animation. But the only way to figure out how to approach the solution is by drawing a storyboard first.

Storyboards are really the first place where a movie comes to life. They're that critical step from the written word to the visual image. And they're fun to play around with — you can certainly indulge your creativity most freely when it's just you and a pencil!

The board you see here is for a musical sequence, "Garbage in the Moonlight," where the roaches introduce themselves to Joe for the first time. This sequence was a blast to make, because I wrote the lyrics of the song as well as drawing the storyboard.

I wrote each line with an image in mind and then gave the lyrics to the composer, who turned them into a song. Once a version of the song was recorded, we fed the music and the images from my storyboard into the computer. We were then able to make what we called a "board-o-matic" — storyboard images edited to the song — which gave us a rough little movie to work from even before we shot a foot of film. It was helpful to see how each image worked with others in a given sequence, and also exactly how long each image needed to last, before we started shooting.

The editor input the "board-o-matic" images into the Avid editing system, and as the shots were finished one by one, he replaced the still images of the storyboard with moving pictures.

The "miniboard" for "Garbage in the Moonlight" follows. We called it "mini" because the storyboard panels were small enough on the page to leave room for other information, like the accompanying screen direction and checklist of possible effects techniques. A box by the bottom left-hand corner of each panel indicated the length of the shot in seconds.

The checklist for techniques lists "Live" for live action (live cockroaches); "Rig" for a special camera rig that had to be constructed for the shot (rig also refers to puppeted action); "CG" for computer-generated animation; "SM" for stop-motion animation; "2-D" for cel animation, matte elements, or both; and "Other" for some weird hybrid path to achieving the shot. Image 1 shows all of this information. In the interest of space, we provide only the image and description of the remaining panels.

"Garbage In the Moonlight"

The basic set-up: five filthy containers of various types in and around Joe's trash can.

Panels 1), 2) and 3) depict the action, but not necessarily the framing.

- ☐ Live
- ☒ Rig
- ☒ CG
- ☐ SM
- ☐ 2d
- ☐ Other...

Five Roaches sequentially pop out of containers in sync to bell chord.
Live Roaches on sticks or SM.
Fun container openings: cork, screw-off lid, beer can pop top, etc.

MUSIC: "GARBAGE....
GARBAGE...
GARBAGE...
GARBAGE

All the roaches scram out of the frame at once (Pop to CG overlay?) Leaving one roach behind who BELCHES, covers his mouth demurely with one leg, and scrams several seconds later

ROACH: "(BELCH) SORRY!..."

ages 1–3

/anted this scene to start out with an old barbershop "bell chord" with singers chiming in. We set up real containers in a trash can d popped their tops off with compressed air. Then we animated the roaches popping up inside. The tab of the beer can is virtual o — and it vibrates like a diving board!

Roach (the lead singer) leaps into frame at head of five-roach "chip", or Whiffenpoof-style close-harmony formation. SM.

Split screen: Joe reacts in surprise.

MUSIC: "THE GARBAGE IN THE MOONLIGHT"

LOW ANGLE of steaming, half-eaten enchilada dripping over the edge of Joe's television. A Roach (Live?) pops up and sniffs it, waving its feelers with excitement

MUSIC: "GIVES OFF A LOVELY SMELL"

ages 4–5

is shot has a live-action actor playing Joe and animated roaches in the same shot. Another ot we did had about fifty CG roaches crawling over Joe's body and face! Image 5 was cut m the storyboard and never shot.

LOW ANGLE of singing CHIP. Synchronized swaying, leg gestures. Art-deco style silhouette (ref. John Held Jr.)

MUSIC: "SIPPIN' SEWAGE WITH MY BABY"

ANOTHER ANGLE reveals the chip singing on top of a Combat roach tray. Two horny roaches run inside.

MUSIC: "IN OUR LITTLE ROACH MOTEL"

CONTINUED The roach tray begins to bulge, buck and spin, throwing the singing roaches off.

Possible 2D "hearts" overlay

MUSIC: "PLEASE DON'T TELL"

age 6

is is the only image in the film that is generated completely in a com- ter, including foregrounds, back- ounds, and lights.

Images 7–8

In this shot, both the roaches and the roach tray are CG objects — a real roach tray couldn't bulge, hop, and spin like that!

A single Roach climbs onto the pull tab of the blinds, gives it a tug, and starts to rise out of frame.

MUSIC: SOLO BASS LINE

PULL OUT. The Roach rides up as the blind rises, revealing the brick wall inches away...

MUSIC: 4 BARS

CONTINUE PULL OUT. Several sheets of filthy cardboard rise into frame, suggesting a grade-school quality theatrical set of an ocean scene. Two levels of waves move back and forth, islands, a newspaper moon.

MUSIC: 4 BARS

ge 9

s was a real tab pull, but the roach that pulls cord (and the cord itself) are CG objects.

Images 10–11

I wanted to zoom from a close-up to a wide shot in this scene, but tracking live-action camera movement with a computer is a very difficult process, unless you have a very expensive computer system called "motion control," which we didn't. As a result, we shot most (but not all!) of our CG shots with a stationary camera.

Images 12–13

I pictured this scene where the roaches put on some cheesy high-school musical production to entertain their pal Joe, using real cast-off material. Everything in this little diorama is real and moves except the insects. The ship was a container of sesame noodles with chopstick masts. This shot worked so well we let it take over the time of Image 13, which we never shot.

Image 14

A reverse angle of Joe listening. "Listening shots" are very useful — they're great for covering dialogue changes or bridging difficult cuts.

Image 15

Again, everything in this shot was puppeted except the insects.

Image 16

A listening shot — Joe's getting sleepy.

Images 17–18

We used a rig in this setup because we wanted Joe's dirty sock to visibly smoke like the cartoon image of a bad smell. Rigs can be pretty simple. In this case, it's a burning cigarette pinned behind the sock.

Image 19

Live action.

Image 20

CG animation.

Silhouette of solo roach doing an Al Jolson impression on the pull tab of the window shade

MUSIC: "MY BUG?"

Joe's light goes out.

Joe's light goes out.

38. EXT NY SKYLINE -NIGHT

WIDER The city skyline, featuring Empire State, Con Ed, Met Life, and Chrysler buildings.

One by one, in sync with sung words, they turn off.

MUSIC: "PLEASE BE MY BUG"

Time-lapse photography

39. EXT NY SKYLINE - NXT MORNING

The same setting at sunrise.

Time-lapse photography

Images 21–29

This was one of the most difficult shots in the film and is a perfect example of how important storyboarding is to a project. Because it contains CG animation, 2-D compositing, live-action camera movement, and time-lapse photography, it took an incredible amount of coordination to pull off. A set had to be built to accommodate the one-story crane up to Joe's roof. Then the second part of the shot is time-lapse footage taken from the roof of my apartment building, depicting the skyscrapers of Manhattan going out, followed by the dawn, which took months to shoot.

Both live-action shots were married together, the bug was animated in the tab pull, and the time-lapse image of the dawn was digitally touched up. This ends the first "act," and I think it's one of the coolest shots in the film.

Screen-grabs, storyboards, and photos courtesy John Payson, Peter Wallach, Blue Sky Studios, Chris Wedge, Geffen Pictures, and MTV Networks.

Credits: **Writer and Director:** John Payson **Executive Producer:** Abby Terkuhle **Cockroach Wrangler:** Ray Mendez **MTV Short: Stop-Motion Director:** William R. Wright **Feature: Storyboards:** John Payson, Dan Shefelman, and Jeff Wong **Visual Effects Producer:** Mike Turoff **Stop-Motion Animation:** Peter Wallach **Computer-Generated Animation:** Chris Wedge **Composer, Roach Songs:** Kevin Weist **A.C.E. Editor:** Peter Frank

9 ▪ Kinestasis and Collage

Don't be turned off by the title of this chapter. Don't be turned off, either, if you are a digital animator and figure that this chapter deals with film techniques.

There are three *really* important reasons to read on.

First, here you will learn about the abilities of an animation stand. Every animator must have a solid grasp of the capabilities, moves, and special effects that can be achieved with the precision tools of film animation. You will need this foundation when, in the next chapter, the focus turns to equivalent techniques within computer animation.

Second, this chapter will clue you in to a very powerful conceptual model of how to think about animation design per se. You will discover four distinct ways to select and sequence images and shots. You will be gaining the basic vocabulary of an animation director.

The third payoff is the introduction you are about to have to a little friend called the *cue sheet.* Read on so that you make a good impression and establish a solid relationship with the basic control matrix common to every single animation technique. The term cue sheet is often used interchangeably with *exposure sheet* or *shooting log.* You will be seeing lots more of this device as we go deeper into both film and digital animation.

KINESTASIS

A host of awkward names have been attached to the technique of making movies from still images: still-scan, konograph, photoscan, FILMIGRAPH, photomontage, kinestasis. We'll use the last of these, as it appears to be the most frequently cited, and the word *kinestasis* accurately exposes the semantic roots and essential characteristics of the technique itself: *kine* = "moving," *stasis* = "stillness."

Whatever the name, an operational definition for this kind of animation would be as follows: A series of still pictures becomes animated through (a) variations of movement across the pictures and (b) variations in the succession between them.

You may have experienced this kind of animation without ever knowing it. To the casual moviegoer, kinestasis is often considered a "normal" filmmaking technique, a fluid flow of images that incorporates panning or zooming effects and thus produces much the same perceptual experience as when seeing a live-action film.

Please try the following project. Its requirements are minimal—you're only asked to draw a few straight lines—but it will clearly illustrate the creative choices and the surprising power of kinestasis.

Project: A Photograph Becomes a Movie. Imagine that you are a live-action filmmaker standing, hidden, in a position that allows you to see exactly the same scene as pictured in Figure 9.2. Imagine also that you have a camera equipped with a zoom lens, allowing you to frame and film the smallest detail or the entire scene. One small limitation: The gear you are filming with won't let you zoom, pan, or otherwise move the camera while the camera is actually running, though you can recompose the image between shots.

As you've noted, of course, there are two prints of the same photograph in Figure 9.2. This is because this project requires you to make two different movies from the same single image.

The first is to be a narrative film. Study the picture until you see a story of some kind. Using a ruler and pencil, draw directly on the first copy of the picture at least five separate frames that capture the essence of the story or relationship you have seen. These individual frames should be sequenced and numbered. Part of your problem here is to determine an order

9.1 Camera moves and effects: A bleary-eyed Thomas Edison is rephotographed in six different ways using a professional Oxberry animation stand. Clip *(A)* shows a direct zoom into Edison's face. Clip *(B)* is a pan—the photograph is moved under the camera in precisely controlled increments. Clip *(C)* shows a complex camera movement in which the stand's compound moves both horizontally and vertically to produce the effect of a diagonal tilt. Clip *(D)* is a fade-out. In clip *(E)*, the series begins out of focus. An in-camera dissolve is shown in Clip *(F)* as Edison's image cross-fades into his signature. Note that the 16mm clips are not actual size. *Photo courtesy Edison National Monument.*

A

9.2 Double feature in a photo: In choosing and marking off the five different sets of images in copy *(B)* and copy *(C)* of the photographs, try to use the rectangular frame that is standard in all forms of film-making. Note that in this project it is permissible to use the same area of the photograph in two or more individual frames that you select. The small copy of the picture, *(A),* illustrates these ground rules.

for the images you've selected, and part of the challenge is to actually compose the individual frames themselves.

The second part of the project asks you to use the second photograph in making an altogether different movie. This film cannot be a story in any way. Rather, you should select as your topic another kind of visual element or nondramatic quality in the photograph. Choose and arrange in sequence at least five completely new shots. Base the selection around something like your feeling of the scene's mood, a study of one element in the natural environment pictured in the photograph, or some purely intuitive connection that you can make.

In essence, the decisions you've been making are exactly the same kinds of decisions that are made in creating a kinestasis film. Were you, in fact, to actually film the individual frames you've marked, you'd be producing a simple kinestasis film. In full-scale kinestasis, there are many additional effects that could be added to either of the sequences you have selected, and we'll be discussing these options later in this chapter. But before we get to the nuances of different filming techniques, before we become encumbered in refinements, it will be valuable to break down the kinds of choices that I suspect you were quite easily able to exercise as you marked up the photographs.

B

C

FOUR DESIGN GENRES

When all the fuss and glamour and all the techniques and tools are taken away, the craft of filmmaking boils down to the execution of two tasks. The first is to capture images that are forceful. This is the task of "filming" as it occurs in live-action production. Editing the selected images is the second task. The filmmaker must carefully order his or her individual shots so that they lead the viewer to see a particular relationship among the images. Sometimes the relationship that the filmmaker wants to show comes in the form of a story and sometimes it is something entirely different. All this corresponds to what you were asked to do in the preceding project.

Now, in animated filmmaking, the control of the artist is extended far, far beyond that which is normally available in live-action filmmaking. Consequently, animators are able to design or structure entire sequences with a precision and purity of concept that reveals the four basic genres of media design.

The most familiar basis for choosing and ordering images is that of a "real-life" process or story. This is termed the *narrative genre.* Decisions follow a story line. A different kind of logic rules what has come to be termed the *documentary genre.* Here the criteria for selecting images and sequencing them has more to do with the nature of the subject matter than with an individual story or event. A film about a city, for example, calls for a different approach than a film that is set in the city. Totally different sets of criteria are at work when one operates in a pure *design genre.* In this case images are captured and ordered around a formal quality that links one image to the next. I'm referring here to qualities such as balance, tone, color, shape, texture, perspective, composition, and graphic and photographic elements. These are elements of "pure" design rather than elements that are generic to a story or to the subject matter per se. Sometimes, however, the criteria for selecting and joining images reside beyond the logic of each of these genres. Sometimes it is just a gut feeling, an intuition, that guides the process of designing. A combination of images feels "right," it "works"—although it may be impossible to say just why. I call this the *intuitive genre.*

Certainly intuition plays a large role in fashioning any powerful film sequence, be it a documentary, narrative, or pure-design genre. In fact, all manners of design can operate

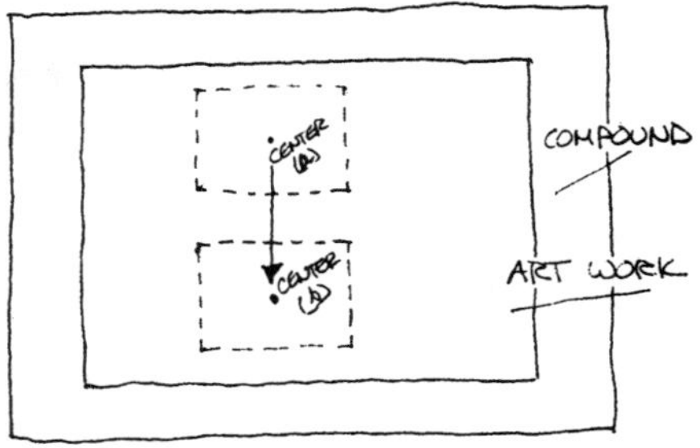

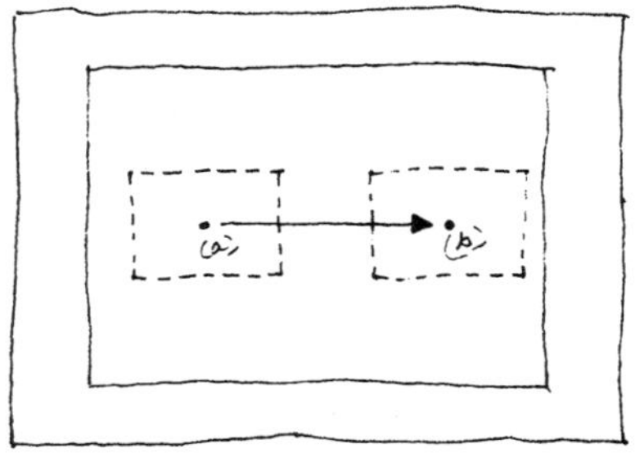

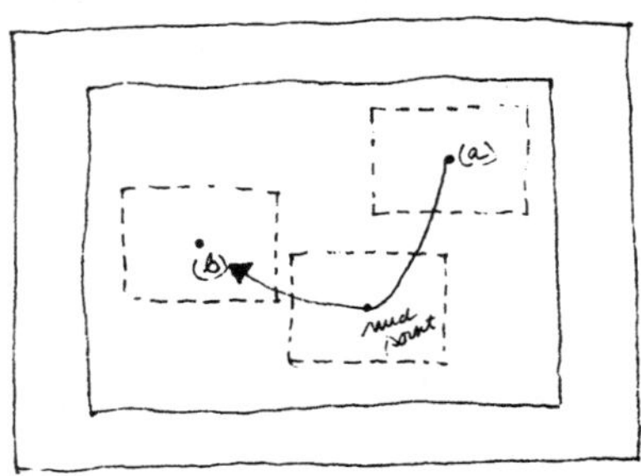

9.3 Compound moves: These drawings place you in the camera's perspective, looking directly down onto a compound that holds the artwork. Three basic kinds of compound moves are suggested: vertical, horizontal, and complex. The dotted outlines represent what the camera is framed to film, the initial and final positions of each movement. Note that the entire compound surface can move and that the artwork itself can be moved across a stationary surface.

9.4 Themes and variations: Tokyo's Tsukjii fish market provides a rich location for testing the different criteria through which images can be selected and ordered. This series of photographs was selected from color slides photographed by independent filmmaker Michael Lemle. The first set, *(A–F),* shows a narrative structure. The sequence *(G–L)* that follows shows a design structure, based primarily on geometric shapes, textures, and parallel

composition in adjacent frames. The third sequence, *(M–R),* is structured in a documentary mode—an establishing shot leads to close-ups of hands and then to a series of portraits of shopkeepers. The final sequence of six images, *(S–X),* was selected by the photographer because it "felt right" in summarizing the Tsukjii market as he experienced it. *Courtesy Michael Lemle.*

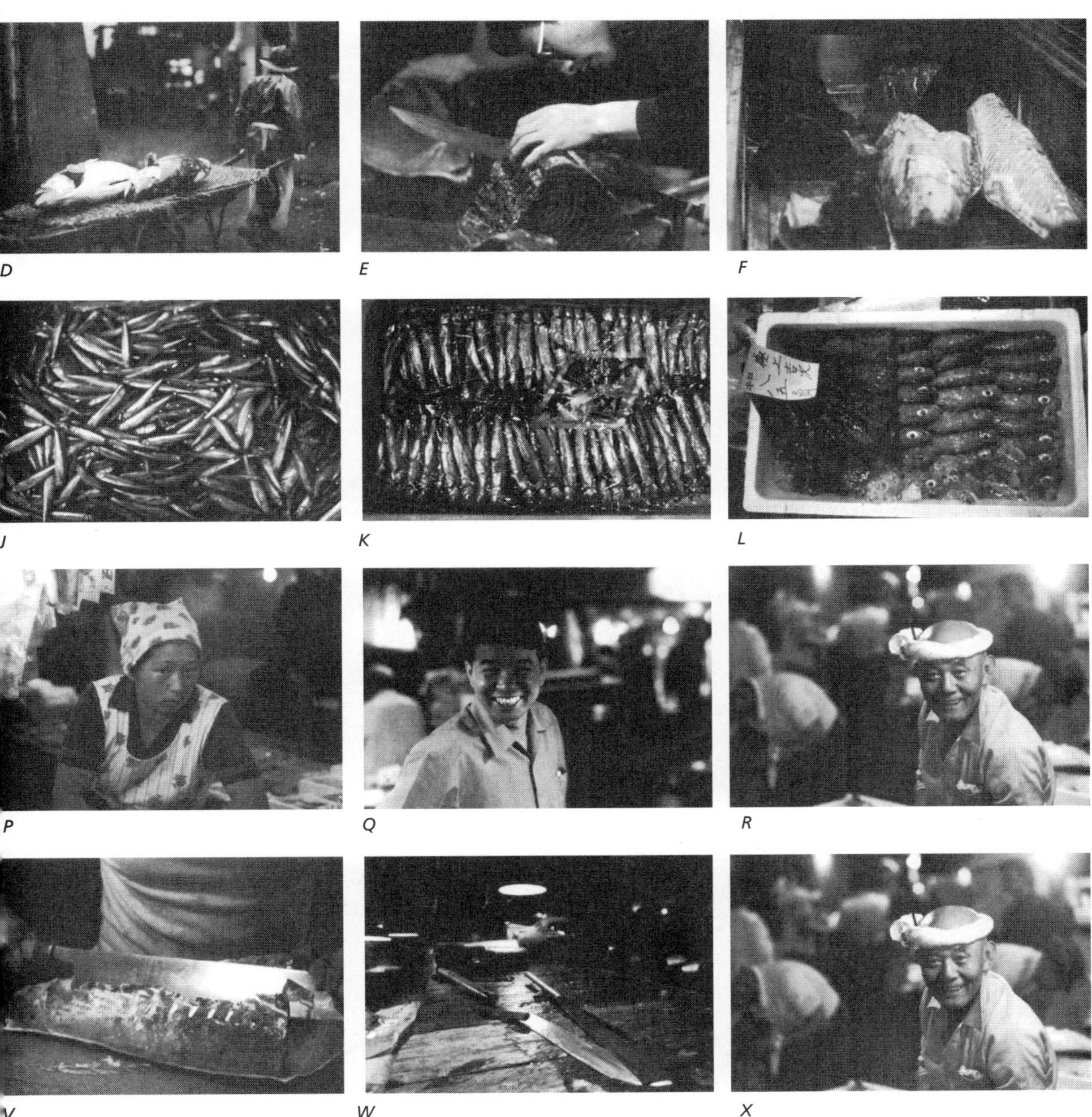

D *E* *F*

J *K* *L*

P *Q* *R*

V *W* *X*

simultaneously within one film, even within one scene. The genres have been isolated here only because breaking things into parts is necessary for exposition. When you lay out a kinestasis film, these categories or "logics" are employed simultaneously.

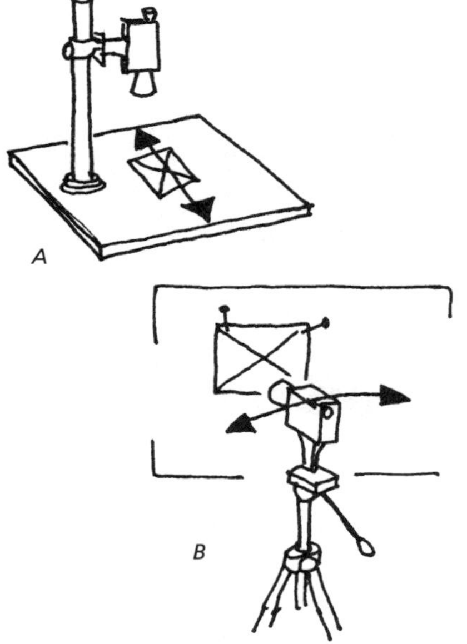

9.5 Kinestasis setups: A copy stand, tripod, or animation stand holds the camera above the artwork. In *(A)*, the artwork itself moves to create pans, tilts, etc. A horizontally mounted camera, *(B)*, allows easy panning by the camera while the artwork remains stationary.

IMAGE SOURCES

One of the most satisfying characteristics of working with this technique is that you have an opportunity to use some of the most powerful images that have been created by our species. You can, for example, make a film from reproductions of photographs made by the world's best photographers. There are books full of these. Old family snapshots are another good source of material. So are drawings, paintings, and other pieces of artwork such as watercolors, pencil sketches, pastel and charcoal studies, silk screens, lithographs, and so on.

Whether your source materials are new or old, original or borrowed, you'll discover that an enormous flexibility exists in the possible ways you can alter an image before filming it, by painting on photographs, for example, or mounting two or more slides within one frame, or tracing a silhouette from a photograph, or altering images with filters and focus variations.

As a general rule, anything that is reasonably flat can be used in making a kinestasis image, including three-dimensional objects such as tapestries or wall hangings, oversized art in the form of posters or wall murals, or specially screened photographs that have been printed on clear acetate sheets for use in conjunction with brightly colored papers or gels.

Whatever the actual source and style of images, during the process of designing and filming animated films, all materials that go under the camera become "artwork."

FILM MOVES AND EFFECTS

Imagine that a sophisticated super 8mm camera is fixed to a tripod and points at an 8- by 10-inch photograph. How could you bring movement to the photograph by moving just your camera?

Tracking/Panning/Tilting. The camera itself can move toward, away from, or parallel to the artwork during the process of filming. Figure 9.1 shows these and other camera movements and effects. If you're using just a tripod, however, it may be very difficult to calibrate accurately the distance and speed of such movements. The design of copy stands or animation stands accommodates such camera movements far more easily and accurately than a simple tripod.

Zooming. Most of today's cameras have excellent zoom lenses built into them. Registration of the zoom movement, in and out, during frame-by-frame filmmaking is tricky.

Focus. Another way to alter an image and to give it life is by selectively changing the focus on your lens. One can go in or out of focus. This can be a good way to bring together images that have no logical connection built into them.

Filters. In addition to the built-in filter on most super 8mm cameras, it is possible to place accessory filters or other devices in front of the lens while filming. Special multifaceted lenses can produce kaleidoscopic effects, for example. Shooting through an irregular glass surface, like a Coke bottle, will produce weird forms of movement within stationary pieces of art.

In-camera Effects. Relatively sophisticated cameras produce a number of in-camera effects that can be forcefully incorporated into kinestasis films. In a *fade-out,* the picture goes to black over a specified number of frames. Or the screen starts black and the image smoothly *fades in.* A fade-out that is superimposed over a fade-in gives the effect of a *dissolve*—one image disappears or "melts" into a new image. Two images can be simultaneously seen on top of each other if your camera has the ability to expose the same piece of film twice. The technique of *superimposition* requires a backwinding mechanism and a variable shutter control, both described fully in Chapter 19. Superimposition is used in creating a dissolve effect.

As suggested earlier, panning or tilting the camera is particularly difficult to control with the kind of accuracy required in animation. For this reason, most animation setups are designed so that the body of the camera itself is never moved. Instead, the subject or artwork moves. Incremental movements of artwork under a fixed camera are termed *compound movements*—named after the movable surface of full-scale animation stands. Figure 9.3 places you in the position of a camera as it points down on a compound.

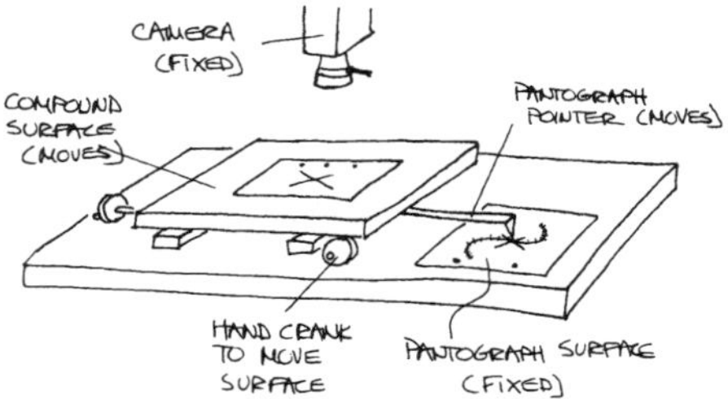

9.6 The pantograph: The pointer is connected to the compound surface and is positioned to indicate the center of the camera's field of view. The pantograph surface generally bears a field guide onto which a camera movement can be charted. Note that the pantograph's needle moves and the surface remains stationary. This is the reverse of the stand where the camera remains stationary and the surface moves. This drawing suggests that the pantograph is to be used with an expensive animation stand.

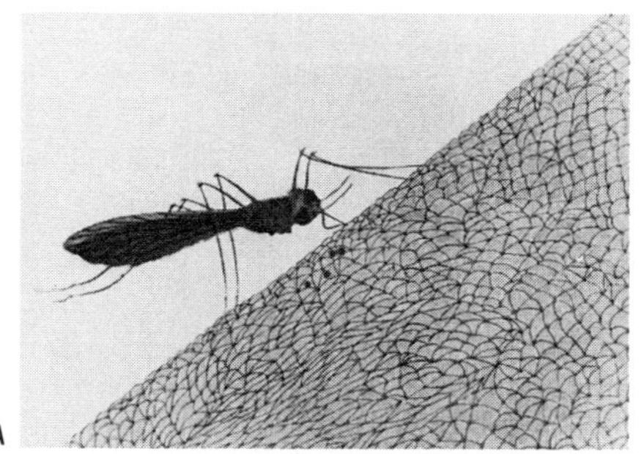

A

B

9.7 Approaches to kinestasis: The first two frame enlargements, *(A)* and *(B)*, sample moments from *Cosmic Zoom,* a film by Eva Szasz and Tony Ianzelo. This film consists of a continuous zoom from the microuniverse of a human cell to the macrouniverse of galaxies. Photos *(C)* and *(D)* are from *City of Gold,* a 23-minute film produced primarily from old photographs showing Dawson City and the Canadian Yukon at the height of the gold rush. Directed by Wolf Koenig at the National Film Board of Canada, *City of Gold* was one of the first kinestasis films. Frame *(E)* is from *Enter Hamlet,* a takeoff on the famous "To be or not to be . . ." soliloquy by Shakespeare. Each word is represented by a humorous drawing by animator Fred Mogubgub. Frame *(F)* is from the film *Secrets* by Phillip Jones. This movie explores the changing moods achieved through graphic variations of the same basic image of a woman's face. *Photos (A), (B), (C), and (D) courtesy National Film Board of Canada; photos (E) and (F) courtesy Pyramid Films.*

SPECIAL EQUIPMENT

Cameras best suited for kinestasis techniques share these general features: the ability to film extremely close up, reflex focusing, a built-in light-metering system, a zoom lens, and both single-framing and continuous-run filming speeds.

Figure 9.5 shows two standard camera mounting arrangements. Both positions enable the camera operator to easily and quickly focus on different-size fields or to reframe the composition of the scene between exposures. If one wants to use a great deal of movement within the filming of a single image (tracking, panning, zooming, in-camera effects, and compound moves), then it's nice to know about two specialized pieces of animation equipment, the pantograph and the viscous damped compound.

Pantograph. A key feature of an animation stand is the control it provides the filmmaker in moving the surface beneath the camera. This surface is called the *compound* or *bed* of the stand. The most precise way to chart a complex movement is by using a *pantograph* (Figure 9.6). The animator draws a single line on a sheet of paper that fits a specially designed platform. This line represents the path that the camera is to follow as it moves across a static piece of artwork. During the process of filming, the animator simply places the sheet of paper into its special mount alongside the compound itself. A special pointing needle lines up the center of the camera's field with the starting point of the line that has been drawn to guide the camera. Now, as the operator moves the compound so that the pointer follows the line on the pantograph, the camera can trace a complex movement over the artwork. The operator pauses to make two exposures at points on the line that have been indicated by the animator.

Viscous Damped Compound. Kinestasis filmmaking can be photographed in "real" time when the animator uses the continuous running feature of the camera to execute preplanned moves across the artwork. With the right equipment, a *viscous damped compound,* the move itself can be just as smooth and accurate as that achieved with a pantograph and frame-by-frame shooting. Figure 9.8 illustrates how this tool is made by mounting the flat surface of the stand in a traylike device that contains a thick, viscous lubricant. This dense fluid smooths out manual movements of the compound. After rehearsing the direction and speed that have been selected for a particular movement, the operator can simply turn on the camera at full speed and proceed to reposition the artwork while the camera is running. The damped compound facilitates a fluidity of all movements—jerks and jumps and other inconsistencies of movement are eliminated.

C

OTHER VARIABLES

We've discussed the conceptual dimensions of choosing and sequencing images. We've reviewed ways in which the animation camera and stand can add movement to an individual frame and how it can join different images through in-camera effects. We've looked also at simple camera setups and at the sophisticated and specially designed tools required in this kind of animation.

In planning an entire kinestasis film, there are a few other variables that you should know about. All of these must be employed with a purpose that matches the basic structure of the overall film. Lest all this begin to sound vague and abstract, here's a quick summary of the kinds of variables I am talking about here.

D

Duration is one variable, an obvious one. The camera can film any image for any length of time, ranging from approximately one twenty-fourth of a second (one frame) to as long as you wish, including several full seconds or even minutes. Because your camera will generally be able to focus on areas that are smaller than the boundaries of the single image you are filming, that image can suddenly become the primary source for a number of different secondary images. This is the process of *reframing.* The photograph of a landscape, for example, can generate additional images that present parts of the overall picture, a tree, a road, or some other detail.

E

CUE SHEETS

Your materials are assembled and the movie camera is loaded. You've studied each image and each movement. The first

F

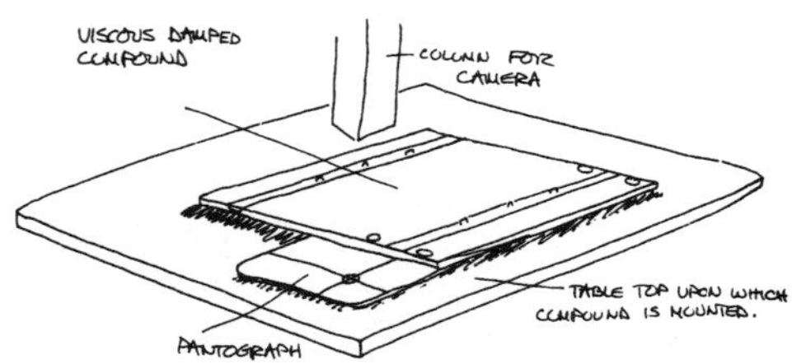

9.8 The viscous damped compound: The line drawing is based on the Oxberry Corporation's Media Pro Animation Stand. The compound is entirely made of metal and resets in a metal carriage containing the thick lubricating substance.

sequence starts well. The process of filming is slow, of course, because it takes time to work with precision. Unless you have a photographic memory, however, as the shooting proceeds you are bound to arrive at a point when suddenly you can't remember what's next or how to shoot it. You've lost your way.

That's a generous scenario. Chances are you will get lost quickly and hopelessly in making a kinestasis film unless you have some sort of notation system that can be used while planning the film and, more important, during the process of filming itself. Otherwise you'll screw up.

The way *not* to become lost is to make yourself a map that stores all the details of your film and helps you to work fast and accurately while at the camera. Figure 9.9 shows a portion of a handmade *cue sheet* (also called a shooting sheet or exposure sheet). You can modify your own shooting sheet to fit the capabilities of your own equipment.

The importance of the cue sheet will begin to sink in when you realize that it tracks visual composition, duration, and movement. As a matter of fact, the cue sheet becomes the one place where all categories of design should be recorded, including:

- identification (labeling) of individual pieces of artwork
- sequence of images
- framing or composition within each image
- camera movement
- compound movement
- duration of each shot
- in-camera special effects (dissolves, superimpositions, filters, etc.)
- sound track information—voice, music, and sound effects
- anything else, including questions, that you will want to remind yourself about when you go to shoot

9.9 Film and digital record keeping: There is simply too much detail involved in *any* animation technique for animators not to employ some notation system that keeps track of things. This page is a celebration of this need. Here, in *(A)*, is a chunk from a handmade shooting sheet that details one way the narrative images in Figure 9.4 might be animated.

Computers require exceedingly accurate and systematic records of image content, sequence, duration, frame number, level, and effects — which can include a wide range of options. A sampling of application windows from five different graphics programs illustrates the digital equivalent to exposure sheets: *(B),* Premiere; *(C),* After Effects; *(D),* Ray Dream; *(E),* Director; and *(F),* Flash.

You can see at a glance that these recordkeeping mechanisms share the same characteristics—little boxes that indicate specific frames; rows that indicate stacked levels of images that are to be seen at the same time. Each program also has a controller (not shown here) that specifies a frames-per-second play rate.

SHOT #	SOURCE	FRAMING	DURATION	CAMERA MOVEMENT	SPECIAL EFFECTS
1	A	FULL FRAME THEN ZOOM-IN	5 sec hold 3 sec zoom	AFTER HOLD, ZOOM IN 3 SEC TO C.U. MAN IN BKG.	DISSOLVE LAST 29 FRAMES
2	B	FULL FRAME ZOOM-IN	4 sec	ZOOM INTO HAND AT SAME RATE AS #1. HOLD AT END	29 FRAME DISSOLVE IN (OVER ZOOM)
3	C	FRAME VERY TIGHT HAND/BRUSH	2 sec	NONE	NONE
4	C	FULL FRAME	3 sec	NONE	NONE
5	D	E.C.U. ON FISH ZOOM BACK WITH PAN TO FULL FRAME	6 sec	WITH ZOOM OUT MOVE TO FULL FRAME IN 6 SEC.	NONE
6	E	E.C.U. MAN'S FACE	18 FR.	NONE	NONE

SHOOTING SHEET

A

Project: Trapeze

20 | Name | Comment

Name		Duration
Black matte — Background Matte	[1]	0:00:02:01
Sample Targa.m... — Movie	[2]	0:00:03:29
Closeup — Movie	[1]	0:00:02:11
Cross position — Movie	[1]	0:00:02:28
Fall Forward — Movie	[2]	0:00:04:00

B

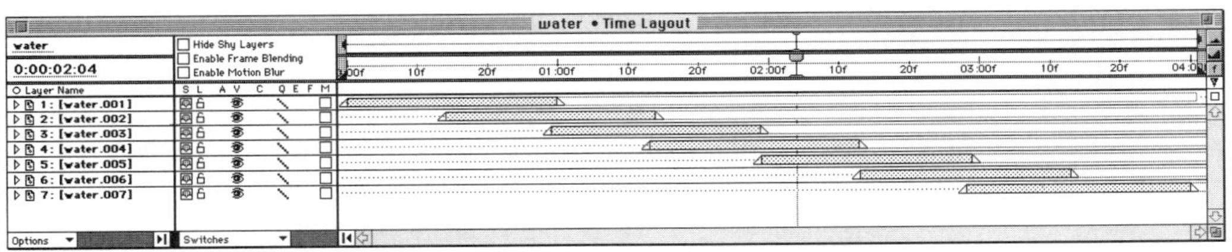

C

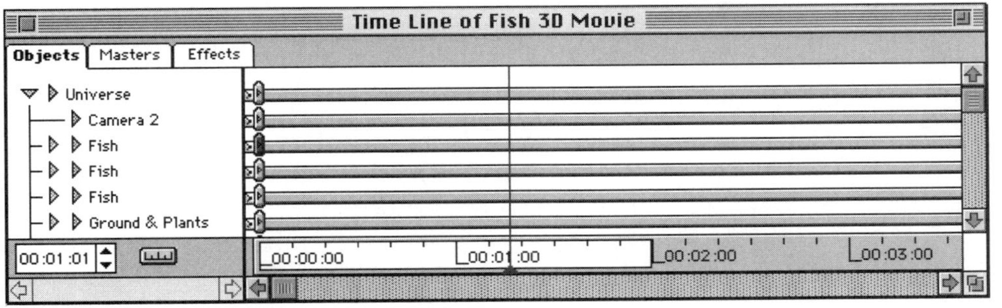

D

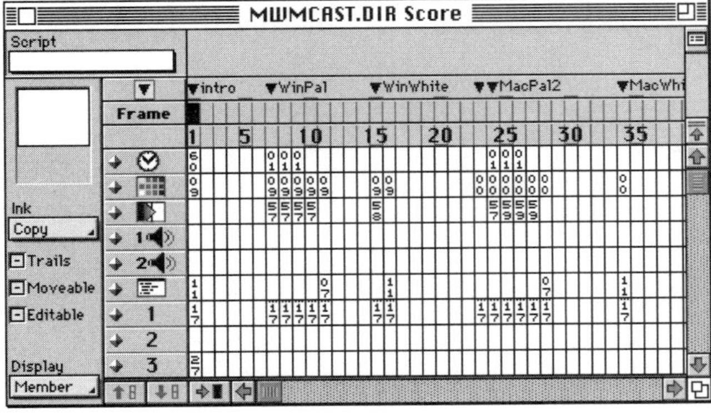

E

F

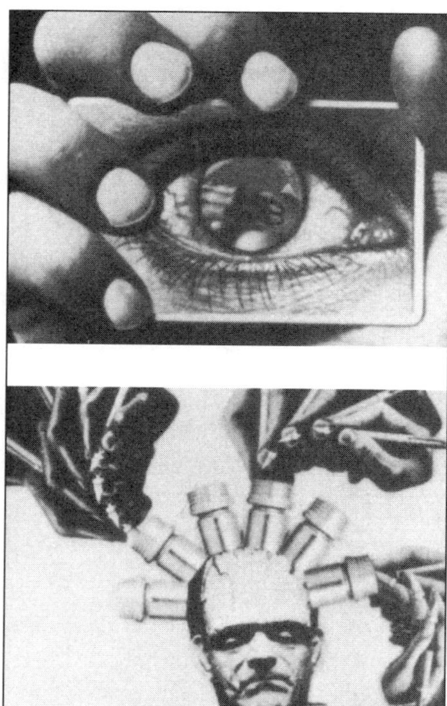

9.10 Approaches to collage: Two frame enlargements from Frank Mouris's *Frank Film. Courtesy Pyramid Films.*

COLLAGE

Collage animation is a technique in which bits of flat objects—photographs, newspapers, cloth, pressed flowers, postcards, and so on—are assembled in incongruous relationships for their symbolic or suggestive effect. The technique is much like kinestasis in the kinds of images used and in the various ways that they can be given movement by means of the animation camera and setup. But the feeling that a collage film engenders in the audience is distinctive. There's often a special quality of zaniness in collage. There's often a feeling of being inside a visual maelstrom.

There seem to be two basic styles of making collage films. The *impressionistic* style is the more familiar. A blitz of imagery fills the screen. The effect can be likened to that of a kaleidoscope, or we might call it an image bomb. The animator creates a flow of images (usually a very rapid flow) through his or her creative use of duration, association, proximity, and, of course, the selection of the images themselves. Full-screen images can be used for this effect or just parts of images, image shards.

Collage techniques have also been used effectively in what I think of as a *narrative* style. In these films a story is acted out, but it's almost always a very surreal kind of story. Terry Gilliam's animation in the television series *Monty Python's Flying Circus* provides a good example of how collage techniques can be used in a rich narrative style. Cutout images combine to create weird characters and weird landscapes. The movement of the characters is often realistic, although the logic of these changes defies description.

If you are drawn to this style of collage, I suggest that you reread the chapter on cutouts. It will give you ideas on how to create joined figures, how to design backgrounds, and how to determine speed and distance as you film under the camera.

10 ■ Motion Graphics

WRITTEN WITH JENNIFER TAYLOR

The superstar of animation software is After Effects, an application published by the Adobe Corporation, a longtime leader in computer graphics. After Effects is a powerful creative tool that allows artists to produce professional-looking 2-D animation and special effects on a desktop computer.

In the few years of its existence, After Effects has achieved a preeminent position within the animation industry—from advertising commercials to broadcast graphics to title sequences to animated shows on TV. So pervasive is the impact of this software program that it has single-handedly relaunched an entire animation technique: motion graphics.

10.1 35mm optical bench: One of a number of computer-linked optical printing machines used in the French and English animation sections at the National Film Board of Canada. *Photo by author.*

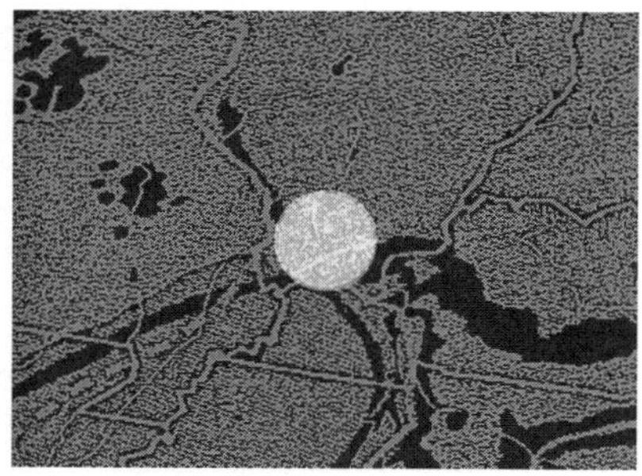

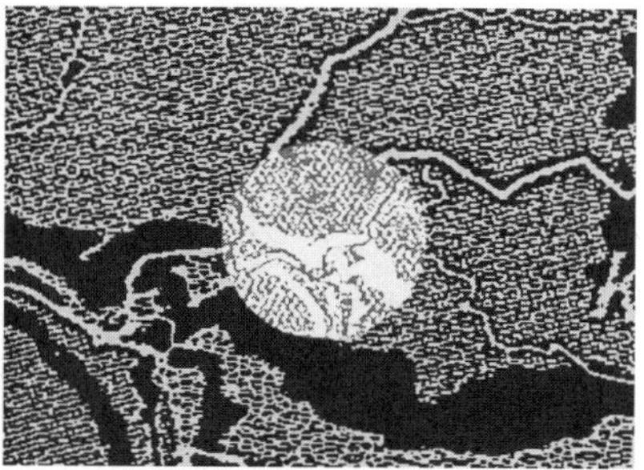

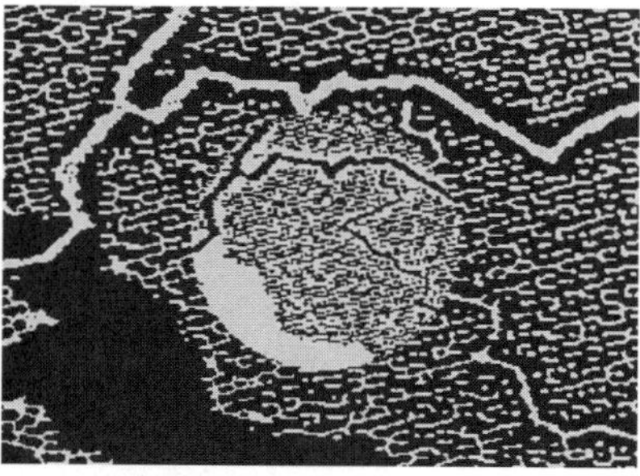

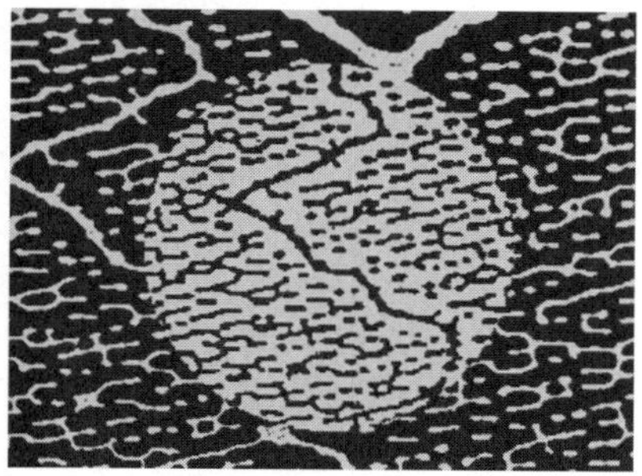

10.2 Zoom: This series of frames demonstrates a "zoom" or "scale up" done in AfX. Taken from *Where on Earth Are You?* by Jennifer Taylor. *Images in this chapter courtesy Jennifer Taylor, unless otherwise noted.*

The roots of motion graphics go back to the 1950s and 1960s, when Hollywood turned to a group of graphics designers like the legendary Saul Bass and asked them to design title sequences for feature films. Movie titles, of course, consist mainly of words: the name of the film, its stars, director, producers, and other key talent. The skill set of the graphic design profession was new to the feature industry: typesetting expertise, logo design, a layout sensibility refined in poster and magazine design, and, in particular, familiarity with a wide variety of photographic techniques such as solarization, audio screens, kodaliths, and duotones. The enriched visual vocabulary of graphic designers began showing up in feature film *opens,* and in Hollywood a number of *title houses* established themselves —postproduction labs/animation studios that specialized in design and use of technically demanding 35mm production tools like the optical printer shown in Figure 10.1.

In the 1980s the television industry began to develop very expensive digital tools that made it easier and faster to do the same kinds of animated graphics effects refined in feature titles. Machines with exotic names like the Paintbox, the Squeeze-Zoom, and the ADO (for Advanced Digital Operations) cost hundreds of thousands of dollars and rented out at $500 per hour and up. In the late 1990s a new generation of such machines (with new names like the Harry and the Flame —and new, higher price tags!) continue to give frame-at-a-time control over image making. With these high-end digital tools, live action, animation, and special effects are all composited into a seamless whole. The results are found most often in feature films, television advertising, and, more recently, high-budget TV series.

After Effects brings to the desktop computer almost all of what was until recently the exclusive province of those expensive, super-high-end digital compositors. It is, in effect, both the 35mm optical bench of film technology and the Flame of digital compositing technology.

It's useful to think of After Effects as a place where the basic raw materials—picture, story, and sound—are fused into a final piece. You can, for example, import artwork created in Photoshop or Illustrator, along with sound created in SoundEdit 16, and marry them into a single digital file that can become a QuickTime movie or be exported to videotape—or even film. Another way to look at After Effects is to see it as an all-in-one animation camera stand, nonlinear editor, and post-production facility where you add titles, sound, and special effects. That's one cool package!

Because After Effects offers such a wide array of creative possibilities, it's impossible to cover in this volume all the different things that animators can do with it. However, these pages can provide a useful fly-over that looks at major components. The case study by Jennifer Taylor (Figure 10.10) will provide a closer look at the interface and some of the nuances of technique. Yet even as you are guided through the major landmarks of After Effects, remain curious about the possibilities that have yet to be charted. Think outside the box.

A

B

C

D

MOVING WITH, THROUGH, AROUND, AND BETWEEN STILLS

It is easy to use After Effects as a virtual camera that will simulate complex navigation across a flat piece of artwork in a way that mirrors the animation stand "moves" used in traditional filmmaking. All the techniques talked about in the preceding chapter on kinestasis and collage animation can be achieved faster, cheaper, and with more accuracy by using After Effects. In fact, it is easiest to explain the powers of After Effects by referring to familiar film terminology.

Pans. When shooting film, you create pans by moving the camera or the animation stand frame by frame to give the feeling of moving across a still image. To create a pan in After Effects, you simply move the artwork left, right, up, or down in the program's *composition window.* You tell the computer where you want your pan to start and stop by setting key frames at your start and finish points.

Zooming. In film, zooming in is achieved by changing the distance, frame by frame, between the camera and your artwork. In After Effects, you can simulate a zoom by *scaling* or resizing your artwork over time. When you scale something up,

10.3 Character animation: These elements from *Choose Your Poison*, by Jennifer Taylor and Randy Lowenstein, show: *(A),* one of the characters as scanned into the computer; *(B),* the character's alpha channel — created in Photoshop; *(C),* the background; and *(D),* the final scene with three characters laid on top of the background. *Stills courtesy Jennifer Taylor and Randy Lowenstein.*

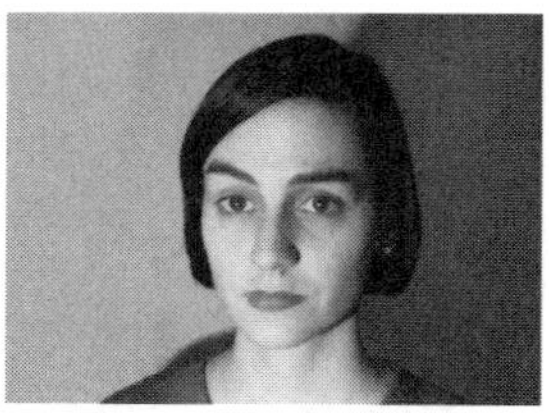

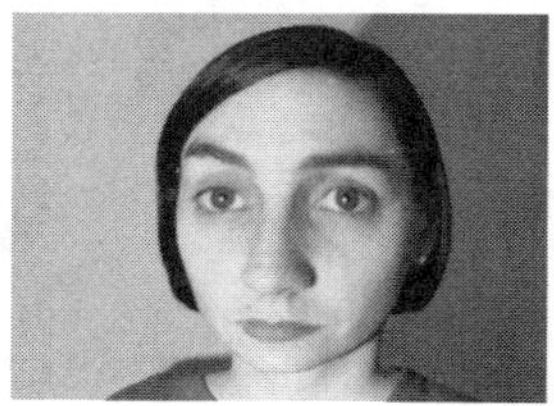

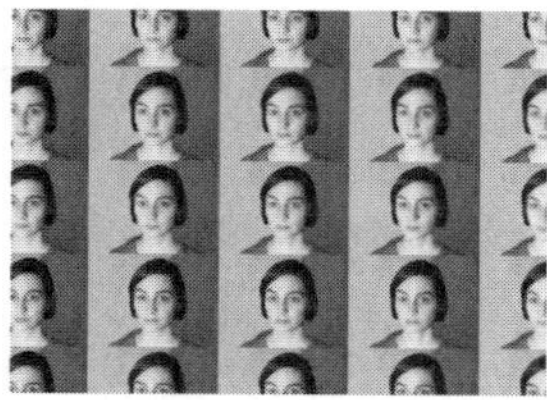

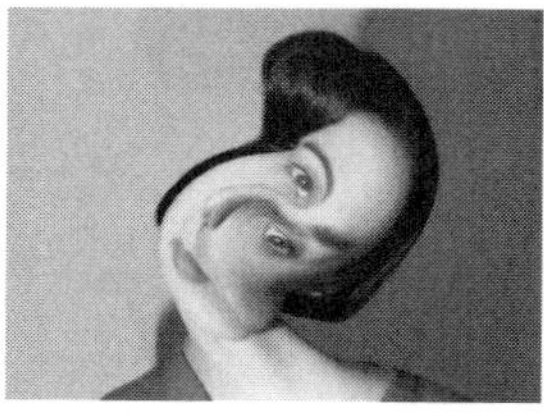

10.4 Special effects at a keystroke: The top photo has been manipulated using different filters that come with After Effects. Because the goal here is to illustrate the full power of such filters, their application to the portrait of Jennifer Taylor has been pushed to its most drastic extreme. Bear in mind that you can also make subtle changes and adjustments. As a matter of interest, the original source image was "photographed" with an inexpensive digital still camera.

this means you're making it larger, and when you scale something down, you are making it smaller. Hence, the effect of zooming in toward your artwork is achieved by scaling up the artwork. This makes sense if you think in terms of things becoming larger as they move closer to you.

In order to fill an NTSC TV (television) screen, an image must have the dimensions of 640 pixels width by 480 pixels height (referred to as 640 by 480). If you want a zoom-in to end with the screen filled (requiring a 640 by 480 image), then the original artwork on which the zoom is planned must be scanned in larger than 640 by 480. This is so the image doesn't pixilate as you scale it up. Remember, scaling up equals zooming in. If you scale something up more than 100 percent of its original size, it will pixilate. If you are planning on zooming out on an image, the image also needs to be larger than 640 by 480 if you want it to end up filling the screen—you just do the "camera" move in reverse. When scanning the image into the computer, you must look at both ends of the zoom and make sure you have enough resolution. See Chapter 24: Computer Hardware for more about scanning images, resolution, and color depth.

Fades. Traditionally, in-camera fades are made by slowly adjusting the amount of light coming through the lens aperture. To fade out (to black) you close the lens down over, say, 24 successive frames. To fade in (from black) on the new piece of artwork, that process is reversed. In After Effects you can fade in and fade out by manipulating *key frames.* Think of key frames as markers of precise moments when a selected image will be manipulated in one way or another. When making fades, a first key frame marks the moment when the computer will begin to adjust the opacity levels. More on key frames will come later.

If an image is set at an opacity of 0 percent, then it is not opaque at all. In other words, you can see through it. If an image has an opacity of 100 percent, then it is fully opaque and can't be seen through at all. It follows, then, that if you program After Effects to slowly bring your image from 0 percent opacity all the way to 100 percent opacity, the resulting effect will be that of watching an image go from being perfectly transparent (clear) to being completely opaque (solid) in the amount of time you have designated by choosing two key frames. In After Effects, 0 percent to 100 percent opacity appears as a fade-up from black—because the black background (a default setting) will show through your image when it is not fully opaque. To the viewer's eye this looks exactly like a fade-up from black in traditional filmmaking.

Dissolves. Dissolves are created on film by overlapping the fading out of one image with the fading in of another image. When this effect is achieved in-camera, there are three steps: The initial image is faded from full exposure to black over, say, 48 frames. The film is wound back within the camera, with the aperture fully closed. A new image is next placed under the camera and the lens is slowly opened over the same 48 frames. The effect is a lovely two-second cross-fade or dissolve.

To dissolve two images in After Effects, simply lay one image over another and slowly diminish the opacity of the top layer over time. This way, instead of revealing black when the top layer is clear, you will reveal the image layer beneath it.

LAYERING

Here is another cameralike effect. Yet it is so revolutionary—the dense flow and overlapping of multi-image collage is so powerful—that the layering possibilities of After Effects require special discussion

Superimposition. To superimpose two images on film, you must shoot one image first, rewind the film, and shoot the second image onto the same piece of film. A superimposition is like a dissolve where the midpoint of the transition is extended —that is, when two images are both clearly visible. In After Effects it is easy to layer many, many images. Four, six, ten, or even more different visuals can coexist within the same frame, each moving independently, fading in and out on its own schedule with perfect control over the legibility of each image. The effect can be breathtaking. In the traditional world of film, such multimedia effects could only be achieved using an expensive optical bench and many trial-and-error passes to achieve the same thing.

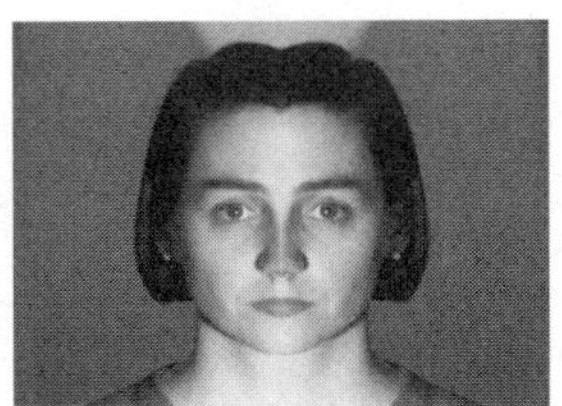

Character Animation. Animators will delight in a related capability: One can make a background and then add an opaque character on top of your background in a way that simulates the techniques of classic cel animation. In other words, After Effects can serve as a pile of acetate sheets onto

10.5 Flying type fonts: These titles were created in Photoshop and then animated in After Effects. The title moves into frame from the top, and the animator's name moves into the frame from the left side. Each chunk of text has only two key frames: a position start point and a position end point.

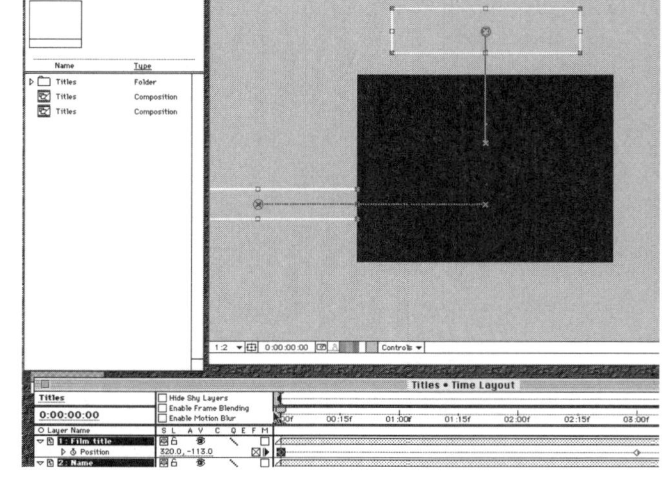

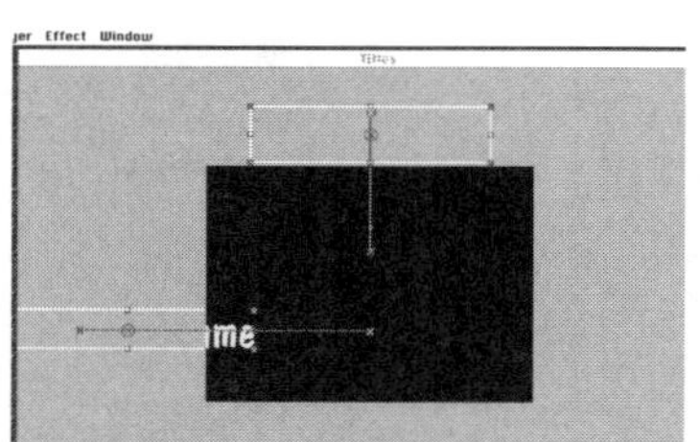

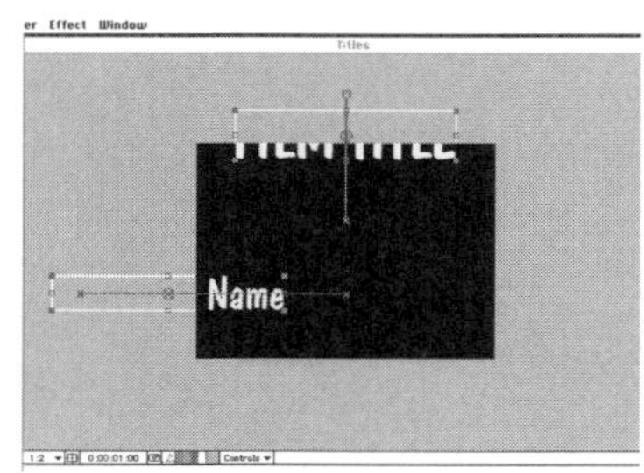

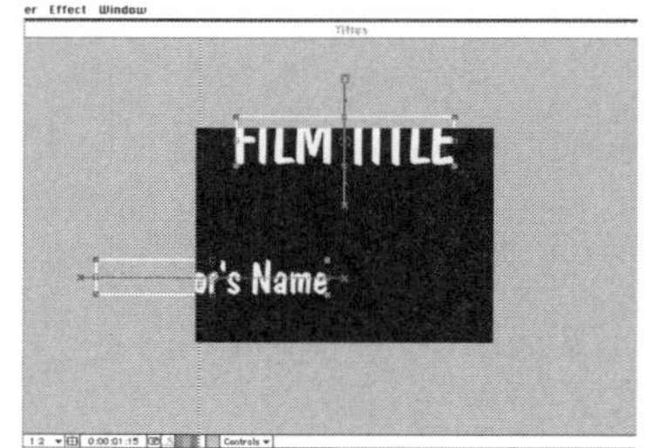

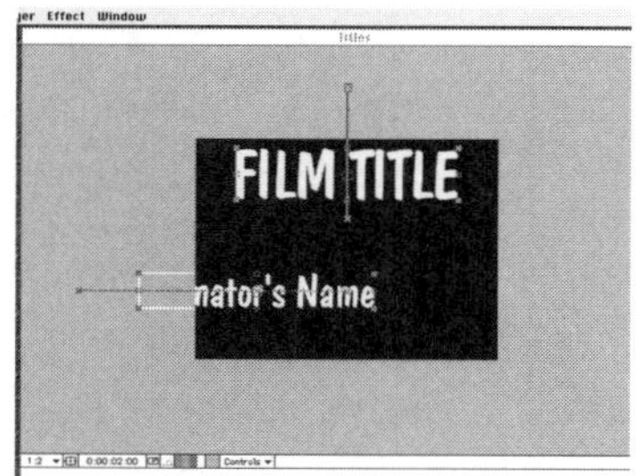

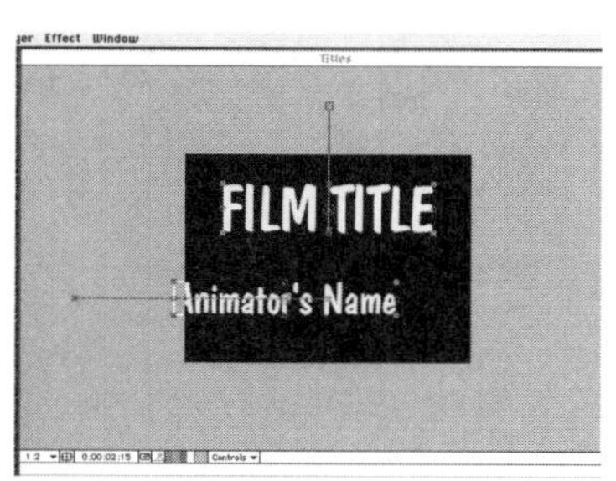

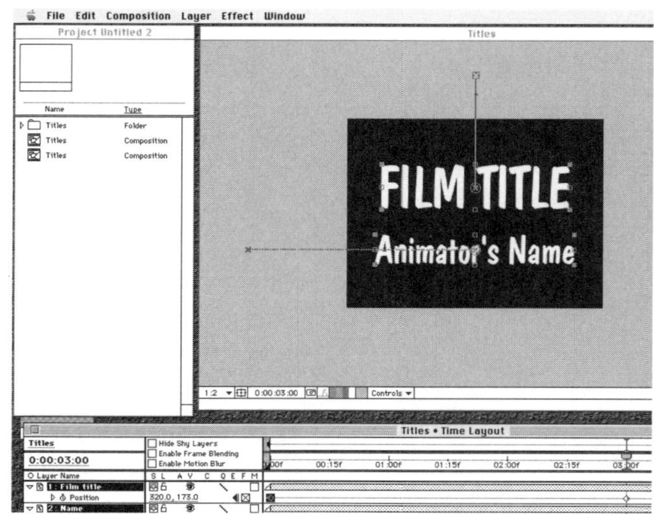

which individual drawings have been mounted and are displayed, in sequence, over the same background.

The multilayered construction within After Effects uses a computer graphics entity called the *alpha channel.* An alpha channel is an "extra" channel in a color image that deals with transparency. The closest direct parallel between traditional animation and the alpha channel of digital animation is to think of the latter as the clear acetate cel onto which a figure is painted. Alpha channels can be built for each level in After Effects and will designate certain areas within that image as being transparent, and other areas as being opaque. Thus the image that is paired with the alpha channel can either "float" above the background or "ghost" with its transparency set somewhere between 0 percent and 100 percent. In After Effects there is no limit on the number of alpha channels you can stack on top of the background image.

A

B

C

10.6 Cutouts: In this scene from an independent production titled *Choose Your Poison*, After Effects is used to animate a cutout arm so that the character looks like he is brushing his teeth. *(A)* and *(B)* show two key *extremes,* or frames. *(C)* suggests the motion that the audience will see when the sequence is "framed up" correctly. *Courtesy Jennifer Taylor and Randy Lowenstein.*

SPECIAL EFFECTS IN AFTER EFFECTS

After Effects comes complete with a host of effects that you can apply with wild abandon to your drawings or other source images. You can apply these effects to individual images, entire sequences of image files, or even whole compositions. Figure 10.4 provides a tiny sampling of such special effects that take little effort on your part and can range from adjusting the contrast of your image to swirling that same image into a big unidentifiable smear. These nifty capabilities are much like Photoshop filters that can be key-framed and therefore animated. In fact, many Photoshop filters can be imported into After Effects. There are also third-party plug-ins such as Final Effects that offer even more special effects for After Effects.

TITLES

One of the instant rewards of using After Effects is that you will have the ability to make titles for your films quickly and easily. You can make the titles themselves in Photoshop or After Effects. If you want a quick job, After Effects will do. For best results, however, use Photoshop or one of the more specialized

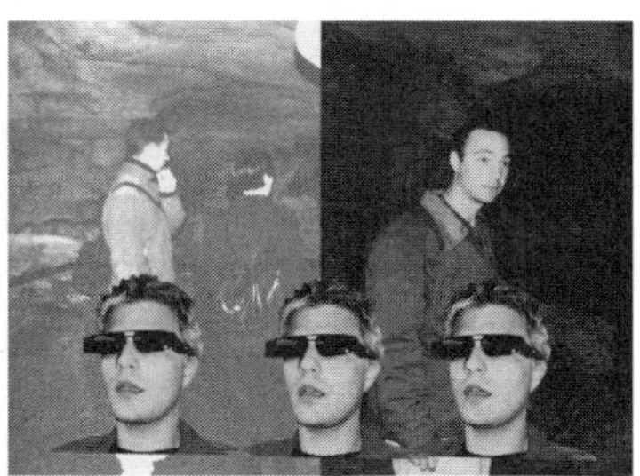

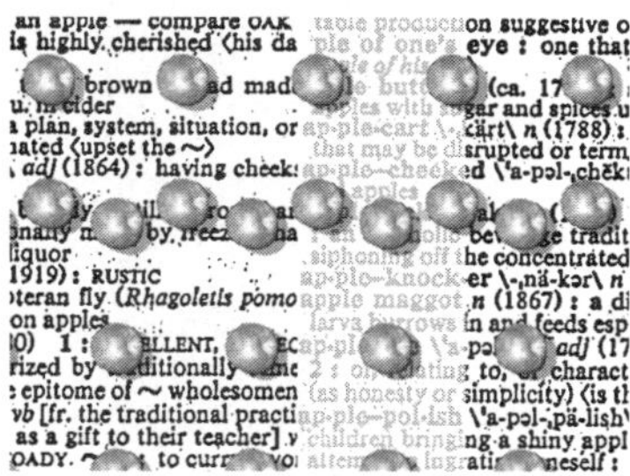

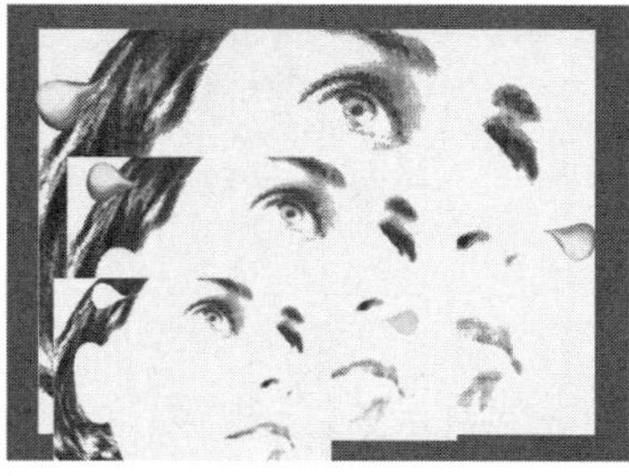

font-creation programs that are available for Mac and Windows computers. Once you have the text, you can create professional-looking scrolls or fades within minutes. Again, using a computer rather than shooting on film takes less time, costs less money, and allows for more experimentation. It's so easy to experiment that you may want to try both of the traditional methods of presenting titles: by fading them in and out and by scrolling them, or you can get fancy and invent something brand-new!

CUTOUT ANIMATION

Using After Effects, you can create digital cutout animation that works very much like traditional cutout animation. There is a huge latitude in how you approach cutouts. Elements can be very simply made and manipulated. For example, you could scan photo head shots of yourself and your family. Bodies could be as simple as one or two exchangeable positions, maybe drawn directly into the computer. For a look at the more complex end of the production spectrum, refer to the *Blue's Clues* case study in Chapter 5.

A FIVE-STEP CREATIVE PROCESS

Like other top computer graphics software packages, After Effects is a complex program with plenty of bells and whistles and more functions than one animator will ever employ. As you would expect, it takes a long time to master a program as rich as After Effects. Fortunately, computer tutorials and a comprehensive user's guide accompany the floppy disks and CD-ROMs that hold the program itself. The local bookstore may have—and can certainly order for you—specially written

10.7 Collage: After Effects offers a rich design palette for combining print, photographic, and handmade imagery. Here are five stills pulled from a video piece titled *Jen, Randy, Nick and Craig Go to Howe Caverns*. The piece is about four people looking for an adventure in nature. They disobey the rules at a tourist attraction and are turned to stone. These highly layered After Effects images combine photos, photocopies, scanned objects, text made in Photoshop, artwork made in Painter, and even frames pulled from video. *Courtesy Jennifer Taylor and Randy Lowenstein.*

texts that are devoted to After Effects or one of the other major software tools described in this volume. Be prepared to spend a number of days messing around with a program like After Effects before you attempt to undertake a project with it.

All that said, it's still true that there are really only a few things you need to know to get started. Once you understand how the program operates step by step, you can spend months exploring.

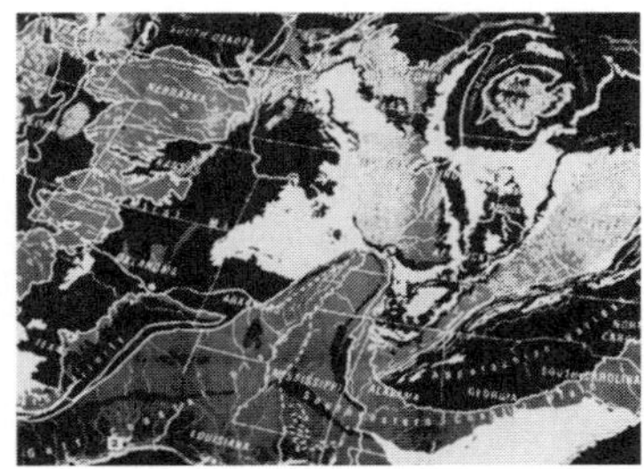

CREATING AND GATHERING ELEMENTS

Before you click the After Effects logo on your hard drive, you should create or collect all the artwork that you'll need for your project. You can create artwork in programs such as Photoshop, Illustrator, and Painter. You can also scan artwork or photographs into the computer. Don't forget that it's easy to acquire images from one of many clip art collections that are already digitized.

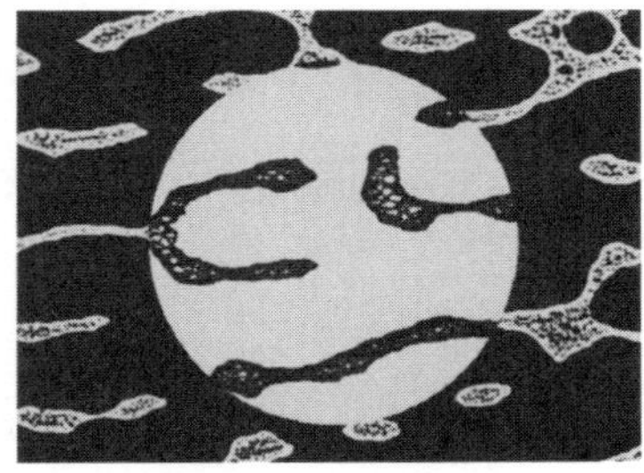

No matter what source materials you choose, make sure you gather them all neatly in a folder that is clearly labeled. It is important not to move these materials to new folders once you begin working. This is because the first time you access a particular file, After Effects remembers the specific place or folder where that source file was stored. The software will be constantly accessing the file at this location. If you throw away one of those files, or place it somewhere that the program cannot find, you will wreck the animation itself.

IMPORTING ELEMENTS

When you open the After Effects program, you will be presented with an empty *project window.* The first thing you need to do is import the artwork from its neatly stored folder some-

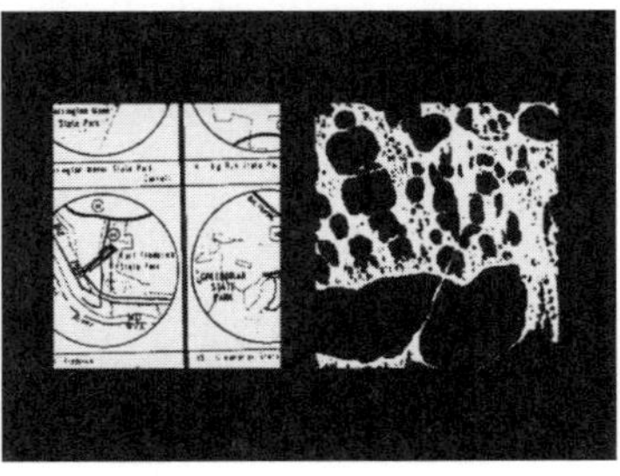

10.8 Xerography meets After Effects: These stills are taken from *Where on Earth Are You?,* a personal film by Jennifer Taylor. It's an abstract animation that leads viewers through a tongue-in-cheek "search" across the globe to answer the question posed in the title. A careful dissection of maps ends up with the conclusion "I think I'm lost." The animator created a hybrid animation style using After Effects to breathe new life into her particular style of traditional xerography animation.

where on the computer's hard drive. Simply choose Import from the File menu. After Effects will ask you where the source material is located, and you will direct it to your source folder. Once an element has been imported, it appears in the project window in the upper left hand corner of the screen. Continue importing your artwork until all of the materials that you wish to use in your new project are listed in the project window.

COMPOSING LAYERS

You compose your project using two production tools that come in the form of windows or frames on the computer screen.

The *composition window* is the canvas where you spatially compose or stage your piece, as if you were looking through the lens of a camera. When you drag your elements into the composition window, they become "layers" of that composition. What you see is what you get with layers stacked from top to bottom, as if in a pile.

The corresponding *time layout window* is where you work out the timing of your film. At one end of this chartlike window there is a list of all the elements that appear in the screen over a chunk of time—which is stretched under the corresponding timeline of seconds. The time layout window provides by name, from top to bottom, the layers that appear in the composition window.

Using these two windows in unison, you will create the flow of images on and off the screen. For example, you can make changes (in size, opacity, etc.) in the composition window, and dictate how long it will take these changes to occur in the time layout window.

ANIMATING WITH KEY FRAMES

Not only is the time layout window where you order the layers and dictate their lengths, it is also the working space where you animate your layers and apply special effects to them. The whole concept behind animating in After Effects lies in placing key frames in the time layout window. A *key frame* is a marker that represents any given change in the nature of a layer, such as a change in opacity, size, or location.

Key frames are often placed in sets of twos, with one key frame as a starting point and one key frame as an endpoint of a particular effect or movement. For example, say you want to move an object into the center of the screen in one second. First you would place a key frame at the start of the layer and set the position of your object outside the composition window. Then you would place a key frame at the one-second mark in the timeline and set the location of the object at the center of the frame. Placing key frames is your way of telling After Effects where you want the object to move and how long that movement should take. If you wanted, you could use the same key frames to instruct the computer to fade in this same object as it slides into its final position.

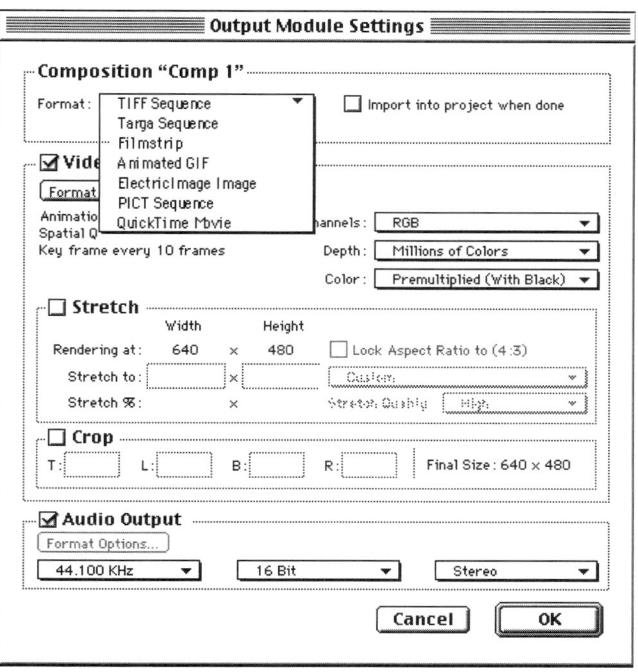

10.9 Exporting AfX: The Output Module Settings window allows you to save an After Effects file in several different formats, so you have flexibility and are able to distribute your work through different media forms, including videotape, disk, and the Internet. *Courtesy Art Bell.*

After Effects approaches key frames in much the same way that they are used in traditional character animation to plot out the "extremes" of a particular piece of animation. Look ahead to Chapter 14 to see how an animator makes key drawings or poses and then "in-betweens" the drawings that are required to flesh out the interval between poses. In After Effects, the computer will provide the in-betweens by taking the object at one key frame and moving it with just the right number of incremental positions required by the number of frames playing each second to the location specified by the next key frame. This is not as complex as it may sound. Still confused? Check out the case study!

SHOW TIME!

When you have animated the images for each scene in your animation, the completed project can be preserved and disseminated in a number of forms.

First you can save the file as a *QuickTime movie.* Because the QuickTime playback module is shared by so many different motion graphics/animation applications, you can be pretty sure that anyone who has a computer will have a copy of Quick-

Time stored away in the Systems folder of their hard disk. This means that any person can just click on the icon representing your completed movie and it will play on their computer screen, with sound included.

You can also export the completed After Effects animation to a videotape. Figure 10.9 shows the pull-down menu choices of the different ways you can save and subsequently share your new masterpiece. Note that in order to output to tape, you will need special hardware that allows for video output.

THE LEARNING CURVE

One of the best things about After Effects is that once you have begun to master the program, it can really speed up your production rate. You simply get more done in less time. When it takes less time to make something, you're also more willing to go back and make changes or go out on a limb in the first place. For that reason, After Effects creates an atmosphere that is more conducive to experimentation than film. Also, nothing is permanent on the computer, unlike film. You are operating in a nonbinding, nonlinear world. After Effects can open the door to a faster, freer, and simpler way of doing animation.

So don't be too frustrated or discouraged the first time you get your hands on After Effects and see the complexity of the program. Just plunge into the tutorials that come with the program. Or simply give yourself a few days of messing around.

You *will* begin to figure things out. The new tools *will* become familiar. And because you can work fast and try out different possibilities, the chances are far greater that you will concoct and refine your way toward a piece of animation that stands out as truly excellent.

10.10 CASE STUDY: Using After Effects

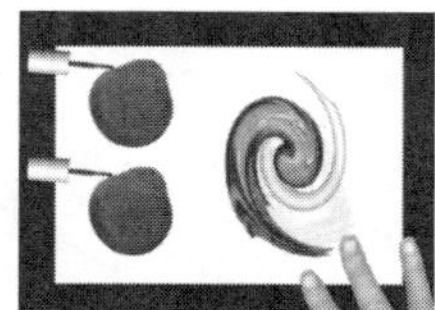

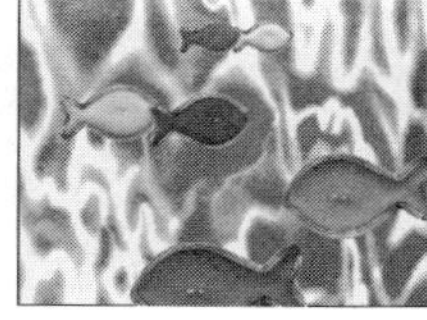

A

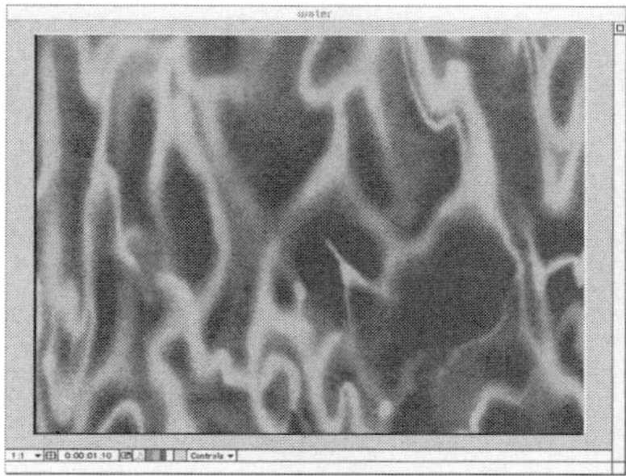

B

The frames above sample a ninety-second video portrait used in a play titled *The Miss Alphabet City Beauty Pageant and Spelling Bee*, written by Emmy Laybourne and performed with a five-member cast at New York City's Catch A Rising Star nightclub. In this tongue-in-check video portrait, a thirteen-year-old girl introduces herself in the best way she knows how: by showing us her favorite possessions. Animator Jen Taylor describes the step-by-step process behind one of the scenes.

Artwork

The goal was to surprise the viewer by, first, showing what seemed like cutout fish swimming through the frame, and then revealing these shapes to be candles, as shown in the third and fourth panels in *(A)*.

I started off by placing the single fish shaped candle I had onto the surface of my scanner and getting a 72 dpi scan. The inside cover of the scanner is white, so that the area around my subject matter was white on the resulting image, although some shadowing showed up. The scan (a PICT file) was brought into Photoshop and given an Alpha Channel by selecting "Inverse" from the Select menu, turning the "white" background that surrounded the fish-shaped candle into the "clear" surface of the Alpha Channel. (Forming the perfect Alpha Channel required some careful tweaking. I needed to get a very clean and smooth-edged fish image that could later be placed over a background.)

As you can see in *(A),* this one image was reused to create additional fish of different sizes. I gave the multiple fish different colors too, although you can't make this out very well in these black-and-white frame-grabs.

The background seemed a challenge, but was easy to do once I studied the problem awhile. The first step involved using Photoshop to create a series of seven abstract "water" images that resembled each other and could be cross-dissolved to simulate the motion of rippling water.

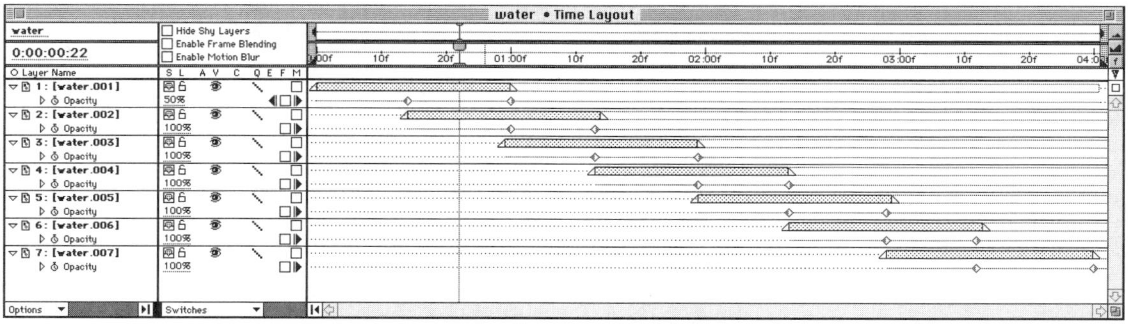

C

Key Frames

A Key Frame is a marker that represents any given change in the nature of a layer, such as a change in opacity or size. Key frames are often placed in sets of two, with one key frame as a starting point, and the other as an end point. In this way, you can animate a change in, say, position or transparency over time.

Animation

It required two different steps to build the animation of the scene where the fish/candles are swimming through the frame.

To create the rippling background, the seven different background stills were imported into the Project window of After Effects and placed into a *composition,* which I called "Water." Basically, all I did was use the Time Layout window to dissolve one water background image into the next. By experimentation, I found that slow overlapping of fades created the effect I wanted. The dissolving background became a self-contained element that could be cycled behind the fish. *(B)* shows one of the water frames and *(C)* is a close-up of the Time Layout window

D

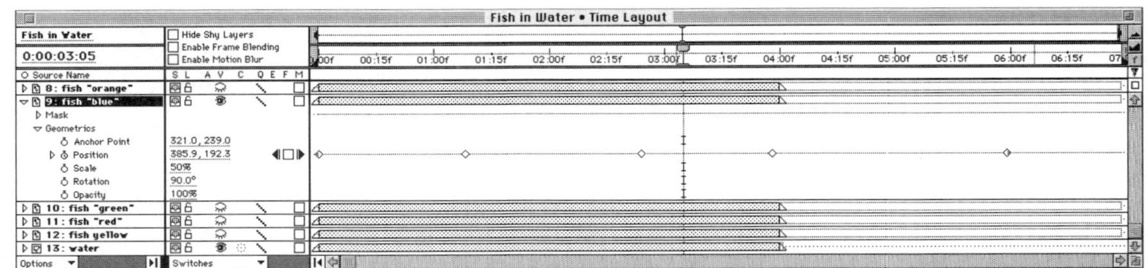

E

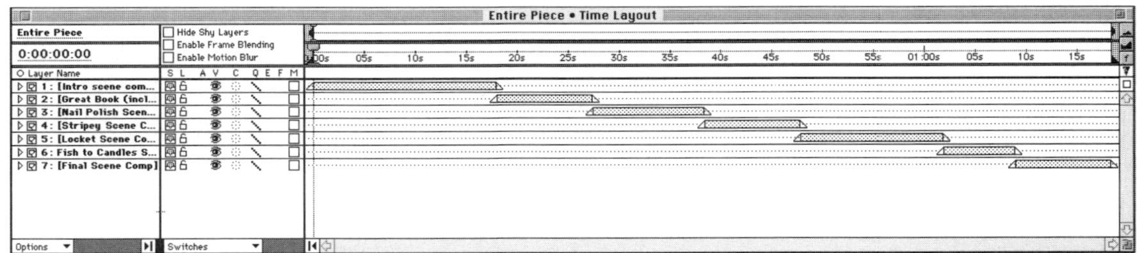

F

that shows how I key-framed the dissolves. Their opacities are set to go from 100 percent to 0 percent, fading into each other.

A new composition, called "Fish in Water," was opened up to hold the entire scene. Into this I placed the completed "Water" composition and then I imported the six fish with their Alpha Channels. The fish needed to swim across the screen, so their positions were key-framed. This was achieved by positioning the different fish objects within the Composition window. First I placed a key frame where the fish was to enter, then I plotted out a point-to-point path for each fish shape to follow until it was off-screen, *(D).* The speed with which the fish followed their path was set using the Time Layout window, *(E).* It all came together visually in the Composition window, as seen in *(H).*

Entire Piece

Name	Type
1. Intro Scene	Folder
2. Great Book	Folder
3. Nail Polish	Folder
4. Stripey Project	Folder
5. Locket Project	Folder
6. Candles Scene	Folder
7. Final Scene Proj...	Folder
Sound	Folder
Entire Piece	Composition

G

The Movie

The candle scene (which combined the animated background with the animated foreground) was just a few seconds long. I placed this composition into its proper order with the other compositions that went into the ninety-second piece, (G). *(F)* shows the Time Layout window for the combined piece, which was then saved

Project Window
Once an element has been imported, it appears in the Project window in the upper left-hand side of the screen. The Project window lists by name all of the artwork, compositions, and audio tracks in your project.

Composition Window
The Composition window is the canvas where you spatially compose your piece, as if you were looking through the lens of the camera. When you drag your elements into the composition window, they become "layers" of that composition that you can actually see. The layers are stacked from top to bottom as if in a pile.

Info Palette
The Info Palette displays numeric values that describe things such as coordinates and color values.

Time Controls Palette
The Time Controls Palette allows you to play your movie-in-progress in the Composition window. You operate its controls much like a VCR's.

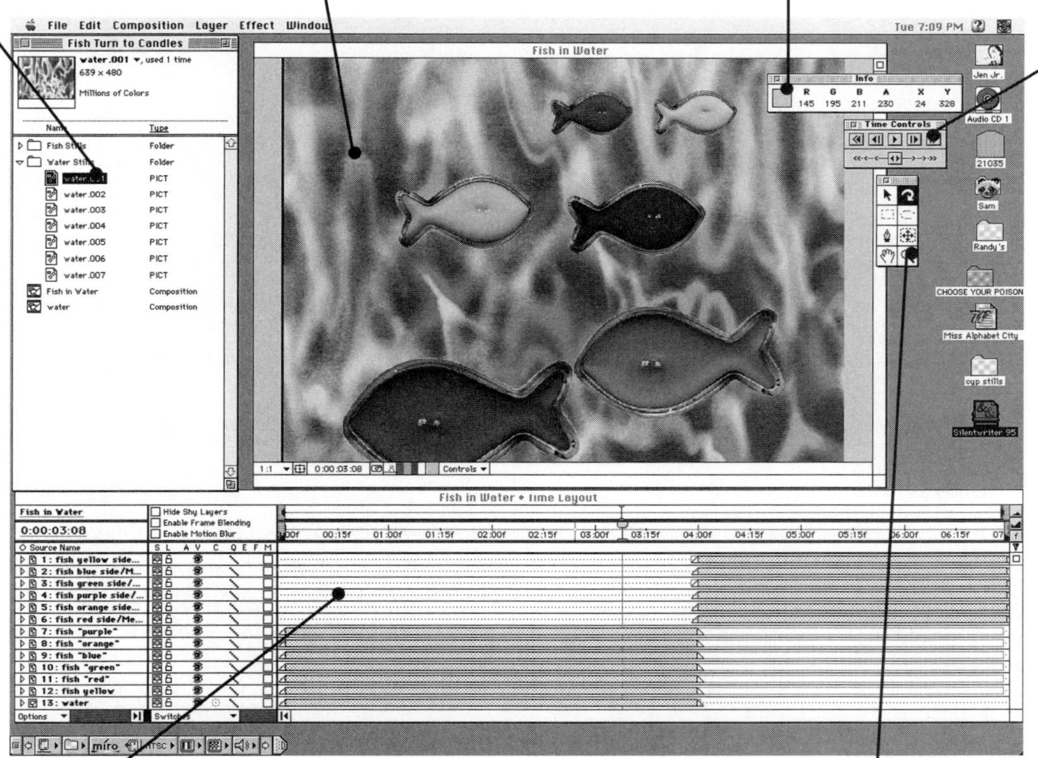

H

ime Layout Window
he Time Layout window is where you map out the timing of your film. It lists by name, rom top to bottom, the layers that appear in the Composition window. This is where you rder your layers, dictate their lengths, and apply special effects to them. Key frames mark hanges in the nature of a layer and are often placed in sets of two, marking the start and nd points, as they do here.

Tools Palette
The Tools Palette contains tools used in the making of masks. It also holds tools for rotating layers and magnifying objects.

(under File menu) as a QuickTime movie. (PS: The "reveal" shot of the burning candles was another challenge: I ended up drawing the "flames" in Photoshop and attaching these to the scanned profiles of the candles in their glass holders. I was lucky in that the reflection on the glass is just the way it came out in the original scan.)

11 ▪ Sand and Paint-on-Glass Animation

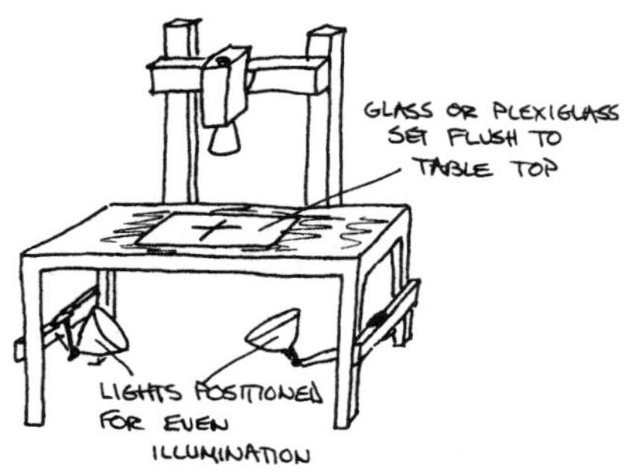

11.1 Basic bottomlighting setup: A fixed camera with fixed lens over a fixed, rear-lit surface: The sketch represents a basic setup for doing sand or paint-on-glass technique.

The brilliant work of animator Caroline Leaf created two new techniques in animation. This chapter describes her approach and samples the extraordinary effects of her films.

THE BASIC SETUP

The way to animate with sand or paint is almost identical. A stationary camera is mounted above a piece of frosted glass or Plexiglas. A special light or series of lights is positioned beneath the compound so that the translucent white surface is illuminated evenly. (See Chapter 22 for bottomlighting gear.) The material is placed on the glass. In one technique a fine sand is used, and in the other ordinary water-based inks or paints are used in combination with an agent that keeps the liquid from drying out. Because of the grueling demands of the animating process, it is valuable for the animator to have a foot pedal or similar release for exposing single frames of film in the overhead camera.

As you can see, there is nothing particularly special about the tools or setup. When the camera is loaded, work can begin. But beware the apparent straightforwardness of these techniques. Animating sand and paint is extraordinarily demanding. Hour after hour the animator sits at the working surface and makes minute changes in the art, exposes a frame or two, and then repeats this operation again hundreds and even thousands of times. If there are any mistakes, an entire sequence must be re-created from scratch.

Someone has observed that either technique requires the precision of a watchmaker, the endurance of a long-distance swimmer, the concentration of a mathematician, and the vision of a great artist. Each of these qualities is evident in Caroline Leaf's work.

DESIGNING IN SAND

Figure 11.3 carries frame enlargements from three of Caroline Leaf's films. Within these images perhaps you will be able to sense some of the special qualities that reside in animation with sand. Consider these elements as you study the samples.

Texture and Tone. Variations in the size of the sand

11.2 Metamorphosis in sand: These fourteen frames indicate the bold and fluid style with which Caroline Leaf sets a scene. In this example from *The Owl Who Married a Goose,* the two prime characters are introduced and their relationship is established. The seven-minute film is based on an Eskimo legend and the sound track consists of sound effects and dialogue recorded by three Eskimo collaborators on the project. *Photos courtesy National Film Board of Canada.*

Animation and Direction: Caroline Leaf
Advised by: Co Hoedeman
Design: Nanogak
Editing: Pierre Lemelin

Sound: Jeela Alilkatuktuk, Paul Anglyou, Martha Kauki, Samonee
Rerecording: Jean-Pierre Joutel
Production: Pierre Moretti

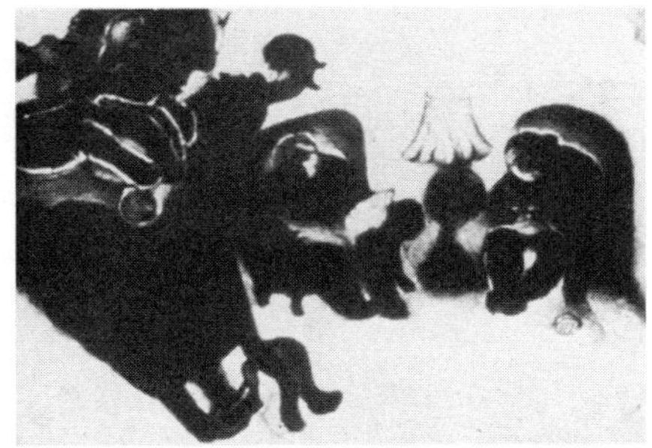

A

B

C

D

grains or in the camera's distance from the surface affect the textural quality of the image. Caroline Leaf prefers to use an extremely fine white sand. By spreading it thickly or thinly with her fingers as she works, she can achieve a full range of gray tones. The full black tone is created, of course, by piling sand deep enough so that no light shines through it toward the camera. Conversely, any area that is not covered with sand will be photographed as pure white.

Positive/Negative Fields. The rich graphic power of sand animation is achieved, in part, by designing characters and objects in high contrast: white whites or black blacks. Using this flat, sharp-edged imagery, you can manipulate the positive and negative relationships. See, for example, how the shape of a bird (Figure 11.3) that has been used as black on a white field can suddenly be transformed into a white shape on a black field. Such transitions of figure/ground relationships are powerfully exploited in Caroline Leaf's films.

Planning and Execution. It is almost impossible to previsualize the form and pacing animation will follow when one is animating with sand. Through experimentation the animator develops a sense of the visual effects that can be achieved through various styles of moving the sand under the camera. Caroline Leaf works on a surface about the same dimensions as this page. Generally she uses her fingers to manipulate the sand, but occasionally a sharp point is employed to etch a thin line. Informal sketches are sometimes used to plan ahead for the next sequence or to record the appearance of an earlier sequence (Polaroid photographs could be used in recording what has already happened.)

But basically animation is improvised as the work proceeds. Great concentration is required, particularly in maintain-

11.3 Sand animation: Frame enlargements *(A), (B),* and *(C)* are from Caroline Leaf's *The Metamorphosis of Mr. Samsa.* Based on a Kafka story, the film demonstrates the animator's abilities to achieve a full tonal range and sharp detail with sand on a rear-lit surface. Enlargements *(D), (E),* and *(F)* are from Ms. Leaf's *The Owl Who Married a Goose.* Note the positive/negative reversals of figure and ground and note the emphasis on silhouette forms. The production still from *Sand* (also titled *Peter and the Wolf*) *(G)* reveals the way sand is piled up on the glass surface. This film was Caroline Leaf's first work in the sand technique. The photograph of Ms. Leaf, *(H),* was taken in her office at the National Film Board of Canada's production headquarters near Montreal. *Photos (A) through (F) courtesy National Film Board of Canada; photo (G) courtesy Phoenix Films.*

ing a steady rate of manipulation so that there will be a consistent flow in the finished sequence.

The laborious and painstaking nature of this technique does have its dividends. In order to keep one's mind alert, the animator invests each image with tremendous thought and care. He or she is absolutely conscious of every manual change, and this attention to the expressive quality of the smallest detail is clearly evident in the finished footage. Figure 11.2 gives you a taste of the improvisation process and of the significance that just a few details can yield.

Here is some advice that will help when you try the sand technique. Because the bottomlighting shines so directly into the camera, it is best to run some tests to determine the proper exposure. Whites should be their whitest and blacks their blackest. A sheet of clear glass can be used as the compound surface provided you rig a system that diffuses the light evenly. Try ordinary tracing paper to diffuse the lighting. It works best to judge evenness by eye. Also experiment with different field sizes until you find a scale that feels right to you.

As with all animation techniques, watch for variations that open up new boundaries. For example, you might try sand animation with toplighting only. Or instead of using regular sand, try gold, silver, and multicolored glitter. Substitute other sandlike materials: peas, seeds, rice, flour, and coffee. The material you select will dictate much about the best way to light and to animate.

E

F

G

H

DESIGNING WITH PAINT-ON-GLASS

On the basis of a single film, universal recognition has been achieved for a distinct animation technique that involves working with paint on a glassy surface. Nominated for an Academy Award in 1977, Caroline Leaf's *The Street* represents in one step the discovery, exploration, refinement, and perfection of the technique.

The Street is actually not the first work in this mode. In fact, Caroline Leaf herself worked with ink on a glass field in an earlier film titled *Orfeo* (Figure 11.4). Other animators also experimented with similar effects prior to the completion of *The Street.* However, the scope and level of artistry within this

10-minute film clearly identifies and proclaims it as the source of a new technique.

The frames reproduced in Figure 11.5 can't do real justice to what is going on in this film, to the magic that makes it what it is, a masterwork. All the same, I'd like to describe a few of the major effects that can be achieved through this technique, and the most direct way to do this is by discussing how Ms. Leaf has designed *The Street.*

Caroline Leaf works small. Her painting surface is roughly 6 by 8 inches and she uses just her fingertips in applying the paint and then reworking it, frame by frame, as a particular scene develops. Because of the relatively small scale, the images themselves must be boldly rendered—there is not a lot of room for detail and this seems to facilitate a visual metamorphosis in which one image, scene, or movement is transformed into another.

Paint-on-glass animation has a distinctive look. Water-based inks and tempera colors are animated over the bottomlit field, and thus colors have the quality of illumination rather than of reflection. Water-based colors present a major problem in that the artwork dries out quickly. To keep the paint or ink workable over extended periods, a wetting agent is required. Caroline Leaf uses Colorflex, a medium that is commonly used in painting animated cels.

Two additional structural elements warrant special attention: *story* and *choreography.* Both these elements may be more particular to the personal style of Ms. Leaf than they are generic to the basic technique of painting on glass. Be that as it may, *The Street* forcefully displays these important elements.

Caroline Leaf's films depend heavily on a narrative structure. The story line is developed through the music, sound effects, and dialogue tracks. During the year and a half that it took to create *The Street,* Ms. Leaf followed a sound track that was recorded at the outset of the production and then was carefully analyzed before the shooting process began. For me, a good measure of the artistry of *The Street* resides in the way that dialogue and sound effects are used as primary sources for constructing images and guiding transformations.

Caroline Leaf choreographs her films with distinctive flair. It is astounding to most people that she animates all "camera moves" (as they would be called in live-action filmmaking). A zoom, for example, is accomplished through

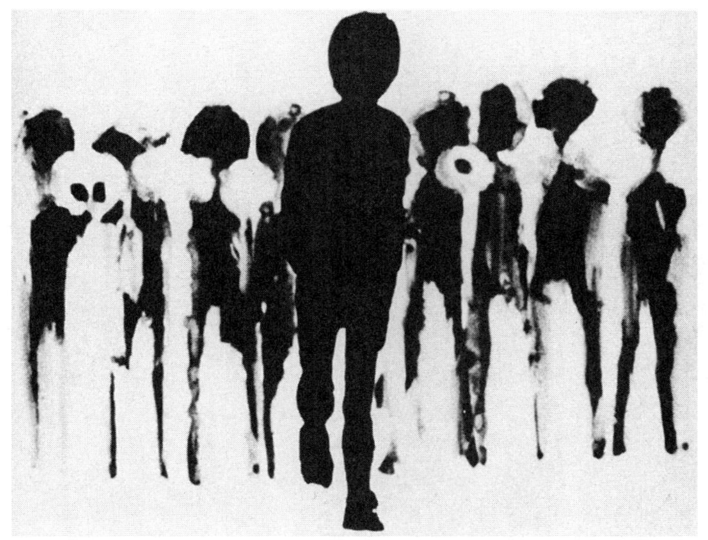

11.4 Ink-on-glass: Two frames from *Orfeo,* a simple and powerful telling of the myth by Caroline Leaf. *Courtesy Pyramid Films.*

manipulations in the artwork rather than by actually moving the animation camera. As a result, the movement within the frame is free in a way that could never be accomplished through real camera moves. In Leaf's work the viewer's point of view is totally fluid.

But the choreography goes further. As they move within the frame, *The Street*'s characters establish dynamic and dramatic spatial relationships between themselves and with their setting. Each scene is well "blocked," as someone in the theater might describe the positioning of actors on a set. Figure 11.5 shows key frames from one sequence in *The Street.* I hope that by studying it you will be able to get a sense of the animator's structural design. And here is an activity that will extend that appreciation through induction.

Project: An Animated Through-Move. Create a simple scene by employing ink or watercolors on a rear-lit background. Or work with sand. Work in just enough detail so that the viewer can recognize the location—if the set is an exterior, for example, show a few trees. The "problem" of this assignment will be to animate the scene so that the viewer will have the same visual effect that would be achieved if a live-action camera mounted on a dolly was pushed through the set while film was being exposed at normal speed.

I admire Caroline Leaf's work. However, I want to temper my celebration of her art by noting that I don't believe that

11.5 Key frames: By sampling the flow of key frames within a sequence, the complexities of choreographic and dramatic flow are revealed. With great economy of form and movement, see how much information, characterization, and detail Caroline Leaf is able to invest in this series of paint-on-glass images from *The Street. Courtesy National Film Board of Canada.*

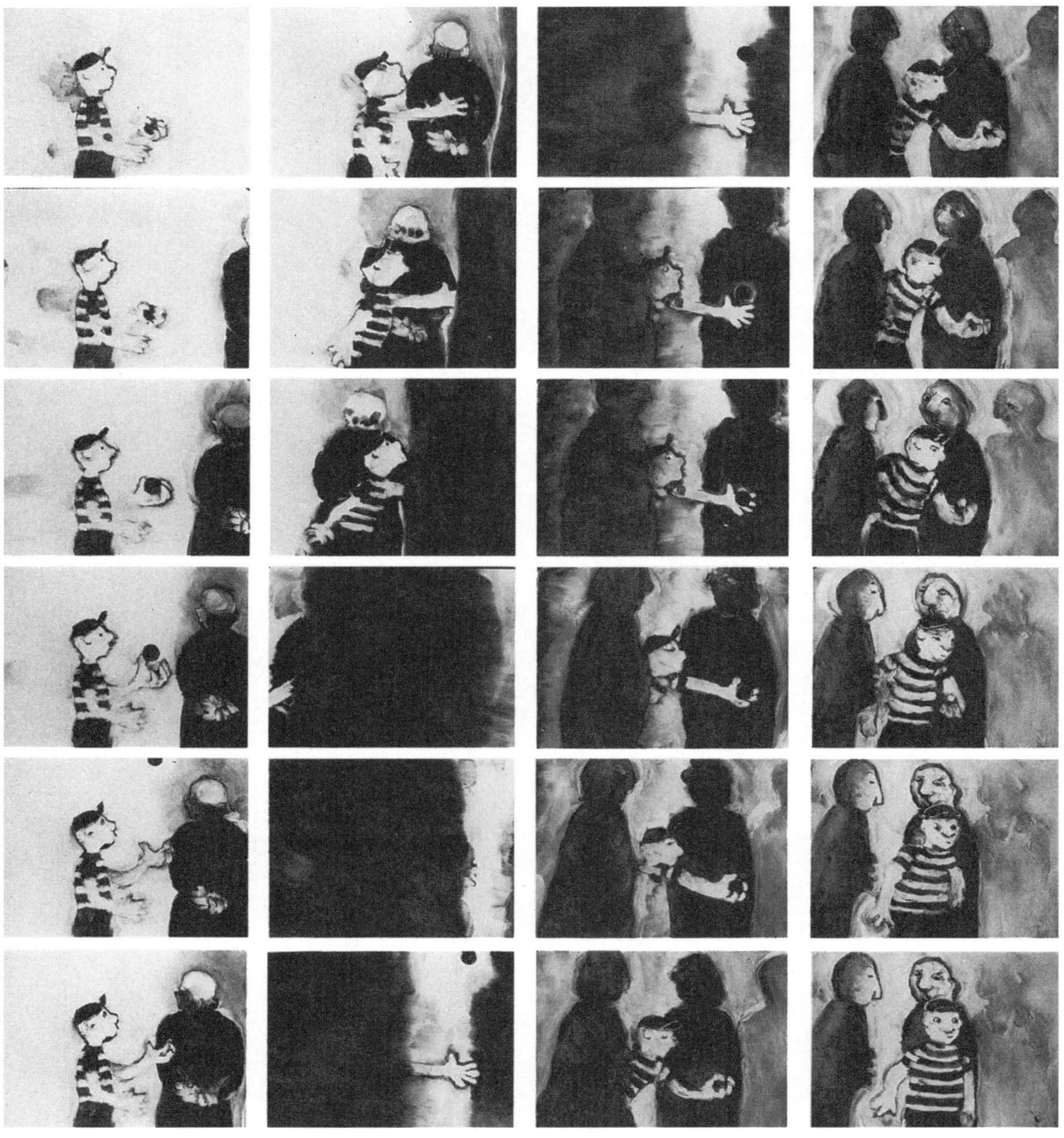

this animator's mastery of sand and painting-on-glass techniques in any way precludes or discourages the work of others in these styles. In fact, many animators have worked with sand and paint since Ms. Leaf pioneered this technique, and their work shows that each animator brings his or her particular gifts to each technique. If Ms. Leaf's particular gift is in telling stories through brilliant choreography and metamorphoses, other animators display different sensibilities for using sand or paint to explore different relationships of image to sound and other fresh dimensions of what is possible. Finding your own vision and fresh techniques is what it's all about. Still, I believe it is also valuable to study others' work. And nowhere is there more nourishing fare for an independent animator than in the movies and techniques of Caroline Leaf.

11.6 Caroline Leaf at work: *Courtesy National Film Board of Canada.*

12 ▪ Clay, Puppet, and Stop-Motion Animation

12.1 Clay animation: Independent animator Eli Noyes was one of the first to use clay in animation. The four frames below and the 35mm clip on the opposite page sample his stylistic range.

Animation of three-dimensional

objects has long fascinated filmmakers and film audiences alike. "Trick photography" in films was one of the most important attractions for early movie audiences. When the animation of small objects became familiar, enterprising filmmakers found new sources of magic through allied techniques such as pixilation and time-lapse animation.

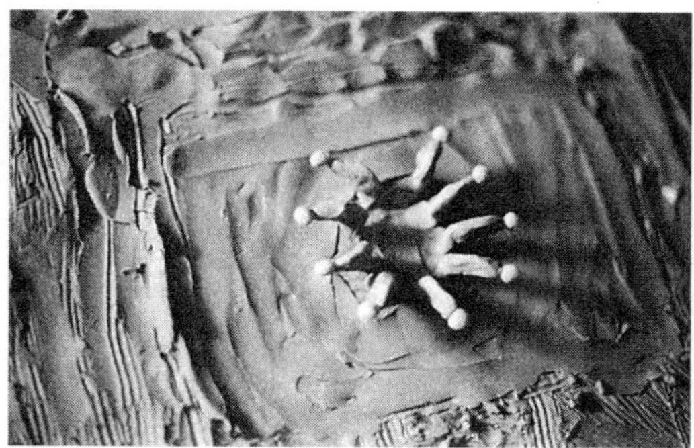

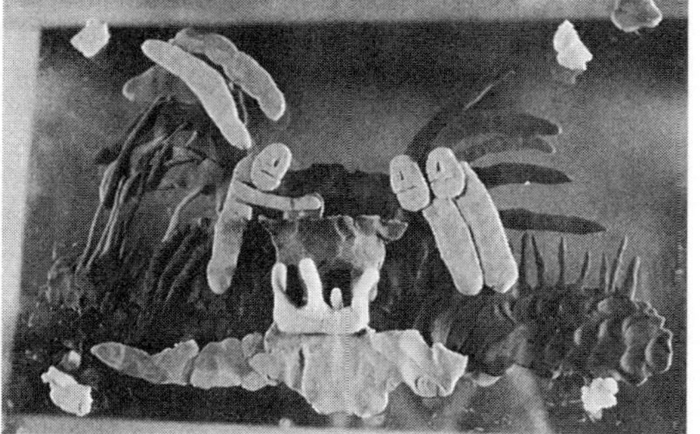

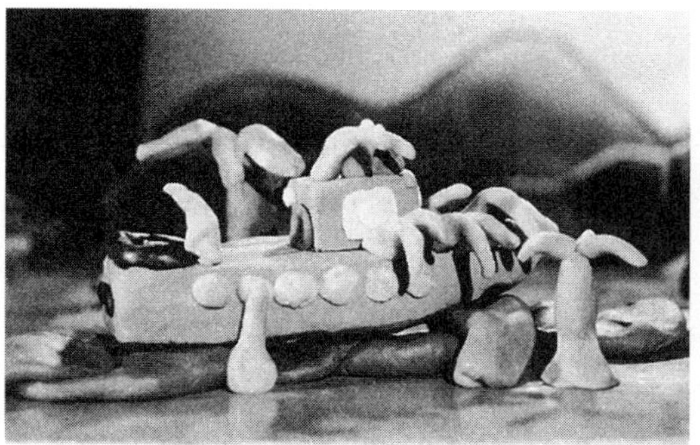

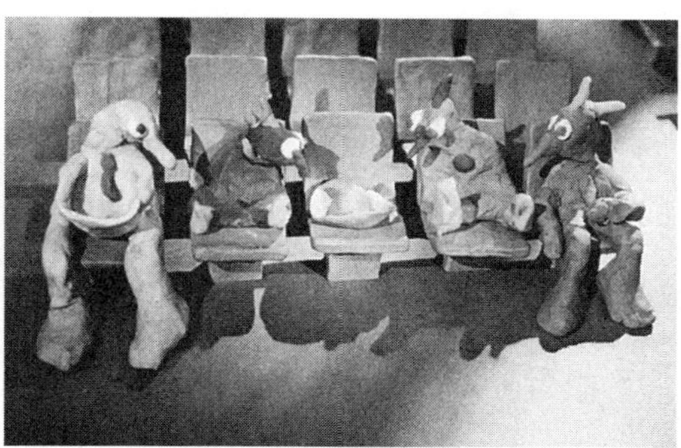

The ultimate refinement of three-dimensional animation is probably represented by puppet techniques that originated and were refined in Eastern Europe. This chapter will introduce puppet techniques through clay animation—a direct descendant of puppet films that had its genesis in North America.

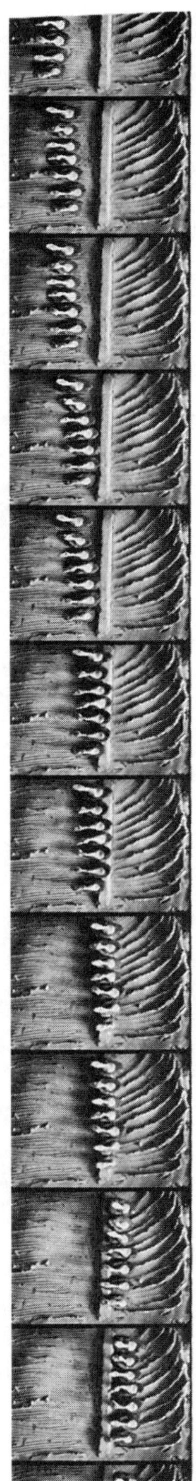

CLAY

Few animation techniques more fully exploit the medium's power of metamorphosis than does working with clay. A three-dimensional object can form itself and then transform itself through endless variations. Clay lets us witness transformations in very concrete terms. The images are palpable. They have texture, shape, and weight.

Standard modeling clay works well. It doesn't dry out under hot animation lights, is easily shaped, stores well, and comes in a variety of colors. A number of synthetic substitutes, such as plasticine, also work well.

I recommend that you work large. It's far, far easier to manipulate relatively big clay models than to work with small ones. Generous size also gives you room to establish significant detail. Sheer bulk helps in maintaining the shape of your clay characters or background. However, you must be careful to design characters so that they will be able to stand independently over long periods of time. For this reason, the legs of animated characters are usually short and their feet large. Be certain to check the practicality and flexibility of the clay figures you want to use. Birds, you'll discover, are difficult to keep together and long flower stems are simply impossible.

There are dangers in working on a large scale, however. One is tempted to fill the entire surface of the character or the background with too much detail. Doing this will give you two headaches, one technical and one aesthetic. The constant handling of the clay figures between camera exposures is bound to flatten and obliterate fine detail. More important, complicated detail will detract viewers from seeing the essential movement and timing by which the characters and the story are revealed.

In shooting clay animation, mount the camera horizontally, or else point it down from an angle of about 45 degrees. Artificial lighting is required, and by using different lights at

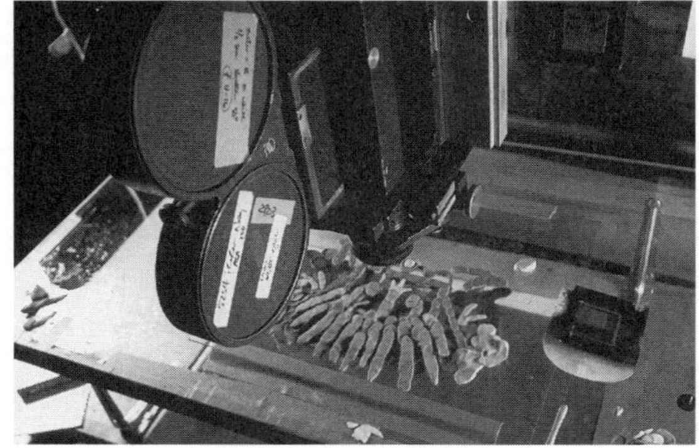
A

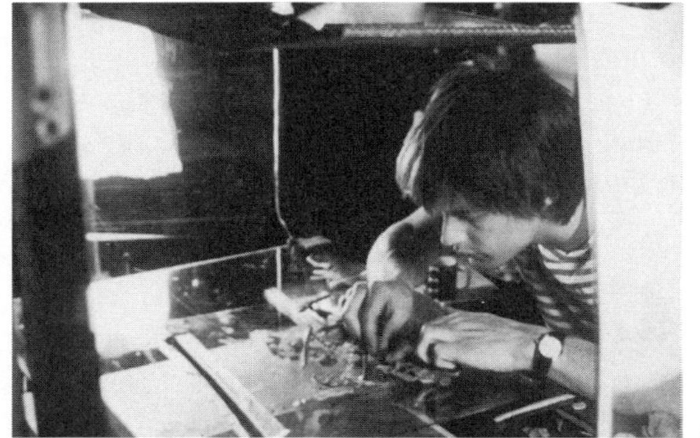
B

C

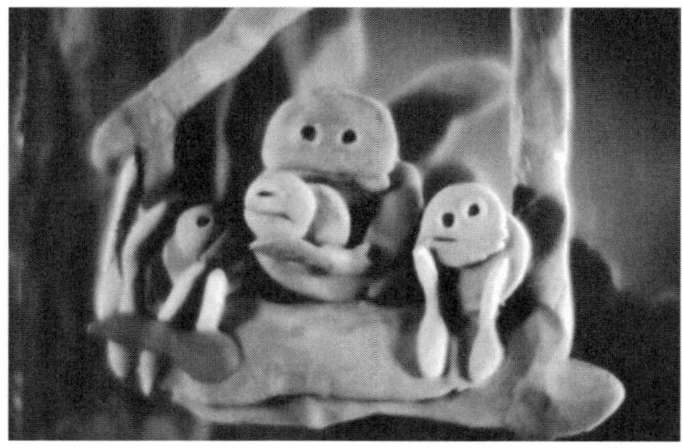
D

12.2 Multiplane clay production: In his award-winning independent film *The Fable of He and She*, animator Eli Noyes invented a form of multiplane production. He arranged several levels of glass beneath his 35mm camera, *(A)*. The top level was used for the active character animation. It was on this level that Eli incrementally animated the simple pieces of clay that comprised his characters, *(B)*. The floor of Eli's TriBeCa loft, *(C)*, became littered with pieces of brightly colored plasticine. Because the camera lens had a very sensitive field of depth, the second plane is slightly out of focus while the top plane is sharp, *(D)* and *(E)*. This helps establish a dramatic sense of depth and foreground. Eli used large sheets of precolored Pantone paper to create a "limbo" background and sky. *Photos courtesy Isaac and Abby Noyes.*

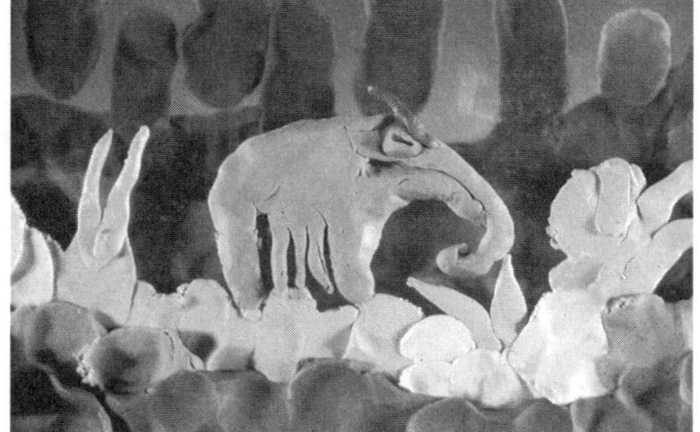
E

different locations, dramatic effects can be heightened. The camera can also be mounted directly above the clay, and if the lighting is strong and comes from a sharp side angle, the surface can be seen in sharp relief. With this setup, the contrast between highlights and shadows and ridges and troughs encourages experimentation with abstract patterns.

Each technique of animation seems to have its

A

B

C

12.3 Claymation®: The Portland-based studio of Will Vinton has made a huge contribution to the technique of clay animation, refining their own particular aesthetic to the point where it was appropriate to register the name "Claymation." *(A)* samples Vinton's first major breakthrough, a short film titled *Closed Mondays* that won an Academy Award. *Courtesy Pyramid Films. (B), (C),* and *(D)* sample a more recent film titled *Will Vinton's Claymation Easter Special*, which won a pair of Emmys, including one for outstanding animation show and one for individual achievement in animation, awarded to animator John Ashlee Prat. The production still, *(E),* shows the scale and setup. *Photos © Will Vinton Studios.*

D

E

own distinctive style of movement. As suggested earlier, clay techniques tend to lend themselves to transformation; that is, one shape turning into another. In order for the metamorphoses inherent in clay animation to be convincing, the action needs to appear relatively smooth. Changes between one position and the next should be very gradually revealed and, where possible, the evolution of one shape into another should be anticipated and exaggerated. Any extraneous source of movement should be eliminated.

Project: Spontaneous Generation. Clay animation can also be used for "character" animation—the art of creating highly anthropomorphic, humanlike characters and stories. This next project will help you get a feel for the tolerances of the materials and the possibilities inherent in transformation. Begin with a fist-sized ball of clay or plasticine. Place it on a simple background—a tabletop will do. Animate the clay as it rolls into the middle of the frame. Then, in exactly 10 seconds of film (240 frames) make your lump turn into five different objects. Plan ahead so that one shape will naturally transform itself into another. Each transition should be as smooth as possible. Shoot on twos. At the end of the 10 seconds, have the lump of clay leave the screen.

PUPPET ANIMATION

History's seven leading puppet animators have been identified as George Pal, Bretislav Pojar, Ladislas Starevitch, Jivří Trnka, Hermina Tyrlová, Zenon Wasilewski, and Karel Zeman. Their names aren't exactly household words. For most of us, in fact, the names are terrifyingly difficult to pronounce, even to read. With the exception of George Pal, these master animators have done their work in Eastern Europe.

Although the technique of puppet animation has been used widely throughout Western Europe, North America, and Japan, it is no coincidence that the major figures of the genre are Eastern Europeans, for within these countries there is a long-standing tradition of puppetry. National film production centers in Eastern Europe, Central Europe, and the former Soviet Union have produced puppet films that exhibit levels of artistry, technical proficiency, and narrative power that are capable of capturing the most profound human experiences.

PUPPET CONSTRUCTION

The basic principles of single-frame movement and photography have been discussed. So let's focus attention on the "stars" of puppet films and study what they look like and how they are made.

Animation puppets share certain basic features. They are freestanding and able to support their own weight; they have movable body joints that can hold any position; they are all well executed, with remarkable detail; and they can be viewed from 360 degrees.

The anatomy of a puppet begins with an *armature,* an inner construction that allows the puppet to be both sturdy and flexible. There are three basic systems for creating an

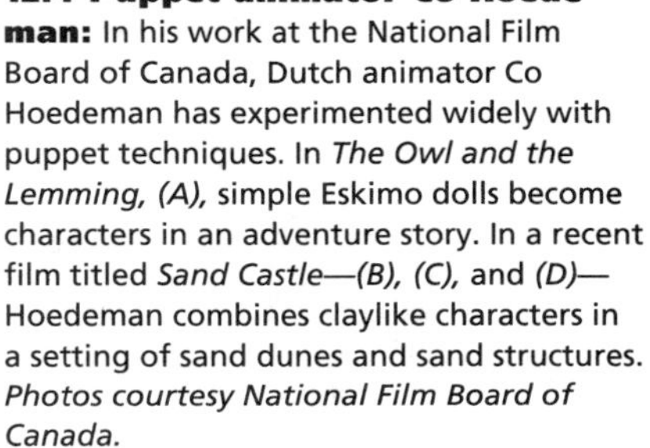

12.4 Puppet animator Co Hoedeman: In his work at the National Film Board of Canada, Dutch animator Co Hoedeman has experimented widely with puppet techniques. In *The Owl and the Lemming, (A),* simple Eskimo dolls become characters in an adventure story. In a recent film titled *Sand Castle—(B), (C),* and *(D)—* Hoedeman combines claylike characters in a setting of sand dunes and sand structures. *Photos courtesy National Film Board of Canada.*

A

B

C

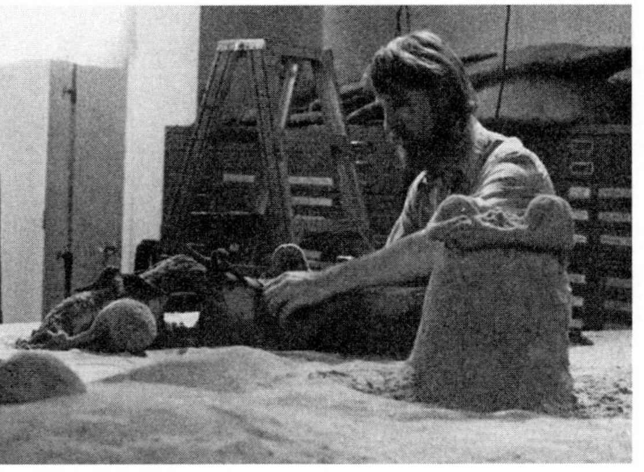

D

A

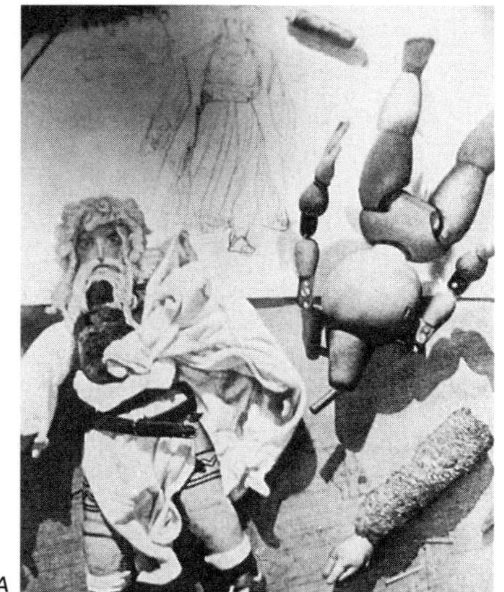

B

12.5 Puppets by Jivří Trnka: The craft of Czech master animator Jivří Trnka is shown in *(A)*. Note the original sketch, the armature, and a finished puppet. Another Trnka, *(B)*, puppet is seen with the lighting and set that complete the film's scene. *Photos collection R. Bruce Holman.*

armature. A wooden body can be crafted that has snugly fitting joints, but eventually the wood wears smooth and no longer holds its position. Flexible wire can be folded to create a basic body structure, though the armature must be padded and clothed, and the wires will eventually break from constant manipulation. Probably the best armature device is also the most sophisticated, a series of metal joints constructed from rods and combinations of ball joints and slip joints that are held together under tension. Figure 12.6 shows examples of various armature designs.

Armature systems help to define the different body types commonly used in puppet animation. The oldest of these is a simple wooden toy, a carved doll that can be moved with precision. Contemporary animators have used the modern equivalents of traditional wooden dolls. Any toy store can provide plastic-jointed dolls that can be used in puppet animation. Such modern toys can be used as they are or modified in varying degrees by an ingenious animator.

Another traditional body type is made by covering an armature with a padded and costumed body. These are often referred to as Czech puppets because of their association with the films of Jivří Trnka and other Czech puppet masters.

Finally, there is the molded body type. The most recent and technically advanced of puppets, a molded body puppet is created by placing an armature within a plaster mold, usually made from a clay or a plasticine model, which is then filled with a flexible rubber or plastic compound. More specifically, foam rubber is injected into the plaster-of-Paris mold. After the body has "cured," the puppet is taken out of the mold and finishing touches are applied.

SET DESIGN

Puppet animation techniques require special worlds for these special characters. The animator must build mini-sets that are appropriate for a particular story. Like a regular theatrical set, the stage and environment of the puppet film may need flats, backgrounds, props, and other details. Here are a few requirements that are unique to designing a set for puppets.

Scale. The set must match the size of the puppets. Select

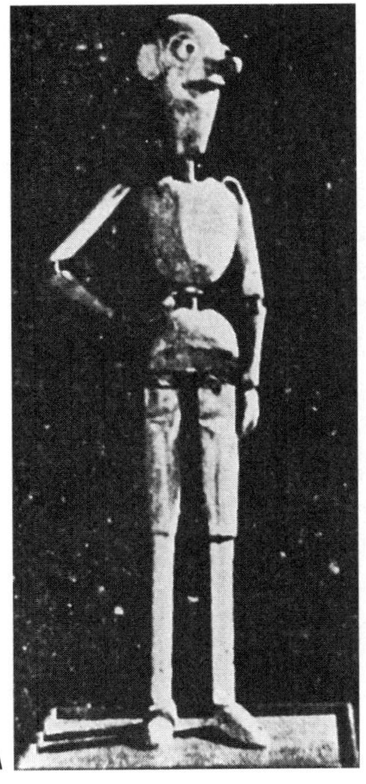

A

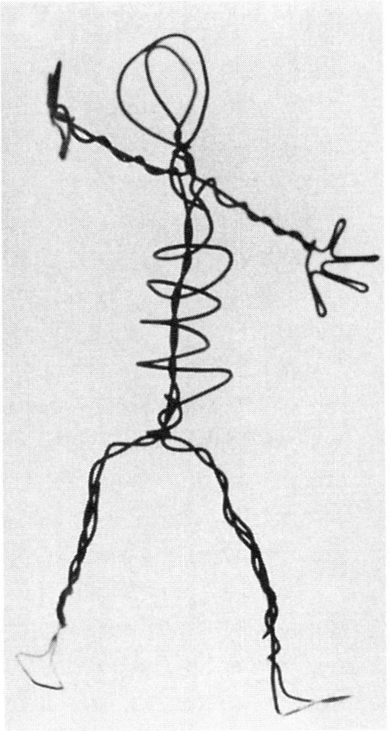

B

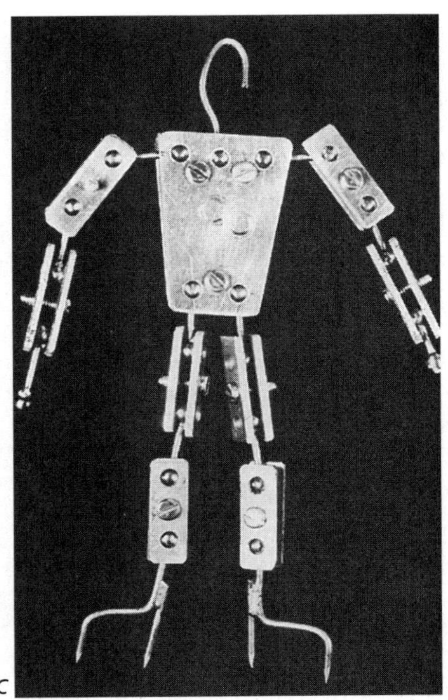

C

12.6 Armatures: The wooden foundation of a Trnka puppet, *(A)*; a simple wire armature, *(B)*; and a metal armature, *(C)*. *Photo (C) courtesy The Yellow Ball Workshop; others from collection R. Bruce Holman.*

a scale in which characters and settings can be easily manipulated. As a general rule, don't try to animate a puppet that is much smaller than 6 inches.

Stability. Not only must the construction of the sets be very solid in order to stand up to prolonged use, but the set designer should be sure that "disturbable" items are few. Here are a few things that can provide migraine headaches because they are so easy to move unintentionally: billowy curtains, tables in the middle of a room, real or artificial plants, shag rugs, or any surface that shows dust easily.

Camera Access. Before designing a set, the animator should carefully consider what kinds of camera positions will be used. Will a high angle show the floor? Is it possible to shoot through a window? Can a wall be easily removed in order to get a reverse-angle cutaway? Can the camera get extreme close-ups? Is it possible to track within or past the set?

Lighting. Here is an opportunity to create special lighting effects. Various combinations of color, intensity, and placement can totally reshape an existing set. There should be lots of experimentation.

A

B

C

D

E

PLOTTING MOVEMENT

There is no really satisfactory way to plot or plan the movement of an animated puppet. The best approach involves some careful observation and simple arithmetic. If you are trying to execute a puppet's walk, for example, time the duration of the overall movement with a stopwatch. Measure the distance to be covered in the walk and measure the average pace of the character. Divide the distance by the number of steps and determine the number of separate moves within each pace. Figure 12.7 illustrates the procedure.

You will still have to use your intuition in applying this system. There must be adjustments because puppet characters will be called upon to perform moves with subtle variations of tempo and emphasis, so that each individual puppet's mood or personality will be revealed through its movement.

If you can imagine the painstaking process of walking a puppet across a set, then you'll agree that the supreme requirement of the puppet animator is patience. But spontaneity is also required. The animator must have the patience to work slowly over long hours with great concentration, and he must have the spontaneity and ingenuity to project into the puppet those details of movement that bring a recognizable personality, a distinctive character to each of the handmade movie stars.

Project: Gulliver's Desktop. As a way to explore some of the characteristics of puppet animation, find a child with plenty of toys and borrow a doll for a few days. If you can, get one of those plastic dolls with full-jointed bodies—ankles, knees, hips, waist, arms, neck, and head. Film a set of tests that shows the doll walking, running, sitting down, and waving. Try various physical activities so that you can find the particular capabilities and limitations of the doll.

12.7 Plotting movement: In this example, the action calls for the Lego character to drive through the gate, past the camera, and out of the frame to screen right. If the real distance to be covered is 12 inches and the shot takes two seconds, then each position of the driver would need to be ½ inch farther along the trajectory of planned movement (24 positions divided by 12 inches—shooting on twos). Photo *(A)* suggests how one would visualize the plan. To exaggerate the drive past, however, spacing between positions might be varied—less space at the beginning and more at the end. *(B)*, *(C)*, *(D)*, and *(E)* show the positions 5–8. Note how the character's face turns toward the camera.

Now for a short movie. Mount your camera on a tripod so that it points at a desktop. Light the area in a dramatic way and distribute a normal sample of desktop paraphernalia: Scotch tape, pencils, books, a lamp, a coffee cup, whatever. Shoot a film in which your doll awakens to find itself lost on a giant's desk. Have the doll explore the surface and its implements. You might end this sci-fi saga with the giant's hand suddenly appearing and lifting up the terrified doll.

STOP-MOTION ANIMATION

Beginning in the 1970s, animators began adapting computers so that they could provide the control mechanisms that drive animation stands. It wasn't long before *computer control,* as it was called, moved over to the high-end world of object animation, thus giving birth to a domain known as *motion control animation.*

This is the technique used when you take the animation camera off its stand and want to move it, with frame-by-frame control, across a stage area. Sometimes the camera is mounted on tracks. Sometimes it is attached to a long jib arm that can move over and above the area being shot. The movement of the camera is interrupted so that the animator can make changes to puppets or objects in the scene before resuming with the exposure of individual frames. When the scene is projected, the camera movement is absolutely smooth and appears as if it were shot in one live-action move.

It takes a specially built computer (not your desktop variety!) to guide the camera's path through space with such a degree of precise, minute control that an identical shot can be achieved time and time again—even if the camera starts ten feet above the stage surface, swoops down following a winding path, and ends up panning to look back at the position from which it started.

The Nightmare Before Christmas and *James and the Giant Peach* are two features directed by Henry Selick, who used the 35mm stop-motion technique. Most often, replaceable character models were switched out from the beautifully constructed sets and carefully repositioned between each exposure on the motion control rig. Figure 12.9 gives you a look inside a stop-motion studio.

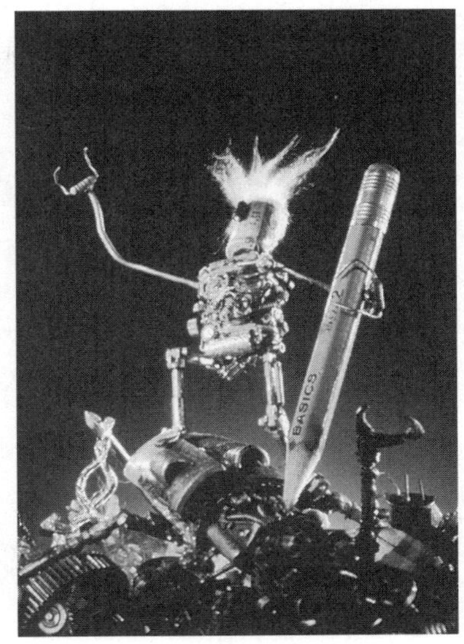

A

B

12.8 Mr. Resistor: *(A)* and *(B)* sample a series of independent films that Will Vinton Studios has produced featuring a character named Mr. Resistor. In this recent work you can see stop-motion director Mark Gustafson leading the Vinton Studio in new directions. Such ceaseless exploration and willingness to try new forms is the hallmark of all the really great animation studios. *Mr. Resistor © Will Vinton Studios.*

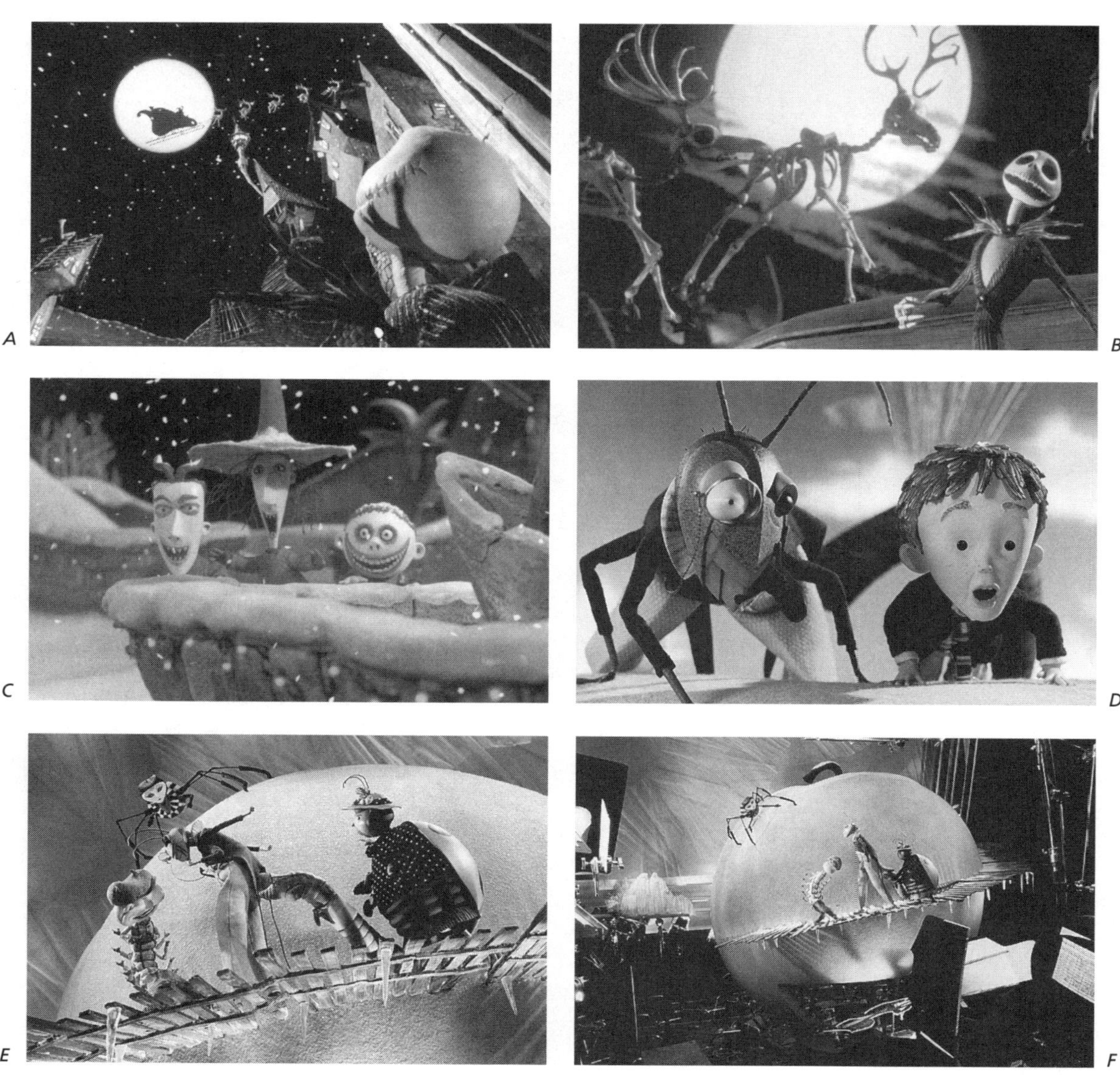

A B C D E F

12.9 The world of Henry Selick: Two feature films have taken three-dimensional stop-motion animation techniques to new heights. Both were directed by Henry Selick. The frames from *Nightmare Before Christmas—(A), (B),* and *(C),* © Touchstone Pictures—suggest the rich atmosphere and dynamic camera angles that characterize Selick's work. Production stills from *James and the Giant Peach—(D), (E),* and *(F),* © Disney Enterprises, Inc.—show the detail of puppet fabrication and the scale of production sets and characters. *Photos courtesy Tim Burton, Touchstone Pictures, and Walt Disney Pictures. All Rights Reserved.*

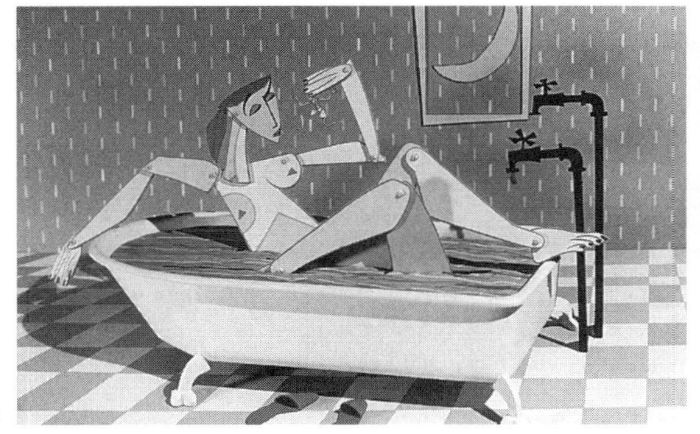

G

H

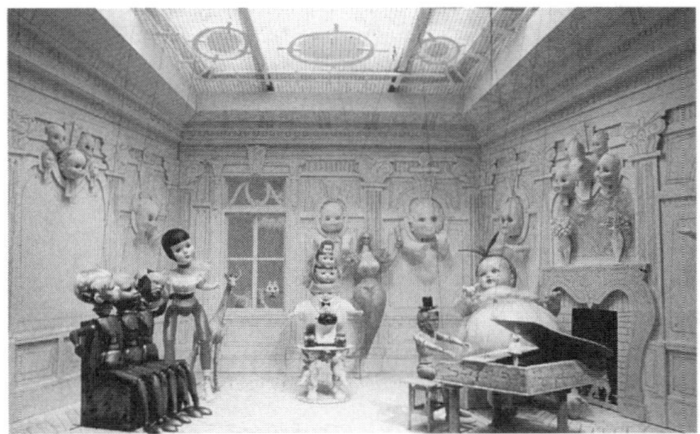

I

J

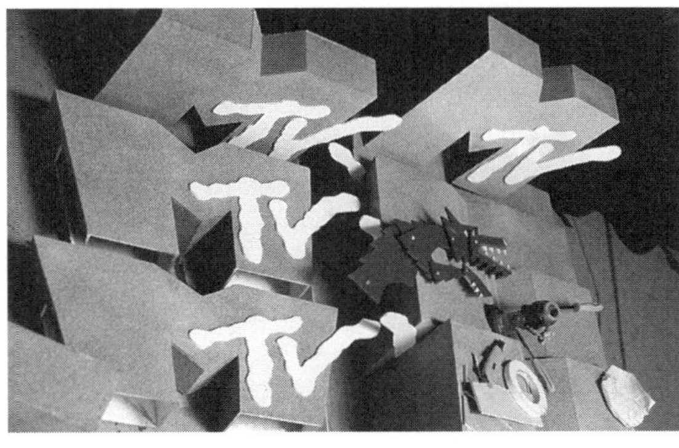

K

L

Henry Selick developed his craft, in part, through a series of short station IDs and other broadcast graphics that he did for MTV. Frames from three different IDs show the range of aesthetic styles possible with stop-motion techniques: *(G)* is *Bathtub M*, *(H)* is *Contortion M*, and *(I)* is *Dollhouse M*. The production still, *(J),* shows Selick and co-designer Ron Davis holding up a character that appears in *Contortion M* and *(K)* shows some of the replaceable block M's that were used in the piece. In *(L)* Henry Selick is seen animating test footage for *Dollhouse M. Photos courtesy Henry Selick, Twitching Image, Inc., and MTV Networks.*

13 ▪ Rotoscoping

WRITTEN WITH MICHAEL DOUGHERTY

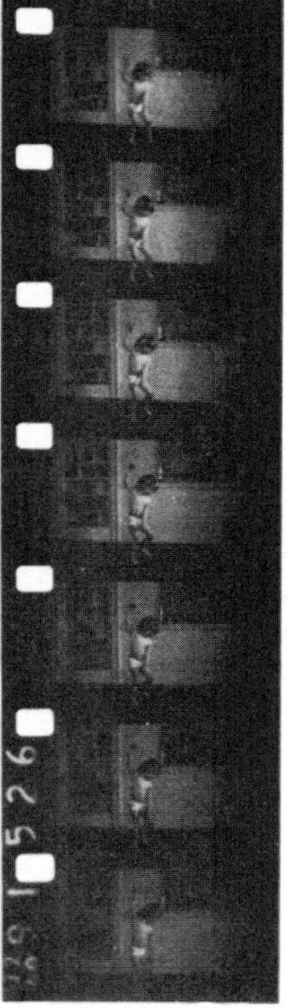

A

Rotoscoping is a form of animation that creates a very fluid, lifelike movement because each image is traced—frame by frame—from live-action reference footage. To some, this animation technique is restrictive because the movement is so closely linked to "reality" and there is not much room for exaggeration or caricature. Yet with creativity and experimentation, rotoscoping can yield animation that is breathtakingly beautiful—and even quite abstract.

Prior to the computer, the process of rotoscoping required a very accurate set of filmmaking tools that pretty much limited the technique's use to feature films and expensive TV commercials. The technique originated in the 1940s as a way to combine live action with animated images onto a single piece of 35mm celluloid film. This old-fashioned process is almost never used anymore. But understanding its principles will be useful in approaching digitally based rotoscoping.

The original process involved projecting a piece of live-action film, one frame at a time, onto a piece of glass. A sheet of tracing paper was placed onto the glass surface and this provided a screenlike surface onto which an animator could draw. Sometimes the live-action footage would be used as a source for the animator to study, for example, the precise movements of a galloping horse. Sometimes the animator would need to have accurate access to each frame so that he or she could draw a character *into* the existing scene—for

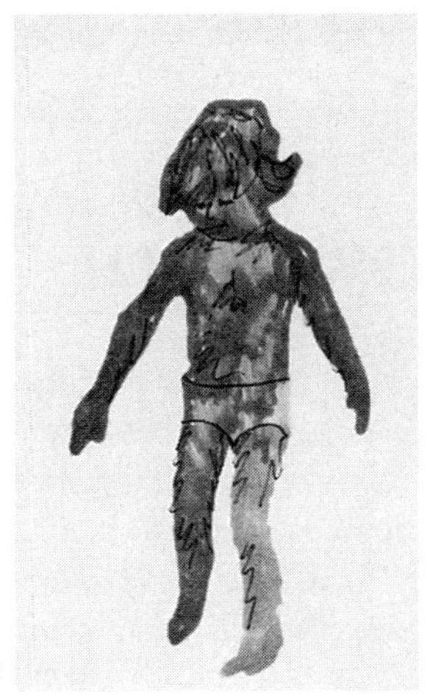

B

C

example, drawing a cartoon bird that would perch on a human actor's outstretched finger.

In classic rotoscoping, once the drawings had been transferred to cels, the new animation and the old live action were combined on a single strip of film. This was done in one of two ways. The resynthesis could be achieved by placing the appropriate cel from the animation sequence over the rear-projected frame of the live-action film on the animation stand's compound surface and then rephotographing each frame. The alternative technique was to photograph the animation sequence by itself and then combine it with the live-action footage in a process involving mattes and optical printing techniques.

Sound complicated? It was. But the resulting visual effects were deemed worth all the labor and expense. And because rotoscoping opened up such imagination-grabbing vistas, independent animators working in the 1960s and 1970s concocted simple, homemade alternatives to the expensive rotoscope gear. These animators tended not to be interested in combining animation with live action, but rather in applying an artist's hand to imagery that was generated from live-action sources. Two of the jerry-rigged approaches developed by independent animators are shown in Figure 13.2.

But it's a new day. Inexpensive computer techniques have firmly bypassed the old film technology and digital rotoscoping now blossoms as a vital new artistic frontier. It's a

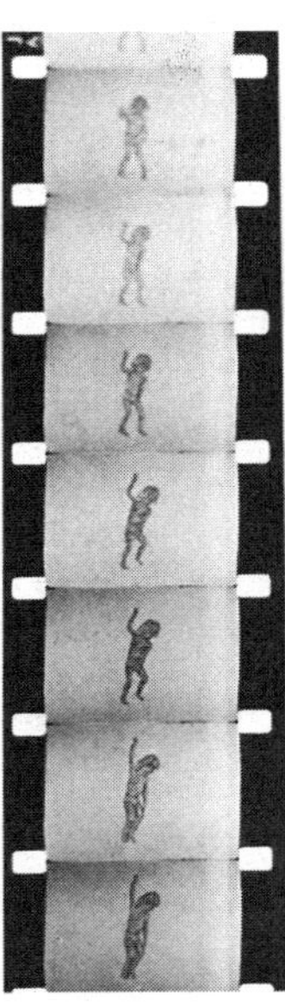

D

13.1 Rotoscoping steps: A section of live-action footage, *(A),* is selected and alternating frames are traced off a rear projection of surface, *(B)*. The animator may make a series of tests, *(C),* to determine what graphic style or styles best suit the original material and the nature of the film. A complete set of images is created and filmed. The resulting footage will bear similarities to the original footage, but it will have a distinctive, animated look, *(D)*. The illustrations are from *Family Spots,* by the author.

A

CEL OR TRACING PAPER
FROSTED GLASS (FOR REAR PROJECTION)
45°
MIRROR
PROJECTOR CASTS A SINGLE FRAME

B

13.2 Tracing setups: If rotoscoping from a viewer, *(A)*, select a model with as large a viewing screen as possible. Also see if you can use a brighter light source so that the illumination of the selected frames will be bright and sharp. Be careful, however, that the light isn't so hot it burns the film in the viewing gate. Precut tracing paper to match the dimensions of the viewer screen. Setup *(B)* illustrates how a film projector can be used in rotoscoping. Note that one can adjust the rig to get an image of almost any size. Old projectors can be modified for rotoscoping so that a single frame is advanced and so that the image is very bright as it appears on the drawing glass surface.

wide-open horizon with plenty of room for animation homesteaders to stake out their own claims.

SOURCE MATERIALS YOU CAN FALL IN LOVE WITH

If you want to try out digital rotoscoping, the first step is the most important one: You must either find or create live-action reference footage that will sustain your interest through the long and demanding process of transforming it into animation.

Of course, anything is possible. You can rotoscope from a TV show or from a home video. However, we recommend that you look for short clips of live action that contain lots of movement. Consider these rotoscope playgrounds: sports footage, dancing, action that forms a natural cycle (like turning in a circle or swinging), or footage that incorporates sweeping camera moves and dynamic zooms. Common to these suggested sources is a complex physical action that is both eye-catching and *really* difficult to animate using traditional methods.

The best suggestion on how to get cool source material for rotoscoping is to simply shoot it yourself, using a digital camera. This way you can skip the digitizing process altogether (see page 165). One of the cheapest and most user-friendly digital cameras is the Connectix QuickCam (see the case study in Chapter 4: Animating Objects). These small eyeball-shaped cameras hook up to the back of your computer and can capture video or single images in either grayscale or color, depending on the model. The image and sound quality won't be the best, but it's good enough for the purpose of rotoscoping. The only disadvantage of QuickCams or similar cameras that sit on your desk is that they have to be directly connected to the computer in order to be used. So unless you have a notebook computer you can take on location, plan on shooting source videos in the same room as your desktop computer. Keep in mind, too, that original clips should be short enough for both you and your computer to handle. Three to ten seconds is an appropriate length. Don't plan on trying to rotoscope an entire basketball game unless you're sitting in jail and facing a life sentence!

DIGITIZING/VIDEO CAPTURE

Let's assume you've got a very nifty chunk of source material. The next step is to transfer the video you want to use from a camcorder or a VCR to your computer—a process called *digitizing* or *video capture.* This can be tricky since the type of computer you have will affect how well the video is digitized. Most Power Macintosh AV (Audio-Video) computers can achieve digitizing very easily by just hooking up a VCR to the proper jacks, without adding any additional cards or parts. With Windows computers you may have to purchase and install video cards in order for the machine to handle the video. Such cards will also enable you to hook up a VCR or camcorder to your computer. Before purchasing the cards, it's important to make sure your computer has enough RAM, a speedy processor, and enough hard disk space to handle the large chunks of data you'll be using. See Chapter 24: Computer Hardware.

To capture or digitize source material, you'll also need the right software. One of the best packages for digitizing and editing video for both Macs and PCs is Premiere from Adobe. Premiere has an easy-to-learn interface and lots of extra features that, as an animator, you will enjoy. These include dissolves, wipes, creating titles, and editing sounds. It's no surprise that Premiere works extremely well with other Adobe software packages like Photoshop and After Effects.

Before we can get on with the heart of the rotoscope process—where one creates custom imagery with frame-by-frame reference to the source footage—it is necessary to quickly discuss *compression.* A short video clip—say, 5 seconds long—creates a huge digital file that can easily take up 5,000 kilobytes (5 megabytes) of storage (that's three and a half floppy disks). You don't need a pocket calculator to know that these file sizes are going to hog space on your hard drive and slow down the computer's CPU as it tries to handle such a gigantic wad of data. To handle this problem, computer scientists have invented compression algorithms—mathematical models that reduce the size of digital files by simplifying or eliminating some of the data. To get the idea of compression, imagine how much smaller the black-and-white copy of this 5-second video clip will be than its color version. Similarly, a frame size that is one quarter as large as a full 325 lines of TV resolution is going to take up less space. And if you can take a

13.3 Rotoscoped variations: Four variations on a single frame show the creative breadth of rotoscoping. From *Odalisque*, a film by Maureen Selwood. *Stills courtesy Maureen Selwood.*

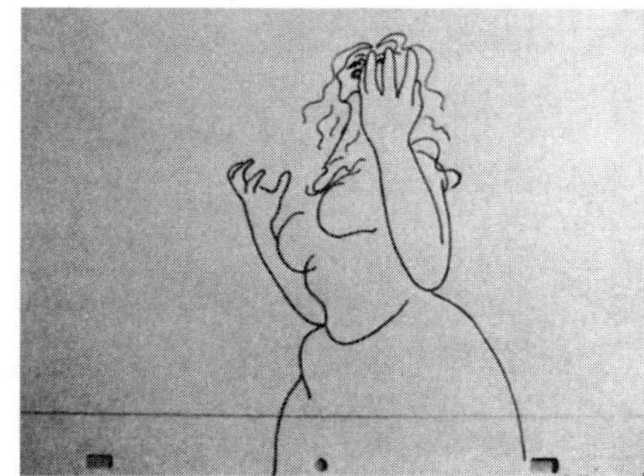

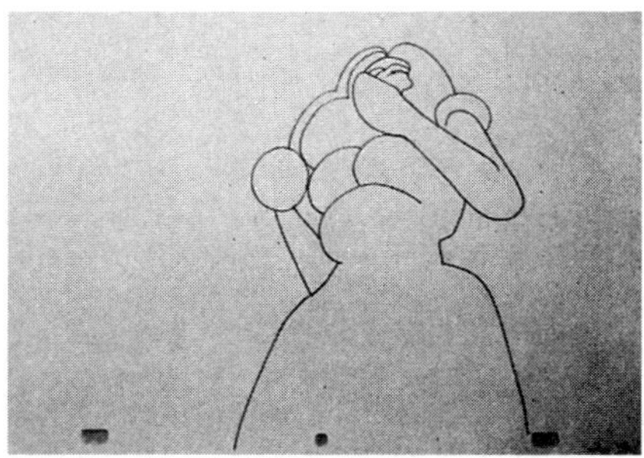

A

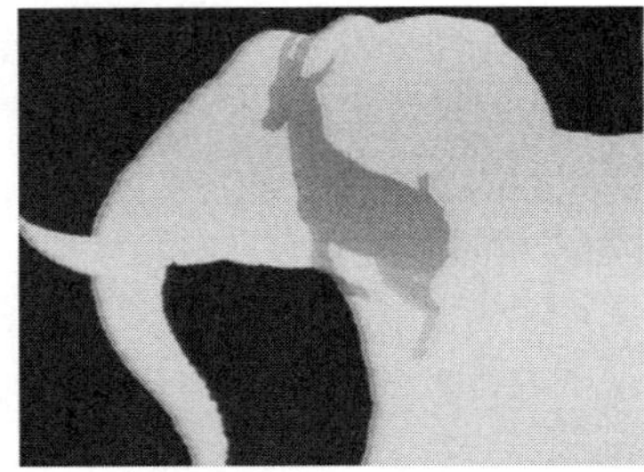

B

C

D

E

range of, say, five distinct shades of gray and treat them as one, you will reduce storage requirements of a file that started out with 256 shades of gray.

Video compression is an ever-expanding frontier in digital animation. Or should that be ever-reducing? Technological advancements in compression make it easier for you and your computer, but at a price. A decision awaits you. Here you must engage in a give-and-take proposition in which you must find the balance between quality and size. If you want or need to trace from a high-definition clip, you shouldn't compress it much, but the resulting file will take up lots of space and slow down the rate at which images can play back. But the more you compress the file, the grainier your image will become (and the clip's sound quality will become muffled). In Chapter 23: Computer Software, you'll find a discussion of different compression schemes that can be used in making QuickTime digital movies. Each compression program has its own set of benefits and corresponding uses.

Rotoscoping is a technique that can take advantage of the biggest compression factors. Since your video source footage is only being used as a reference that you will draw over and it won't actually appear in the final product, a grainy or blurred image isn't a problem. It is best to choose the compressor that will give you an adequate image for your particular project, and will significantly shrink the size of the file.

TWO DIGITAL ROTOSCOPING METHODS

Figure 13.5 provides a detailed case study of how a single piece of live action can be used in the rotoscope technique.

13.4 Rotoscoped TV IDs: Here are five frames from a series of ten-second rotoscoped channel IDs that Kit Laybourne and Eli Noyes did for Nickelodeon. *(A)* and *(B)* are from a piece titled "Menagerie." Public domain wildlife footage was the source for silhouette tracings that lifted animal outlines from backgrounds. The resulting sets of imagery were then combined and colorized using high-end video tools. The same effects could be achieved today using After Effects or Premiere. In an ID titled "Skating," a set of "loose" tracings was done with crayons and colored pencils onto white sheets of animation paper—*(C), (D),* and *(E).* The images were reversed in video postproduction, resulting in some of the looks sampled here. *Courtesy Noyes & Laybourne Enterprises, Inc., and Nickelodeon.*

Actually, you learn about *two* techniques using two different digital tools.

Surely one of the all-time great graphics applications is Photoshop from Adobe. This amazing program is sometimes billed as an image-processing application. In truth it is much, much more. The first portion of the case study shows how one can use Photoshop for rotoscoping.

But there is another all-time great graphics application that can be used for rotoscoping. This is Painter from Fractal Design. A breakthrough computer paint program in its own right, Painter can open up a video clip and let you draw or scribble directly onto each frame using a variety of digital painting tools that mimic the range of visual effects one can achieve using traditional art media such as charcoal, oil paints, watercolors, crayons, and pastels.

Each method has its own advantages and disadvantages, which you will best discover yourself through trial and error. By altering each frame separately in Photoshop, you use up less memory than attempting to alter and design around an entire video clip as you would using Painter. However, by using Painter's built-in video and animation features, you can skip the often time-consuming process of exporting a clip as individual frames, tracing in Photoshop, and then reassembling them again in Premiere. Whichever method you decide on—or even if you find another hybrid of software programs—digital rotoscoping allows you to create an almost endless variety of animations based on a few short live-action video clips.

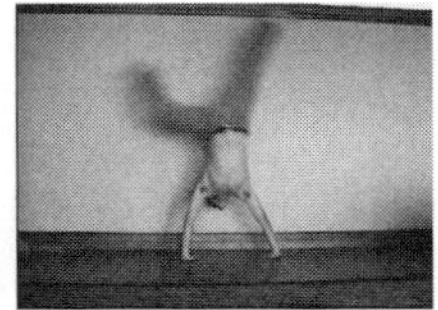
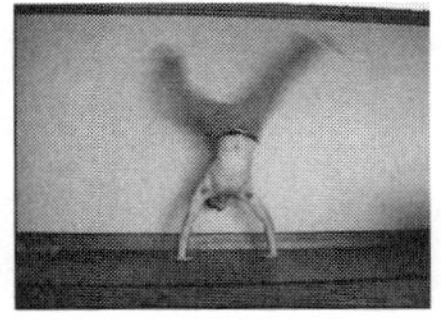

A

B

Kit Laybourne and Michael Dougherty asked Sam Laybourne, Kit's son, to help out in creating some interesting footage for a rotoscoping case study. The source footage was shot on Sam's Hi-8 camcorder. We shot against a white living room wall so that the background wouldn't be distracting when it came time to trace over the short video clip. But that's getting ahead of ourselves. Here's how we did it, first using Photoshop and Premiere, and second, using Painter.

E

The Digitized Source Footage

The five frames in *(A)* sample the dynamic action we chose—the full shot was two and a half seconds long. It was digitized by hooking up a Hi-8 deck to the computer, and using Premiere to turn it into a 320 x 240 (less than full-sized) QuickTime movie. The QuickTime clip, *(B)*, can be viewed frame by frame to study the action and begin thinking about what the rotoscoped animation version might look like.

We used After Effects to export the clip frame-by-frame as a series of sequentially numbered PICT files.

Approach I: Drawing the New Frame in Photoshop

One by one, each PICT file was opened up in Photoshop. Using the layer function, *(E)*, a blank "tracing" layer was placed on top of the source PICT and used to draw on. This is a lot like placing a sheet of tracing paper on an image you want to copy. We could have used any of the brushes, pencils and colors from Photoshop's tool palette—*(C)*, *(D)*, and *(E)*. *(F)* shows a black line being traced against Sam's image.

F

Once the drawing/tracing was completed on the new layer, the original PICT layer from the live-action footage was deleted, leaving only the newly sketched frame, *(G)*. The new image was saved with a new name and a corresponding frame number, i.e., Roto 00074.

G

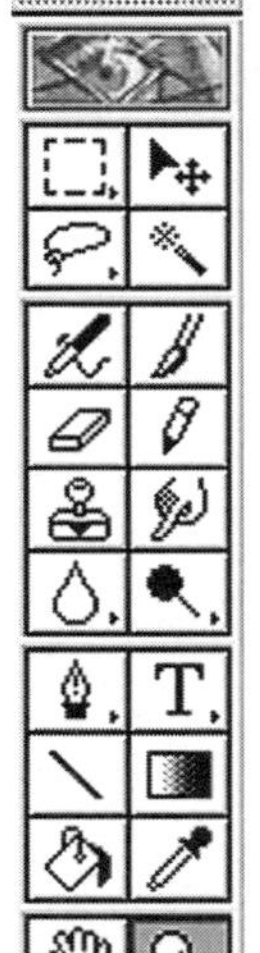

C

The same procedure was repeated for every frame *(H)*. Well, not quite. To save our time and efforts, we did the process for *every other* frame—following the age-old animation dictum of working "on twos." Once all the desired live-action frames had been traced and saved, we were ready to import all these sequential PICT files into Premiere.

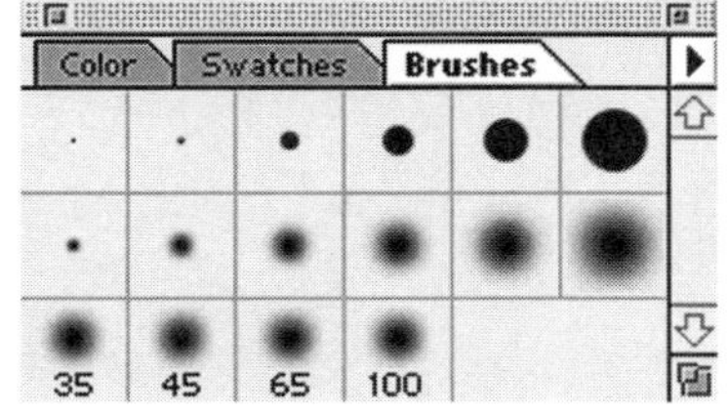

D

H

Construction Window

01:22 0:00:01:26 0:00:02:00 0:00:02:04 0:00:02:08 0:00:02:12

V I D E O

A T B S1

I

Folder: TAB/Rotoscope

56	Name	
	sam00066/roto Still Image Duration: 0:00:00:02	[1]
	sam00068/roto Still Image Duration: 0:00:00:02	[1]
	sam00070/roto Still Image Duration: 0:00:00:02	[1]
	sam00072/roto Still Image Duration: 0:00:00:02	[1]
	sam00074/roto Still Image Duration: 0:00:00:02	[1]
	sam00076/roto Still Image Duration: 0:00:00:02	[1]

J

Reassembling in Premiere

Premiere is a versatile video editing application that not only allows you to edit video and QuickTime clips, but also lets you assemble a series of still images together to make animation. All the PICT files were dragged (as a bunch) or "imported" into the Project Window in Premiere, *(J)*. A setting under Premiere's Preferences menu lets you determine the length of each imported still image. In this case, we chose a duration of two frames — because we had only traced every other frame but wanted the final rotoscope to have the same overall length as the original clip.

Next we dragged the PICT files from the Project window to the Construction Window, *(I)*. This process had to be done one PICT file at a time, kind of like splicing on new frames to a filmstrip one by one, or exposing a new frame if shooting on a camera.

Creating the Movie

Our end product, a QuickTime movie of Sam's rotoscoped cartwheel, was rendered by Premiere once all the PICT files had been brought into the construction window and placed in their proper order. Note that in order to create smaller file sizes, we compressed the frame size by rendering the clip at 320 x 240 screen resolution (half-size).

The end result, *(K)*, is a new QuickTime clip with drawn animation that mirrors the same motion and timing as the original live-action video footage.

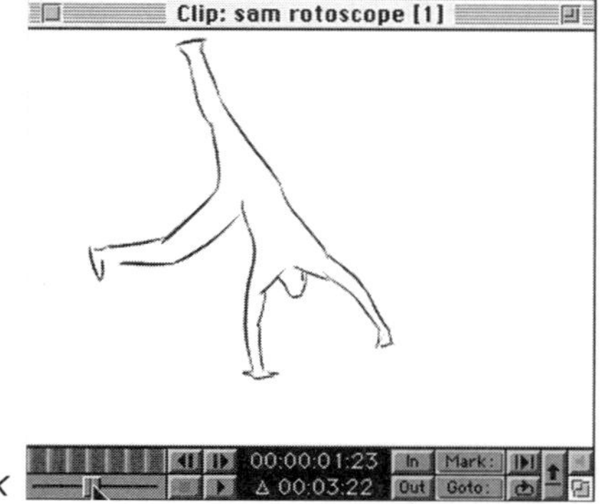

K

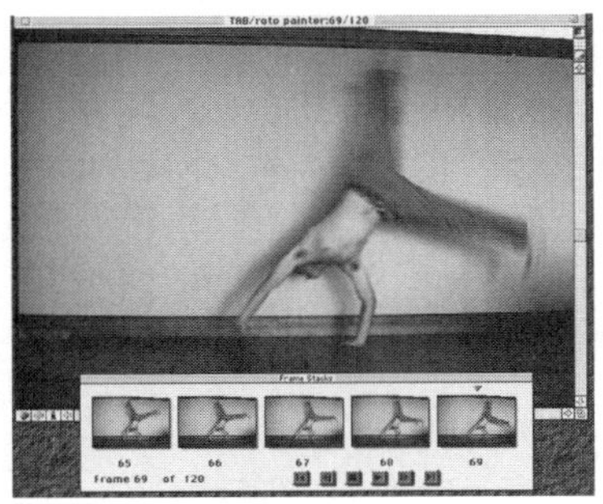
L

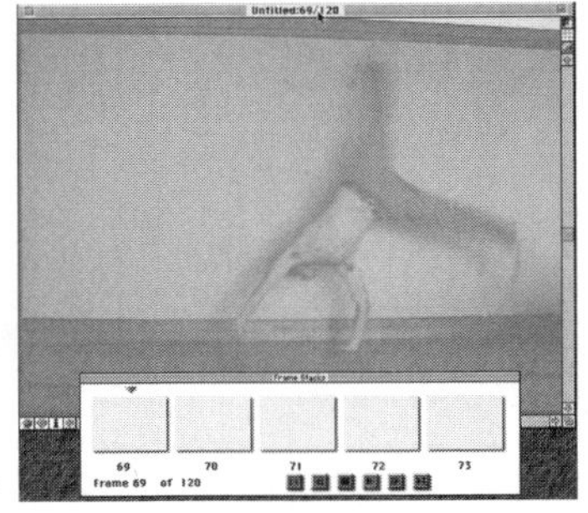
M

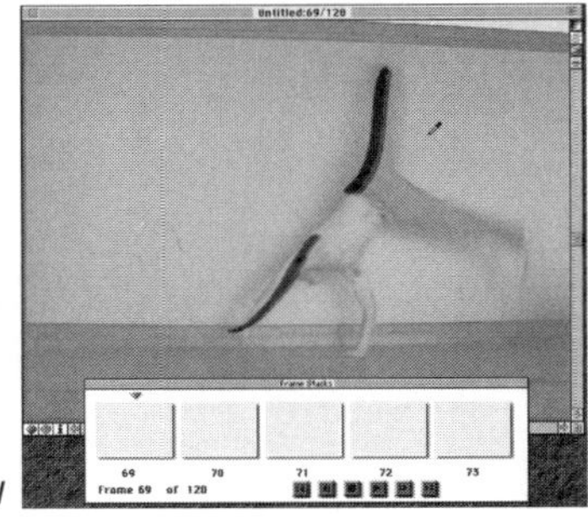
N

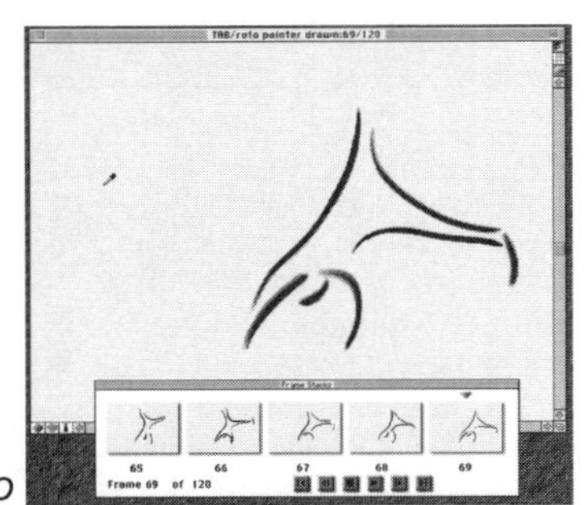
O

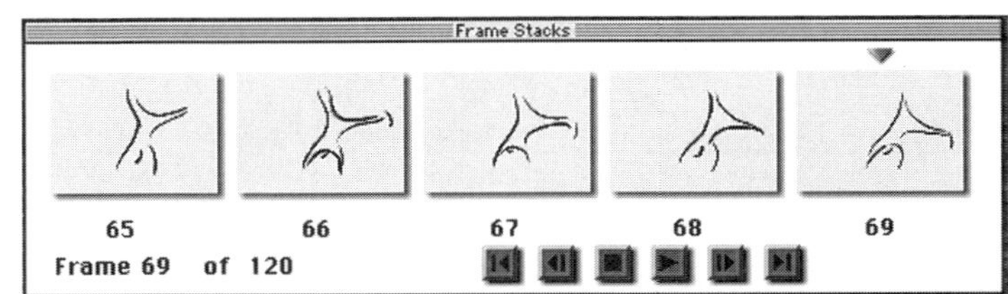

P

Approach II: Rotoscoping with Painter

Here's an alternative technique for rotoscoping, which may seem easier. To start, you import the QuickTime clip directly into Painter, the paint/draw application from Fractal Design. We opened our video clip just as we would any other file. Painter turned the clip into an image stack and calculated the number of frames that composed the file. We were able to view the footage frame-by-frame via the Frame Stack window, *(L).*

Painter will let you draw or scribble directly onto each frame using a terrific variety of traditional art tools, including brush strokes, marker lines, colored pencils, airbrush, pastels, and more. Or, with the tracing paper option, you can work on a transparent level that is placed over the original. Choose MOVIE instead of IMAGE in order to create a new file in Painter that is the same size as your original captured video, and has the same number of frames. As you can see, this function in Painter lets you see the frame you are working on plus the four frames that come before or after it.

We used the Movie menu and selected Set Movie Clone Source so that we could view both the source frame and the tracing layer. With the tracing layer set as Frame 1, Painter knew that we wanted to use the video frames as source and the tracing layer as a new clip where all the changes were saved. Once this step was completed, the original footage appeared as "ghost" images, *(M),* that were slightly faded but still suitable for tracing over.

We worked quickly in drawing a slightly abstract set of action lines over the figures of Sam *(N).* As you can see, we took a simple design path. But we could have taken advantage of Painter's ability to mimic traditional media effects, including smudges, smears, and splatters.

Before long we got to the final frame. The original video clip was removed and the image stack was Saved As, creating a new QuickTime clip, *(P).* Some weeks later, Kit imported the QuickTime clip into his After Dark screensaver program. Now Sam cartwheels for his dad many times a day.

14 ■ Line and Cel Animation

Traditional American animation is cel animation. The technique has taken its name from the transparent sheets of celluloid that bore the likes of *Snow White and the Seven Dwarfs, Cinderella, Bambi, Pinocchio, Peter Pan,* and the special universe of *Fantasia.*

It is no coincidence that the classic animated feature films are almost all the work of the great Disney studios. More than any other individual, Walt Disney guided the development and refinement of cel animation techniques: its recognizable cartoon style, its perfect synchronization of movement with music and voice (called Mickey Mousing), its assembly-line system of production, its division of labor and massive logistics management. The Disney features are the penultimate achievement of cel animation techniques.

American cartoon animation, or *character animation,* as it's now called, created not only a brand-new art and industry but also the first truly international audience for animated films. Cartoons have entertained viewers for almost three quarters of a century, and some contemporary psychologists might argue that the fantasy world of full-cell animation has reshaped the imaginations and the inner imagery of successive generations of children the world over.

Walt Disney is credited, and rightly, as being the father of classic American animation, but the body of work created with this technique was the business of entire studios of artists, filmmakers, businessmen, and technicians. While tributes are being handed out, here's just a beginning list of individual ani-

14.1 The look of line animation: Nine frame enlargements suggest the breadth of drawing style and subject matter that have been used in animating with registered sheets of paper. Note that the last three in this series are from a single film, Ryan Larkin's *Street Musique.*

Mildred Goldsholl, *Up Is Down. Courtesy Pyramid Films.*

Geoff Dunbar, *Lautrec. Courtesy Films, Inc.*

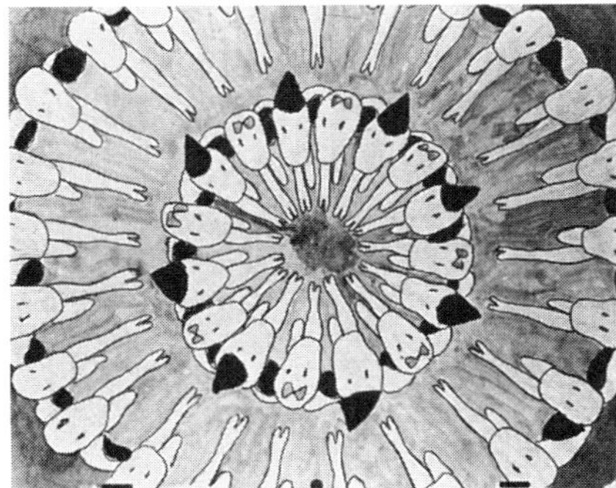

Sally Cruikshank, *Fun on Mars. Courtesy Serious Business Company.*

mators, those facile artists and storytellers and designers who contributed to the artistry and impact of full-cell animation: Walter Lantz, Oskar Fischinger, Winsor McCay, Max Fleischer, Paul Terry, Otto Messmer, John Hubley, Ub Iwerks, Art Babbitt, Bob Clampet, Tex Avery, Milt Kahl, Chuck Jones, Shamus Culhane, Frank Thomas, Tissa David, George Dunning, Grimm Natwick, Preston Blair, Ollie Johnson, and many, many more.

In "big-time" Disney-type animation every movement is conveyed through acetate cels that lie stacked on top of each other under the animation camera. Each acetate cel has been painted by hand. The outline of the character is applied to one side and the interior colors to the other side. The number of layers of cels is determined by the number of moving parts in any particular scene. For example, if a character is motionless on the screen for a few moments, except for his mouth, cel techniques allow the animator to use one cel to show the body and head, and another set of cels to animate the various positions of the moving mouth. A background scene, of course, would rest under the stack of clear cels. In a nutshell, that's the cel technique.

OVERVIEW

The step from any other technique to full-cel animation is big. It's enormous, in fact. Although I don't want to scare you off from this most complicated of animation techniques, I do think it is important to point out that you are now entering a rugged domain. It's an area filled with as many new concepts, tools, and techniques as have been introduced in all the preceding chapters. The road beginning here is most challenging. The journey is not a short one. To get to the highest levels it's necessary to begin a lot further back. In this case, you'll be starting with ordinary sheets of paper instead of acetate cels and with simple pencil lines instead of intricately drawn and painted artwork.

This chapter is complicated enough to require its own overview. It begins with an important discussion of the surface area on which animated characters are created. Next there is discussion of how to control movement and timing within a series of drawings. The problem of animating a bouncing ball is given here. It's a classic assignment and an important building block for technical refinements that follow. Central topics rele-

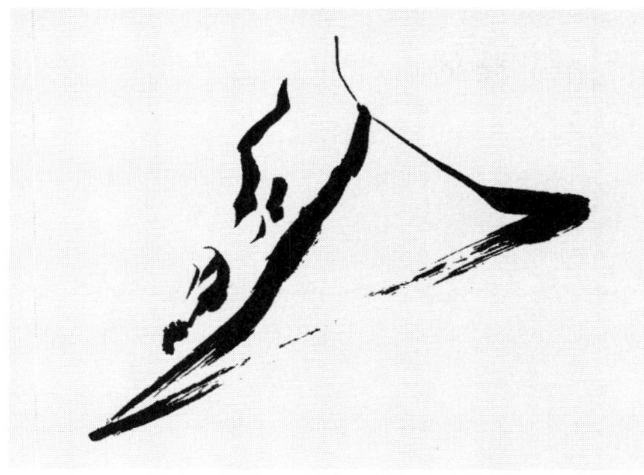

Marcel Jankovics. *Sisyphus. Courtesy Pyramid Films.*

J. P. and Lillian Somersaulter. *The Light Fantastic Picture Show. Courtesy Films, Inc.*

Grant Munro and Ron Tunis. *The Animal Movie. Courtesy National Film Board of Canada.*

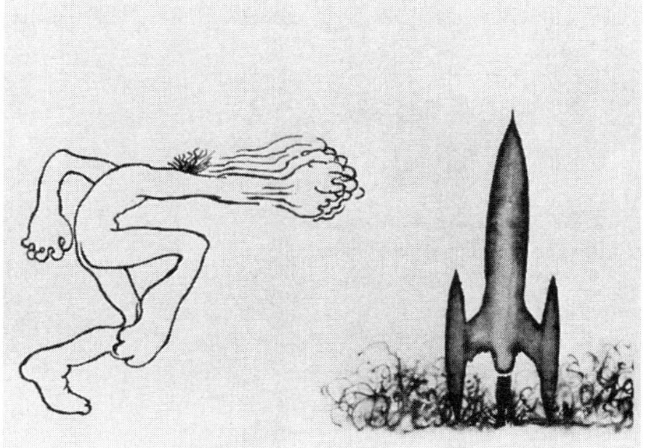

Ryan Larkin. *Street Musique. Courtesy Learning Corporation of America.*

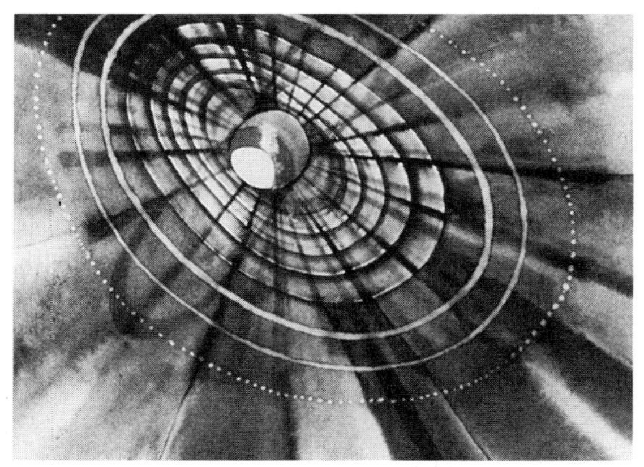

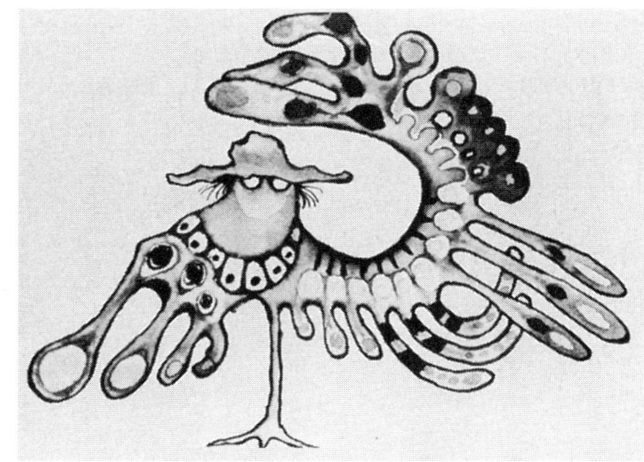

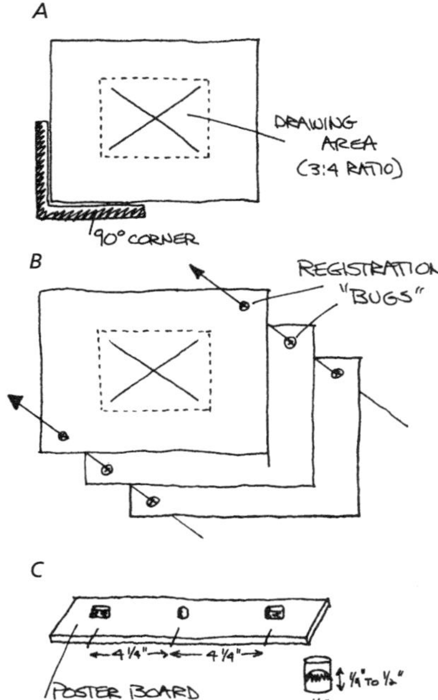

14.2 Registration devices: A wooden molding, a 90-degree metal rule, or a similar structure is fastened to a wood or Plexiglas drawing surface, *(A)*. You can select any corner you prefer, but note that the area used for drawing should not come too close to the edges of the paper sheets you use. In *(B)*, a simple pair of "bugs" is traced on successive sheets. These marks must be aligned in filming as well as drawing. Using a standard two-hole or three-hole punch (or a variation of your own), a sturdy registration peg system can be made with dowels and papers punched to match it. Measurements in *(C)* conform to the three-hole standard punch.

vant to the aesthetics of animation follow. The terms of the discussion are neither vague nor abstract but perfectly concrete. You will design a character and then make it walk across the movie screen.

If you can get this far, you're practically home. Creating your first "walk" is probably the most difficult animation problem you will ever face. Once the technique is mastered, you'll have reached a wonderful plateau in which the refinements come easily and the parameters of your growth become wider than ever.

LINE ANIMATION

Line animation is the technique of using registered sheets of paper with simple artwork to create a series of drawings that are then photographed by the animation camera. Line animation has only recently been recognized as a distinct and identifiable technique in its own right. An older cousin to this kind of animation is the *pencil test,* a familiar term around classic character animation studios.

The traditional pencil tests of the great animation studios were simply a stage in the development of full-cel sequences, whereby the initial drawings of an animator were quickly photographed right off the paper sheets used in developing such preliminary sketches. The pencil test was, literally, a test, a way of checking the flow and style of a series of drawings before they were laboriously (and expensively) transferred to transparent acetate sheets.

Today's independent animators often go no further than this paper stage of animation. In so doing, they design their movies in ways that make it possible to redraw the entire image (including background) on each subsequent sheet of paper. In the *line-animation* technique, the paper drawings aren't "rough" drafts to the final film. They *are* the finished drawings—fully articulated movements through finely rendered artwork in watercolors, charcoal, felt-tipped pens, or other graphic media. As we'll see, today's independent animators have made other modifications in paper-cel techniques that give their drawings the maximum impact with a minimum duplication of drawings.

The frame enlargements in Figure 14.1 and elsewhere in

these pages should clearly indicate that enough significant work has been done with simple paper sheets to justify classifying line animation as a distinct technique.

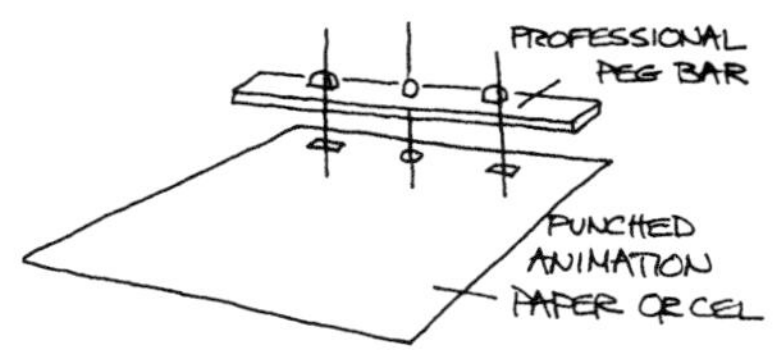

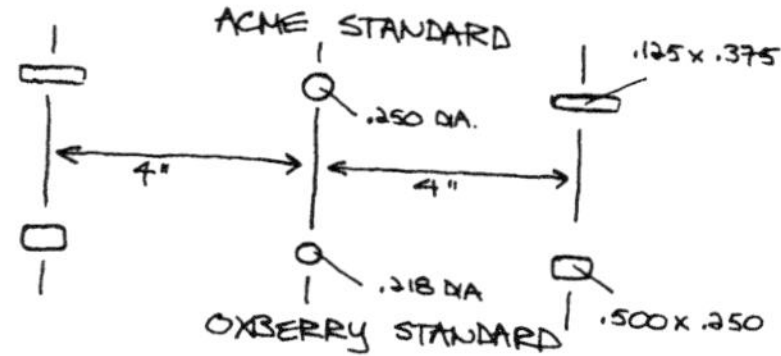

14.3 Professional peg dimensions

REGISTRATION

A first requirement of line and cel animation is to keep all drawings in accurate alignment so that the projected image won't be jerky. A *registration system* is a way of establishing that the individual sheets of paper on which the drawings are made can be easily and precisely aligned. Here are four possibilities.

Ninety-degree Corners. It is possible to use the squared corners of paper sheets or index cards in developing an accurate registration system. The animator simply lines up the edges of the sheets, which should, of course, have the same dimensions. A corner of a wood or metal picture frame can be used to make a bracket to assist in the alignment. If you are using corner registration, work with the thickest paper stock available, index cards, or heavyweight bond.

Registration Bugs. When using a light table or tracing paper, individual pages can be lined up visually by superimposing a "bug" drawn at opposite corners of each sheet of paper. This kind of registration system is recommended only for those situations in which the artwork is particularly large, unwieldy, or cannot accept a punched registration system.

Round Punch. The standard two- or three-ring punch that is used for school loose-leaf notebooks can be adapted to provide a serviceable and inexpensive registration system. A *peg bar* must be made that holds the sheets during the drawing and filming of paper cels. A simple way to do this is to glue ¼-inch-diameter wooden dowels, approximately ¼ inch long, to a piece of heavy cardboard (Figure 14.2). Sand the top edges so that the pegs won't rip the paper sheets.

Punch the outside holes at a distance of 4¼ inches from the center hole. In this way you can use standard three-ring notebook papers, and you can always have your sheets punched to commercial specifications should you decide to shoot the drawings under a camera that accepts only professional registration peg systems.

Professional Peg Systems. Different standards have

14.4 Standard field size and peg bar arrangement

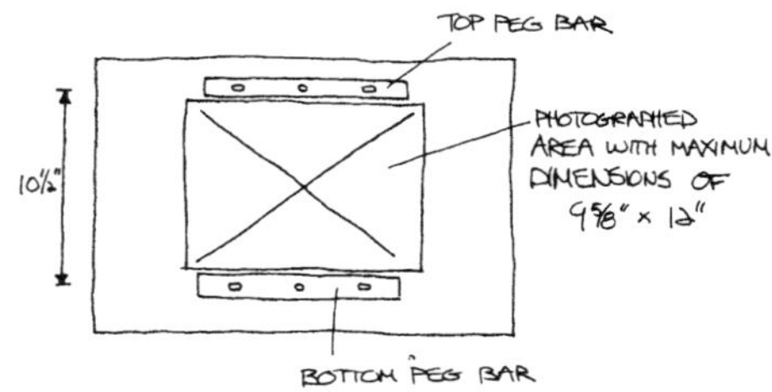

evolved for the placement and specifications of professional peg systems. The standard in widest use seems to be the Oxberry System, named after the firm that manufactures much of the professional-caliber animation hardware. Specifications of the Oxberry pegs are to be found in Figure 14.3. There is another peg system that is used commercially today, the Acme system. Its dimensions are also given.

While these professional systems are accurate, they are also expensive. Metal sheets with a set of mounted metal pegs cost about $35 each. Paper has to be specially punched. Chapter 21 will give you more information about registration equipment and costs.

FIELD SIZE

While there's no universally accepted standard for registration, there is one for determining the area that lies below the camera. Figure 14.4 shows how the total field area and the pegs are placed on the compound bed of an animation stand. The dimensions cited are standard for the industry. This means that if you have your paper cels photographed on a professional stand, the camera operator will always be able to line up your artwork in the same way you've designed it.

In Chapter 25 you'll find an accurately reproduced *field guide*. A *field* is nothing more than the standard rectangular frame that the camera "sees." As you'll remember, there is a common aspect ratio for the film image. It is roughly 4:3, width to height, unless a special "wide-screen" format such as CinemaScope is being used. Camera fields are sized in common animation parlance with a simple number. A #8 field, for example, is exactly 8 inches wide. A #5 field measures 5 inches in width. The function of the field guide is thus to specify the area that the camera will be set up to photograph. A greatly reduced model of the standard field guide is shown in Figure 14.5. You may want to detach the guide in Chapter 25 and adapt it to whatever registration system you're using. This tool will be immediately helpful in the projects to follow.

There is a standard relationship of the field (and peg bars) to the animation camera. The center of the field guide is matched to the center of the camera's frame. This means that all directions for movement from the central position can be indicated by N (north), S (south), E (east), and W (west).

More now about the field guide itself. Its measurements are based on width increments of 1 inch. The smallest-size field is called a #1 field and it is 1 inch wide by slightly less than ¾ of an inch high. A #12 field is the largest photographic area that is normally used. Its dimensions are 12 inches wide by a little less than 10¾ inches high. Here are a few important things to know about the standard field guide.

Aspect Ratio. Regardless of field size, the dimensions of the frame must always assume the ratio of 4 to 3 (width to height). This is roughly equivalent to the standard format used

14.5 Standard field guide: In Chapter 25 there is a full-scale section of the standard field guide that you can detach and use in your own work. The circle beneath these coordinates: N-2/E-3. The heavy line shows a #4 field (4 inches wide).

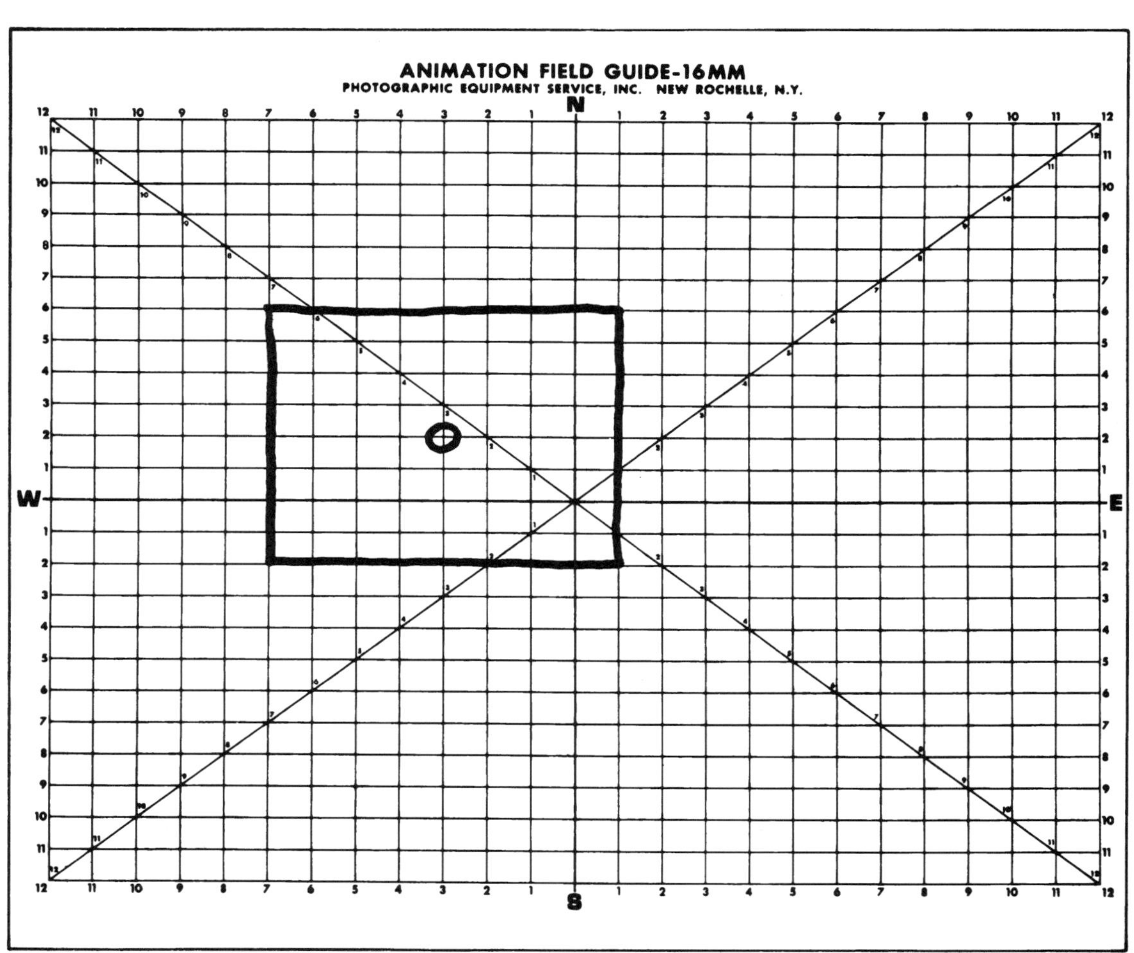

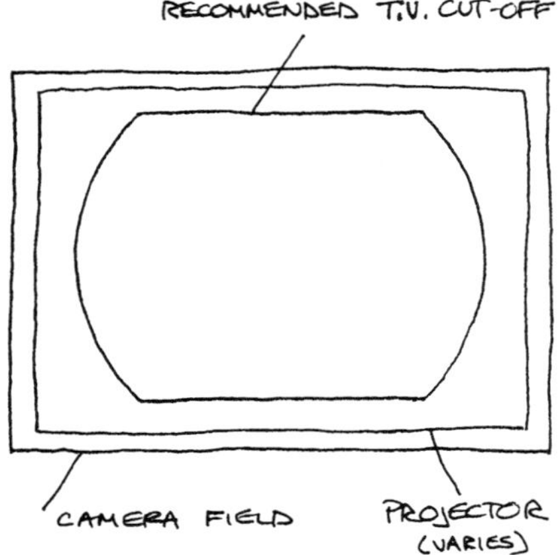

14.6 Projection and TV cutoff: The degree to which the edges of the field are masked during film projection will vary with the specific projector being used. Similarly, television sets differ in how much of the full picture they cut off. This illustration suggests rough limits for both projection and TV cutoff. Special field guides are available to indicate the recommended "TV safe" area for various field sizes.

in super 8mm, 16mm, and 35mm filmmaking, except for Cinemascope and other "wide-screen" configurations. The actual standard aspect ratio is 1.33 to 1.

Recommended Field. Typewriter paper, tracing tablets, quality drawing papers—almost every imaginable type and quality of paper stock—can be purchased in the common 8½- by 11-inch format. So often do animators use this size paper for their work that a practical field size has evolved that you may want to incorporate in your work. It's a #10 field, slightly smaller than an 8½- by 11-inch sheet, used horizontally.

Coordinates. When the standard field guide is *zeroed out,* that is, when the center of the camera frame coincides with the center of the field guide, it's possible to specify any point within the field by stating the numerical value of its coordinates. This allows animators to specify precisely and universally the position of any point under the camera. For example, in Figure 14.5 the tiny circle would be indicated by N-2/E-3. If you wanted to begin a sequence with a small field located off center, you first mention the coordinates of the field's center and then the size of the field itself. In our example, N-2/E-3 is a #4 field.

This standardized system of coordinates permits an animator to provide very exact instructions for complicated movements. For example, the following would be immediately understood by an animation cameraperson: Initial position: #8 field 0-0. Final position #4 field N-2/E-3. Track and zoom in 40 frames.

Edges. Here is a practical note that is based on painful experience. Although super 8mm and 16mm reflex cameras are *supposed* to show precisely what the camera is seeing when you look through the viewfinder, often there is a slight discrepancy. For most forms of filmmaking this error is insignificant. But in animation it matters very much. The precise location of frame lines is often critical. So be cautious. Always leave a little extra room between the edges of your drawing and the edges of the paper sheets you are using.

The Safe Area. When a film is broadcast over television, part of the picture area is always lost. As a matter of fact, some of the picture area is normally lost in 16mm or super 8mm projection. In any event, if you think your film might end up on television, you should design it so that the action takes place within the *safe area.*

THE BOUNCING BALL

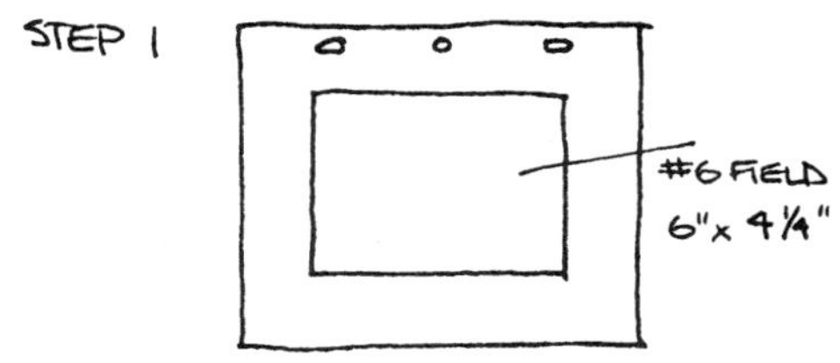

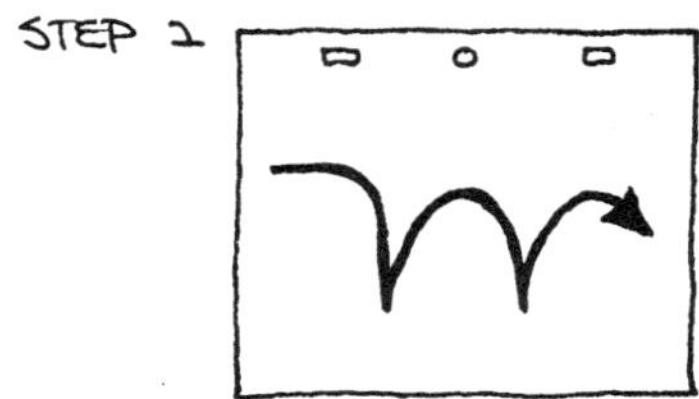

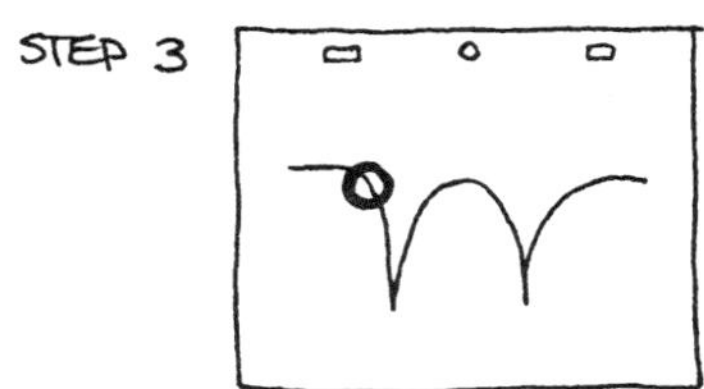

14.7 The bouncing ball

It's easy to despair in the face of field guide coordinates, registration pegs, aspect ratios, and safe areas. In fact, there's probably something wrong with you if you're not inundated with details. However, all this information can be effectively assimilated if you put the information into real operation. Complete the following project. The challenge is to use paper sheets to animate a bouncing ball. This is one of the two classic exercises in animation. The other is the walk. As best you can, try to follow these instructions. You'll make mistakes—anticipate that at the outset. But you'll be familiarizing yourself with the basic process. And that will help you to understand the more detailed explanations and refinements yet to come.

Step 1: Field. Using a field guide and whatever registration system you have (the edges of the sheets will do), draw the outside perimeters of a #6 field. Center the field so that the coordinates are 0-0.

Step 2: Path. Using either tracing paper or a light table to help you see through more than one piece of paper, place a fresh sheet over the one showing your field and draw a path a bouncing ball might take if it were to enter the frame from the left and then hit the ground twice as it moves across the field and disappears off the right side of the frame.

Step 3: Ball Size. Select a size for the ball you will animate. Don't make it too small or too large—something in the area of 1 inch in diameter ought to work well in this project.

Step 4: Timing. In this exercise you should try to pace the drawings so that the ball will hit the ground at intervals of exactly 1 second. You will be shooting on twos, so twelve drawings are required to make 1 second of screen time.

Step 5: Draw It. Without further explanation, try the problem of the bouncing ball. Use the path diagram as a guide and use a different sheet of paper for each new position. Try to have twelve drawings between each bounce of the ball. Remember, there should be two bounces as the ball moves across the screen. Number each sheet of paper in the upper-right-hand corner. Attacking this problem will require about thirty-six pieces of paper and several minutes of your life, but it will prepare you for the discussions to follow and for all the fine points of animating with line drawings and finally with plastic cels.

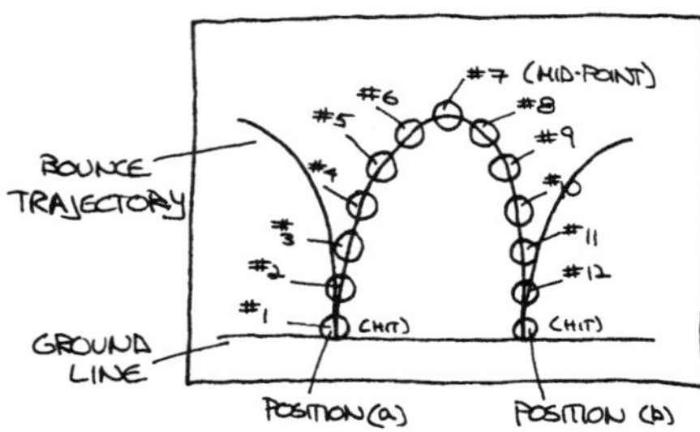

14.8 A mathematical bounce: A 1-second duration means there are 24 frames between each contact of the bouncing ball with the ground. Shooting on twos would thus require the animator to produce 12 drawings that start with position A and end *just before* position B. The ball hits only once per second and here we start each second with that hit.

Step 6: Preview. After you've completed the problem, hold the stack of drawings in one hand and flip through the pile with the other hand. This will let you preview your work.

CONTROLLING MOVEMENT AND TIME

It is, of course, the amount of change in position between one drawing and the next that creates the illusion of movement, but how can you tell how much to change each drawing from the ones preceding and following it? How is it possible to plan ahead so things end up where you want them and at the right

ANIMATION TIMING CHART

FILM FORMAT		SUPER 8MM				16 MM		35 MM	
PROJECTION SPEED (FRAMES PER SEC.)		18		24		24		24	
RUNNING TIME AND FILM LENGTH		FEET + FRAMES		FEET + FRAMES		FEET + FRAMES		FEET + FRAMES	
SECONDS	1	0	18	0	24	0	24	1	8
	2	0	36	0	48	1	8	3	0
	3	0	54	1	0	1	32	4	8
	4	1	0	1	24	2	16	6	0
	5	1	18	1	48	3	0	7	8
	6	1	36	2	0	3	24	9	0
	7	1	54	2	24	4	8	10	8
	8	2	0	2	48	4	32	12	0
	9	2	18	3	0	5	16	13	8
	10	2	36	3	24	6	0	15	0
	20	5	0	6	48	12	0	30	0
	30	7	36	10	0	18	0	45	0
	40	10	0	13	24	24	0	60	0
	50	12	36	16	48	30	0	75	0
MINUTES	1	15	0	20	0	36	0	90	0
	2	30	0	40	0	72	0	180	0
	3	45	0	60	0	108	0	270	0
	4	60	0	80	0	144	0	360	0
	5	75	0	100	0	180	0	450	0
	6	90	0	120	0	216	0	540	0
	7	105	0	140	0	252	0	630	0
	8	120	0	160	0	288	0	720	0
	9	135	0	180	0	324	0	810	0
	10	150	0	200	0	360	0	900	0

SUPER 8MM – 72 FRAMES PER FOOT
16 MM – 40 FRAMES PER FOOT
35 MM – 16 FRAMES PER FOOT

14.9 Animation timing chart

moment as well? Is there a way to speed up the process of drawing? And how can you get a real feeling of character or personality into a particular movement?

If we were to undertake a structural analysis of the bouncing-ball problem, we would divide the distance to be covered by the time required for a particular movement. If you looked at a single, complete bounce of the ball, you would have a series of drawings that were symmetrically placed. The plotting of the balls in Figure 14.8 has been determined through straight arithmetic.

In real life, a ball accelerates as it approaches the ground and decelerates as it reaches the top of its bounce. This change of speed has to be designed into the animation. More than that, this nuance in the reality of a bouncing ball must be accentuated. There is a way to help decide where to place each drawing of the ball. The key lies in the analysis of the time for any given movement.

By now it should almost be second nature for you to change time in seconds into time measured in drawings. The process becomes almost unconscious once you've done it a few times. The camera shoots on twos. Each drawing will become 2 frames of film. Therefore twelve drawings are needed for 1 second of finished film.

To always know the number of frames comprising a given amount of "screen time," mount a *time chart* close to the work area where you draw and shoot your animation. Such a chart is shown in Figure 14.9. If you'd like, you can cut this out of the book, put it into an ornate rococo frame, and hang it over your drawing table.

Back to the bouncing ball. The given time in the problem is 1 second between bounces. The second-to-drawing computation is easy; shooting on twos, you need twelve individual drawings to fill the 24 frames of 1 second. Figure 14.10 shows the same bounce trajectory as used in Figure 14.8. Note, however, a different selection of positions for the twelve individual balls. This new spacing plan takes into account the unique way in which a ball really bounces: slower at the top of its arc, faster as it approaches the ground. Look at the ball's positions on the curve. Although these points are symmetrically placed on either side of the bounce (after all, things go up and down with the same flow), the positions of the six ascending and the six descending locations are asym-

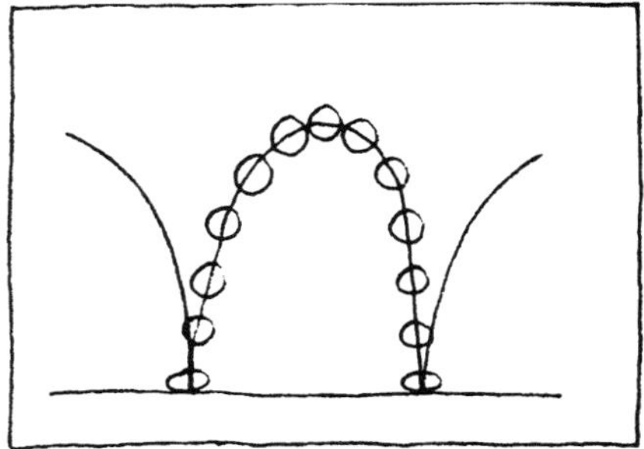

14.10 An exaggerated bounce: Field, trajectory, and ball size are identical with those in Figure 14.8. Note, however, the different placement of individual positions on the arc. This bounce will look and feel "right," but even greater exaggeration would read correctly when filmed.

metrically arranged. This arrangement will accommodate the varying speed of the ball.

Study Figure 14.10 carefully. The slower you want the ball to appear to move, the closer must be the positions of consecutive drawings. Remember that in filming and then in projecting, these closer images will show less change over the same period of time than would wider-spaced images. Speed is always relative. To make the movement appear faster, the positions are spaced farther apart.

The series of positions in the arc actually exaggerates the positions of a real bouncing ball. In Figure 14.11 you'll see frame enlargements of a bouncing tennis ball from an actual film taken at live-action speed. Compare the difference in spacing in the real version with the animated version.

The spectacular squashing of the ball as it bounces summarizes the difference between an animated movement and the real thing. As gross as the squash may appear in the spacing guide (Figure 14.10), its effect will "feel" right in the finished movie. Exaggeration is the single most important quality in giving "personality" or "character" to an animated movement.

Here's another fine point that you can study through experimentation. Movement is always perceived relative to the size of the objects and background, as well as to the speed of other objects. Different-size balls will seem to move at different speeds. Smaller balls look faster than larger ones even if the different balls are traveling the same trajectory for the same period of time. A similar effect exists with field sizes. The same set of drawings will be perceived differently according to the composition and size of the field in which they are placed.

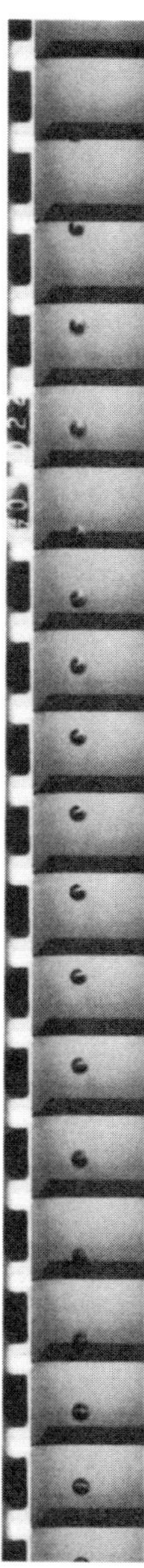

14.11 A live-action bounce: This enlargement of a 16mm film clip shows the positions of an actual rubber ball as it bounces in real time. In this example, the ball appears to be hitting at roughly a half-second interval—each 10½ frames.

EXTREMES AND IN-BETWEENS

If you have tried to animate the bouncing ball, you have probably found yourself beginning with ball #1 and then working steadily through the remaining drawings in numerical order. Somewhere in the process, I suspect, there may have come a moment when you realized that you must plan ahead so that things would come out right. In this case, the thirteenth drawing would show the ball hitting the ground as it initiates the second bounce. Getting to the right position at the right time is a universal problem in animation. With those techniques that demand precise timing, planning ahead is absolutely essential. The best way to animate fluid movements that must arrive where you want them when you want them is to employ a process called *extremes and in-betweens.* The process generally begins not with one of the actual drawings but with a schematic sketch that outlines the course of a movement and the relative positioning of the entire set of drawings. This is called a *spacing guide.* A sample has been introduced already in Figure 14.10.

There is a second notation/planning device that animators often use before they start drawing. It's called a *break-down count,* and it gives the animator an order in which to do the individual drawings. In the bouncing-ball exercise, the ball was to hit the ground at the beginning of each second. This is called a *24 beat,* one hit every 24 frames. As we've seen, the sequence requires twelve drawings. The animator begins by writing out the numbers of all the drawings required in the first beat (Figure 14.12).

The numbers representing the beats are circled. In our example, #1 and #13 represent the start of the first and second beats. These become a first set of *extremes* and should be drawn first. Next, the midway points between these extremes are located and drawn. In the example in Figure 14.12 this is #7. The midway points between the new extremes—#4 and #10—are drawn next.

Whatever the length or speed or quality of a movement, the extreme positions should always be drawn first. Because the light table enables the animator to see through at least three sheets of paper at the same time, extremes become very concrete guides to where to place and how to draw the remaining images—called *in-betweens.*

Only after the extremes have been drawn does the animator finish up the series by doing the remaining drawings. In our exercise, the in-betweens are numbers 2, 3, 5, 6, 8, 9, 11, and 12. While independent animators generally draw these themselves, in large studios the chief animator often does just the extremes and an assistant animator finishes off a particular movement or scene by drawing all the in-betweens.

Project: Bouncing Ball Revisited. To assimilate all this information and master the awkward process it requires, you should undertake another bouncing-ball problem. Use a #8

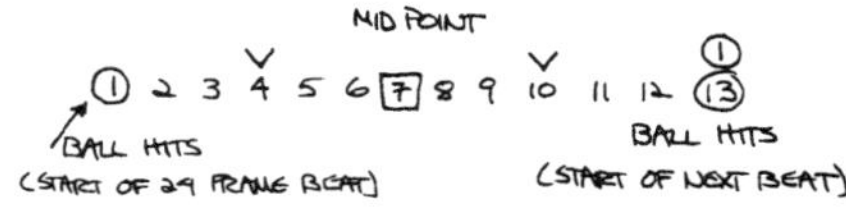

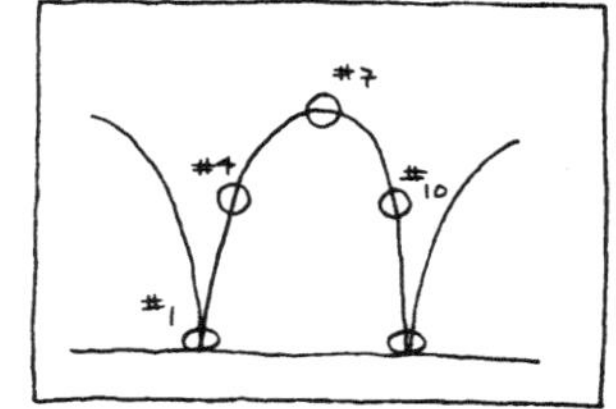

14.12 Breaking down the beat

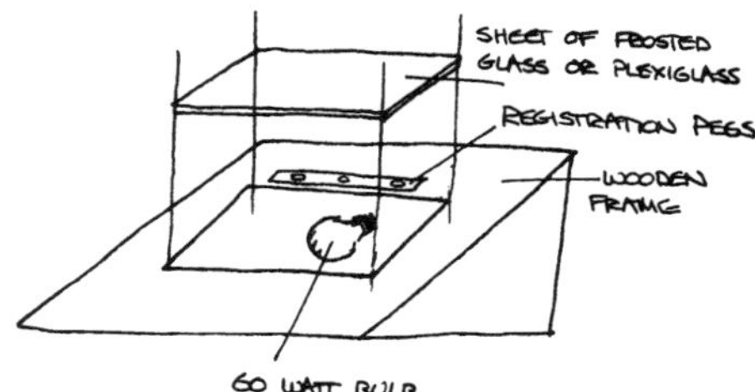

14.13 A rear-lighted drawing table: The sketch suggests the features of an animator's drawing table. The raised angle makes it easier to draw. Light tables can be easily jerry-rigged. For example, a sheet of glass with tracing paper on its underside can be suspended between two piles of books, allowing a desk lamp to light the surface of the glass from underneath.

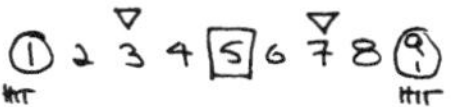

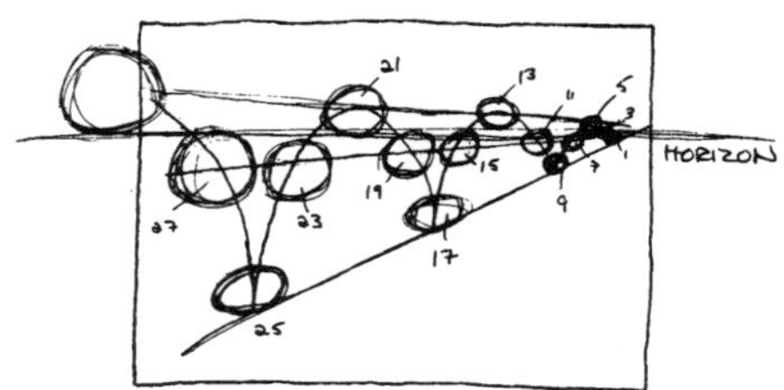

14.14 A perspective bounce: This sketch suggests the loose working style that many animators use in roughing out a spacing guide and in breaking down a given time-distance problem.

A

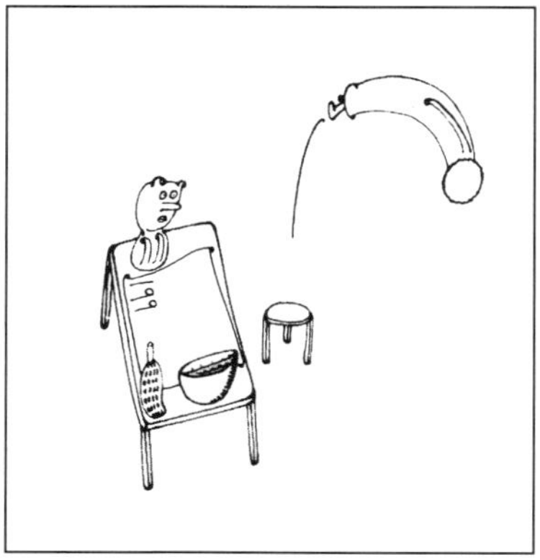

B

C

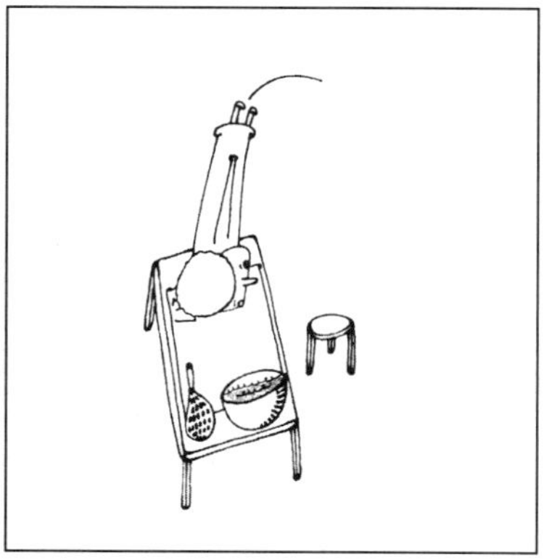

G

H

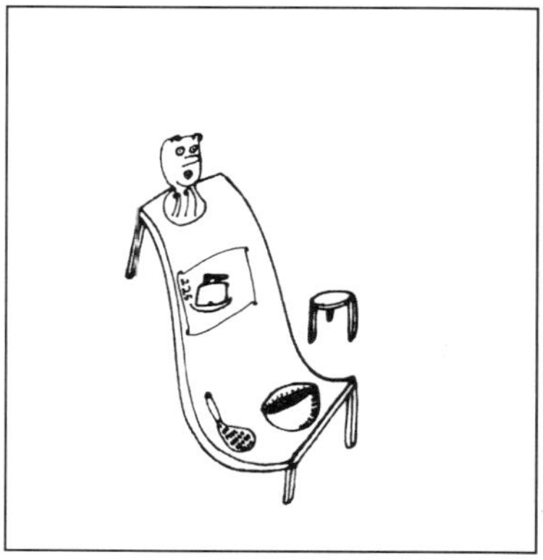

I

14.15 Kathy Rose's rules: Lest the process of extremes and in-betweens or the values of motivation, anticipation, and exaggeration become too strictly interpreted, study this sequence of key frames from *The Doodles* by Kathy Rose, *(A)* through *(L)*. Ms. Rose breaks all the rules. Some 6,000 drawings comprising the movie were all drawn upside down on paper. Voice, sound effects, and music were added after the filming. In all her films, Kathy Rose does the animation "straight ahead," very rarely using extremes or in-betweening. Most of the work is done using a basic storyboard, with a great deal of improvisation. *Photos courtesy Kathy Rose and Serious Business Company.*

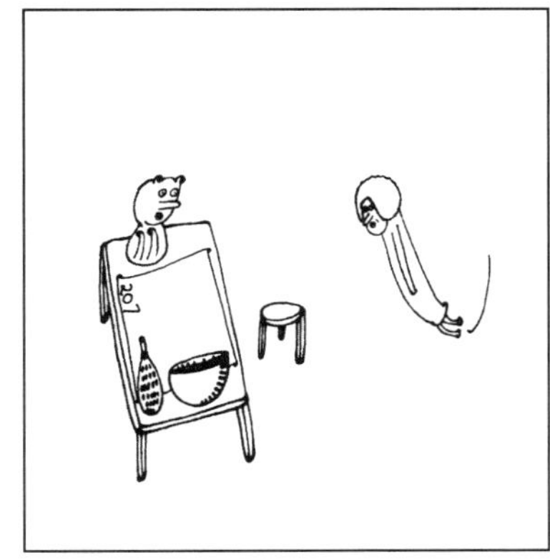

D

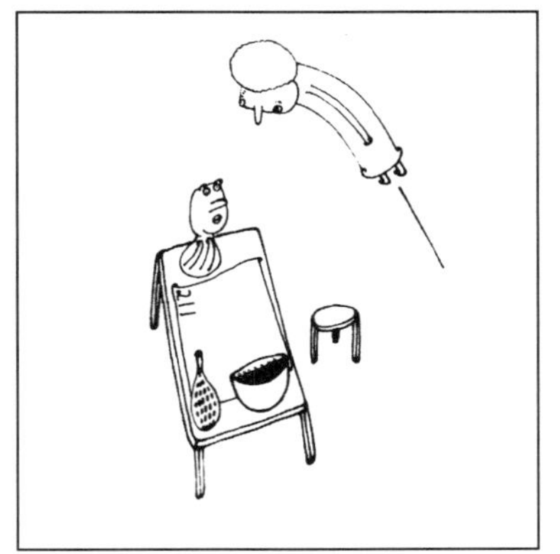

E

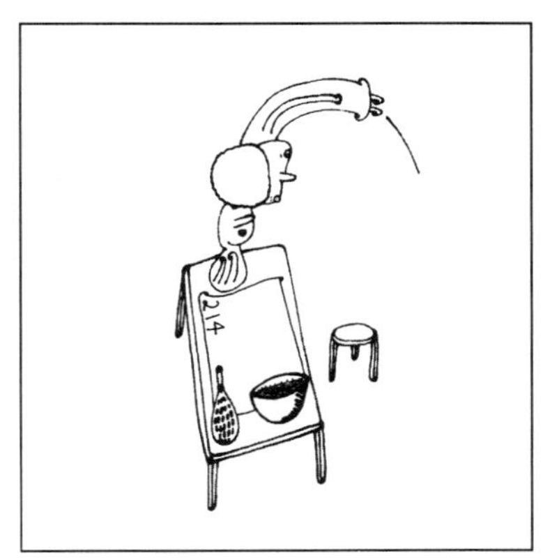

F

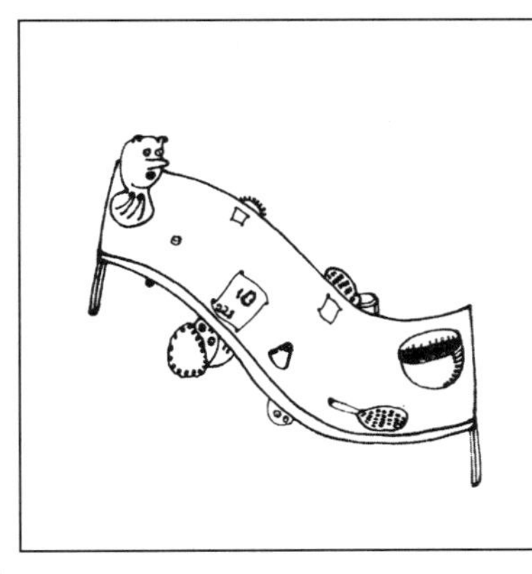

J

K

L

field, have the ball bounce on a "beat" of every 20 frames, and use a very small-size ball, about half an inch in diameter.

First make a spacing guide and select a trajectory within the field. Next, make a breakdown count in order to determine the order for drawing extremes and in-betweens. After you've completed the animation, film the registered sheets of paper, shooting on twos.

If you feel up to it, try this more difficult bounce problem. Select a large field and have the ball hit three times as it moves from deep in the distance toward the foreground. Use a beat of 16 frames. Have the ball begin small and end up very large in size. In fact, have it fill the screen in the last drawing. Film this shooting on twos. Figure 14.14 provides a rough planning chart for this more difficult problem.

14.16 Animating a walk

Step one: Character design: Easily drawn circular shapes are used as the foundation of the character. All nonessential details should be eliminated. The rear leg is shaded to help distinguish right and left legs.

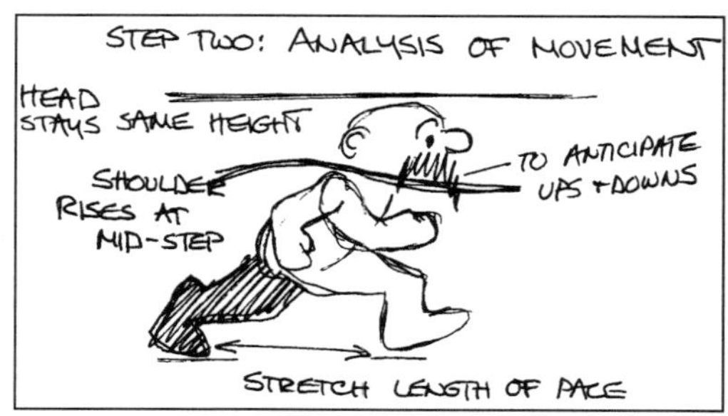

Step two: Analysis of movement: Spend time studying how different kinds and degrees of movement will best express the distinctive personality you wish your walking character to exhibit. This is a critical step.

ANTICIPATION, EXAGGERATION, AND MOTIVATION

In reading all this, and in trying it, you may begin to feel that the technique of animating with registered sheets has become its own end; that figuring out and then executing a series of drawings is the highest goal and most prized competency an animator can attain. But in animated filmmaking, as in all other art forms, technique is meant to serve the primary goal of expression. How something is realized is never more important than what is being realized. Execution should never be of more importance than content.

As you are laboring with your spacing guides and extremes, try to keep part of your mind focused on the following kinds of questions. Is this movement right for the character? Does it have the same qualities as real movement? Does the movement reveal subtle qualities about the character or about the story? In the final analysis, does the movement "move" the mind and the feelings of the viewer?

Learning how to block out and animate a bouncing ball is only a means in that by mastering the process you'll be able to bring to life a drawing of your own design and make it move in ways that are unique and fitting to it.

Anticipation and *exaggeration* have become special concepts in character animation. Wherever and whenever possible, the animator tries to give a character a distinct physical movement in anticipation of what is to be a major movement. If Porky Pig is to walk toward screen left, that

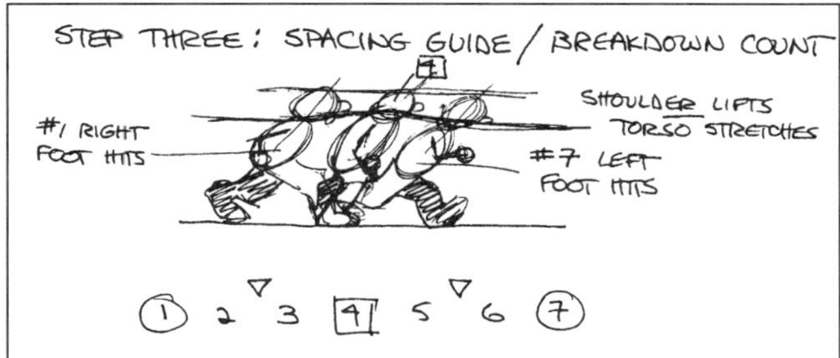

Step three: Spacing guide/breakdown count: A foot is to hit the ground each half second or every 12 frames. This happens to be the timing of a "normal" walking step.

Step four: Extremes: In animating a walk, it is necessary to simultaneously see through at least three sheets of paper. Common tracing paper will allow this, or you can use a simple light table like the one shown in Figure 14.13. In drawing extremes, remember that if the character has weight, the position of whichever foot is on the ground will remain fixed throughout an entire step (in this case, for six drawings).

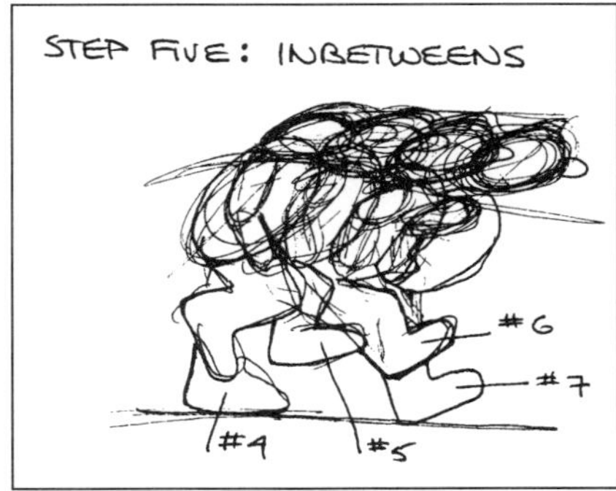

Step five: In-betweens: This ought to be an easy step. A registration system is required, of course, in steps four and five as well as in the eventual filming of the drawings.

movement is first prefaced and anticipated by a small movement toward screen right. You've probably noticed how cartoon characters always seem to pause miraculously in midair before they begin their fall after walking over the edge of a cliff. This pause is "anticipation." And then the fall itself is often exaggerated. Wile E. Coyote chasing the roadrunner seems to fall off cliffs that are at least five miles high. And the impact of the fall causes the poor coyote to disappear into a deep hole. Exaggeration again.

So central is the use of extreme anticipation and exaggeration in classic American cartoon animation that the matched concepts have taken on more informal names. When talking about the aesthetics of their craft, animators will speak generically of *stretch* and *squash.* The terms are synonymous with *anticipation* and *exaggeration.* The bouncing-ball problem should have introduced both of these quite clearly.

One of the animator's most creative acts is the study of real movement. Hours must be spent watching how a friend walks, studying the mannerisms of animals, peering with concentration at one's image in a mirror. Yet understanding how and why things move as they do is only part of the creative process. If a movement is to work, it must be properly *motivated.* The animator has to convince the audience that the only

Step six: Cleanups: Depending upon how roughly you work, it may be important to redo each drawing quickly so that the clarity and simplicity of the original design (step one) is maintained consistently throughout the walk. This is also a good time to check the details of movement and characterization that you've designed in step two (anticipation, exaggeration, motivation, stretch, and squash).

14.17 Spacing guide for walk project

possible kind of movement for a given situation is precisely that movement the animator has created. Developing motivation for a particular sequence and movement becomes another central aesthetic concern.

THE WALK

Needless to say, a ball is a lot easier to draw than even the simplest cartoon character. The classic bouncing-ball project is often followed by another classic, the walk. Animating a walking character for the first time just has to be one of the most difficult and challenging problems you'll ever set for yourself.

The illustrations in Figure 14.16 clarify some basic principles about animating a walk. Note that the process of planning the drawing is exactly the same as the process used in executing the bouncing-ball problem.

Project: Your Very Own First Walk. This project assigns you the same specifications as the sample walk just presented: a #10 field size, a 12-frame beat, and a figure of approximately the same height and width of gait. But this time it's *your* character, a creation of your own mind that will move with the style you draw into it. The degree of exaggeration and anticipation that you select for your walking person is for you alone to decide. You can, if you want, even have the character look a little like yourself.

Here's a simple narrative framework for this walk sequence. Your character enters the screen from the left and

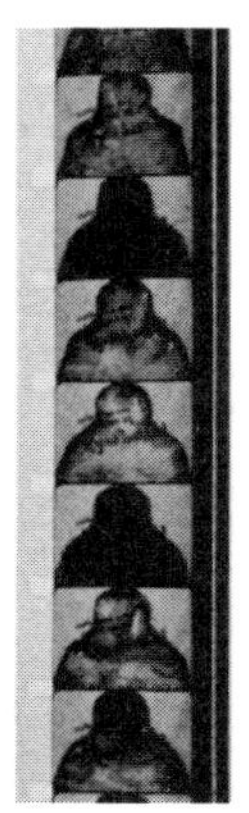
E

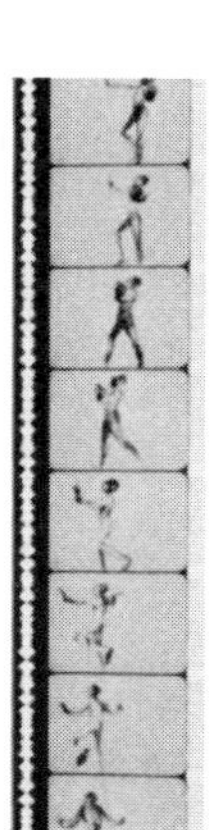
F

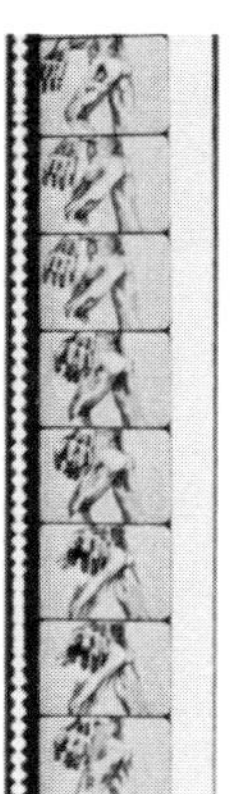
G

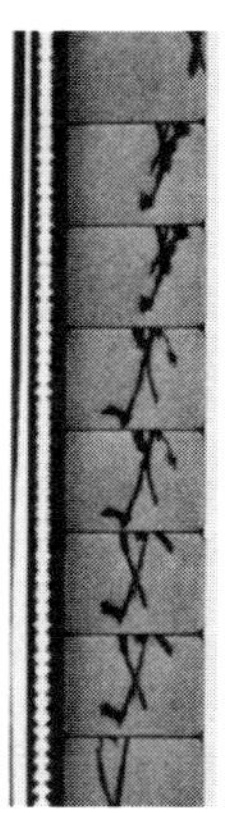
H

I

A

14.18 Ryan Larkin's <u>Walking</u>: In Ryan Larkin's deceptively simple film, the screen shows us beautiful studies of people in motion. Done with sheets of paper, not acetate cels, *Walking* explores different walking gaits, different kinds of walkers, different angles of observation, different moods, and different—boldly different—styles of animation, *(A)* through *(N). Stills courtesy National Film Board of Canada and Learning Corporation of America. Film strips by the author.*

C

B

D

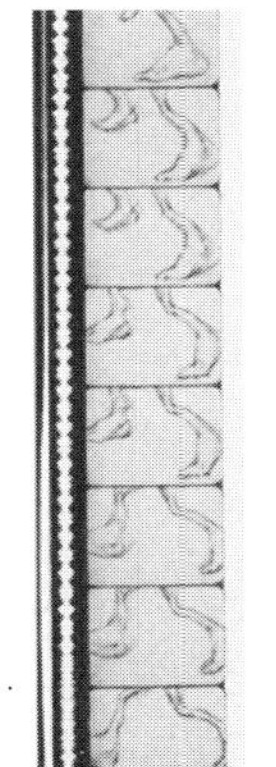

J

K

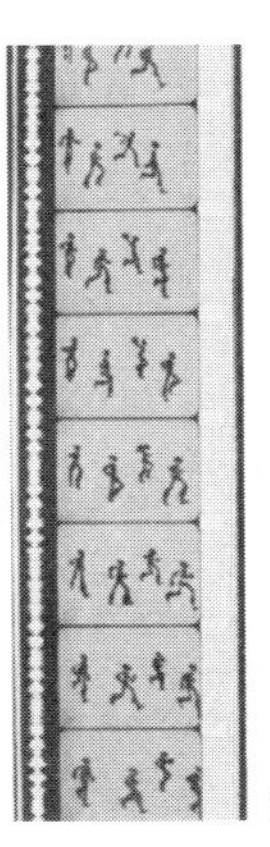

L

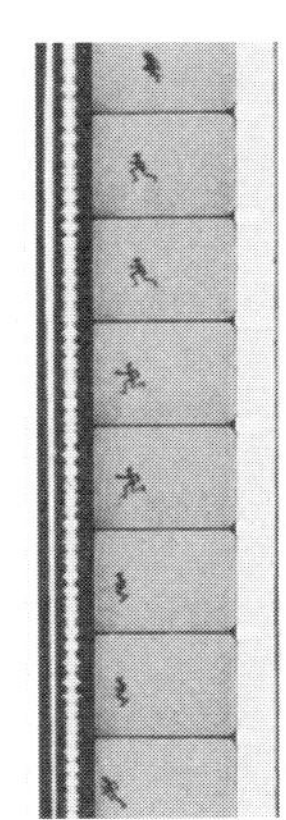

M

N

takes two full paces (four steps) into the middle of the field. Plan this so that when the character comes to a standstill after the second full pace, he or she will still be within the given field size. At this point, have your character pause for ½ second and then, in another ½ second, turn the character so that it faces the camera. Let there be another pause, this time for a full second. Finally, raise your character's arms in a move that takes ½ second. In the last drawing of the sequence have a "bubble" pop onto the screen beside your character with this written inside: "I'm terrific." This project should require in the neighborhood of thirty-five drawings on separate sheets of paper. Figure 14.17 gives a scaled-down suggestion of how your spacing guide might look.

EASING IN AND EASING OUT

It's the moment for an import digression. In animating a pan, or any other movement, it is necessary to move *gradually* up to the full speed of the movement. A character can't suddenly appear to walk at full speed any more than someone in real life can suddenly jerk into full movement. A transition is required.

In animation, this transition from stasis to a steady rate of movement is called *easing in* and *easing out*. Complex formulas have been invented to help the animator chart the initial movements in this transition. At this level of exploration, however, it is enough to improvise easing in or easing out of a pan by gradually increasing the distance between positions during the initial phase of filming.

CYCLES AND HOLDS

Animators love shortcuts. It's not surprising. Shortcuts allow one to make the longest film for the least labor. Shortcuts stretch the impact of an animator's art. You'll grow to love shortcuts too!

In doing the walk problem, you can use a single drawing for a full twenty-four frames at the pause before the character lifts his or her arms. Such a moment is called a *hold*. This shortcut enables the animator to avoid drawing the stationary figure twelve times.

A *cycle* accomplishes the same thing, increasing the number of times a single drawing can be used. But a cycle does this with movement. In the walk, you were asked to draw two full paces, which was accomplished in twenty-four drawings (the "beat" was twelve, which required six drawings for every step). A cycle allows you to accomplish the very same visual effect with just twelve drawings, half the total number normally required.

The principle of a cycle is based on the ability to design a series of drawings so that the last drawing leads smoothly back to the position of the first drawing. This means, of course, that the same set of drawings can be used again and again as long as you want the sequence repeated.

In order for the last drawing in the cycle (#6) to lead smoothly into the first drawing (#1), the drawings must be modified so that there is no forward movement. The character's head stays in the same position, although the legs continue to move as before. This means that the character's feet must "slide" somewhat between drawings so that the walking figure always returns to the same initial position.

No matter what kind of character or object you are animating (a car, an animal, a spaceship) or how many drawings comprise the cycle, the movement will appear correct if the last position leads smoothly back to the first position.

To get a feeling of *forward movement,* a separate background is made, which will pass under the walking figure while it remains in exactly the same place. To achieve this effect, it is necessary to see through the top paper sheets to a background sheet. This is most effectively done with clear acetate sheets, as opposed to paper sheets. However, by using a rear-lighted surface under the camera, it's possible to see through more than one page of the paper sheets (Figure 14.19).

Try to cycle the animated walk you designed in the last project. Create the effect of forward movement by panning a background beneath the walking figure, using rear lighting, or by simply advancing the drawings of the cycle under the camera without a background of any kind.

14.19 Cycling with paper sheets: The sketch shows an animation disc that has been mounted into a table and then lit from beneath. In addition to the frosted glass or Plexiglas insert in its center, the disc bears two sliding peg bars. In filming a cycle, one set of pegs remains stationary (it bears the cycle of drawings) while the other set of pegs (bearing the same background) is moved incrementally between exposures.

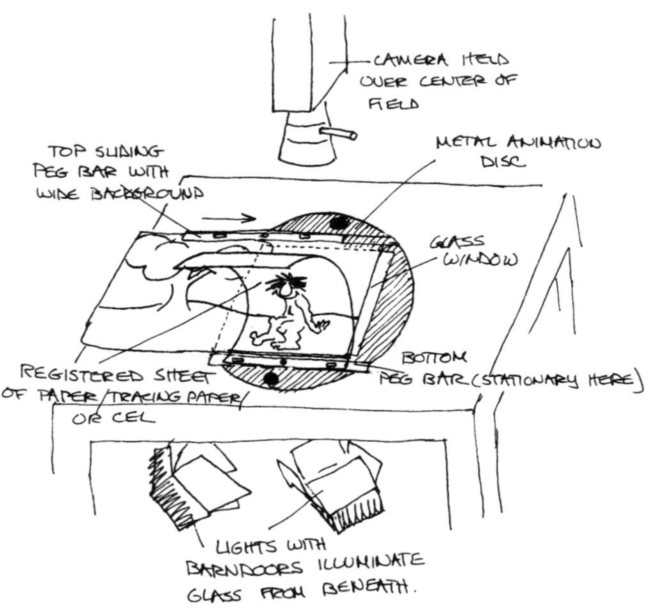

MOVEMENTS OF THE TWO LEGGED FIGURE

HERE IS A COMPARISON OF THE VARIOUS TWO LEGGED FORWARD MOVEMENT CYCLES -- I HAVE DRAWN ONE-HALF OF EACH CYCLE BELOW -- REVERSE HANDS + FEET FOR THE OTHER HALF -- THESE CYCLES CAN BE USED AS "REPEATS" - (THAT IS THE DRAWINGS MAY BE REPEATED OVER + OVER IF THE FIGURE REMAINS CENTERED ON THE SCREEN AND THE BACKGROUND MOVES.

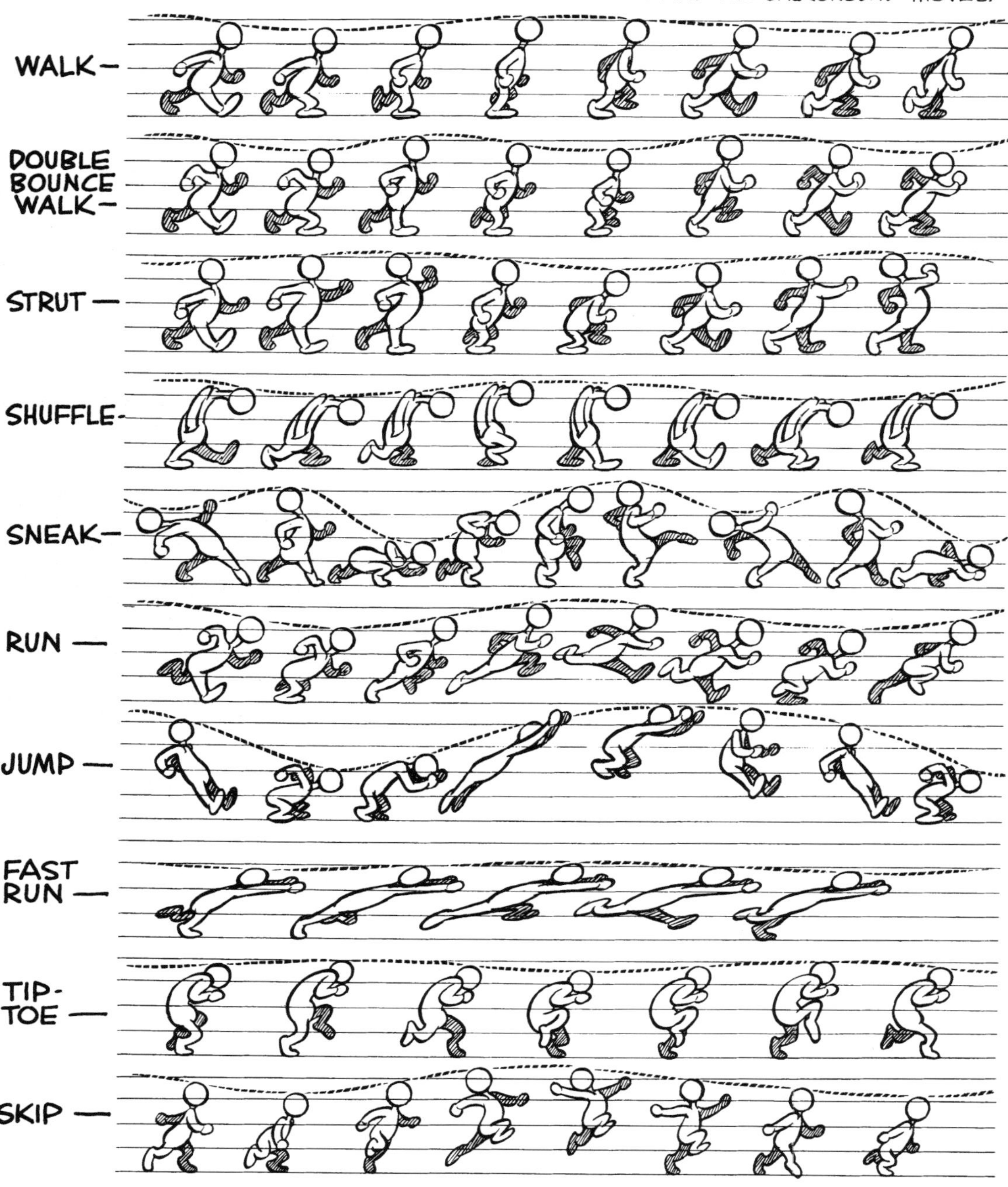

14.20 Cycling someone in motion: As only a master animator could do it, these variations have been created by Preston Blair. Mr. Blair worked for many years at the Disney studios, where he contributed to *Pinocchio, Bambi,* and *Fantasia.* (In the latter he designed the hippos.) Later, working for MGM, Mr. Blair designed and animated *Little Red Riding Hood* and directed cartoon shorts. *Material here is gratefully reproduced from* Animation—How to Draw Animated Cartoons, *by Preston Blair, published by Walter Foster.*

NUMBERING

You have already discovered how important it is to consistently number the separate sheets of paper that hold your drawings. This is especially important when you get into cycles and holds. The more complicated your film, the greater the need to keep accurate records of how many times to expose a particular drawing or how many times to repeat a particular cycle.

With the paper sheets of line animation, the numbering system can be relatively simple. Informal notation devices can be used. But the problem becomes much more complicated once you start to use layers of paper cels, or when you incorporate a camera or compound movement with the use of paper sheets. A sophisticated system for numbering is discussed later.

REFINEMENTS AND SPECIAL EFFECTS

Although we've not yet reached the use of acetate cels, I'd like you to know that once you have mastered the walk, you've dealt with animation in its most complete and most elegant state. The differences between line techniques and cel techniques are differences in execution, not conceptualization. In fact, cel techniques are designed to make simpler, easier, and faster the sort of complete animation that we've been working through thus far.

Line techniques yield "full" animation. Each element of the image can be modified in the twelve drawings required for each second of screen time. Unless one chooses to film separate drawings for each frame, there is no way to pack more

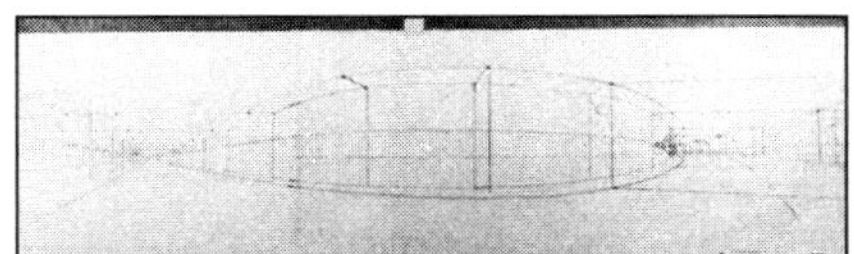

A

B

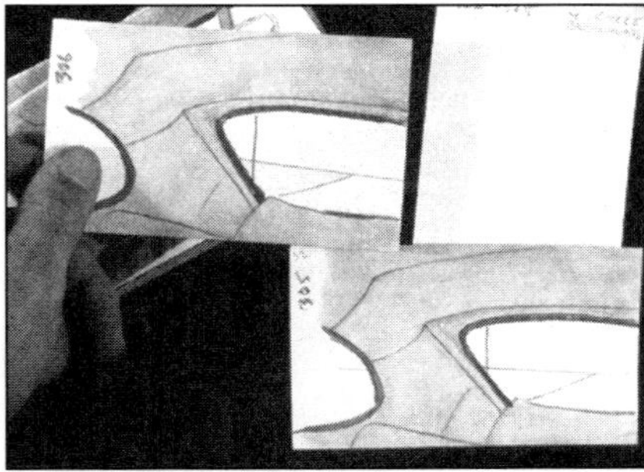

C

D

14.21 Precise planning: These four photographs show a sophisticated spacing guide, *(A),* planning elements, *(B)* and *(C),* and final art used in Al Jarnow's *Autosong, (D). Courtesy Al Jarnow.*

14.22 Cel/line hybrids: George Griffin demonstrates a few innovative techniques that combine line and cel animation. In *(A)* an acetate sheet with lines and a paper sheet with textured color are cut so that a third level of material can show through a *Candy Machine* character's eyes. Also from *Candy Machine, (B)* shows how a paper lightning bolt has been mounted on an acetate sheet and then placed on top of other artwork. George Griffin stands beside an extended pan background that combines cel and paper levels, *(C)*. In his hands the animator holds cutout paper figures used in this scene from *The Club.*

A

C

B

movement within a second of film. You've reached the saturation point. Here are some quick notes on simple modifications of line-animation technique through the use of registered paper sheets.

Multiple Layers with a Light Box. A rear-lighted surface allows you to see through a number of sheets of paper. This helps in doing in-betweens once you've drawn extremes. If you use *tracing paper,* you'll be able to get a clearer illumination for more layers of paper sheets.

Die-cuts. One paper cel can be used as a "frame" for those cels placed under it. The effect is quite pleasing—a frame within a frame.

Cel-over. An easy way to show a very simple background

without redrawing it is to create a single *acetate cel* with a few background lines drawn on it. This cel is placed *over* the sheets of paper before each shutter release. A set of top pegs and bottom pegs are helpful to ensure good registration.

Special Papers. Don't overlook the creative effects inherent in different kinds of drawing paper. You can take good advantage of variations in texture and color.

Graphic Media. There are also creative possibilities in using different marking tools. Experiment with watercolors, felt-tipped pens, pastels, crayons, colored pencils, charcoal, acrylic paints, drawing inks, and so on. Each dictates a special kind of drawing. Discover, if you can, which drawing media best fortify and extend the unique character of your own personal drawing style.

Dissolve Series. If your camera has the capability to execute in-camera dissolves, you can produce very powerful effects by "dissolving" between drawings. When used appropriately, this technique can save you the work of doing in-between drawings. For example, a 12-frame dissolve (which lasts ½ second) requires only two drawings, while full animation during the same ½ second would require six drawings. Experiment with dissolves of different lengths.

Multi-image. With registered paper sheets you can easily create multi-image movies. For example, the camera can be positioned so that there will be a large field that accommodates two different sets of drawings. A masking device can be used to create a regular frame for different cycle drawings.

14.23 Use of line and texture: A sketchbook's black pencil drawings come to life in George Geersten's *The Men in the Park.* Animation is sporadic but intense. Facial expressions change. A figure of a man rises and shambles off screen. There is a blur of nearby traffic. Note the texture of the lines. *Courtesy National Film Board of Canada.*

CEL ANIMATION

Believe it or not, cel animation was invented because it saves work. The transparent surface of the plastic sheet makes it unnecessary to draw *all* the parts of a scene every time there is to be a change in *one* of the component parts. In turn, this means that the imagery of animation can become far more detailed and finely rendered than would ever be possible if every single element of the drawing had to be completely redrawn each time a change is required, as it is in line animation. The term *cel* comes from celluloid, the chemical substance of the early plastic sheets. *Acetate* is the proper chemical name of the sheets in use today, but the old name has stuck. Today,

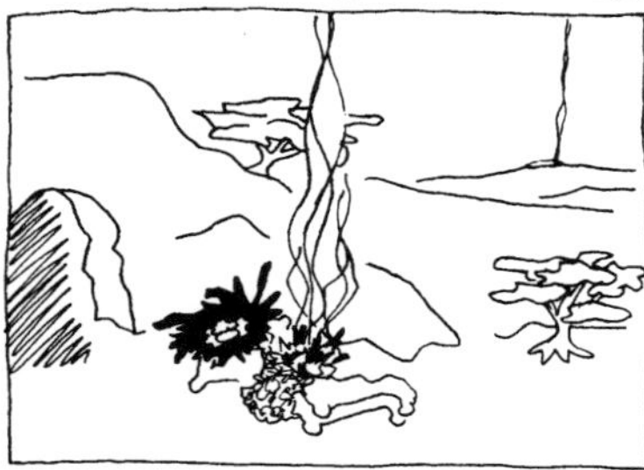

14.24 How acetate cels save work

the word *cel* refers to transparent plastic sheets, regardless of material.

The heart of the process of working with cels has already been introduced: registration, extremes and in-betweens, cycles and holds, stretch and squash. The discussion that follows deals with the unique characteristics, potentials, and problems of using acetate sheets, and of the system of animation that has developed around them. Let's begin with the cel itself.

The standard animation cel is a transparent plastic sheet that is .005 of an inch thick and measures 10½ by 13 inches. The sheet is slightly larger than a #12 field and it is usually punched to fit one of the two standard registration systems, Oxberry or Acme.

The surface of the cel is very soft. It scratches, bends, rips, and smudges easily. It collects dust like a magnet. For these reasons, cels are handled with great care and usually the animator will wear cotton gloves while painting or handling them.

If you try to remember that the function of the cel is to save work, it becomes easy to see how the cel is used. Whenever and wherever possible, an animator will save the labor of redrawing. An example or two will make this clear.

Imagine that we want to animate that famous moment in prehistory when one of our ancestors discovered fire. We'll first create a scene, perhaps the mouth of a cave. Our character might be sleeping when a huge lightning bolt zaps the ground beside him. Let's say that this dim-witted caveman fails to stir during the first smashing bolt. He snores on. A second blast manages to burn off his fur clothing. He continues sleeping peacefully. Eventually, however, our ancestor starts shivering in his sleep from the cold. A third lightning blast awakens him and he notices a burning branch. His curiosity is aroused. He crawls over for a look, a very close look. He burns his nose and discovers the meaning of fire. It's an elevated moment—one small sniff for man, one giant waft for mankind.

The drawings in Figure 14.24 present the opening scene of this drama, plus quick sketches for key moments within the story: the first zap of lightning, the burning clothes, the awakening, the moment of glorious discovery.

The largest saving of work comes right at the outset. Using cels, one never has to redraw the background. The same drawing can be used throughout the entire sequence. The opening moment, for example, requires just one layer of cels for our sleeping friend. If the animator wants to have him snoring, then there might be a cycle of, say, three or four cels that would show the body moving slightly as our caveman sleeps. These few drawings could comprise the first five seconds of the film.

But greater economies are in store thanks to cel techniques. During the first strike of lightning, the animator could use a second layer of cels to show the lightning itself as it blasts onto the scene. No need to redraw our hero if the lightning is carried on a second level of cels.

In the sequence where our ancient relative sleeps peacefully as his loincloth is burning off, the animator might use one cel for the naked body, while a different cel level shows, in stages, the fur being singed away. Executed as a close-up, this sequence might bear further refinements. The great discoverer's face might reveal a sensual smile as his body is warmed for the first time by a flame. In this case another cel level could be used to show the changing facial expression. Without the use of cels, the entire body and background would have to be drawn twelve times for each second of screen time. With cels, only the burning cloth and perhaps the eyes and mouth would need to be redrawn for each exposure.

Skipping now to that moment of triumph, the animator could design the sequence so that the caveman's body remains motionless once he has crawled over to investigate the burning branch. Only his head and nose would actually be extended with suspicion toward the novel sight. At this moment three cel levels would be used: one for the body; one for the head and nose, stretching and sniffing; and one for the burning branch, the latter probably a cycle of three or four cels. The original background would still be used.

A maximum of four layers of cels can be used at any given time. The animator must study and plan the action carefully in order to break down the overall image into no more than four layers.

Each acetate cel has a small degree of optical density even though a single plastic sheet appears transparent. If you hold one cel over a picture you may not be able to make out the slight fading of color. But if you hold three or four cels against a background, your eye will easily detect a major change in color intensity and in the definition, or clarity, of the image.

Any change in the number of cels covering a given background will be noticed. Hence, if there is a scene in which four cels are used—even for a short sequence—four cels will have to be used throughout the film. In our example, three cel levels are required for the finale. This means that two *blank cells* are required for the first part of the scene in order to provide a consistent optical density. Otherwise the background will appear to change intensity and color.

Wherever possible, of course, the animator uses previ-

A

B

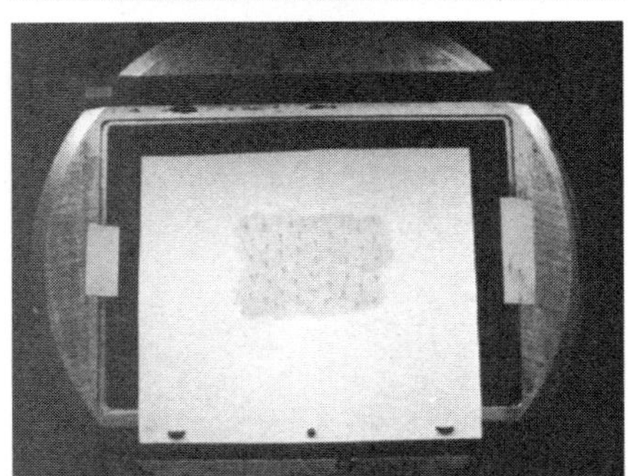
C

D

14.25 Cel levels: Two cel levels, *(A)* and *(B)*, combine with a background painted on paper, *(C),* to compose a frame, *(D),* in Janet Perlman's *Lady Fishbourne's Complete Guide to Better Table Manners.* Frame enlargements, *(E)* and *(F),* show other moments in the film.

E

F

ously prepared cels or opts to hold particular cels under the camera for as long as possible. The latter is termed a *held cel.* And if there's a way, the animator will always keep one level of cels for use with a cycle of drawings that can be repeated again and again. When there is full movement, of course, everything that moves must be redrawn. For such movements the animator generally consolidates all the action into one cel level. This is simply for economy and drawing convenience.

As noted before, animators usually first try out complicated movements with a pencil test—a film of the first rough sketches that have been done on paper. It is after the animator has been able to review this rough cartoon that drawings are transferred to transparent cels. The necessity and value of the pencil test is easily understood when one learns how much work goes into preparing a cel.

First, the pencil drawing is covered with a sheet of acetate and the outline of the character is traced. At this stage the "roughness" of the pencil drawing is cleaned up and a smooth continuity between drawings is developed. The traced line is usually varied in thickness to provide emphasis to the character it portrays. This operation is called *inking.* In large animation studios there are teams of assistants called *inkers* who perform only this task. It's a demanding job and it requires great precision.

In order to keep the lines of the drawing as strong as possible, colors are applied to the reverse side of the acetate sheet. This work is done by another set of assistants called *opaquers.* Their job consists of carefully painting the correct opaque paints on the appropriate parts of the character. Figure 14.26 illustrates the processes of inking and opaquing. They are every bit as laborious as they look. And the materials are expensive too.

When there is no need to redraw a background for every scene in the finished film, it becomes practical to fashion detailed and delicately rendered backdrops against which the animated action will take place. In classic American cartooning, there are individual artists, and even entire production departments at the larger studios, who do nothing but design and execute backgrounds.

Backgrounds are fashioned with care. A fine-quality paper stock is used and art materials are also of the best quality. Watercolors are the medium most often used. All back-

14.26 Inking and opaquing: By flipping over a completed sheet of cel art, it is possible to see how the process works. Because the black outline of the character has been inked onto the front of the cel, the process of opaquing on the rear side can be somewhat imprecise. This frame is taken from *The Dog Who Said No,* an animated segment done for *Sesame Street* (The Children's Television Workshop) by animator Derek Lamb.

grounds, however, must be styled to fit the graphic quality and look of the characters. Backgrounds must also, of course, conform to the needs of the story. And, as pointed out earlier, it is important that the background not be so busy as to obscure or draw attention away from the movement of story and characters.

One of the most interesting effects of cel animation is experienced in *traveling scenes.* The use of transparent cels, combined with the principle of a cycle, allows the animator to create a long, exciting chase with very little work. The secret lies in a *panning background,* an extra-long strip of background artwork that is moved underneath the cels and joins up with itself so that the pan can go on forever.

Project: A Long Way to Tipperary. Using commercially punched cels or a makeshift set of acetate sheets that have somehow been registered, copy (ink) the outline of the walk cycle you designed earlier. After the inking process, turn each cel over and color in the character using paint or felt-tipped pens. For this project you will be making a cycle of six different cels.

Design a panning background. It will be the same height as the field on which you are working, but it will be much, much wider (Figure 14.27). On this background draw and then paint a road upon which your character will eventually appear to be walking.

Once the background and cels have been completed, place them under the camera. The cycle drawings are placed over one set of pegs, and the background will be constantly moving between exposures. In order to provide a smooth reg-

14.27 A panning background

istration for this panning movement, punch the background drawing and place over a *sliding peg bar,* a set of pegs that can be moved between exposures to carefully established positions. The sliding peg bar is usually located at the bottom of the animation disc. Chapter 21 will show such equipment in detail. It's possible, however, to use a common ruler in calibrating and controlling the movement of a background.

With some experimentation you'll discover how the speed of the walk is a factor in determining how much the background is panned between exposures. The movement of the character's legs must visually match the speed with which the background passes underneath the walker.

An illusion of depth can be created with the use of a second and closer moving background. An additional element can be placed over the cel sequence. It is moved at a relatively faster rate than the distant background. The combined effect mirrors the perceptual experience of real life, in which objects nearest one appear to pass more quickly through the field of vision than objects in the distance.

The background you've designed can itself be a cycle. If your drawing is long enough, you may be able to reconnect it with itself so that the left edge of the drawing joins the right edge.

SPECIAL PROBLEMS

Working with cels is difficult and time-consuming. It requires special care and it is expensive. Employing cels is certainly worth the trouble, however, when you want to present dazzling backgrounds and full movement of characters. There is a lushness to cel animation that cannot be equaled by any other technique or production shortcut. It *is* rigorous, however, and filled with special problems. Here are a few of them.

Reflection. Acetate sheets will reflect light into the camera lens. You must be particularly mindful of glare from the side lights, even if they are mounted at a 45-degree angle to the artwork. The glass platen that holds the acetate sheets flat on the compound should control such reflection. Problems nonetheless persist.

Shadow Board. Even with a platen and angled lighting, the cels can act as mirrors and reflect an image of the camera itself from the cels back onto the film's surface. To remedy this, a black, nonreflecting surface can be rigged to mask the reflections of the camera body or lens metal. See Chapter 22.

Newton Rings. It is important that there be just the right amount of pressure on the cels piled beneath the camera. If there is too little pressure, the cels will wrinkle, causing ephemeral shadows that are very distracting. Too much pressure produces an effect known as Newton rings. This is partially caused by the thickness of the paint on the cels, and treatment of the problem may involve a very light dusting of the cels with face powder so that they won't adhere to each other under the pressure of the platen.

Cel Paints. As suggested earlier, there is often an appreciable loss of definition and color caused by the density of the

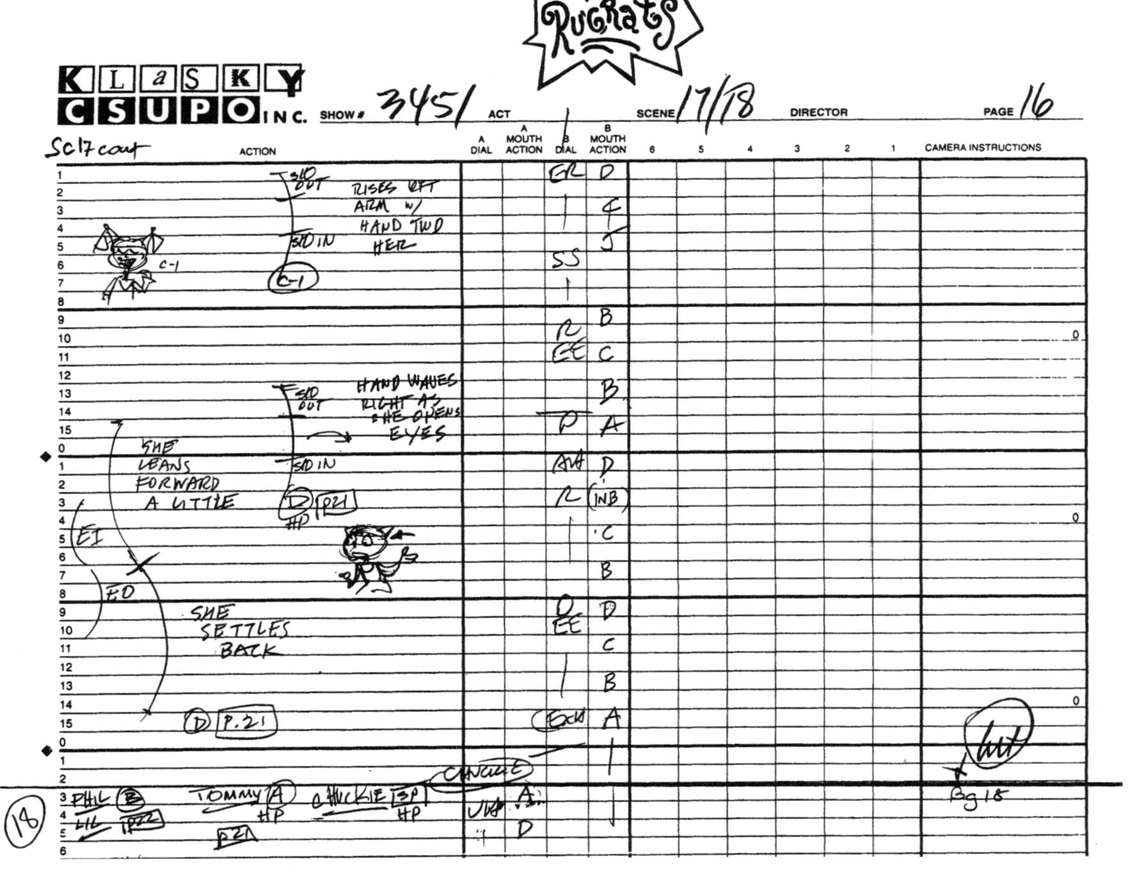

14.28 An exposure sheet: Here is an exposure sheet (also called a cue sheet, a dope sheet, or an x sheet) from the TV series *Rugrats.* Across the top, from the left: space where the animator and/or sheet timer describes the action (often with a quick sketch); A and B dialogue tracks with the words themselves entered under "Dial" and the mouth action references next to the dialogue — character Angelica is completing the phrase "(aniv)ersary party"; six columns that the animator will use to note different cel levels; and finally, a column for camera instructions. In Chapter 25 you will find a blank, appropriately simplified exposure sheet to use in your own movies. *Courtesy Klasky/Csupo Studio and Nickelodeon.*

acetate itself. This cel thickness causes a 5 percent loss of light and this shows up most clearly with color paints. If a red is used on the top cel and then the same red is used, say, on the fourth cel layer, these reds will have a different hue in the final film. This problem is solved by using special sets of commercially prepared paints in which the same colors are supplied in four different hues to compensate for the change in color caused by the cels themselves.

Dust. Tiny pieces of dust will settle on the glass platen during filming and these, unfortunately, show up as specks on the finished film. Dirt and dust particles also tend to adhere to the surface of cels, helped along by static electricity. Polaroid filters can be used to optically remove the glare caused by specks of dust and dirt (see Chapter 19).

Shooting Complexities. The more complicated the arrangement and number of cel layers, the easier it is for the camera operator to screw up during filming. There is simply too much to remember, too many changes to make, and too many variables for any individual to keep track of without some kind of written aid. And this leads us to exposure sheets.

EXPOSURE SHEETS

Full-cel animation requires a fastidious and precise system for recording the sequence and order of various cel layers, the number of exposures given to these layers, and the movements of camera, background, or compound. In practice, all this information is written down by the animator on an *exposure sheet.* Figure 14.28 provides an example of this record-keeping device, and in Chapter 25 you'll find a full-size exposure sheet that you can detach and have duplicated for your own use. Only a few comments about the exposure sheet seem necessary. *Dial* is a special term that refers to the consecutive numbering of each and every frame within the finished movie—24 frames to the second, 40 to the foot (in 16mm), 1440 frames to the minute, and so on. In order to keep a running count of frames, the animator adds the appropriate prefix numbers before each zero. Cardinal numbers reading from left to right (4, 3, 2, and 1) refer to the order of *cel layers.* If possible, the animator positions those cels that require the most frequent changing at the top of the pile. The *background* (Bkg) is usually

A

B

C

14.29 Character design: In a large animation project such as a TV series or feature, many different animation directors and animators will be working on the same characters. Hence "model sheets" or "character packs" are compiled and updated from time to time. Sampled here are three of four standard elements, all taken from the *Rugrats* TV series. The *size relation* layout, *(A),* shows the relative heights of different characters. Each character will be drawn in a set of typical *poses, (B),* such as these for Angelica. There is also a *turnaround* for each character — here for the show's protagonist Tommy Pickles, *(C).* The other standard element is a *mouth chart* — see Figure 14.31. *Courtesy Klasky/Csupo Studio and Nickelodeon.*

A

given a letter reference or a short descriptive term in order to differentiate between the various backgrounds used in a single film and even in a single sequence.

The exact timing and position of each cel is indicated by the individual cel's number, and this is placed on the exposure sheet in a box beneath its appropriate cel layer. Usually the animator employs both a letter and a number in marking individual cels. The letter is chosen to indicate the subject matter of the cel series. *H* might be for head, *F* for feet, *HA* for hat, etc. The number beside the letter indicates the order of cels for each layer (F1, F2, F3, F4, etc.).

The exposure sheet has space for *camera instructions* such as fades, dissolves, superimpositions, and for camera/compound movements such as pans, zooms, and spins. In addition, the animator can requisition space on the sheet to indicate where a special sound exists on a prerecorded track or to note for further reference any item that might easily be forgotten or wrongly executed.

B

C

D

E

F

G

H

I

J

K

14.30 Inside Klasky/Csupo: The Los Angeles–based animation studio of Klasky/Csupo has produced a string of TV hits including *Duckman, Rugrats,* and *Aaahh! Real Monsters* — which is also being produced as a feature film in conjunction with Nickelodeon and Paramount Studios. Klasky/Csupo's next series, *The Wild Thornberrys,* is scheduled to air on Nickelodeon in 1998. The exterior walls of the company's main office, *(A),* sport some super graphics that preview characters from their newest show. *Rugrats* director Norton Virgien works on storyboards in *(B).* A typical animator's desk and drawing disk, *(C).* Character Designer Sharon Ross, *(D).* Sketches of backgrounds are pushpinned to a wall, *(E).* Actress Cheryl Chase records the voice of Angelica, *(F).* Frames, *(G), (H),* and *(I)* are from *Rugrats, Duckman,* and *Aaahh! Real Monsters*, respectively. Founders Gabor Csupo, *(J),* and Arlene Klasky, *(K). Photos by Vince Gonzales. Courtesy Klasky/Csupo Studio and Nickelodeon.*

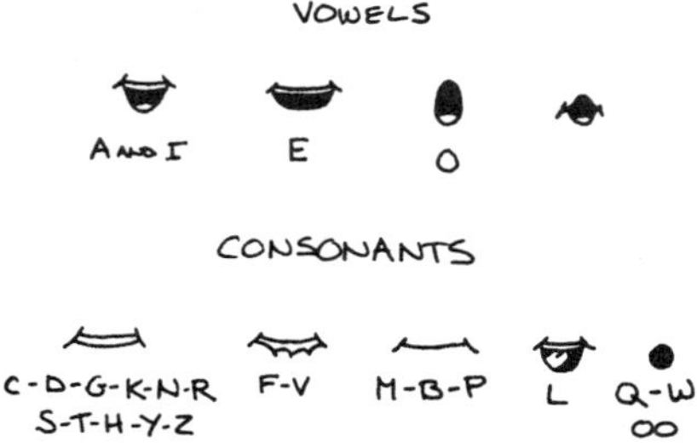

14.31 Common mouth shapes: This simplified set of vowel and consonant shapes must be styled to your particular character as well as the kind of emotional delivery the lines are being spoken with. Some animators like precise and fully articulated lip synchronization while others feel a "loose" technique is quite adequate. An alternative to such shape charts is for the animator to sit facing a mirror so that he or she can study the visual movement of his or her own lips in saying particular phrases with particular inflections.

Treat this aid informally. Modify the standard form to suit your own particular needs, the capabilities of the production tools that you will be using, and the particular techniques being employed.

LIP SYNC

You can create animated characters that appear to talk, whose mouth movements precisely match the voice track. It's not easy, but you can attempt and eventually master the process if you have the required equipment for analyzing recorded voice tracks (see Part III: Tools).

In order to create lip sync for cel animation, a *bar sheet* must be filled in with a frame-by-frame analysis of each vowel and consonant sound in a given piece of dialogue. With this information, the animator is able to match each utterance with the appropriate mouth shape.

Both track analysis and lip-sync drawing are laborious processes, as you can imagine. Doing it right requires patience and perseverance. A real mastery of lip-sync animation requires a control of head and body movement and of gestures appropriate to individual words, to the character's personality, and to the dramatic requirements of the film as a whole. If you are interested in pursuing this specialized technique, obtain one of the advanced manuals on the subject and be prepared for a lot of work.

To get you started, however, look at the chart in Figure 14.31. It has been created over the years as a standard simplified guide for animating lip movements. In application, modify the style of each given mouth shape so that it matches the graphic style of your cartoon character.

THE STUDIO CELL SYSTEM

Your understanding and your appreciation of the technique of cel animation can be extended by studying the assembly-line hierarchy developed at the large animation studios in the heyday of character animation in the 1940s and 1950s. Incidentally, the same system, or one close to it, is being used today in the production of most contemporary animation features and,

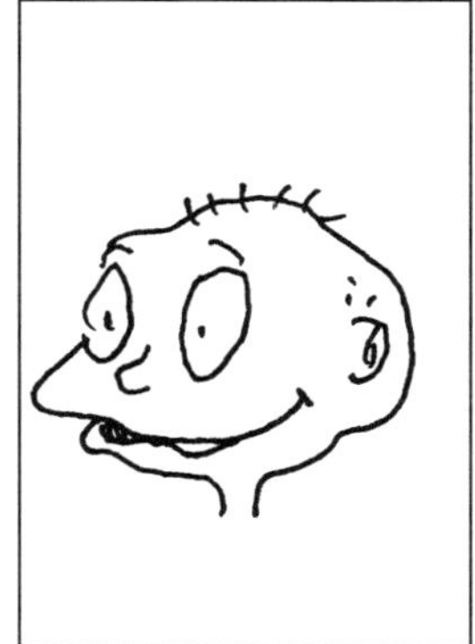

14.32 Same sound but a different shtick: Although all the utterances of the English language can be simplified down to nine or ten basic mouth shapes that humans use in forming the sounds of our language, animators can exhibit a broad stylistic range in showing how a particular character enunciates. Here are corresponding mouth action poses for two of the *Rugrats* characters, Angelica and Tommy Pickles. *Courtesy Klasky/Csupo Studio and Nickelodeon.*

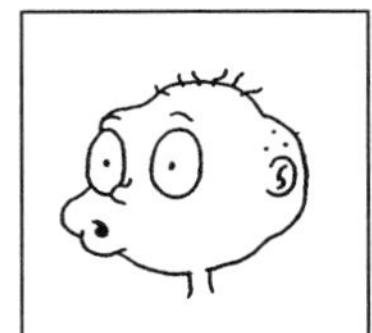

to a lesser extent, in the production of animated series and TV spots. In doing your own independent animation, of course, you have to cram these separate jobs into a single job—yours. Anyway, here's a rough breakdown of the studio system of traditional cel animation.

The *producer* is the individual responsible for the overall development of the animated film. He or she selects personnel, raises capital, manages expenditures, arranges for distribution, and much more. The producer is everyone's boss. The *writer* is assigned or chooses the subject, defines the characters, and begins shaping story line and dialogue. The *storyboard artist* breaks down the story into component scenes. The storyboard provides a visual system for making a detailed analysis of the film's development. The storyboard includes a detailed study of character appearance and movement, a detailed workup of backgrounds, and a delineation of scenes and sequences within scenes. The producer and all "creative" personnel (writers, designers, animators, directors) work with the sudio's sound crew, the "actors," or voices, and the musicians and composers in recording the full sound track for the finished movie, including music and dialogue. Special sound effects can also be placed on the track at this point or added later in the final editing and sound-mix process. A *track analysis* is then made. A specialist will make a bar sheet from the sound track. All words, beats, and music are identified with the utmost detail and precision. Working with the bar sheets and the layout/design models, the *director* begins filling in an exposure sheet for every moment of screen time. Decisions made at this point shape the remaining work; the timing of movement, the stretch and squash of characterization, and the amount of time devoted to

A

B

C

particular bits of action. Working from the layout drawings and in consultation with the director, the *background artist* creates each background that is required in the film. When full-cel animation is being used, backgrounds are lavishly executed pieces of art. Different *animators* are now assigned to different scenes or different characters within the overall movie. If the film is short, the animation may all be done by one individual. It's the animator's job to create the extremes for the cels that make up the film. These are drawn roughly on paper sheets. The animator is given the background against which he or she is to bring the story to life. The *assistant animators* prepare the in-betweens, also completed on paper sheets. Generally, the assistant smooths out the actual drawings of the animator and refines the animator's working sketches. At this stage the completed paper drawings may be filmed in a pencil test.

Once the animator and director are satisfied with the pencil test, the drawings go to the inking department, where *inkers* trace the outline of every drawing onto an acetate sheet. At this stage decisions are made on how and where to break down the drawings in order to save work, time, and expense by using different cel layers. Now the *opaquers* flip over the acetate cels and paint in the colors of the characters. There is always at least one *checker* to be sure that all the cels and layers and exposure sheets and bar sheets and backgrounds are properly executed and marked and registered prior to the actual filmmaking. Working from the exposure sheets, a *camera operator* or team of operators films the entire movie. Then a *film editor* cuts together the pieces of exposed film and matches the images to the sound track. The editorial department will also prepare the original footage for duplication and printing at the laboratory.

That's the general process for making a cel animation film. But the work isn't over once the film is made. For if everyone involved wants to make more movies, then the finished film must generate revenue. It's the American Way. So there are always a host of other people involved in the studio animation system—accountants, publicity people, film programmers, distribution agents, and more.

Digital tools, combined with the Internet, are beginning to change the traditional studio system. Digital ink and paint techniques (see Chapter 15) are making it possible for a dozen computer stations to replace rooms filled with ink and opaque

D

E

F

G

H

I

14.33 John and Faith Hubley: Animators John and Faith Hubley are designing a sequence from their independent feature, *Everybody Rides the Carousel, (A).* Frames from a few of the over twenty-five films they have conceived, directed, and produced follow: *(B), Adventures of an Asterisk; (C), The Hole; (D), The Hat; (E), Cockaboody; (F), (G), (H),* and *(I), Everybody Rides the Carrousel. Photos courtesy The Hubley Studio.*

desks. On a movie like *Space Jam,* small animation studios across the United States collaborated with Warner Bros.' Feature Animation group in Burbank, California.

TV AND THE LIMITED-CEL SYSTEM

The costs are tremendous for full-cel animation, and the process is slow. It requires a lot of people, most of whom must be highly trained and experienced. The equipment and supplies are costly, too. Thus it is not difficult to understand why the preservation of the full-cel system can be sustained only by the vast budgets allotted to feature film production and theatrical release by large studios like Disney, Columbia Tri-Star, Warner Bros., Dreamworks-SKG, Fox, Nelvana, Nickelodeon Features, and Tooniversal.

Just as the use of cels was originated in order to save production time (and money!), the relatively modest budgets

A

B

C

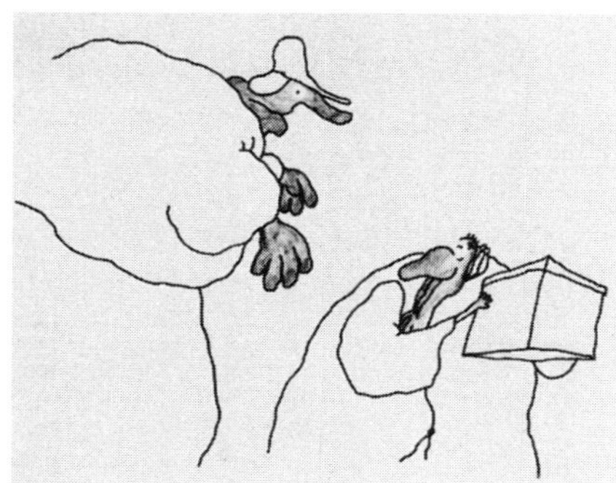

14.34 The look of cel animation: Seen in proximity with each other, these nine frame enlargements indicate the wide range of graphic styles and subject matters that have been used with cel animation techniques.
(A), John Leach, *Evolution, courtesy Learning Corporation of America; (B)*, Stephen Bosustow, *The Legend of John Henry, courtesy Pyramid Films; (C)*, Paul Driessen, *An Old Box, courtesy National Film Board of Canada.*

of animated series for TV have forced the evolution of a distinct genre of character animation. This is often referred to as *limited-cel animation.*

To the purist, limited-cel animation is animation with limited power and effectiveness, as well as limited technique. In planning their films, directors of limited-cel productions try to minimize the number of drawings that appear in a minute's worth of show. Typically, a character's face will be frozen while only the mouth moves in talking, or a body will be still while an arm performs a task. Limited animation stresses the use of cycles, as you'd expect. Wherever possible, frames are panned or held while the voice, music, and sound-effects tracks attempt to drive the story forward. The storyboard artists and layout teams are required to recycle as many library drawings and backgrounds as possible, hence stories tend to share the same locations, angles are consistently drawn from an identical perspective, and close-ups are heavily featured.

What is described above is limited-cel animation at its worst. And you won't have to flip through too many cable and network channels to find plenty of examples. Just check out Saturday morning programming for kids.

In recent years, however, it seems that TV animation has come of age, with new creative juices. While production techniques and production values are still "limited" vis-à-vis theatrical animation, there have been a number of enlivening breakthroughs in the aesthetic realm, through shows like *The Simpsons, Liquid Television, Pinky and the Brain, Beavis and Butt-head, Dr. Katz, King of the Hill,* and of course the willfully eclectic and groundbreaking work of Nickelodeon cable's Nicktoons, which are sampled in Figure 14.35. What distinguishes the best work for TV is not necessarily the technique execution, but rather the concept and writing level. As new generations of creators bring their own characters to TV, these animators (individuals and small studios) innovate storytelling forms, animation styles, and art direction in ways that make the most of the budgets available for television.

THE SIMPLIFIED CEL SYSTEM

The aspect of cel animation that requires the most tedious work, takes the longest time, and is the most expensive is the

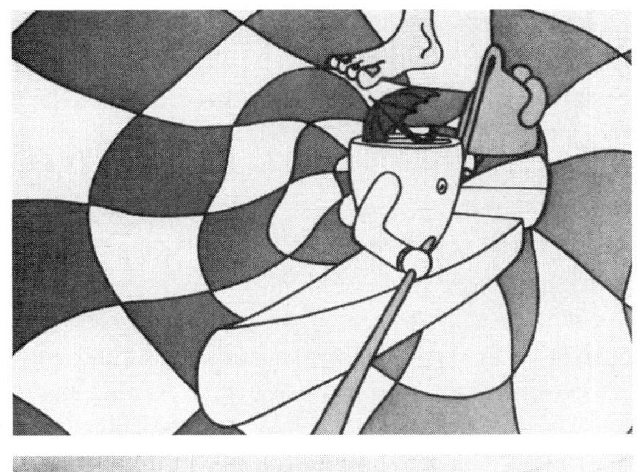
D

E

F

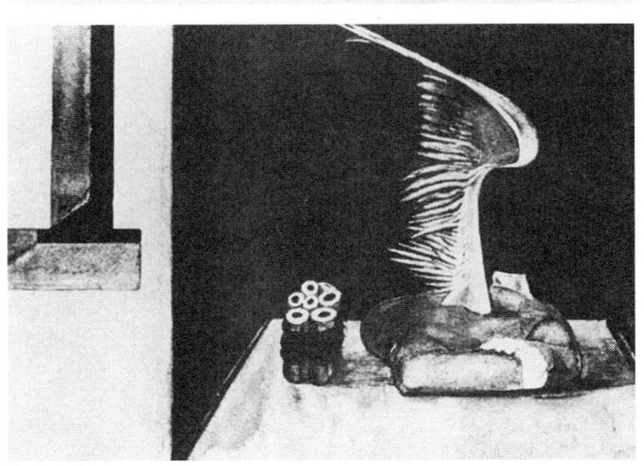
G

H

I

(D), Vincent Collins, *The Mole and the Music, courtesy Phoenix Films;*
(E), Zdenck Miller, *The Mole and the Music, courtesy Phoenix Films;*
(F), John Halas and Joy Batchelor, *Animal Farm, courtesy Phoenix Films;*
(G), Walerian Borowczyk, *Game of Angles, courtesy Pyramid Films;*
(H), Bruno Bozzetto, *Opera, courtesy Films, Inc.;*
(I), Zlatko Grgic, *courtesy National Film Board of Canada.*

process of inking and opaquing. In looking for ways to simplify the cel process, some animators have simplified the graphic quality of their animation rather than the amount or quality of the movement itself.

The husband-and-wife team of John and Faith Hubley have done just this and their work's quality and integrity, as well as its technical structure, have served as an important model to many aspiring independent animators.

Up until John's untimely death in 1977, the Hubleys produced films of exceptional quality at the Hubley Studio, their New York–based production company. John Hubley was a Disney-trained animator who went on to become a major figure at UPA Studios, where the first major reaction to and competition with the Disney style of animation began in the late 1940s.

The Hubleys and other character animators since have developed a number of techniques that produce the full-cel look without all the full-cel work. We might call these techniques "simplified" as opposed to "limited" cel animation. The best trick of the simplified cel system is its new way of mounting paper drawings on a regular acetate cel, thus eliminating the requirements of inking and opaquing. Using standard registered paper sheets, the animator carefully draws the character. A background is prepared as in full-cel animation. The same coloring techniques are generally used for figure and background. Watercolors, crayons, or felt-tipped pens can be used. In this simplified cel technique an outlining of the character is not required.

The paper drawings are now mounted on cels. A thin layer of rubber cement is applied to the back of the paper sheet. Then each paper sheet is quickly placed on top of an acetate cel bearing identical registration holes. The paper cel is smoothed out and allowed to dry. In the final step, the sharp point of an X-acto knife or a similar implement cuts away the surplus white paper from the outline of the character. The completed cel can now be placed over its background.

Such simplified cels work well. There are some significant limitations to the technique, however. The number of layers one can use is confined in practice to three or, better, two. This technique, like all others, must be tailored to the style of the individual artist. It works better for some artists than for others. The appearance of paper figures creates its own unique "feeling" and "space." It is well suited for some stories and types of films, but not for others.

There are many more production shortcuts that ingenious independent animators are developing for their work in basic cel and character animation.

Drawing directly on the cels, as opposed to "rehearsing" first with a drawing on paper, is one such idea. In this variation markers or grease pencils are often used. It requires a practiced eye and a confident hand to work this way. But the technique may yield a spontaneous and unrehearsed quality that is often lost as pencil drawings are transferred to cels.

Airbrush coloring and painting can speed up the process of creating backgrounds. The airbrush has a characteristic style that can be employed for special effects of one kind or another. Transfer lettering, shading, and lines can all be adapted to cel techniques.

Reframing is often employed. The same set of cels are photographed against different backgrounds using a different frame composition each time. Typically, the animation camera or zoom lens is focused closer to the artwork when reframing.

Ultra close-ups are being used more and more in today's simplified animation. There are two reasons for this. Close-ups require less intricate drawing, and they also work well on the TV screen, which is the prime distribution medium for animated films.

There are other effective tricks and shortcuts for working with cels: the use of cutaways, silhouettes, die-cuts, color xerography, video keying, rephotography, and many, many other special techniques. In particular, the recent innovations in the area of digital ink and paint and digital compositing offer today's animators ways to harness the full creative range of character animation, but to do so with simplified production systems where virtually all creative steps take place in the province of the computer.

A GRADUATION EXERCISE

As a final project, let me suggest that you use full-cel techniques (or simplified techniques as described above) to reanimate one of the characters or stories that you developed in an earlier project. Find a favorite, one that really interests you and

A

B

C

D

E

that feels within reach of your present level of skill. The reason for suggesting that you repeat the same problem is that you'll discover in the process things about the characteristics of either technique. A second try at the same basic material will also give you a chance to redesign the characters and reshape the story. You'll be able to experiment with new variations in timing, emphasis, dramatic flow, and plain old physical movement.

Limit this last film to a maximum length of 15 seconds. In the process of executing the full-cel film, allow yourself to begin sifting through all the projects you've completed. Search for images you enjoyed or transformations that excited you or interesting audio ideas. Begin studying the parameters of your tools. Determine what they are capable of doing and what their limitations exclude you from doing. And as you sift through earlier projects on your sample reel, look for an idea or for a technique that can become your first magnum opus.

14.35 Nicktoons aesthetics: In the 1980s the Nickelodeon cable TV programming service launched a set of "creator-based" animation series whose source was independent artists, not existing toy lines or spin-offs from other media forms. The Nicktoons lineup of shows displays remarkable variety. In *The Ren and Stimpy Show*, the two heroes, *(A)*, display extreme "stretch-and-squash" style as sampled in Ren's poses, *(B)*, *(C)*, and *(D)*. There is an urban, hip-hop look to *Hey Arnold! (E)*, an animated series that has very lush backgrounds, *(F)*, and *(G)*. In *Angry Beavers, (H)*, the styling is reminiscent of the "modern" styling of animation from the 1950s, *(I)*—although the characters are all '90s. *Rocko's Modern Life* is about a Wallaby and friends, *(J)*, with aesthetics that often feature distorted angles that extend to pan backgrounds, *(K)*. The creators of the Nicktoons shows are John Kricfalusi (*The Ren and Stimpy Show*, premiering in 1991), Arlene Klasky and Gabor Csupo (*Rugrats*, 1991, and *Aaahh! Real Monsters*, 1994), Joe Murray (*Rocko's Modern Life*, 1993), Craig Bartlett (*Hey Arnold!*, 1996), and Mitch Schauer (*Angry Beavers*, 1997). *All images courtesy Nickelodeon.*

F

G

H

I

J

K

15 ■ Digital Ink and Paint

WRITTEN WITH GREG PAIR

15.1 Drawings rough, cleaned up, and cleanest: The same cartoon figure seen as it might appear in an animator's rough drawing, *(A),* and after that rough drawing has been given a cleanup by erasing unwanted lines and making sure that interior areas are sealed off, *(B).* If necessary, this could be scanned and painted for the final production. Version *(C)* shows a tracing of the original, done with a black Magic Marker. Notice the stylistic variations in line thickness and quality that can be done with inking.

Whether you are erasing the unwanted pencil lines on a rough drawing or tracing a new version, the first firm rule in digital ink and paint is to make sure that the lines of the drawing all connect with each other. You'll see why this is important as we get a bit deeper into the process. *Courtesy Greg Pair and AMPnyc Deluxe Animation Studio.*

The preceding chapter focused on the basics of classic character animation and the process of drawing extremes and in-betweens—the unchanging process that yields those cartoon worlds we've all come to love. This is the domain of purest creativity: Alone with just a pencil, a stack of blank paper, a peg bar, and a rear-lit surface to work on, the animator brings drawings to life.

Such creativity begets drudgery. That stack of inspired pencil drawings must now follow a rigorous journey to become the full-color scene of a finished animation. In this chapter, the focus turns to those steps that come *after* the inspiration ends . . . and when the perspiration begins! We now enter the world of inking and painting.

In the early days of animation, each cartoon studio had large rooms with rows of desks at which painters (usually women) worked with ink pens to meticulously trace the animators' rough drawings onto sheets of clear acetate. Such "inked" images were allowed to dry and then taken to an even larger room, where more artisans turned over each cel and painted colors onto the back, using thick cel paints that were specially formulated to stick to the slick surface of the acetate. This step was referred to as "painting" or "opaquing." Eventually the ink and paint sweatshops of Los Angeles were replaced by specialized facilities. Until very recently, virtually all the animation on TV was "finished" in one of the Asian countries using the same labor-intensive process.

As you may have guessed from this chapter's title, the traditional ink and paint process has been touched by the miracle of computers. Be warned that this process of technological innovation has barely started. You will see in the following pages that digital animation—at least at the low end—remains a labor-intensive and somewhat awkward procedure. Yet as imperfect as it still may be, the digital miracle that allows drudgery to disappear and artistry to remain is quite potent. To fully appreciate this, let's do a little arithmetic.

Before you is a stack of sixty drawings representing a five-second segment with one character moving against a simple background. Using traditional film techniques, a reasonable estimate of how you might spend your time would include: 20 minutes to set up a camera stand for rear-lit pencil testing; 20 more minutes to shoot the sequence; 40 minutes to get the exposed film to the lab and back after it's been processed; 15 minutes to screen rushes and study the rough animation as it appears in the pencil test; let's say another 30 minutes to revise and redraw portions of the scene; 30 minutes to reshoot pencil tests; 40 minutes to make another pair of lab runs; 30 minutes to cut the new footage into the test footage or animatic and decide that the animation is ready to be prepared as final art; allow 90 minutes to retrace all the drawings onto acetate cels using a Rapidograph pen for inking (at 90 seconds per drawing); 30 minutes to prep materials and work space; 180 minutes to fill in each inked image by painting the rear side (3 minutes per cel); 30 minutes to redraw and color the background; 20 minutes to set up the stand for final, toplit shooting; 40 minutes to shoot the cels after carefully checking each for dust and smudges; and, finally, 40 minutes for a third set of lab runs.

To screen your finished five seconds of animation, you will have spent at least eleven hours. That doesn't count time spent waiting for cels to dry and handling them. Nor does it count three overnights while the lab is processing your pencil tests, revisions, and final footage. And if there are multiple levels of cels (and that's why one uses cels), the inking, painting, and shooting stages will take even longer!

Starting with the same stack of drawings, but using digital tools instead of film tools, you could see the identical piece of finished animation in 4 to 5 hours! The very same day!

A miracle? Not at all. Read on and you'll find out how and why digital techniques have started to replace traditional

A

B

C

film and video techniques. Read on to marvel at how these tools will allow you to make changes in timing and framing without ever having to reshoot (or rescan) a single one of those sixty sheets of paper.

CLEANUP/INKING

The digital ink and paint process always begins with a *cleaned-up drawing:* a sheet of white paper with a firm dark line that outlines the character—no half erasures, smudges, or faint sketch lines of initial placement.

A word of advice here: It is much less time-consuming either to erase superfluous lines on the actual drawing paper or, even better, to return to the light table and retrace the drawing onto a fresh sheet of paper than to attempt cleanup using a computer graphics application. Truth is, the eye and hand are faster at spotting and eliminating unwanted marks than a computer will ever be. The computer cannot distinguish a faint line from a partially erased line. Another good reason why the cleanup stage should be done the old-fashioned way is that the animator can focus his or her attention on making minor adjustments that keep characters "on model" and correcting those unintentional variations that always creep into a character's appearance. And although it may feel like an uncreative or rote activity, you'll find that inking a rough drawing onto another sheet of paper provides you with an opportunity to forget about what the drawing looks like and focus solely on the characteristics of the line itself. Does the character require a strong, solid line or a thin, delicate line? Or one that tapers to reflect the contours and volumes of the object itself? Figure 15.1 starts off our tour of digital ink and paint by looking at the raw output of your five digits.

DIGITAL PENCIL TESTS

The *pencil test* gets its name, as you might expect, from the days when pencil drawings on registered sheets of paper were put under the camera in order to give the animator a quick check on how well the animation was working. The technique is still in wide use today: A light source is placed under the frosted glass or milky Plexiglas surface on which the drawings are placed. This way two or sometimes three levels of paper can be filmed at the same time, with the bright, even light source shining through the stack of levels.

The resulting footage is murky, because the grain of the paper itself creates a grayed-out background against which it is sometimes difficult to make out the pencil lines held on the different sheet levels. Regardless of poor image quality, pencil testing is absolutely essential, for it gives the animator his or her first chance to study the motion of a given scene or sequence. The pencil test can be played forward and backward (on film or videotape), again and again, as the animator studies the timing and smoothness of the action. Such careful analysis usually leads to revisions and these are also pencil-tested so that the animator can be sure the movement is perfect before beginning the laborious, time-consuming, and expensive process of ink and paint.

In digital character animation, pencil tests remain the only way for creators to see whether the frame-by-frame drawings come to life exactly the way they envisioned. When done right, digital pencil tests are far, far less murky than their film/video counterparts. The case study in Figure 15.3 reconstructs the pencil-testing technique.

But there's an even bigger payoff in store. If the scanned rough drawings have been properly cleaned up or inked, then the animator will only need to enter those drawings into the computer *that one time.* The images that constitute the pencil test can undergo a digital transformation that turns them into the images of the final cartoon itself.

But this metamorphosis requires careful attention. It starts with a stack of clean images. To get to the pencil test, those images must make their way into the computer.

SCANNING

The drawings on paper (cleaned up or re-inked) must be scanned correctly into the computer in a way that maintains the accurate registration of one drawing to the next, so that when viewed in quick succession, the image "files" will fall one on top of the other just the way they were designed. In essence, the scanner becomes the equivalent of the animation

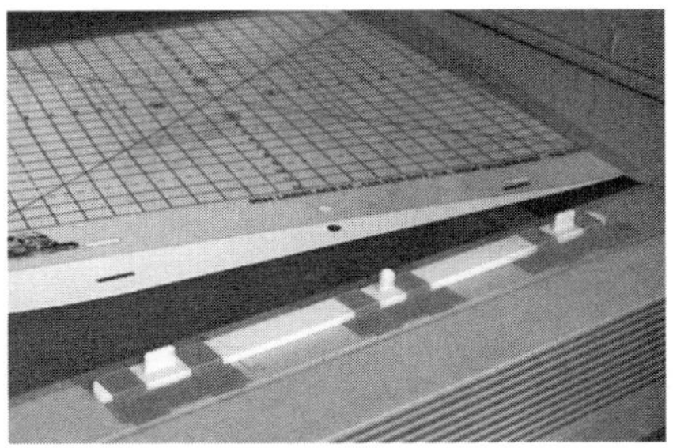

15.2 Setting up your scanner: Purchase an inexpensive plastic peg bar with the same registration pegs as the ones on your animation disc. You will tape this onto the outer casing of the scanner, as shown, following this procedure:

1) With a field guide mounted on the pegs, position the peg bar on the scanner's lengthwise casing so that it is roughly centered along the length of the scanning glass. Through all four steps, make sure the field guide stays on the peg bar and is facing right side up.

2) Slide the peg bar toward or away from the center of the scanner so that the field guide's #12 field (the biggest field) is centered along the width of the scanning glass. If the scanner's surface isn't large enough to accommodate a full #12 field, then note the largest field size it can fit and limit all your drawings accordingly.

3) Level out the top and bottom lines of the #12 field so that they run parallel to the edges of the scanning glass.

4) Carefully hold down the plastic peg bar as you tape it to the scanner's casing. Use a strong tape rather than regular Scotch tape. Try the simple, paper-base tape that is available in art supply shops. Don't use anything with a strong, gooey adhesive like duct or gaffer tape. Paper tape will do just fine. Put pieces of tape across the peg bar at opposite ends as well as in the center. With your fingernail, rub the tape tightly down the peg bar's edges.

To ensure that the paper is lying absolutely flat against the scanning glass, get a piece of foam rubber, about a half an inch thick or so, and tape it to the underside of the scanner's cover. *Courtesy Greg Pair and AMPnyc Deluxe Animation Studio.*

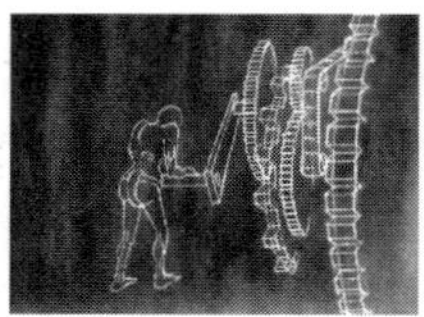

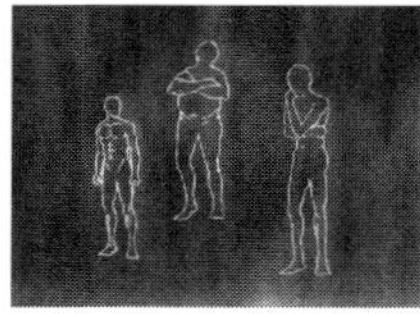

A

Many believe that pencil testing is the most critical step in animation. It is here that the movement and actions of characters and other on-screen elements can be previewed and fixed. In this case study, Randy Lowenstein demonstrates how he uses the well-known application Premiere, from the Adobe Corporation, to compile as a QuickTime file the five different layers that make up a scene from his NYU thesis film titled *The Rule of Whim*, frames from which are sampled in *(A)*. Not withstanding this chapter's strongly argued position that only "cleaned-up" images are suitable for the digital ink and paint process, this example shows how working drawings can be used effectively as final art.

At NYU we have an ancient Lyon-Lamb setup. This is a video pencil testing station that uses a video camera with zoom lens that points down onto a light box with Peg-Board taped to the top. There is a controller that allows the camera to take single frames, which are recorded onto a modified VHS videotape recorder.

I hate this machine!

It takes hours to shoot a stack of drawings. If you want to use more than three levels, the image gets so murky you can't get a good look at your animation. Worst of all, you are locked into the exposure sheet you follow during the shooting process. So if you want to see how the action works on "threes" instead of "twos" — or if you want to try out a longer hold — you have to reshoot the entire thing.

I discovered that using Premiere offers a one-step alternative for pencil testing that not only eliminates all of the problems described above but even provides new options and increased freedom. It doesn't actually save any time the first time you scan in the registered sheets of paper you've drawn, but in the end the time you can save is immense. The best thing about using the computer to pencil test is that once you have your drawings in the computer, you'll never have to touch them again. You can mess around with different timings, experiment with in-camera effects like dissolves and fades, or even play your animation backward, without ever "reshooting" your drawings. And if you want to make changes in just one of the cel levels, you can rescan just those sheets and see how they look with the other levels. And remember, on the computer there is no limit to how many levels you can have. After you've tested all of your scenes, you can even string them all together to make a test of your entire film. If you get really ambitious you can even add sound effects or music to your pencil test. Try doing that on a Lyon-Lamb setup!

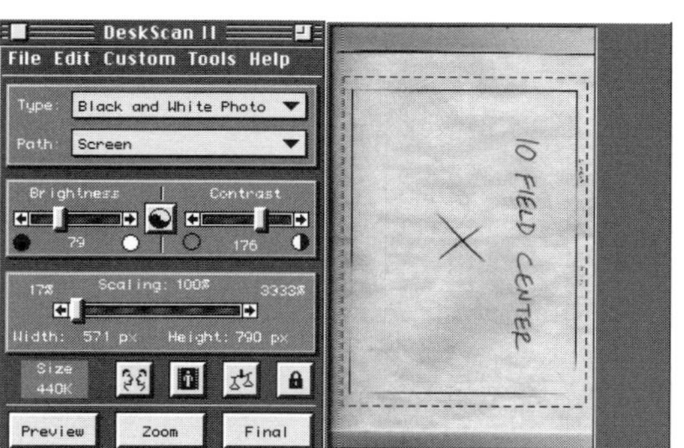

B

Scanning Original Drawings

I choose one "typical" drawing from each stack to adjust the scanner's settings for brightness and contrast. Then I lock the settings so that these settings will not be readjusted automatically for each scan. Of course, the peg bar that positions the drawings is firmly anchored on the scanner's copy surface.

The first image I scan is not of the first image in the pile of drawings for that particular level, but instead, it is one that shows my field guide, positioned on the peg bars, *(B)*. This way if you need to add a single drawing to a particular shot at a later date, all you have to do is rescan your field guide, checking to see that it matches the field guide image from before. The "insert" drawing or drawings will be perfectly registered to the ones you did before.

Now you can proceed to scan into the computer an entire series of drawings.

Naming Your Files and Batch Importing

One of Premiere's great features is that it will display a series of drawings in sequential order. The program learns what drawings live together and their order, through the filename you give them. So it's important to do this right. I like descriptive names to start, such as "big," "multi," "cranky," and "small" — the names I used for the four stacks of drawings in this scene. Note the underscore (_) that is used to link the filename to the number of each drawing in the sequence—Premiere doesn't like periods. I like to use three digits in case I get really inspired and create a cel level with more than ninety-nine drawings!

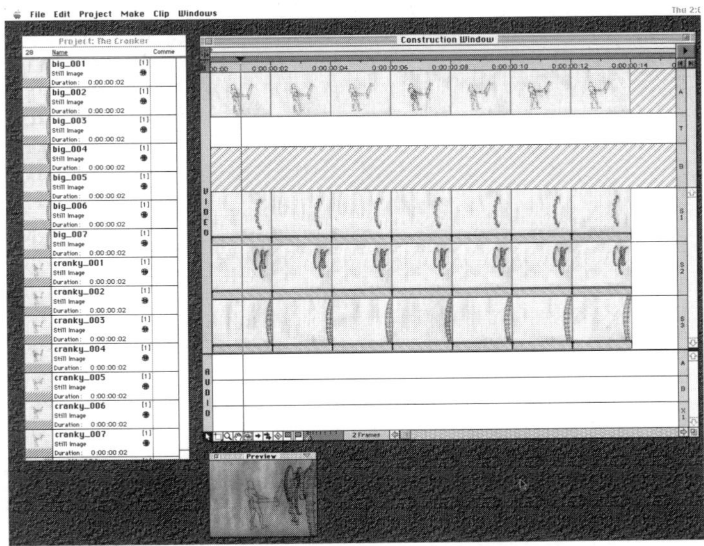

C

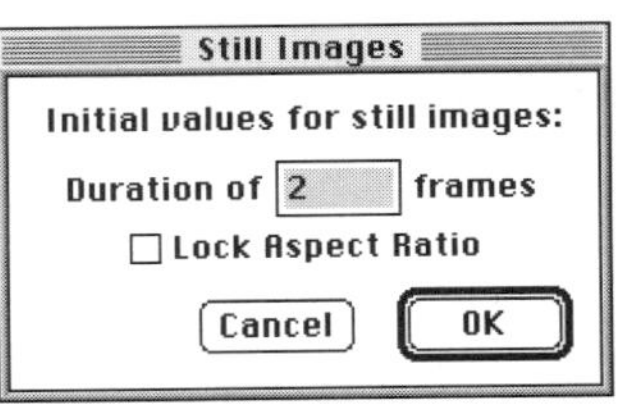

D

An entire folder of scanned artwork files can be imported into Premiere and then dragged as a whole into the construction window. Premiere automatically arranges the files in their order. *(C)* shows the Premiere interface, with a frame-grab of my whole desktop configuration. You can specify the number of frames assigned to each imported image in the Still Image Preferences menu. For example, if you want to test your animation on twos, you tell Premiere that you want each imported still to last for two frames, as shown in the dialogue box in *(D)*.

Working with Multiple Layers

When I do pencil tests, I simply superimpose individual levels of artwork over each other. This is a quick and dirty alternative to creating alpha channels in Photoshop (which float the "subject" on a clear layer). I use the "multiply" function to create an effect similar to that achieved by underlighting on a conventional pencil tester. Each pixel of a digital image has a numeric value based on that pixel's color. The "multiply" transparency mode literally multiplies the numeric values of the two images you are "multiplying." When two colors are multiplied the result is a darker color resembling the color you might get if you placed two sheets of colored acetate over each other. When a color is multiplied with black, the result is always black, and when a color is multiplied with white, the result is always the original color. When this mode is used with two images that consist of dark lines on very light backgrounds (our pencil drawings), the result is a single image with dark line drawings from both images and a light background, i.e., a superimposition of sorts. In order to use the "multiply" function in Premiere, first select the "transparency" option from the Clip menu. Change the setting from "normal" to "multiply," *(E)*. Keep in mind that in order to use the trans-

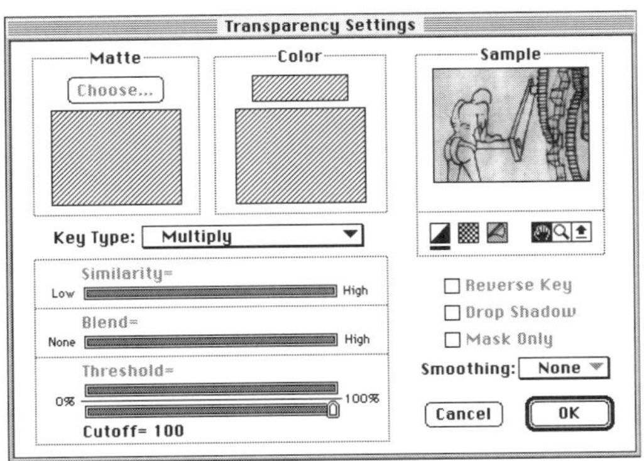

E

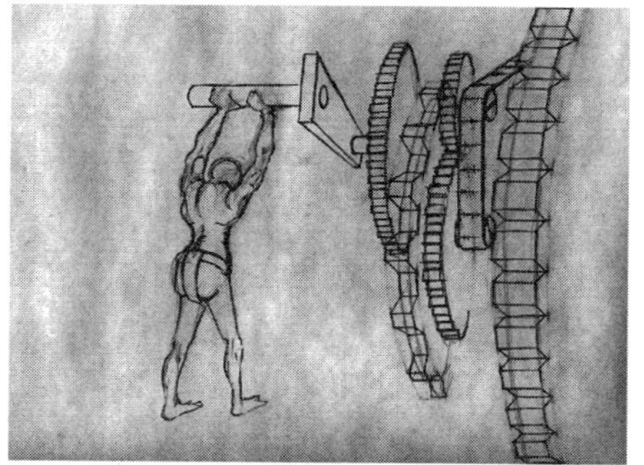

F

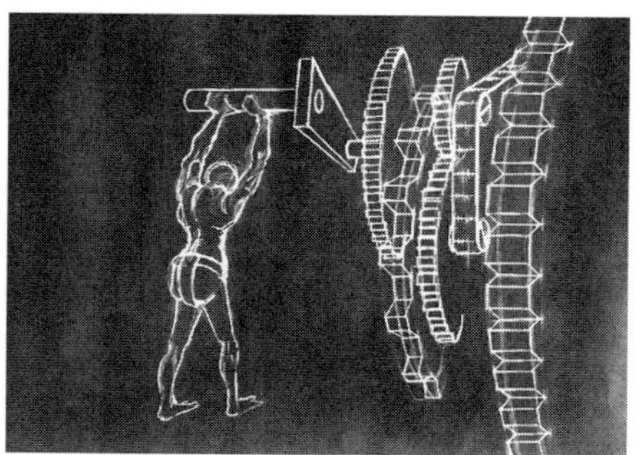

G

H

I

parency setting, artwork that is to be superimposed must be on one of the superimposition tracks, *(C),* not track A or B. The result is a computer-generated pencil test that looks just like a traditional pencil test, but is actually much more flexible and efficient.

Where You Can Go from Here

Once you've finished using Premiere to test your animation, there are many different things you can do to turn your pencil test into a completed film.

The easiest thing you can do is to simply alter your entire pencil test in some way. You can use Photoshop or Painter to add color to your already existing files, or you can use a Premiere filter to do anything from "spherizing" to "solorizing" the entire movie.

In my film *The Rule of Whim*, I wanted certain scenes to consist of white line drawings on a black background. To do this, I simply inverted my pencil test. The dark pencil lines became white and the white paper became black. *(F)* and *(G)* show the same frame that has been inverted.

Another step you can take to make your pencil test into more of a finished film is to create alpha channels for the necessary elements. This way your characters and props will be opaque and float above your backgrounds.

In another film I created, *Li'l Dainty Rudy*, all of the artwork was done in charcoal. Many of the drawings I used to test the animation ended up as final artwork in the film. After I was satisfied with my pencil tests, *(H),* I went back and created alpha channels for each of the drawings, *(I).* This way I was able to maintain a loose sketchy quality of line and at the same time I could place my characters on top of backgrounds. So it's not always necessary to ink and paint a drawing for it to be finished.

And of course you always have the option of digitally inking and painting your pencil test. If this is the route you decide to take, pencil testing in Premiere will give you a great head start. If you don't change your filenames when you ink and paint your drawings, your pencil sketches will be automatically substituted with the new inked and painted versions. The next case study will take up how to ink and paint using Photoshop.

camera and the computer becomes the equivalent of the projector. How is all this done so precisely? Whether you're drawing animation or scanning animation, you use the same registration device: a peg bar (Figure 15.2).

Most scanners have a scanning glass for artwork as large as 8.5 by 14 inches. With the peg bar positioned properly, you can scan up to a #11.5 field. But to play it safe, design your layouts and subsequent animation to fit into a #11 field, or smaller.

Before scanning scenes with a lot of drawings, make one last check to be sure the peg bar is still taped securely. Test it by trying to jiggle the peg bar at its center peg. If you feel the peg bar moving too easily back and forth, then the taping has become loose. It's good practice to replace the taping every so often since frequent use will start to wear it down after scanning hundreds of drawings.

Maintaining accurate registration begins with the peg bar but also depends upon the software of the particular scanner you are using. Most scanner software programs allow you to custom-set the size of its scanning field or *scanning box,* which is the predetermined area of the scanning glass to be scanned and digitized. It is important to make sure that this custom-sized scanning box keeps the same field for each drawing scanned, since most of the common video/animation software applications that you will be using rely on utilizing the scanned artwork's digital dimensions to maintain registration. Premiere from Adobe is one such program that tracks original scanning fields to achieve registration.

Unfortunately there are other software applications that require you to pick a single registration or "anchor" point on each individual image. This makes the peg bar useless and forces you to employ an alternative, visual form of registration. You must carefully place a registration mark (see Figure 15.4) on each original rough animation drawing and maintain it through to the cleanup/inking stage.

When you begin to scan a collection of drawings, start off with a custom-drawn field guide that shows the outside edges of the particular frame (field size) into which this particular scene has been designed. A sample is shown in Figure 15.5. Place this guide facedown on the scanner's peg bar.

Just as there are many makes, models, and manufacturers of scanners, there are also many versions of software that coordinate the activity between a scanner and a computer.

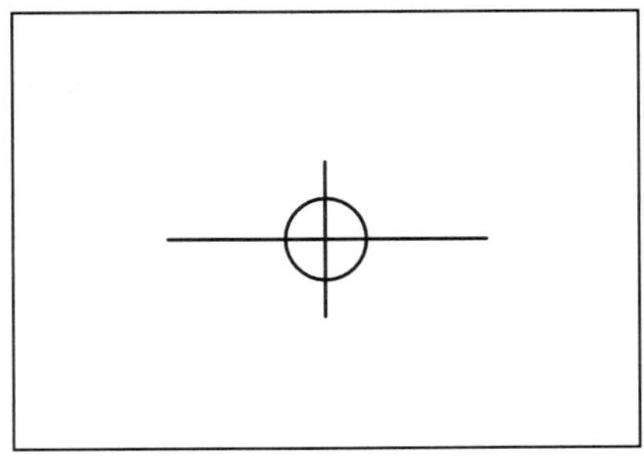

15.4 Registration bugs: Although it's a tough way to go, registration during scanning can be achieved by using visual marks, often called "bugs" as shown here, that are copied and used for each drawing. Don't attempt to draw and redraw such markers, because it's impossible to preserve the necessary precision when tracing the registration mark over and over again. It's easier to use preprinted registration stickers or to make your own registration mark in a graphics application and print out hundreds of them onto sheets of clear label stickers with a laser printer. Both preprinted registration stickers and clear labels can be found with other computer printing labels in an office supply store.

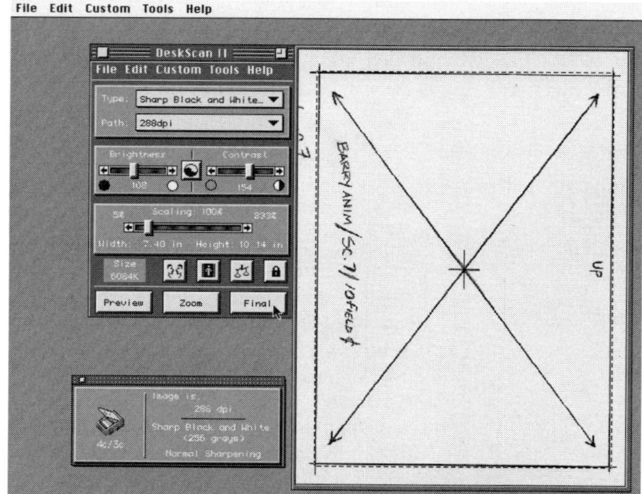

15.5 Targeting Barry: In this chapter, we follow the example of a single scene featuring a character named Barry who was created by Greg Pair and his partners Ted Minoff and Michael Adams at AMPnyc Animation. Note that the field guide shows which side is "up," centers the custom-made field guide to the middle of the scanner, clearly labels the scene, and indicates what field size has been used (here, a #10 field, which measures 10 inches wide). This framegrab shows the software interface for DeskScan II, the software that comes with the Hewlett-Packard Scanjet 4C scanner. *Courtesy Greg Pair and AMPnyc Deluxe Animation Studio.*

However, all these programs have the same basic functions.

By using the preview mode of the scanning software, animators can set the scanning field to fit the field size they have chosen for their artwork. Previewing a scan means that an image is scanned but not committed to the computer's hard drive memory storage. It is necessary to preview each scene's custom field guide in order to see how it lines up, as shown in the software's scanning window.

Once you have adjusted the scanning box of the software so that it lines up with all four sides of the drawn field guide, use the software's lock button to fix the scanning box's dimensions. Remove the field guide from the peg bar and pick out one of the animation drawings from the particular sequence you are about to scan. You need to preview-scan this drawing in order to set the scanner's software settings for (a) brightness, (b) contrast, (c) resolution, or how many pixels per inch (called ppi, or dpi, "dots per inch"), and (d) the type of image that is being scanned (black-and-white line drawing, black-and-white photo, color photo, etc.).

In choosing these settings, the goal is to create a set of black-and-white scans that will require very little time cleaning up extraneous lines and/or unwanted paper texture. If you're scanning rough animation in order to preview (pencil-test) the action and you know you will be subsequently redrawing and rescanning each image, then you can choose brightness and contrast settings that allow all the pencil lines to be picked up and easily seen, even if this means picking up paper texture as well. The quality of the image isn't all that important for pencil-testing, so the ppi/dpi setting can be set to 72, keeping each image's file size small. Whether you're scanning for pencil testing or for final coloring, the type of image setting used should be black-and-white photo rather than black-and-white line drawing, so that all gray-scaled pencil line nuances are picked up.

Now let's look at the slightly different process for scanning final, cleaned-up drawings—whether they are in pencil line or black ink. Your main objective here is to keep the lines in the drawing from getting an aliased look. That's when the lines become "jaggy" or "blocky."

In order to avoid this, the scanner settings need to be set for scans of a much higher quality. First, the ppi/dpi should be set for 288, which is four times more pixels per inch of information than 72 ppi. Of course, each image's file size will be bigger as well. The type of image setting is still set at black-and-white photo (sharp).

Here is where you must develop real finesse in playing with brightness and contrast settings. You want to eliminate as much of the paper texture and superfluous image information as possible so that all you have to deal with in Photoshop are the black lines of the drawing placed against a clean white background. You'll have to play with the settings of your particular software, but what you're trying to get is this: The brightness should be set precisely at the point where the paper texture drops away and is not picked up by the scanning process. Adjust the contrast setting so that the drawing's cleaned-up lines become black. Be careful here. If the brightness is pushed up too high, the whole image gets washed out: The line becomes grayed and the contrast will have to be pushed up even higher to blacken the line. But if the contrast level is pushed up too high, it starts to eat into the lines and makes them aliased.

It's a good idea to try a few test scans of a single drawing from the stack you are about to scan. This will give you the chance to study the effect that different settings will have on the quality of the line. But once you have the settings figured out for that drawing, they should be kept constant for the whole batch of animation drawings.

NAMING FILES

Remember, in digital animation you are dealing with a *huge* number of images that will need to be placed in a certain order. Each image must also receive a timing—whether it is at a constant rate of, say, 12 images per second or whether the particular image is held for five full seconds. As an animator, you must develop and label each drawing with a predetermined naming system, or *code,* so that the order of the drawings matches the timing sheets that are used to assign duration. An example will help to make this more clear.

15.6 Scanning resolution: In a graphics application, like Photoshop, you can open the scanned artwork files in order to view and paint them. It is in Photoshop that this file of "Barry Bertrum" is seen. *(A)* shows an enlarged segment of Barry's line drawing at the "pencil test" resolution of 72 dpi. *(B)* shows the same segment when scanned at 288 dpi. Notice when you zoom in on these two different images, *(C),* that there are more shades of gray in the 288 dpi (right) image, for a more gradual transition from black line to white background, making a smoother (less jaggy) line. This gives you an antialias line. Most tools in Photoshop give you an option to perform its function either alias or antialias. *Courtesy Greg Pair and AMPnyc Deluxe Animation Studio.*

A

B

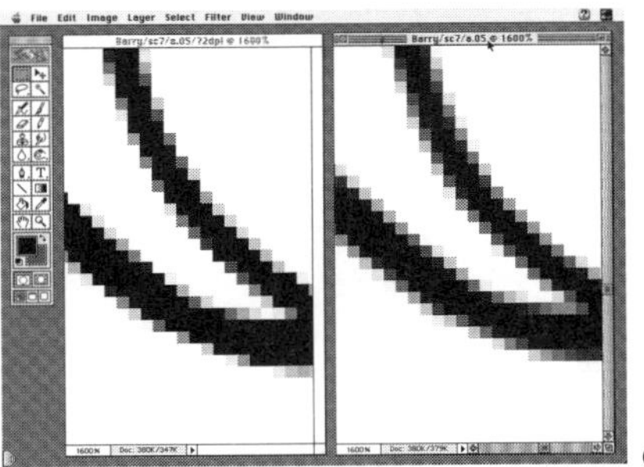

C

A

B

C

15.7 Good scan, bad scan: *(A)* has too little contrast and not enough brightness. *(B)* has too high a brightness and too much contrast, which eats at the line of the drawing. *(C)* is just right. *Courtesy Greg Pair and AMPnyc Deluxe Animation Studio.*

Written out, the drawing in Figure 15.8 would be known as Scene 7/Level A/Drawing #5. That's a pretty long name for a file, so it has been shortened to Sc07/A05. Such a code reference is one that different people working on the same project can easily understand. But the designation is also useful in organizing files on the computer. Figure 15.9 carries this process a bit further, suggesting a folder name.

Always make sure that the drawing number is at the end of its filename. If the drawing is between 1 and 9, put a zero in front of it, like 01 or 09. The filenames of all drawings in the same scene should be similar, regardless of what level they are on. Hence, when scanning all the animation drawings of level A of Sc07, the only part of the filename that you must change for each scan is the end number. Always cross check to be sure the filename matches the drawing number you originally placed in the lower right corner of the source drawing.

Once all the drawings of level A are scanned, then the next level from Sc07 should be scanned, with the filename's level changed to correspond with the new level. Continue this process until all of the drawings from Sc07 have been scanned.

An important note to save you some grief: All drawings in a particular scene should be scanned in one session, without turning off the scanning software. When you quit the software and then go back to finish scanning that same scene later, you must repeat the entire process of realigning the scanning box to that scene's specific field guide. Recalibrating leaves room for human error in setting up the scanning box, and that, in turn, can completely mess up your registration!

PAINTING IN PHOTOSHOP

With all the animation drawings from a scene cleaned up, scanned, and labeled, it's time to "paint" them. As you will see, this is a pretty complex process, but once you get your technique down, the steps become routine and your speed increases dramatically.

A number of computer companies have published software programs that are specially designed to assist in coloring the thousands and thousands of cels that are created in the course of making a TV show or feature film. Such digital ink and paint systems often require very high-end desktop hardware. These are sophisticated, highly specialized packages designed to function within large, complex, and well-financed operations. Unfortunately, such specialized ink and paint programs are beyond the reach of all but a handful of those who will read this book.

What we're going to focus on instead is the digital paint process as it can be done using one of the oldest and best-established computer graphics programs. That program is Photoshop, created and refined over the past decade by the Adobe Corporation. Photoshop is a multilevel, comprehensive editing program that is used not just by animators but also by graphic designers, illustrators, photographers, and other visual artists. Photoshop isn't something you can learn in a day. Successfully using it as a digital planning tool will require lots of patience.

The best method of explaining painting with Photoshop is to approach it step by step. The following case study will demonstrate some of the depth and power of Photoshop.

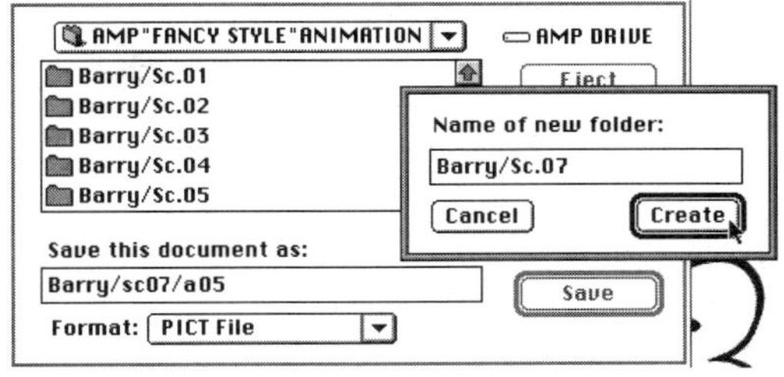

15.9 Nomenclature: Here are interface windows from Photoshop. All the scanned images from a scene are placed into their own distinct folder, which is named "Barry/Sc.07." *Courtesy Greg Pair and AMPnyc Deluxe Animation Studio.*

15.8 Barry's original drawing: Here is #5 in a stack of cleaned-up drawings of Barry's head shot. In the bottom right corner, near the peg hole, you can see the number 5, preceded by the letter A, indicating that this belongs to the A level. If the animator had been drawing a pie that was heading towards Barry's grin, that set of drawings would probably be done as a B level. On the opposite corner of the original drawing, the scene number has been noted, "Sc7". In a movie with multiple scenes involving the same character, it is important to number all scenes. This usually takes place at the storyboarding/layout phase. If a new scene is added, it becomes "Sc7a" or "Sc7.1." *Courtesy Greg Pair and AMPnyc Deluxe Animation Studio.*

15.10 CASE STUDY: Digital Ink and Paint with Photoshop

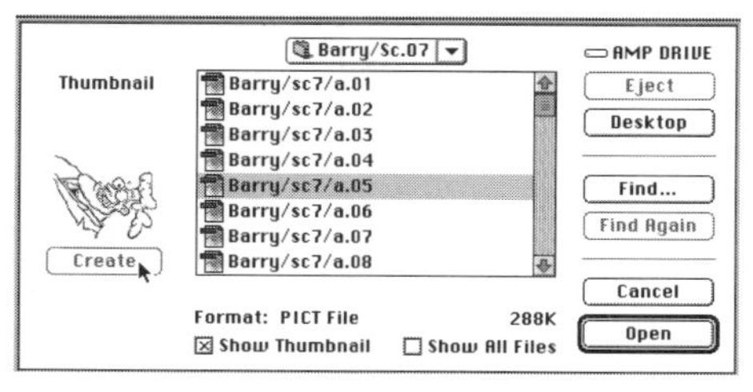

A

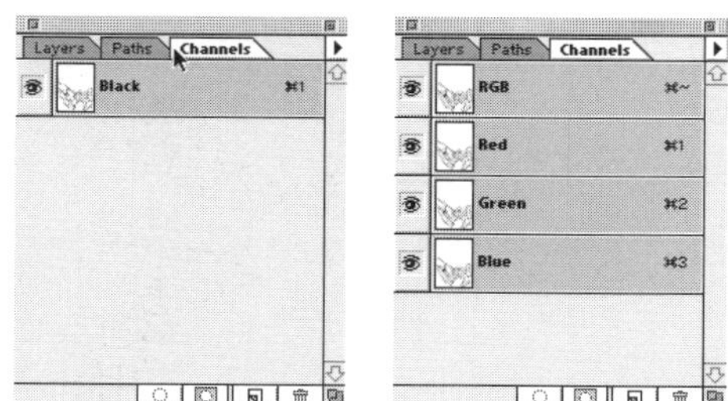

B

C

This case study puts you in the capable hands of digital animator Greg Pair. Stick with him as he works his way through the steps he used to color a scanned black-and-white drawing.

#1 **Scanned black-and-white images become color images.** When you first open a scanned image file in Photoshop, it will appear just as it was first scanned. It's a good idea to click on Create so that a thumbnail picture of the file is made, *(A)*. A couple of adjustments will need to be made before any coloring can begin. First the image's Mode will need to be changed from gray scale to RGB, *(B)*. Note how the "channels" went from just one, "Black," to three channels: "Red," "Green," and "Blue" (often abbreviated as "RGB"). A fourth channel, or "alpha" channel, will be added later for matting purposes.

#2: **Rotating the image.** The entire image needs to be rotated so that it is right side up. That's taken care of at Image > Rotate Canvas > 90 CCW. (In calling out a sequence of menu selections to take place on the computer, the ">" means "go to" and usually requires you to move your mouse into a new portion of the menu.)

#3: **Place color model adjacent to unpainted cel.** Before you begin to paint a sequence, you must have already made a *color model* of the character or object you are working on. This planning step does more than simply allow you to work out the aesthetics of choosing colors, although that itself is very important. In the digital paint process, you work directly from the color model, using Photoshop's *color picker tool* to quickly select the right colors as you need them when painting the black-and-white scan. Note that both the color model and the black-and-white scan can be open at the same time, *(C)*. Just resize and position the color model's window so that it's still easy to work on the b&w scan. Complete step three by using Photoshop's color picker tool (the eyedropper icon) to select the first of the colors that you will be filling into the black-and-white scan.

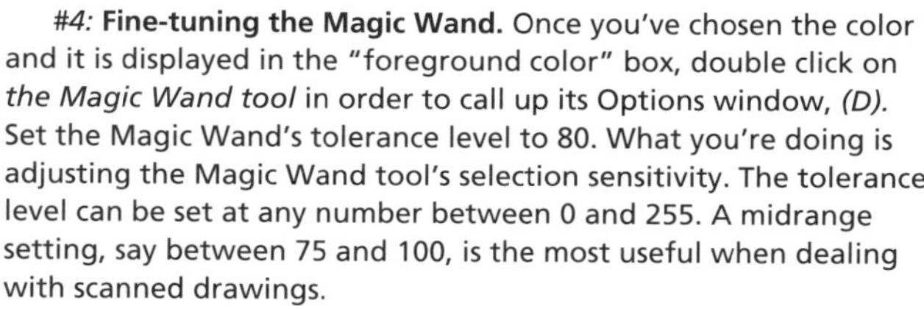
D

#4: **Fine-tuning the Magic Wand.** Once you've chosen the color and it is displayed in the "foreground color" box, double click on *the Magic Wand tool* in order to call up its Options window, *(D)*. Set the Magic Wand's tolerance level to 80. What you're doing is adjusting the Magic Wand tool's selection sensitivity. The tolerance level can be set at any number between 0 and 255. A midrange setting, say between 75 and 100, is the most useful when dealing with scanned drawings.

#5: **Prepping a color area.** On the black-and-white image, use Photoshop's Magic Wand tool by clicking once in the white of the area you wish to fill with the first color you've chosen from the color model. If the white area you clicked into is defined by solid black lines surrounding and enclosing it, then the Magic Wand

E

F

G

H

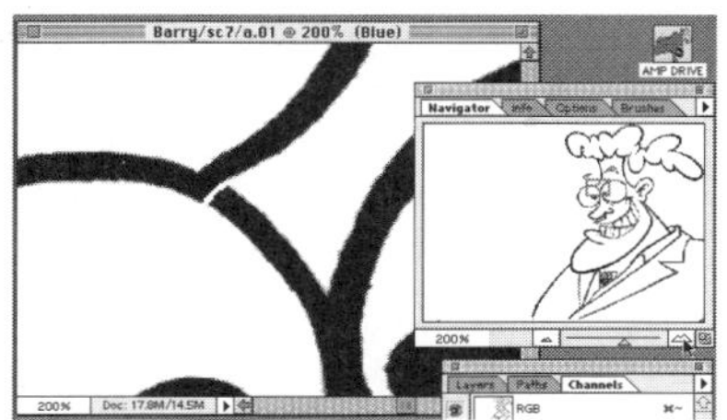

I

tool will only select that confined area—which is what you want! In *(E)* you can see how the area of Barry's face is highlighted by the "marching ants" interface convention of the Magic Wand.

#6: **Repairing leaks: finding the gaps.** But, if there is a gap in any of the area's defining lines, then the Magic Wand selection will "leak" out through that gap and into the adjacent white areas. The leak will spread until it finds itself confined by solid black lines or by the edge of the screen, whichever comes first. Illustration *(F)* shows leaks involving Barry's right eye, two of his bottom teeth, and his shirt collar. If you were to try coloring Barry's face now, your image would look like *(G).* When this happens, you've got some line repair to do!

Chances are you will run across a leak at some point, so we'll explain leak repair in two distinct steps in the process unfolding here. In the black-and-white image, study the moving selection dashes, or "marching ants" as most people call them, that have been created by the Magic Wand. Often you can spot the leak just by looking at the marching ants. If you need to get closer, or if the ants are distracting, use the keyboard shortcut Command D to deselect and employ Photoshop's Navigator window—*(H), (I), (J)* and *(K)*—to zoom and pan around the image in order to hunt down the gap.

L

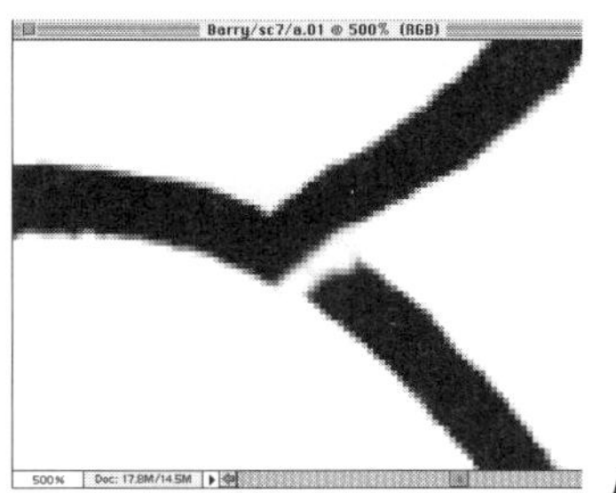

J

K

#7: **Repairing leaks: filling the gap.** Once you've found a gap, you need to use the Paintbrush tool. Double click on the Paintbrush icon so that its Option window is called up, *(L).* Select Paintbrush Options > Mode: Normal > Opacity > 100%, *(M).* Next go up to the menu bar and select Windows > Show Brushes. Here, choose a brush size that most closely matches the line thickness of the scan, *(N).* With the right brush size, you can quickly fill the gap(s) in the line(s) and get back to painting. Once you've chosen a brush size, then choose black as your foreground color to paint with.

M

#8: **Completely checking all areas to make sure there are no leaks.** Once the known gap(s) are painted in, get the Magic Wand tool again and systematically click once on each of the areas that you want to fill in with the same color. If there are no leaks, the area you select should be ready to fill with the color you've selected, *(O).* Are there any other

N

O

areas on the black-and-white image that you want to color with the same color at the same time? You can do that by simply holding down the Shift key as you click the Magic Wand onto each of these areas. This is handy when filling in lots of polka dots, checkered squares, stripes, or anything else that's a repeating pattern or a predominant color on a character.

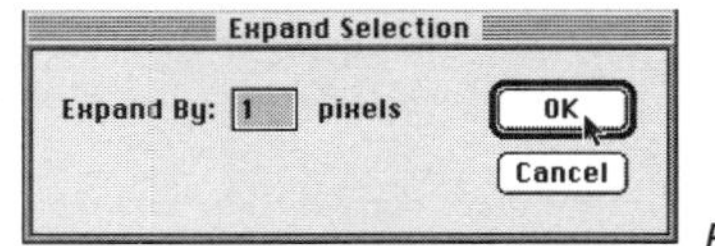

P

#9: **Expanding color areas into surrounding lines.** Once you have selected all the areas that carry the same color, go up to the Photoshop menu to Select > Modify > Expand. The Expand Selection window will pop up, *(P);* type in "1" for number of pixels to expand by. This step expands the selection outward from the center by one pixel, which will push the colors of all the areas you have painted into the black line of the drawing itself. This is necessary in order to make sure that when these areas are filled with color, the whole area is filled right up to and into the black line.

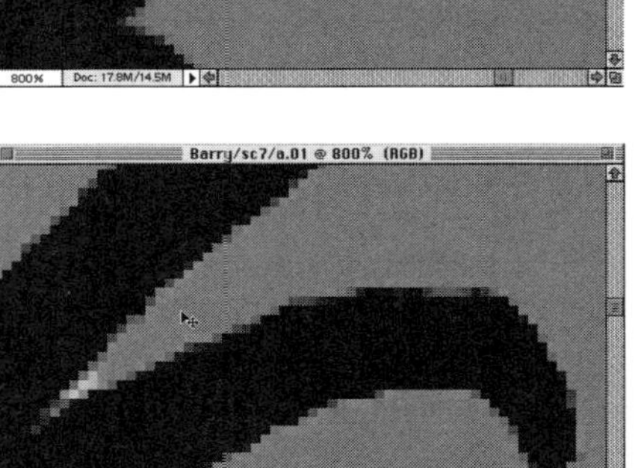

Q

R

#10: **Filling the drawing with colors.** Finally! Now the selected areas are ready to be filled with the color. Go up to the menu bar to Edit > Fill > Opacity: 100% > Mode: Darken. With the Fill mode set at Darken, the black pixels of the lines are protected and only the white pixels and other shades are painted the foreground color. *(Q)* shows a close-up of an inner area painted correctly. *(R)* shows the same inner area that was not expanded adequately, demonstrating why Step #9 is so important. Look closely and you will see light gray/white pixels, which, as you might expect, will be visible even when one is looking at the entire image. This expand-by-one-pixel technique guarantees that the areas will be filled in completely without any inner line aliasing or unwanted antialiasing gray fringing inside the colored areas. *(S)* shows Barry colored and ready.

S

#11: **Resaving.** At this stage, with the image completely painted, it's a good idea to save the file, just in case some unforeseen computer foul-ups come up. The colored image should be saved under a different filename than the original black-and-white one (just to be safe, save that black-and-white file). Here's a fast and simple way to modify the name:

Barry/Sc07/a01 = the original black-and-white image
Barry/Sc07/a01c = the color version of that image

The full filename is maintained, with just a single letter added onto the end in order to denote that this file is a color version of the original black-and-white image. The color files can be kept in the same folder as the black-and-white files or in a new folder called Barry/Sc07 Color. It's that simple!

T

U

#12: **Creating the alpha channel.** The color files you have made in Photoshop all show the colored image against a white background. Now you must replace the background white with a transparent surface that is like a painted image on a clear acetate cel. Putting it another way, you want to keep only the painted image so that it can be composited on top of background art. The white background surrounding the painted image must be digitally cut out. This is easily done using a special digital feature that is the equivalent of the clear celluloid sheets that give cel animation its name. This is a fourth color channel called an *alpha channel.*

At this point, we've selected one of the colored Barry images, clicked on Photoshop's Magic Wand tool, and then clicked on the white area that surrounds the figure we've been painting. If there are any "negative spaces" that must become transparent (for example, the inner space formed when a person's elbows are spread and his fists remain on his hips), then hold down the Shift key and click on those white areas as well with the Magic Wand. Your fully painted character should end up surrounded by the "marching ants," *(T).*

With the white areas selected as such, go up to the menu bar and click on Windows > Show Channels. The Channels window will pop up. Click on the Channels window's pull-down menu and choose New Channel. In the Channel Options window, *(T)*, Photoshop will give you the default setting as shown.

A new, fourth channel should appear in the Channels window, *(U).* In this empty channel, the selected area of "marching ants" should still be seen against white. Go up to the menu bar and get Image > Apply Image. The Apply Image window's options should be set as shown in *(V).* Make sure to "x" the Invert box as well. After you hit Return, the selected

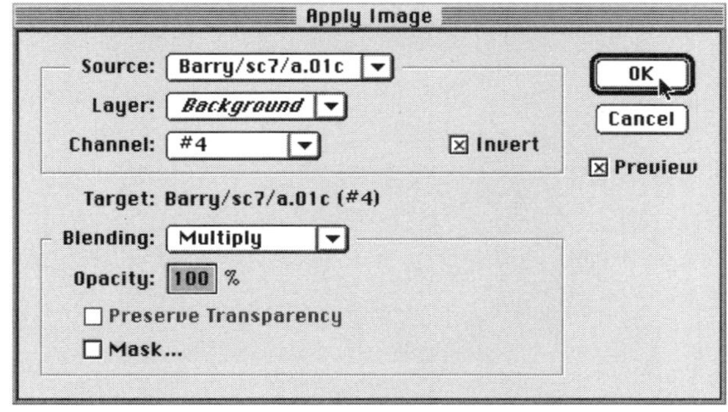

V

W

area will fill with black. The black represents areas of the image that will be ignored when composited with other images, *(W)*. With the white background area of the colored image being ignored, the painted character is now on a digital "cel." After completing the alpha channel, save the file. Do not save it under a new filename, but replace the old, colored file with this new, modified one.

#13: **Final image resizing.** After alpha channels have been cut for each image you have painted, there is one final step that has less to do with the look of the animation (all your cels will now appear correctly registered over the background you have selected) and more to do with file size and how efficiently your computer can play back the finished sequence.

You finish by reducing resolution. Recall that in order to maintain good line quality, it was necessary to scan, color, and alpha-channel the image at 288 ppi. Now that the painting stage is over, the image's size can be reduced to a smaller size that meets standards for television—see *(T)* through *(W)*. A video image on standard broadcast television has the image size of 640 by 480 pixels presented at the low resolution of 72 ppi.

The trick at this final stage is to make sure that the image's new size is what you really want and need. Remember, the image size required for standard broadcast video is 640 by 480 pixels. If you change the image's resolution down to 72 ppi and its height and width drop below 640 and 480 respectively, then you have an image that is too small for the television screen ratio. Sometimes you may find it helpful to "stretch" the image a small bit, say 20 to 30 pixels, in order to bring it up to the minimum of 640 by 480. If that's the case, go up to the menu bar at Image > Image Size to open the Image Size window; then type 480 into the window's Width box for pixels so that it meets the standard.

A surefire way to make sure that the line quality isn't affected at all and to meet the 640 by 480 ratio is to change the image's resolution from 288 ppi to 144 ppi. The end result will mean a larger file memory size than a 72 ppi file, but the line quality is guaranteed to be safe.

#14: **Saving at playback resolution.** Each colored image on its alpha channel background can now be dropped down to this broadcast resolution by once again opening the Image Size window, *(X)*. As before, you will see the image's height and width measured in pixels as well as inches (or other unit of measurement you choose). The image's resolution rate is also displayed. Following the paint and alpha channel process, the image is at 288 ppi. Before typing new numbers in any of this window's boxes, make sure that the two options "Constrain Proportions" and "Resample Image: Bicubic" are both active with an "x" in their boxes. With both checked, all that's left to do is to change the image's resolution from 288 ppi to 72 ppi. The image's height and width pixel measurements should adjust accordingly, by shrinking down to lower numbers, as shown in *(Y)*.

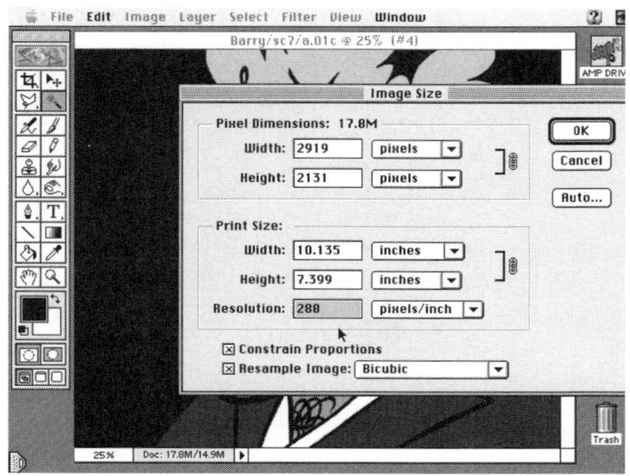

X

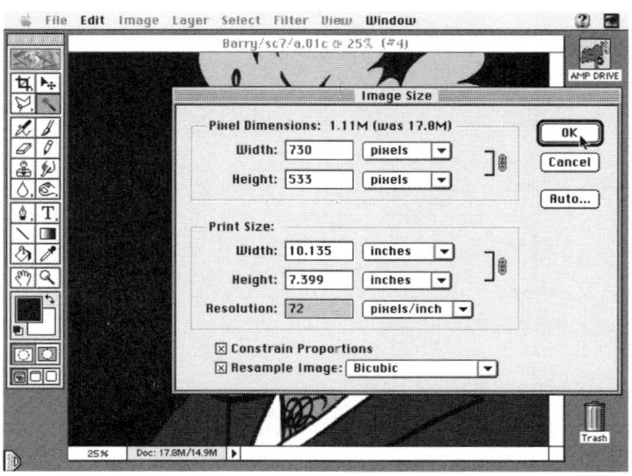

Y

BARRY IN HIS MOVIE

You deserve a payoff for working your way through the digital ink and paint process. Here it is—Big Barry Bertrum, seen in action with a barrel of rather stinky monkeys, from a promo for the Cartoon Network.

By knowing the basic Photoshop methodology, you will be able to assess and master other desktop ink and paint programs, such as Streamline, Illustrator, and Freehand, all from Adobe Corp. Using these programs, you can turn your originally scanned art (which is a master image, meaning the ink lines are stored in individual pixel locations) into a vector image, where your ink lines become real curves that you can modify as line art based on mathematical algorithms. And this knowledge will enable you to evaluate and conquer the next generation of ink and paint programs.

16 ▪ 3-D Animation

WRITTEN WITH DAN SCHRECKER AND DAN MOSS

Three-dimensional computer animation (3-D for short) is probably the most prevalent form of animation seen in media today. As its popularity has skyrocketed, 3-D has forever changed the look of TV, movies, video games, the Internet, and other forms of entertainment.

The ability to create convincing 3-D animation was once limited to high-powered, expensive workstations, but with computer technology on the move—always improving and always becoming more accessible—the cost of 3-D software and the platforms to run it on has dropped dramatically. While higher-end systems remain the ones that are used to make the most professional-looking imagery, it is now possible to create quality animation on a home computer.

16.1 Cobble street: The above image shows the level of realism possible with even "low-end" software. This was created using Specular Infini-D, a program that sells for about $600. *Courtesy Dave Merk and Jen Jeneral.*

If you have no experience with 3-D animation, it is helpful to think of it as a digital hybrid of classic animation techniques and live-action film. All of the basic principles of animation still apply, and the best 3-D animators have usually started as cel or stop-motion animators. Traditional animators' knowledge of movement, weight, and expression of character allows their work to be far superior to the overused "flying logos" and traveling-camera moves so typical of poor 3-D animation. A working knowledge of film production is also a key tool in creating high-quality 3-D work. Just as a live-action director sets up cameras, positions actors, and lights scenes, so too must a 3-D animator.

Because the computer aids in the process, a common myth is that 3-D animation is easier, faster, and even "better" than other forms of animation. While the machine does, in fact, draw every frame of the animation, the entire process is complex and takes some getting used to. As with any form of animation, 3-D requires liberal amounts of patience and diligence, but with a home computer and software costing less than $500, you can create amazing imagery that realistically duplicates our own world or explores new visions limited only by the imagination.

THAT THIRD DIMENSION

Before getting into the details of 3-D animation, a basic grasp of "space" is required. Other forms of animation, such as cel and line animation, generally work in two dimensions. Such "flat" surfaces comprise the Cartesian plane, which is defined by an x-axis and a y-axis. Any line or geometrical shape can be defined as a series of coordinates within this grid.

Obviously, 3-D animation introduces a third dimension. This adds a new coordinate, the z-axis, and you now have three-dimensional space within which to work. Instead of simply drawing a square and moving it horizontally across a frame, as you might do with cutout animation, the computer allows you to take your square, transform it into a cube, and move it *around* the camera. In traditional two-dimensional animation, the dot on a piece of paper could always be located by a field guide showing north, south, east, and west from the center of the frame. The third dimension adds the possibility of forward and back. The z access slices through the center of the frame at 90 degrees from the xy plane.

It probably won't surprise you to learn that the third dimension of space has a big impact upon the process of animation itself. Whereas in other techniques one has some freedom in how to attack a new project, in 3-D animation there are five basic steps that must be followed in this order: *modeling, applying textures, building the scene, animating,* and *rendering.* Although there is some overlap, this sequence is fairly rigid in that it is necessary to complete one process before moving on to the next.

Each step is very time-consuming and it is all too easy to

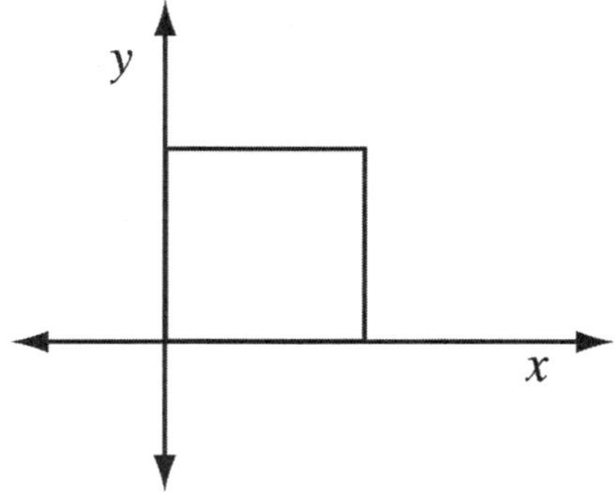

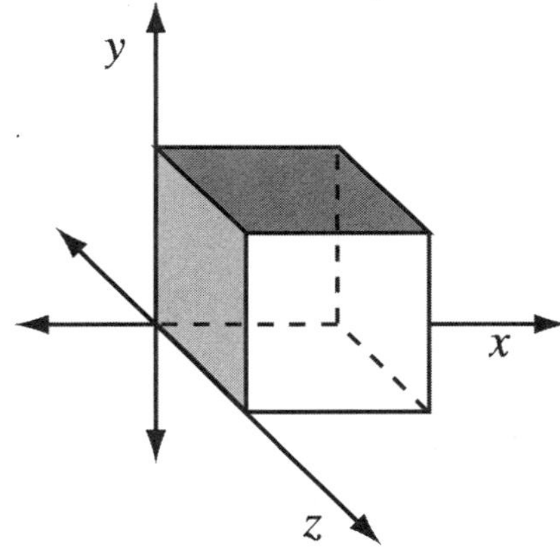

16.2 Space: In order to work in three dimensions, you have to be able to think in three dimensions. By "extruding" a square along the z-axis, a cube is created. *Images courtesy Dsquared, Inc., unless otherwise noted.*

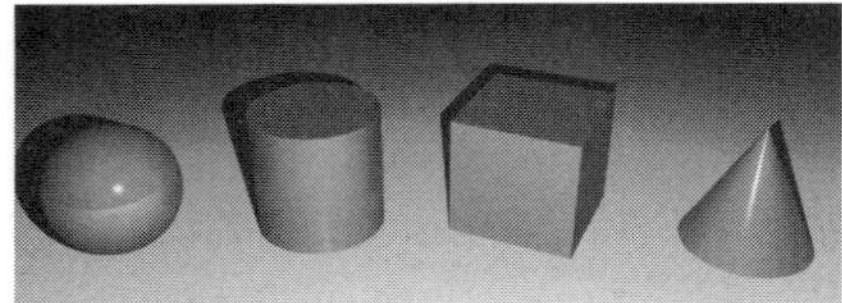

16.3 Primitives: Sphere, cylinder, cube, and cone.

lose track of one's overall idea in the detail of the new creative tools. Hence it is crucial to have a good storyboard in place before you begin, so that you don't end up going backward and losing track of what is important.

In fact, each of the five steps within the creative process can be so elaborate that professional 3-D animators often specialize in only one. As you explore 3-D, you may find that you are drawn to one of these areas more than the others. Consequently, you can learn to tailor your 3-D animations to highlight your strengths and minimize those areas where you feel less interest and/or have less skill.

MODELING

Regardless of what kind of animation you are planning, the first thing you must do is create the objects you plan to use in your animation. In keeping with the live-action metaphor, objects can be thought of as anything that is in front of the camera. This includes actors, props, and scenery. *Modeling* is the process through which you build all of the objects that appear in your scene.

The easiest way to get started modeling is by using *primitives,* which are simple 3-D shapes that come with most software packages. These cubes, spheres, cylinders, cones, pyramids, and planes can be combined to create more complex objects. This process is very much like building with blocks or Legos. You don't actually build the pieces you use; you just put them together. Once you've built an object by combining some primitives, it is likely you will want to link these together so that whenever you move one piece, the rest will follow.

16.4 Primitive temple: This Greek temple was constructed with nothing but cylinders and cubes. The primitives were stretched and squashed varying amounts.

A second way to create a model is by starting in two dimensions. Almost every 3-D package gives you the ability to draw a curve or shape, manipulate it, and use it as a basis for creating a model. The types of curves you can create vary from program to program, and each type has its own set of characteristics. In addition, most software will allow you to import 2-D curves from some other program, such as Adobe Illustrator.

After creating a curve, there are a number of ways to bring it into the third dimension, the simplest being *revolving* and *extruding.* Revolving is the process of taking a 2-D curve and rotating it around an axis so that it acquires some volume.

A good example of this is the construction of a wineglass. In order to make a wineglass, you must first draw a curve that represents half of the silhouette of that glass. By taking this curve and revolving it around its vertical axis, you give volume to what was once a flat shape. When you extrude a 2-D shape, you are taking that shape and pulling it off "flatland" into 3-D space. For example, a cylinder is simply an extruded circle. Shapes can be extruded along a straight line or a curved one.

Don't be too overwhelmed by modeling. It takes a lot of time and effort to master, but wonderful results can be achieved with even a cursory exploration. If you find that modeling is not your strength, you will be relieved to know that many software packages include a library filled with premade objects such as cars, tables, cups, people, animals, and even dinosaurs. In addition, companies such as Viewpoint Datalabs have made a business of selling models to 3-D animators. It is possible to buy models of anything from the Eiffel Tower to a 1956 Chevy Bel Air to the most intricately detailed sunflower.

Polygons Versus Curves. What is a 3-D object actually made of? To the computer, each object is nothing more than a series of numbers that define the object's boundaries. Several distinct methods have been developed to define objects, and different software packages utilize different systems. Some programs use polygons to break up an object's surface into straight-edged geometric shapes, such as triangles and quadrilaterals. The more polygons a model has, the smoother the surface. Curve-based modelers define an object as a series of curved *splines.* Because this type of modeling offers tremendous flexibility and smoother edges, it is the method employed by high-end software, such as Alias.

Although different methods are employed by different modeling software to create the surface of a 3-D object, all renderers *tessellate,* or break down, this surface into polygons. This is done because the renderer uses the small, flat surfaces of the polygons to mathematically compute the color and light information of each point of the object's surface.

TEXTURING

Once you've finished modeling, you are ready to begin the next step in the 3-D process. If objects are your actors, then *textures*

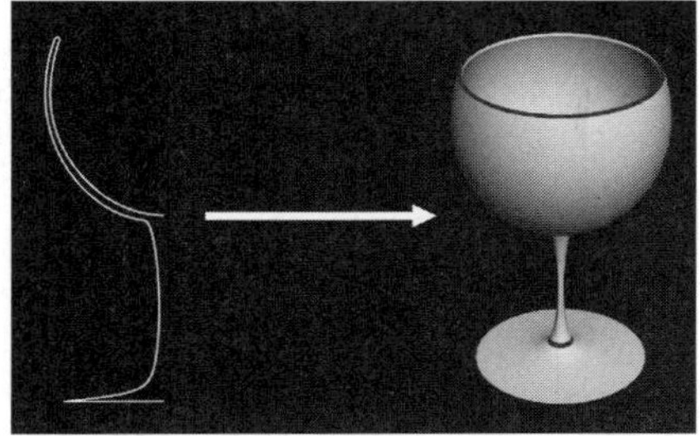

A

B

16.5 Revolving: By taking a two-dimensional curve and rotating it around an axis, you can create 3-D objects. Above, a wineglass, *(A),* and a chess piece, *(B),* were created from shapes made in Adobe Illustrator.

16.6 Extruding: Here, a star shape was extruded along a curved line to create a tube. Note that in order to create a hollow object, the star shape has to have some thickness along its edge.

are their costumes. While modeling defines the shape of your objects, textures define what their surfaces look like. There are two basic types of textures: *materials* and *image maps.* You will find that most projects require both types of textures.

Most programs come with a set of materials that you can simply select from a menu and apply to your objects. There is a wide variety of metal, stone, wood, skin, glass, and liquid to choose from. As in the real world, materials differ from one another in the way that they reflect light. The different ways in which objects appear to the eye, say glass versus metal, is the result of a combination of several different properties. These include color, diffuse shading (how much light it reflects), ambient shading (how dark the shadows on its surface are), reflectivity (how much of its environment shows on its surface), luminence (how much light it emits), and transparency (how much you can see through it). All of the premade materials that come with software packages have these properties already defined. Glass is defined as transparent, partially reflec-

A

B

C

D

16.7 Enhancing realism: *Toy Story* was the breakthrough 3-D computer film made by Pixar and distributed by Disney. This stunning achievement did not come from nowhere. Pixar's Animation Production Group, creatively led by writer/director/animator John Lasseter, spent a number of years working on self-funded short films. Four of these are sampled here. Together they demonstrate how textures such as wood, plastic, metal, and the fabric of a couch help 3-D artists create believable imagery and environments.

Luxo Jr., (A), broke new ground with its ability to imbue inanimate objects with personality and emotion. The film was introduced at the 1986 SIGGRAPH convention and went on to win over twenty festival prizes.

Red's Dream, (B), was completed in 1987 and encompassed several technical achievements. A number of scenes were rendered with procedural texturing techniques, self-shadowing, and motion-blur.

Tin Toy, (C), was Pixar's first work to feature the animation of a human character. The baby's face required the definition of more than forty facial muscles, which were grouped by function to allow the animator better control in creating expressions. Computation of final color images was performed using RenderMan™ from Pixar, for 3-D scene description. This proprietary 3-D application subsequently was released for desktop hardware suites. *Tin Toy* won an Academy Award in 1989 for Best Short Animated Film.

Knickknack, (D), was designed in 1989 as a 3-D stereoscopic film to create a unique three-dimensional visual experience. The film can also be viewed as a traditional two-dimensional work. The cartoon features an original musical score by Bobby McFerrin. *(A) through (D) courtesy Pixar.*

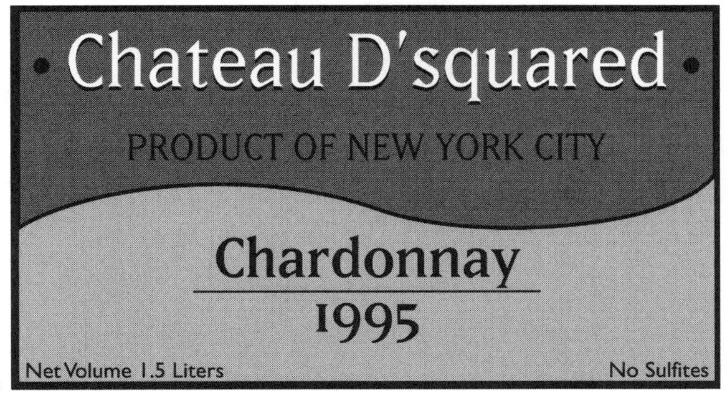

A

B

16.8 Image maps: The wine label, *(A),* was created in Adobe Photoshop, and mapped to the front of the wine bottle in Infini-D, *(B).* There are a number of different ways to place flat images on a 3-D surface. In this case, a "cylindrical" map was used. Other possible types are planar, spherical, and cubical.

tive, and a little diffuse. Stainless steel is completely opaque, very reflective, and barely diffuse. These properties can be combined in an infinite number of ways to produce any desired result.

Image maps are 2-D images that are "placed" on the surface of an object. Think of a wine bottle to go with your wine glass. To make the bottle appear to be made of green glass, you could probably just select a premade material from a menu. But to put a label on the front, you need to first create a label in a paint program, such as Photoshop (or scan one in). You would then import this file into your 3-D software and place it on the front of the bottle. As with other textures, you can define the surface attributes of your image map in order to make it reflect light as an actual paper label would. Image maps (and their cousins, bump maps) can also be *tiled* to cover an entire object (see Figures 16.8 and 16.9). Some software packages also allow you to apply moving images as textures, allowing you to map, for example, a piece of video onto the front of a television model.

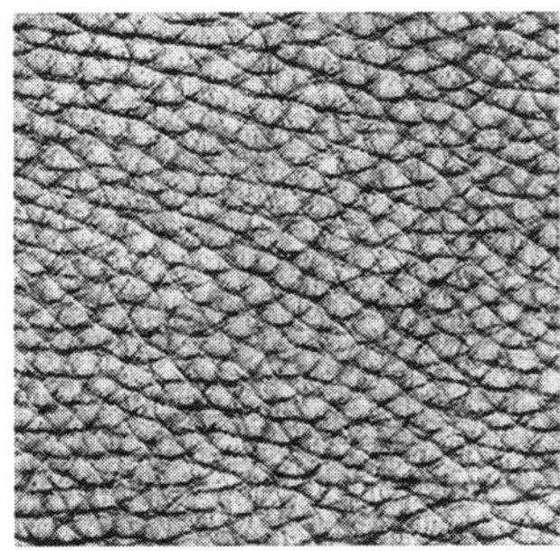
A

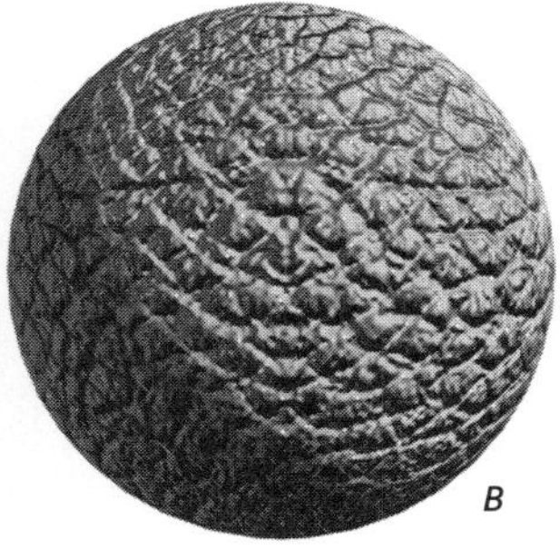
B

16.9 Bump maps: A bump map is a gray-scale image in which the lighter parts of the image appear to be higher than the darker parts, with all of the levels of gray in between. Here, a bump map was used to simulate a leather texture, *(A),* and then tiled to a sphere shape, *(B). (A), ©Form and Function from the Wraptures One CD-Rom texture library.*

A

B

C

16.10 Digital lighting: The drawings above illustrate how the three most common types of 3-D lights affect the same scene: parallel light, *(A),* radial light, *(B),* and spotlight, *(C).*

BUILDING A SCENE

Now that your objects are fully textured, you are ready to build your scene. There are actually three stages to this: *Composition, camera setup,* and *lighting.* It is at this point that any live-action filmmaking and TV production experience you may have had will come in handy. Just as you would place human actors, props, and scenery in front of a camera and light them, you now take your objects, position them in front of the camera, and arrange a variety of lights to illuminate your scene.

The virtual camera that you will employ in 3-D animation is very much like a real camera but is not limited by the laws of physics. It can zoom in and out, dolly forward and back, and vary its focal length, just like an actual movie camera. It can also change size, fly through the air, or automatically track an object moving through the scene. In the end, though, it remains a camera: your eye on the world. Viewers will see only what you show them, so it is essential to compose your frame carefully.

For novice 3-D animators, lighting a scene can be one of the most challenging parts of the whole process. The balance between fully illuminating your scene while still maintaining the proper mood comes only with practice and repeated test renderings. Most software packages come with at least three types of lights: spotlights, which direct a beam in a single direction from a single point; radial lights, which emit light equally in all directions, like a bare bulb; and parallel lights, which, like the sun, emit light equally in a single direction. Each of these types of lights has its proper usage and own set of characteristics (see Figure 16.10). As a 3-D designer, you can control the focus, intensity, and color of each light in order to create the right look. If you find yourself drawn to 3-D production, it is probably worth it to pick up a book on film, television, or theatrical lighting.

ANIMATING

By modeling, applying textures to your objects, and building a scene, you can create a very photorealistic still life. However, in order to create an animation, you've got to introduce the element of time. It's useful to remind yourself regularly that *anything* in your scene can be animated. Objects, cameras, lights, and in some cases textures can all be changed over time in a number of ways, including size, position, and rotation. Many of the techniques utilized in nondigital animation carry over onto the computer. As with cel animation, a 3-D animator builds key frames that define an object's motion. Unlike cel animation, though, no in-betweening is required, as the computer computes all of the intermediate frames necessary for a smooth animation. Objects can be animated along straight lines or curves, and many software packages are equipped to automate basic animation timing methods such as ease-in and ease-out. The more knowledge you have of classic animation techniques, the more polished your 3-D animation will look. The concepts of stretch and squash, anticipation, and exaggeration are as relevant to 3-D as to any other form of animation.

Inverse kinematics is one type of 3-D animation worth noting. IK, as it is called, is the process by which the computer is able to animate a human figure while maintaining proper anatomical structural relationships. If you had a human figure you wanted to make run, it would be a nightmare to first move a thigh, then a shin, and then a foot, all the while making sure these body parts were still aligned properly. IK allows you to simply move a single body part, and have all the adjoining body parts move accordingly within a defined range of motion. It is this type of technology that allowed for the high-quality character animation of *Toy Story.*

A

B

C

16.11 The virtuosity of <u>Toy Story</u>: Both Woody and Buzz Lightyear, the central protagonists of *Toy Story, (A),* incorporate inverse kinematics — which keeps them behaving "naturally" across the wide gamut of expression that director John Lasseter and his team at Pixar have designed and programmed into this brilliant animated feature. Frame *(B)* samples the range of 3-D computer characters in the groundbreaking movie. Each of the toys behaves consistently, yet displays a custom vocabulary of movement and gesture. *Toy Story*'s exploration of lighting effects is suggested in *(C). Courtesy Walt Disney Pictures. © Disney Enterprises, Inc.*

RENDERING

Even though today's computers are capable of doing thousands, and sometimes millions, of computations every second, most are still not fast enough to keep up with the demanding mathematics required by 3-D production. *Rendering* is the process through which the computer takes all of the data that define a 3-D scene, including models, textures, lights, and camera, and creates a 2-D image of that scene. Depending on how realistic the image that the computer must create, rendering a single frame can take anywhere from a fraction of a second to several hours, or even days.

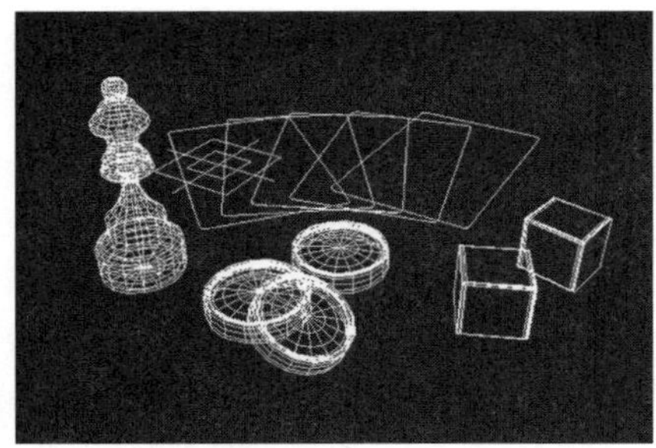

A

B

C

16.12 Rendering modes: The same scene was rendered three times, each time at a different rendering mode: wire frame, *(A),* "fast" mode, *(B),* and ray trace, *(C).*

Although rendering is *always* the final stage in producing a 3-D animation, it is actually something you do constantly throughout the production process. All software packages allow you to view your 3-D scene in a number of ways, and each of these *rendering modes* has its proper place in the process, based on how long it takes the computer to render an image. A *wire frame* rendering allows you to see the shape of your objects and where they are in space, but without any surfaces. Because the computer is able to draw images very quickly when it doesn't have to compute lighting or textures, this is the rendering mode commonly used for modeling, frame composition, and animation. There are different types of *shading algorithms* that allow you to view images with surfaces and lighting at varying degrees of quality. For example, *flat shading* shows lower-quality images with jaggies; they are quicker to render and can serve as a general reference for your image. On the other hand, *smooth shading* eliminates these aliased lines, making it more realistic, but taking much longer for the computer to render. These modes are good for setting up lights and checking your models for any unintended imperfections, a process of rendering known as *ray casting.*

Ray tracing, so named because the computer traces every ray of light, is the most time-intensive rendering mode. Ray tracing allows you to see all of your textures (including reflections, which add considerably to rendering time), lighting effects (including fog and realistic-looking shadows), and in some cases motion blurs. And finally, *radiosity* is becoming the most common highest-quality rendering, which provides more realistic animation than ray tracing. It is mainly used in architectural animation because, at least for now, this mode is too computationally intensive for animation in general. It is not uncommon to set up your animation and allow the machine to render overnight. High-end production houses often have several machines dedicated only to rendering.

THE FUTURE

Because 3-D animation is linked so closely to technology, it only follows that as machines become stronger and less expensive and software becomes more fully featured, 3-D animators will have unprecedented control over their craft. In many ways, 3-D seeks to re-create the physical world as faithfully as possible. The ways in which this will be possible will continue to grow. In the future, animators will have increased control over all aspects of the 3-D process. Modeling will become increasingly sculptural, allowing you to carve and mold objects in a more natural way.

Perhaps the most significant improvement is something that is already widely used by 3-D game developers and other high-end users. Now, 3-D objects have a number of properties, such as size, position, rotation, and texture. In the future, an object's properties will be more dramatic, responding not only to an animator's control but also to the other objects in the scene and physical laws such as gravity and velocity. For instance, to create an animation of a ball bouncing off a brick wall, you would only have to create a ball and a wall (a sphere and a cube) and "throw" the ball against the wall. When the ball hits the wall, the computer would be able to calculate the ball's bounce by taking into account whether it was a tennis ball or a baseball, how hard it was thrown, and what the wall is made of. You will no longer need to fake the relationship.

This is not to say that the concept of stretch and squash will no longer be relevant to 3-D animation. The timing and feel of a piece will remain the result of the animator's vision, regardless of how closely it re-creates the real world. The incredible creative powers of 3-D animation become strikingly evident when one studies, in some detail, the creative process that this technology requires. Toward that end, Figure 16.14 provides a detailed case study. In it, you can see the huge amount of control that the 3-D animator can impose on every aspect of the production sequence. This particular case study goes even further, showing how the tools and techniques of 3-D animation can extend the way we see our world and how a well-worn dramatic situation—almost a tired cliché—can be given new creative force.

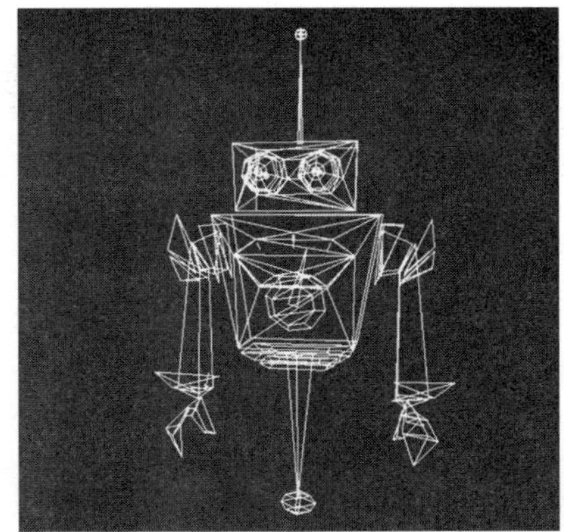

A

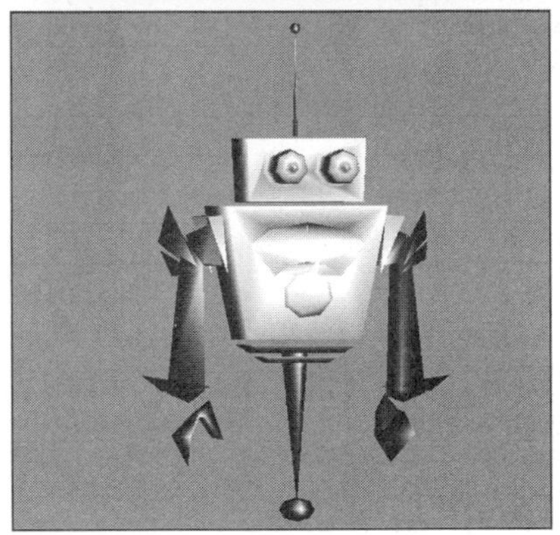

B

C

16.13 Cartoon rendering: Thinkfish Productions of San Francisco has developed a technology that allows 3-D artists to render scenes to look like illustrations. The three images shown here show a model of a robot rendered in wire frame, *(A)*, in a standard renderer, *(B)*, and using Thinkfish's proprietary technology, *(C)*.

16.14 CASE STUDY: Using Martin Hash's 3-D Animation Program

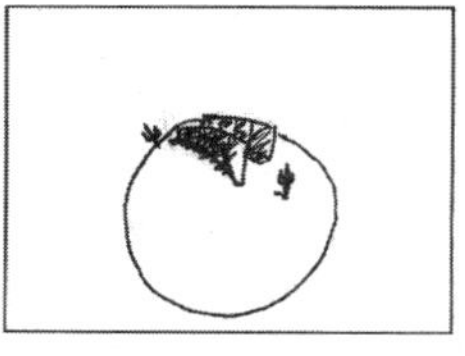

A

1. Extreme high angle of planet in space.
2. Camera begins to move in to show an old western town.
3. Camera eventually stops behind Cowboy 1, poised to draw.
4. After dramatic pause, Cowboy 1 draws his guns.
5. Bang! Cowboy 2 shoots logo-shaped hole clean through Cowboy 1.
6. Cowboy 1 wobbles and falls over, face first.

Shootout is a ten-second motion logo that Dan Schrecker and Dan Moss created for their independent production company, Dsquared, Inc., which specializes in 3-D animation, game design, and Internet projects. This humorous animation, which plays on Western-movie archetypes, was created using Martin Hash's 3D Animation (MH3D), an application that combines a powerful curve-based modeler and superior character-animation capabilities in a relatively inexpensive package (about $600). MH3D is made up of four modules: Sculpture, Character, Action, and Direction. Each of these modules is designed to accomplish a part of the 3-D production process. Objects are modeled in Sculpture, assembled and assigned hierarchies and limitations in Character, animated in Action, and composed into a scene and rendered in Direction. You can move back and forth freely between modules, but the basic sequence of work is similar to that detailed earlier in this chapter. As with most 3-D packages, MH3D is quite complex. The following notes from the two Dans provide an abridged summary of the creative process.

SCULPTURE

MH3D's spline-based modeler provided us with the flexibility we needed to create all of the objects in the scene, including organic elements such as the cowboys and cacti, as well as architectural structures, such as the storefronts. The six frames from our original storyboard, *(A)*, show the setup of the gag that introduces our company's logo.

B

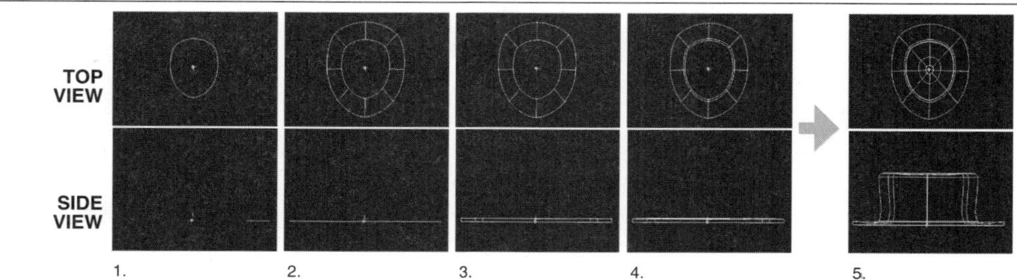

The progression of how the cowboy hat was constructed. The first curve is not even visible from the side because there is no surface at all, *(1)*. As we build the hat up, it becomes visible from the side, *(3)*. When the hat is complete, *(5)*, you can see from the top view that the hat is made from a series of different-sized ellipses.

Before we begin, a note about spline-based modelers. In a spline-based modeler, the surface of an object is made up of patches, not polygons. In MH3D, a "legal" patch (one that renders as a solid surface) is created by any closed combination of three or four control points, made up of two or more splines. This sounds confusing, and it is. With practice, however, you will find that spline-based modeling is very powerful and intuitive.

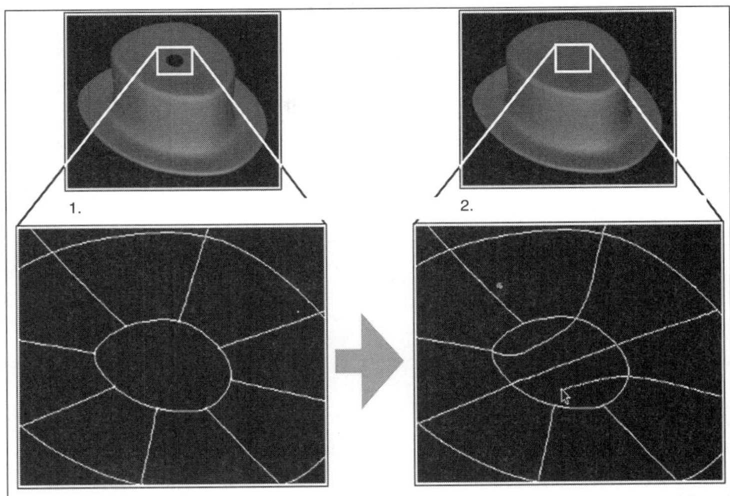

Before the final patch is created, there is a hole in the top of the hat, *(1)*. In order to fill this hole, it is necessary to create "legal" patches *(2)*.

C

Making a hat

Here's a step-by-step description of how we create a cowboy hat. To save time, we model a single, generic hat, which we then duplicate. These two hats can then be individually tweaked to create two seemingly distinct cowboy hats, each with its own unique style.

We begin with a single 2-D curve, an ellipse composed of 8 points, *(B1)*. This curve will serve as the basis for the rest of the hat, which we will build, one surface at a time. By extruding this curve "out," we create our first surface, the bottom of the hat's brim, *(B2)*. Each time we extrude a curve, a duplicate of that curve is created. This new curve then becomes the basis for our next extrusion. In addition, each successive curve creates a new surface between itself and the prior curve. Basically what we are doing is tracing the shape of the hat in three dimensions, using a two-dimensional curve as our brush, and resizing the curve as we go along, *(B5)*. In the end, the hat is just a surface with no volume. This is due to the fact that the ends are still open. This doesn't matter on the inside because the cowboy's head will obscure any gaps. However, if we look closely at the top of the hat there is a hole in the middle, *(C1)*. In order to "fill" the final curve, we must add some lines to ensure a solid surface, *(C2)*.

The finished cowboy hat.

D

Now we want to turn the basic hat into a more stylized cowboy hat. By selecting individual points or groups of points and tweaking their position or rotation, we are able to bend the brim, indent the top, and generally give some Western flavor to something that started out closer to a top hat, *(D)*. The final result is a segment file, "hat.seg," which we will use in the next module, Character.

One feature of the Sculpture module is the ability to glue "decals" onto individual segments. Decals are excellent for animation because they deform with the segment onto which they are mapped. Decals were used simply to make a sign for one of the storefronts. An image map and a bump map—both produced in Photoshop—were combined to turn a flat plant into a sign seemingly built from two boards.

E

CHARACTER

The primary purpose of the Character module is to create what MH3D calls a "figure." A figure, identified by the ".fig" file extension, can be made up of a single segment or multiple segments, depending upon the amount and style of animation required by the figure. Within the Character module, figures are built by adding one segment at a time until the figure is complete. In addition, constraints are defined for segments that will later be animated.

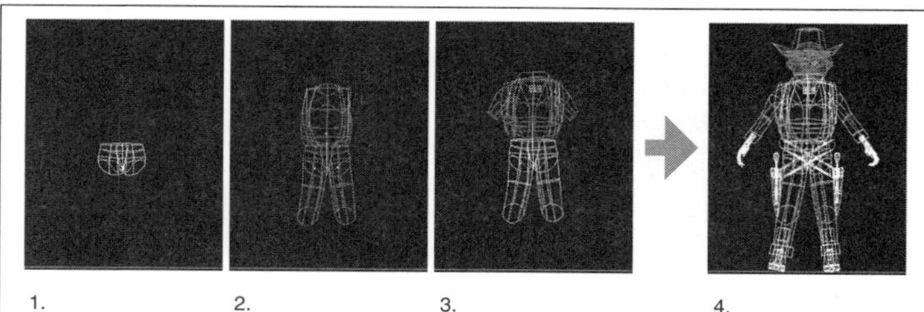

1. 2. 3. 4.

Figures are built one segment at a time. Starting with the pelvis, we add one body part at a time until our cowboy is complete. This creates the hierarchy required by inverse kinematics.

F

BUILDING A COWBOY

The arm bone's connected to the hand bone. The neck bone's connected to the head bone. Although we are working with segments and not bones, the same ideas apply when constructing a human figure. After modeling the cowboy in the Sculpture module, we "cut" him up into the individual segments we require in the Character module. For example, we modeled the entire arm at once, but instead of saving it as a single segment, we divide it up into the biceps, forearm, and hand in order to facilitate future animation.

As we build our cowboy, one body part at a time, we create a hierarchy. Whichever segment is selected when we import a new segment becomes the "parent" of the newly imported segment. We start with the pelvis, *(F1)*. This

1.

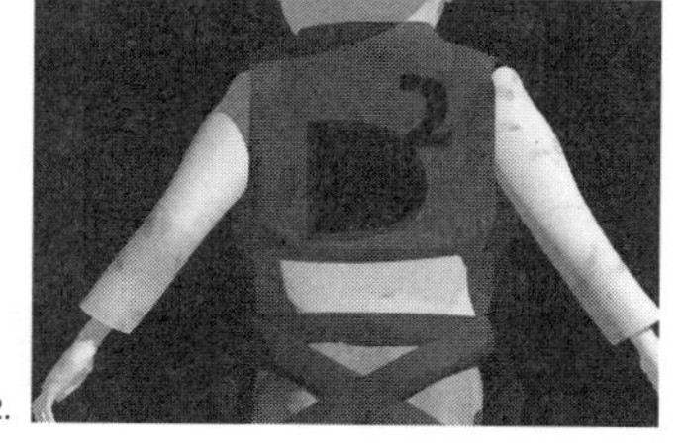

2.

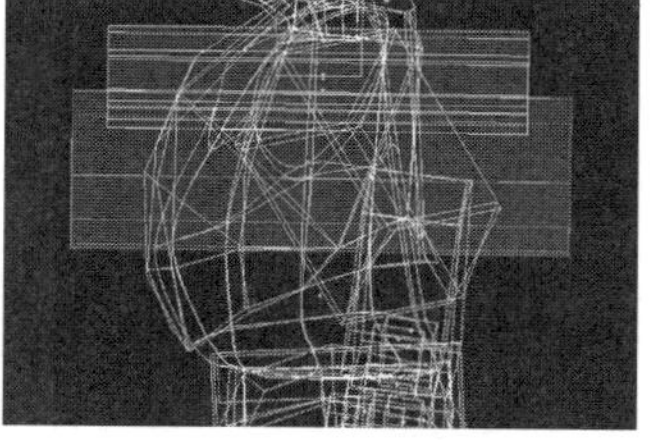

3.

To create the logo in the cowboy's chest, we used what MH3D calls a "cutter" segment or Boolean shape. First, we modeled the logo in sculpture, *(1)*. Then, with the torso selected, we imported it into Character. Next, we positioned the logo so that it cut right through the cowboy's chest, *(2)*. By making the logo a cutter, its volume is subtracted from the segments it intersects and its surface appears on the inside, *(3)*.

G

Although you can apply textures within any of MH3D's modules, we did most of our work in Character. The Materials palette allows you to select any object and edit its surface attributes. In addition, MH3D allows us to combine different materials to create new ones.

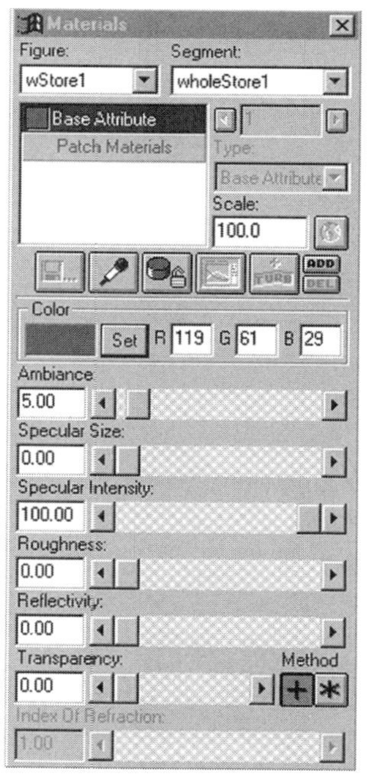

H

will be the top of our hierarchy. Anytime we move the pelvis, the entire cowboy will follow. With the pelvis selected, we import the torso and thighs, *(F2).* With the torso selected, we import the neck, shoulders, and vest, *(F3).* We continue in this way until the entire cowboy has been assembled, from hat down to boots, *(F4).* The reason for constructing the cowboy in this way lies in MH3D's inverse kinematics. In order for the biceps to move the forearm, which in turn moves the hand, it is necessary to link the body parts and establish this hierarchy. Another part of preparing figures for animation is defining the proper constraints to each body part. This assures that a leg won't bend in an unnatural position or the head spin all the way around. Using the constraint panel, we define a range of motion for each segment.

ACTION

Here's where the fun really starts. There are two types of motion possible in MH3D: *absolute*, which determines the position and orientation of an object in space, and *relative,* which determines how an object behaves, regardless of where it is in space. A good metaphor for this is the motion of the Earth. The Earth's orbit around the sun is absolute motion, while its daily rotation on its own axis is relative. In MH3D, the movement of a character's legs and arms as they walk is relative motion, but their actual movement from point a to point b is absolute. The figures constructed in Character are brought into Action, where all relative motion is defined (absolute motion is defined in the Direction module).

MH3D provides three types of animation: skeletal, muscle, and spine. Skeletal is the type of animation that affects a figure's joints, such as bending a knee. Muscle animation affects the position and contour of control points on a single segment. This would be the type of animation used to lift a character's eyebrows or move a character's lips. Spine animation is much like using an armature within a given segment. If you wanted to make a snake slither along the ground, you would use spine animation.

The draw

The animation of a cowboy drawing his guns is an example of skeletal animation. Herein lies one of MH3D's great strengths (and what truly sets it apart from other low-cost 3-D packages). The hierarchy we built in the Character module serves as the basis for the inverse kinematics we will use to animate the draw. In other words, because we've already established links between the character's biceps, forearm, and hand, we need only position the hand where we want it, and the arm will follow.

I

The process of animating the draw is similar to other forms of character animation, except the computer does all of our in-betweening. The animation of the cowboy drawing his gun was created through a sequence of key frames, much like classic character animation. Our understanding of concepts such as anticipation and weight were extremely helpful in creating believable motions. We define our key frames, *(I),* and render a series of wire-frame tests until we are satis-

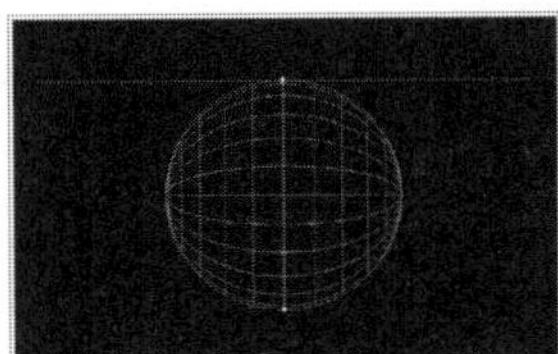
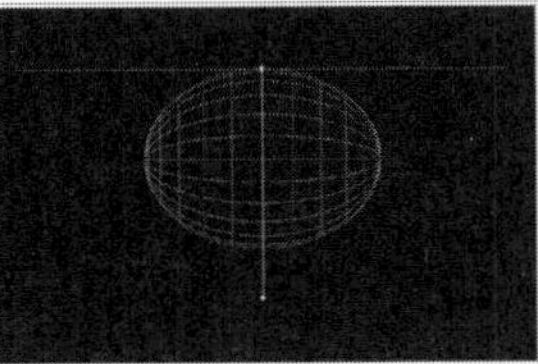
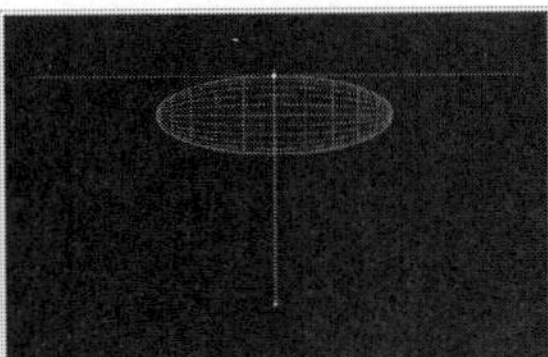
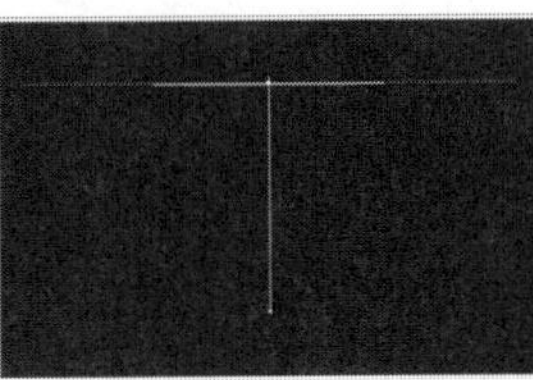

J

fied with our animation. The final product is a twelve-frame animation encapsulated within a file named "draw.act." When we bring this file into the Direction module, we can speed it up or slow it down as we please. In addition, MH3D allows us to apply this action to any character that shares the proper hierarchy, thus allowing us to use it for both cowboys.

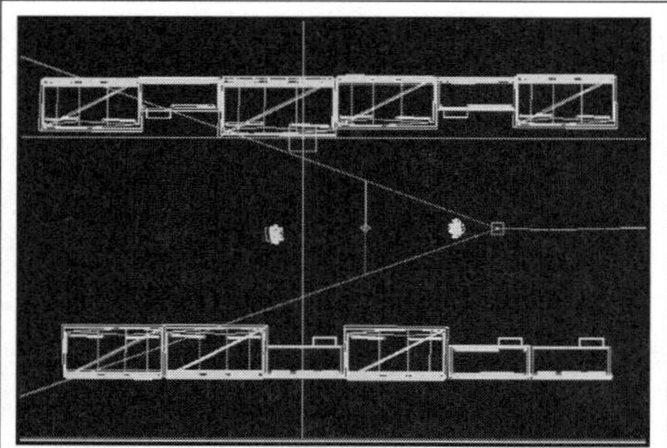

K

This top view of our Western set shows the storefronts, the cowboys, and the camera in its final position. The triangle represents the camera's field of view.

DIRECTION

This is where it all comes together. Lights, camera, action, and figures are combined to create what MH3D calls a choreography. At this point, the process is similar to live-action filmmaking. We build a set, position the actors, light the scene, set up a camera move, and roll the film, or in this case, render the animation.

A basic concept to grasp when working in Direction is that all objects, including cameras, lights, and figures, must be attached to a path. A path can be a single point (which will result in a stationary object), or a curve made up of several points (which will produce an object that moves along its path over the length of the choreography).

On the set

When we begin a new choreography it contains nothing but a default camera and default light. The first step is to put these objects on paths so we can move them around. We will eventually add some more points to the camera's path in order to animate its descent. For now, however, it is easier to keep it on a single point so that we can move it freely about the scene in order to check our composition from a variety of angles.

The first figure we add is the Earth, since it acts as the ground plane and all other objects will be positioned in relation to it. In order for us to have a round planet at the beginning and a flat horizon once the camera is finished moving, we created an animation of the Earth squashing from a sphere into a plane, *(J)*. We did this simply by decreasing the Earth's scale along the y-axis in the Action module.

Next we build our Western set by adding the storefronts and cacti. Finally, we add the cowboys. Because none of these figures will move (absolute motion), they are all positioned on single-point paths. We rotate the storefronts to form a street and the cowboys to face each other, and we're ready for our showdown, *(K)*.

With all of our objects in place, we can light the scene. This is a fairly simple process because we want to emulate natural light. Although MH3D allows us to use five different types of light with a wide array of attributes, *(L)*, we simply want a single parallel light to illuminate the whole scene—i.e., the sun. We add a little bit of color so the sunlight isn't pure white, and render a few tests until the sun is positioned correctly so that our shadows are appropriately dramatic.

It is now time to apply the animations we created in the Action module to the proper figures. The draw animation is applied to both cowboys, the fall animation is applied to the cowboy who gets shot, and the squash animation is applied to the Earth.

We are ready now to animate the camera. In order to do this, we must first add a few points to the camera's

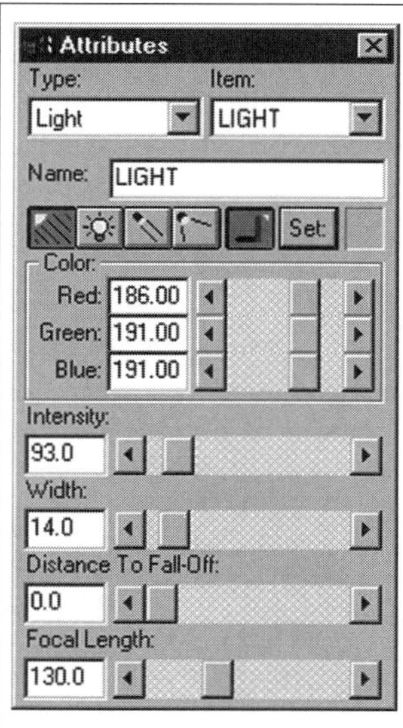

The Attributes panel provides us with total control over how a light behaves, including its color, width, intensity, whether or not it casts shadows, and at what distance the light begins to fall off. To recreate dramatic sunlight, we used a light yellow parallel light that cast shadows and never dropped off.

L

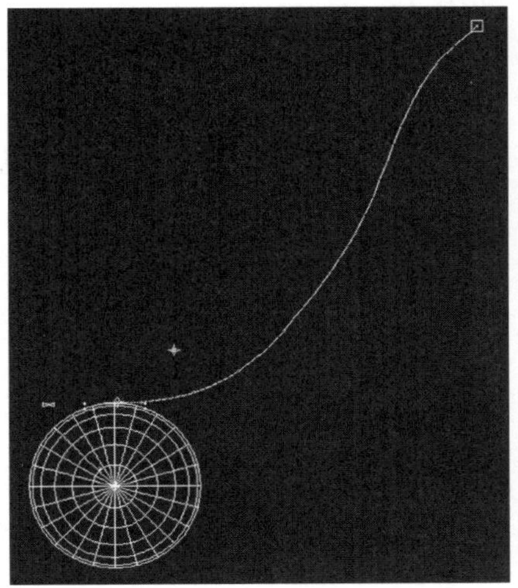

1.

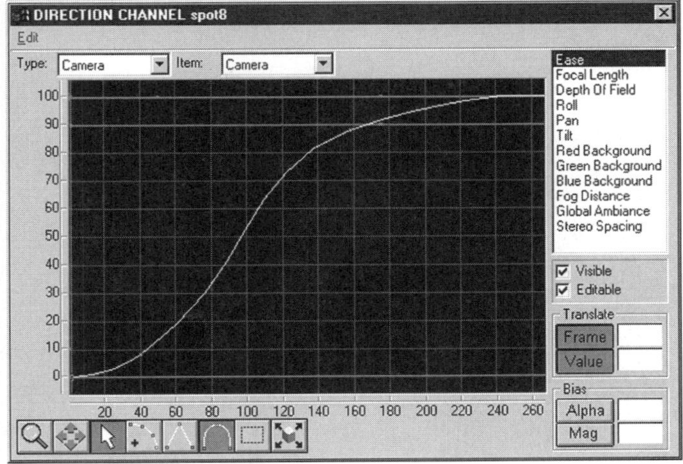

2.

M

path. Each point on the path is positioned individually to create a smooth approach for our camera. To achieve the proper camera move, we created a smooth path that started high in the sky and ended at street level, *(M1).* The timing of this move is controlled through a "channel," *(M2).* The curve here represents the camera's "ease." An ease of 0% represents the beginning of a figure's path and 100% is the end of its path. The S-shape of this curve results in a classic ease-in/ease-out. At frame #240, the curve levels out and the camera holds still for the rest of the animation. Channels are very powerful and can be used to affect every aspect of an animation, including position and scale of figures, camera attributes such as depth of field and focal length, and light attributes such as intensity and color. To check that our path is correct, we render a wire-frame version of the complete animation.

N

A frame from the final render.

After a little tweaking, we're ready. The set is built, the actors are in place, the sunlight is just right, and the camera is hovering a few miles above the set, ready to swoop down into its final position. We make sure our rendering options are correct (ray tracing, full screen, fifteen frames per second), press the Go button, and go home while the machine renders the final animation.

17 ▪ Animation Frontiers

17.1 AAAAH-EEEEE-AAAAH-EEEEE-AAAAH!!!! We swing blindly into the future of animation in the good company an Eli Noyes adventurer. *Drawing courtesy Eli Noyes himself.*

Animation can be viewed as developing in three waves. The first was based on the technology of film: an emulsion coating on a piece of celluloid, which is ratcheted through a projector gate 24 times a second. Films were (and still are) projected to large audiences using machinery invented in the nineteenth century. The first cartoons of Disney and others built huge popularity for 35mm theatrical shorts, then full features, and eventually widened both audience and purpose as animation was discovered to be effective in delivering a broad range of information. Filmmaking tools, once expensive and bulky and therefore available only to a few, eventually became available to everyone through 16mm and super 8mm cameras and related paraphernalia. Toward the end of the first wave—in the 1970s—there was a flowering of independent animation as artists invented and refined new and simpler ways of animating.

This next evolutionary wave in animation was based in video technology. Television became a new distribution medium that, for the most part, delivered animation that was still produced on film. But the needs of TV subtly altered how animators worked. Early animated series like *The Flintstones* and teaching elements on *Sesame Street* led to the development of a simplified aesthetic that looked good within the lower resolution (and smaller budgets) of television. *The Simpsons* and Nicktoons series like *Ren and Stimpy* and *Rugrats* represent the full flowering of film-based animation on the TV screen.

In the 1980s, high-budget TV commercials pushed animation techniques into new digital configurations. Very expensive video tools like the Ultimatte, the Paintbox, and the Harry permitted animators to blend together, or *composite,* images from various sources onto the same screen. Cel animation, live-action video, motion graphics, and archival film merged in a new aesthetic that was named "Blendo" by one of the cutting-edge studios involved in the innovations, (Colossal) Pictures in San Francisco.

Emerging cable networks like MTV quickly adopted the second wave of eclectic, video-based animation. The promotional departments that created on-air graphics for these networks became showplaces for the new hybrids of merged film/video techniques, and shows like MTV's *Liquid Television,* an anthology of beyond-the-fringe, over-the-top animation, showed some of the first stirrings of the wave we're currently experiencing.

Now, at the turn of the century, animation has moved to the computer. The swell of this new wave in animation's evolution started to become clear when—quite suddenly, it seemed—50 percent of American households had computer screens as well as television screens. By the mid-1990s, new computers were routinely equipped with the second or third generation of powerful CD-ROM drives and with modems, satellites, and high-speed telephones providing easy access to the Internet.

This newest wave is still building. Soon cable television services will offer faster digital hookups to the World Wide Web. The Web will respond with new forms of information and entertainment. A new generation of digital, higher-definition television sets (the first appearing before the year 2000) will further blur the line between computing and television. And after that, high-bandwidth switched networks using fiber optics will fuse the worlds of television and computers, of entertainment and information.

But all this is about technological breakthroughs. What about the medium of animation? Are new techniques becoming available? Will there be new opportunities for independent animators and small studios? Are new patterns yet visible?

This chapter offers a look at four frontiers of animation at the millennium. All exemplify the transformational power of the computer. All offer special kinds of lures to capture the imaginations and careers of tomorrow's fearless animators.

• NEW MEDIA •

Written with Eli Noyes

17.2 Living Books logo: This electronic publishing outfit, a co-venture by Broderbund and Random House, pushed the envelope for incorporating original animation within interactive properties. The vision was simple and strong, with a format and interaction model that proved very popular with children and continues to be satisfying in an age with the razzle-dazzle of complex 3-D interactive graphics. *Courtesy Living Books.*

Think of new media as that world where the viewer becomes a user. As someone who has chosen to pick up this book, you are undoubtedly aware that computers demand a subtle but very profound switch from a passive to a participatory mode. The paradigm shift is one of *interactivity.*

You are probably also aware that there already exists a robust configuration of interactive tools operating under the general classifications of *multimedia* and *new media.* We prefer the latter name, partly because the rah-rah, overpromising hype of the early 1990s has marred the term *multimedia,* and partly because the phrase *new media* seems to welcome stuff that isn't yet invented. In day-to-day, operational terms, new media has come to mean any computer-based form in which the user can navigate his or her way through a seamless mix of text, TV clips, still images, audio tracks, and animations.

CD-ROMS AND DVDS

The most exciting new distribution technology to emerge from the digital revolution is the CD-ROM (compact disc—read-only memory). Shiny and silver, like their audio CD cousins, CD-ROMs hold as much as 650 gigabytes of digital information. That's the equivalent of 450 floppy disks! CD-ROM drives that can quickly access information (ten times faster than the first generation) are now being routinely built into every computer coming off the production line. Here is a delivery device that works using off-the-shelf hardware and software and invites the audience to interact deeply.

Production of your own CD-ROM requires the fabrication of many bits and pieces of "media," using many of the same digital tools that are discussed in Chapter 23 and sampled in case studies throughout these pages. The full spectrum of words, images, and sounds are assembled in a CD-ROM so that the viewer (who is now a player) can interact autonomously by clicking a mouse. The means of production is quite artist-friendly and encourages anyone with a modicum of equipment to make their own multimedia products.

The case study for *Ruff's Bone* (Figure 17.4) provides insights into the creative turf of CD-ROM production. It also will introduce you to a software program called *Director* from Macromedia, which is a landmark software application within the domain of new media.

By the turn of the century, a new and improved distribution medium will be in place. The DVD (digital videodisc), has the capacity to store between 7 and 14 gigabytes. A dual-sided DVD holds the same amount of data as almost ten CD-ROMs. The interactive capabilities of this new medium will increase the creative reach of designers of games and other content.

The realm of new media is destined to evolve far beyond the technologies that exist today. High-definition TV (HDTV) and interactive television are still a few years away for everyday use, but you can find a preview of what is to come in the digital program guides and navigation schemes of today's satellite, microwave, and closed-circuit TV environments.

The medium of animation turns out to be particularly well suited to the interactive domain. For one thing, bits of animation can "nest" within larger fields of information, coming alive at the click of a mouse. Animated characters (including animated logos) provide powerful proprietary identities that bring fun and recognition to multimedia products. Because animated characters and worlds compress better than their live-action counterparts, animation has a natural advantage when space and computing power is limited.

Traveling on the backs of such inherent qualities, animation has begun to leave the movie and TV screen behind, making its way into documents, presentations, and toolbars. Animation has jumped into the very heart of the GUI (graphical user interface), making up the icons, windows, and desktop environment that connect each of us to the powers of computing. If you are a young animator looking for a place that invites innovation and values fearlessness, head toward the interactive frontier of new media.

A

B

C

D

E

17.3 Ruff's Bone: The first product for the Living Books line of interactive CD-Roms was *Ruff's Bone,* an original creation developed by Eli Noyes in association with (Colossal) Pictures. Living Books has gone on to develop a proprietary multimedia platform with a full library of interactive titles. *(A)* sets up the story, while *(E)* is the final page and the payoff. *(B), (C),* and *(D)* sample pages of the worlds Ruff enters in his quest for the bone.

17.4 CASE STUDY: Using Director

Ruff's Bone is an original creation for Living Books, a joint venture of Broderbund (a software publisher) and Random House (a traditional book publisher). It was illustrated and directed by Eli Noyes, who developed the project with a group from (Colossal) Pictures in San Francisco. *Ruff's Bone* was created from scratch rather than adapted from an existing children's book, using an early version of Director from Macromedia. Below, creator Eli Noyes describes this process.

Writing

We had to tell a story in about twelve pages. At first I wanted each page to be more absurd than the last. But as the story developed it became clear that a through-line would make a difference. We then tried to "shape" Ruff's emotional state so he gets discouraged before arriving at the Bone Planet. Story line and page content evolved via brainstorming with the creative team. We had to script the opening and closing animation for each page, and come up with "gag" ideas for objects and characters on each page. A sample chunk of the final script is shown in *(A)*.

In the final writing stage, we had to make sure we could tell our story using only fifteen to twenty words on each page. These words had to be carefully chosen for the target audience of preschoolers. To give you a sense of how concepts evolved and word count shrank, *(B)* provides an initial concept sketch and *(C)* shows what it evolved into.

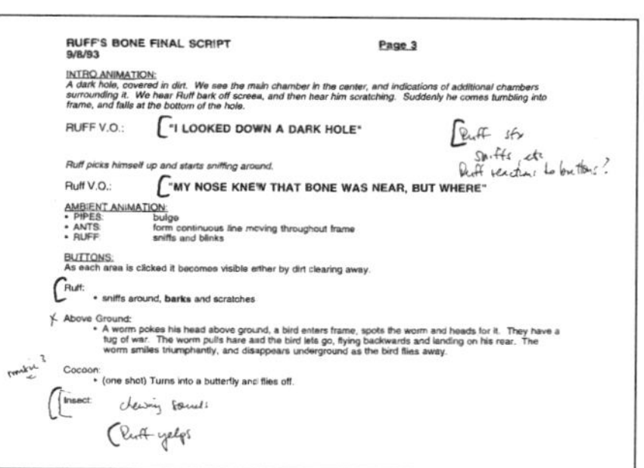
RUFF'S BONE FINAL SCRIPT **Page 3**
9/8/93

INTRO ANIMATION:
A dark hole, covered in dirt. We see the main chamber in the center, and indications of additional chambers surrounding it. We hear Ruff bark off screen, and then hear him scratching. Suddenly he comes tumbling into frame, and falls at the bottom of the hole.

RUFF V.O.: **"I LOOKED DOWN A DARK HOLE"**

Ruff picks himself up and starts sniffing around.

Ruff V.O.: **"MY NOSE KNEW THAT BONE WAS NEAR, BUT WHERE"**

AMBIENT ANIMATION:
- PIPES: bulge
- ANTS: form continuous line moving throughout frame
- RUFF: sniffs and blinks

BUTTONS:
As each area is clicked it becomes visible either by dirt clearing away.

Ruff:
- sniffs around, **barks** and scratches

Above Ground:
- A worm pokes his head above ground, a bird enters frame, spots the worm and heads for it. They have a tug of war. The worm pulls hare and the bird lets go, flying backwards and landing on his rear. The worm smiles triumphantly, and disappears underground as the bird flies away.

Cocoon:
- (one shot) Turns into a butterfly and flies off.

Insect:

A

Design and Character Development

The character of Ruff evolved through a series of rough drawings that became more and more refined as the script took shape. I wanted his world to be rich enough in detail to hold the attention of young viewers, who I knew would appreciate all the little details. The drawings in *(D)* sample the design process that I used with just about every character in *Ruff's Bone*.

The art director part of me wanted to use the gradated backgrounds and the multiple color palettes possible with digital rendering. Unfortunately, you cannot see the beautiful pastel tones we achieved in these black-and-white reproductions. But to give you some sense of it, *(E)* shows the background for the scene in Slim's yard, and *(F)*, the final layout populated with characters and objects.

B

Layouts

I needed to create a layout for each "page." I knew we would eventually place fifteen to twenty short animations on each page, and that the layout would need many inviting things for kids to click on. My fear at the beginning was that layouts would not be interesting enough and that there wouldn't be enough things to point at with the mouse. Hence the backgrounds presented an unusual design problem: They had to tweak the curiosity of kids but they also had to avoid too much clutter. The evolution of one layout is sampled here: an initial sketch, *(G)*; one that is closer, *(H)*; and the final version showing the scene as the young person first sees it, *(I)*, and after all the implied burrows have been clicked on, *(J)*.

The layouts had to create a stage for the little animation scenes that come to life when kids interact with each scene. An outline drawing of the scene in Slim's yard, *(K)*, shows how I laid out the action.

C

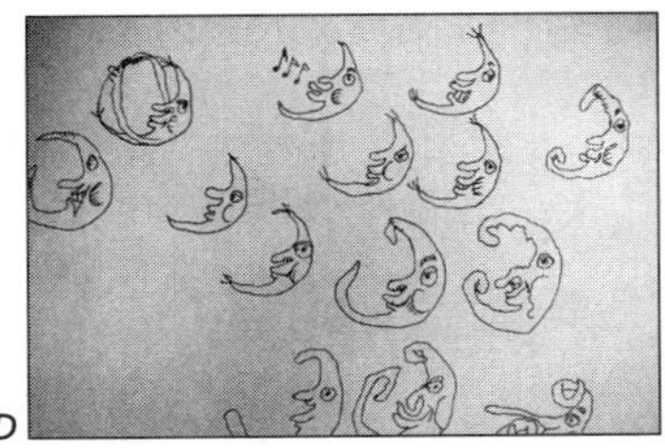
D

E

F

G

Sprite and Space Allocation

We created *Ruff's Bone* using the 3.0 version of Director. This version had a limited number of "cast members" or "sprites" (only 1,000) that could be used for each Living Books page, but that number has more than quadrupled in current versions. Each frame of animation counts as a cast member. Each flashing word of text counts as a cast member. Every piece of audio counts as a cast member. We found that it was very easy to fill up our quota for each page, and that if we spent a lot of frames animating one "gag," we would have to skimp on the next one, possibly using audio only to pay off the mouse click. The screen-grab in *(L)* shows a few of the sprites appearing on one of the Living Books pages.

We also found that we were limited in the size of objects that could move on the screen. Our rule of thumb was that nothing larger than a quarter of the screen could move at a time, since most home computers do not have the horsepower to refresh large parts of the screen at a rapid rate.

Voice Recording

Once the script and basic layout of each page was fixed, it was time to record voices. It was important to have tracks before we did the animation itself in order to make sure gestures and mouth movements matched the spirit and content of the track. Voices for *Ruff's Bone* were drawn from the ranks of employees at (Colossal) Pictures, who delivered a great range of vocaleese. Oftentimes actors would improvise lines that filled gaps in our script. Our sound designer enhanced the voice track with an array of comical sound effects. Our music director/composer created all the music for the CD-ROM and modemed it to us from Minneapolis.

If you look closely at *(L)*, you can see that some of the sprites show mouths, which were laid onto other sprites in Director. This is how the animators created lip sync. It's easy to move and recombine sprites in Director; hence the process involved trial and error until the mouth synced with the already recorded bark track. Note that sprites are not shown proportion-

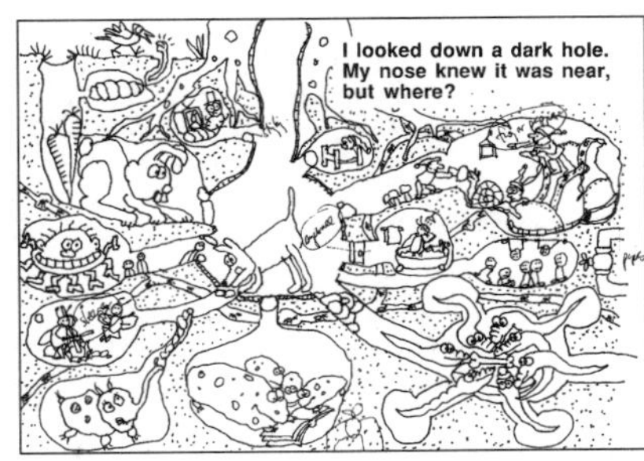

H

I

J

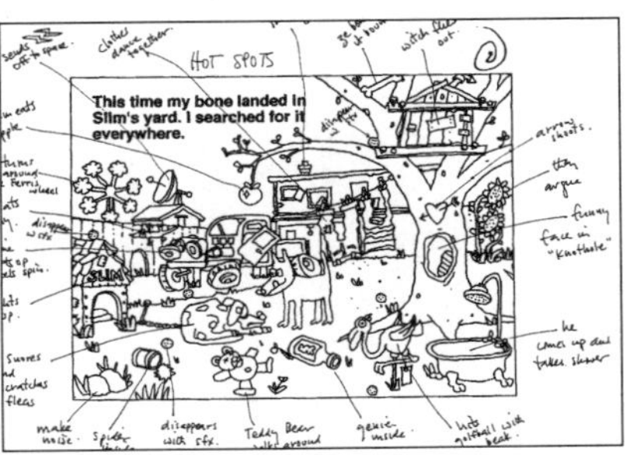

K

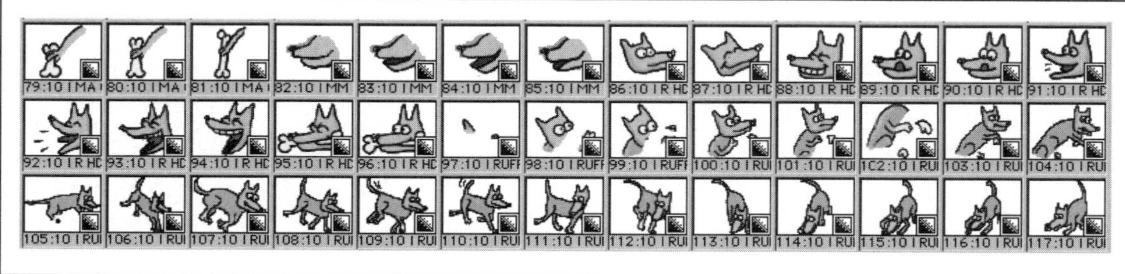

L

M

ally to each other. The program automatically enlarges each sprite to the degree possible in these tiny thumbnail reference images.

Animation

In the particular configuration of hardware and software version we worked with, Director did not play animations at a full 24 frames a second. The rate varied depending on the size of the image and the speed of the computer, but tended to hover around 8 frames a second. This meant that the animators had to rethink the way they worked: Fewer drawings meant less work for them, but each drawing had to count, and the motion had to be simplified and stylized for the lower frame rate. The sprites representing the complete set of drawings for one animated moment is shown in *(M)*.

Animation was done in a traditional way: by hand with a pencil on bond paper with pegs. This was pencil-tested on a frame-by-frame video recorder and, after revisions, inked onto bond paper. Each frame was then scanned into Photoshop for coloring. Animator Beth Sullivan is shown at work in *(N)* and *(O)*.

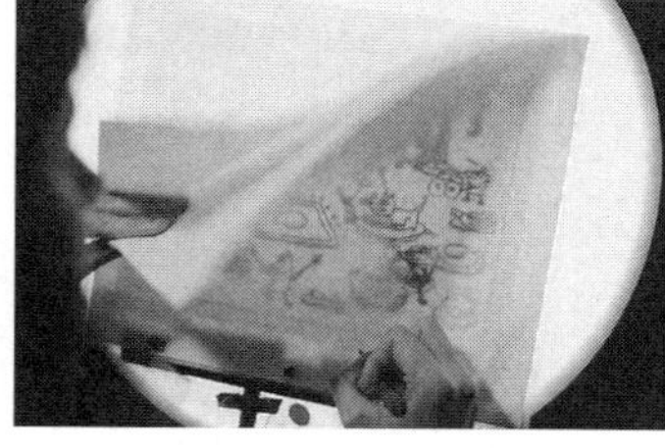

N

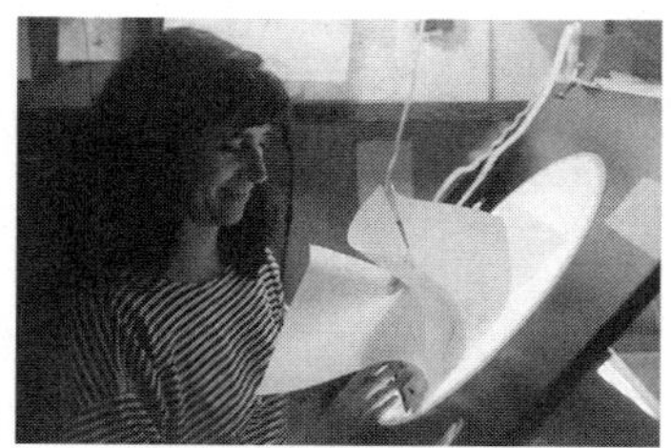

O

P

Color Design

It requires labor and ingenuity to establish distinct color designs of each page. The Director software limited the palette of each page to 256 colors. This may sound like a lot, but when colors are gradated in backgrounds and on characters, what seems like one color can easily eat up a lot of different shades. Also, antialiased lines are in fact combinations of black and, around the edges, pixels of black mixed with whatever color it goes against, *(P)*. So even black lines ate into our palette budget. Our program-

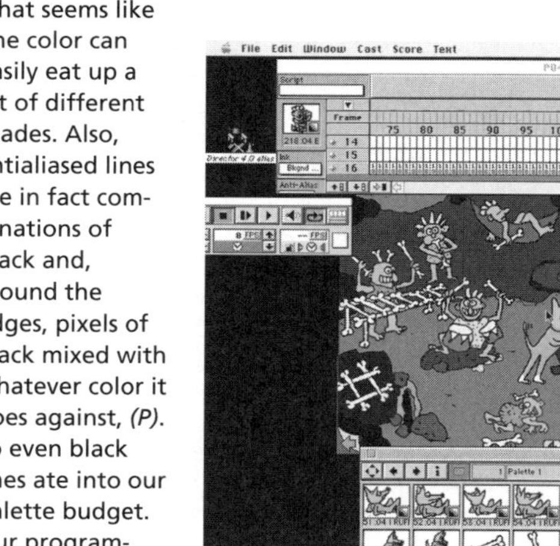

Q

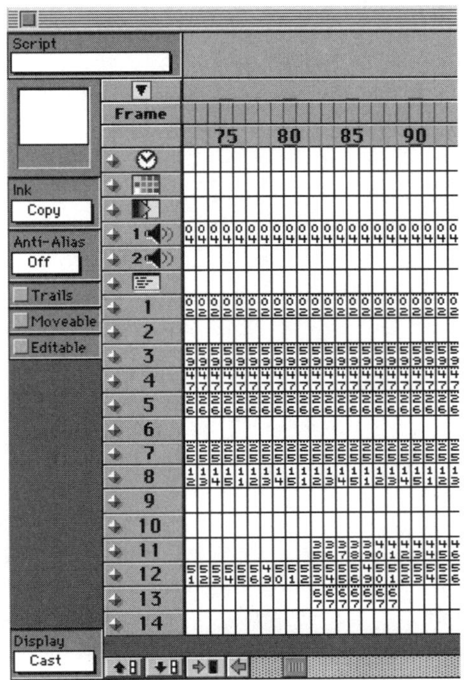

R

mers spent hours tweaking the color palette of each page to get maximum richness.

S

Registration, Timing, and Scene Building in Director

Once animated pieces are drawn, scanned, and colored, they must be entered into Director's cast list. They must be precisely aligned with the background so when the young user clicks on an object, the appropriate sprites will play back and look as if they are married to the page. Director provides the animator with the computer equivalent of a cue sheet, where all of these images can be entered along with the sound effects that they trigger.

(Q) shows four of the major tools of Director: the Score; the Controller; the Cast; and the Stage. *(R)* provides a closer look at the score for this particular screen. Don't be too overwhelmed: It took me weeks to learn my way around Director. Happy to say, my background as an animator (cue sheets, serial images, analyzing audio tracks, etc.) made it much easier to learn a multimedia application like Director.

Viability Testing and Debugging

In the good old days of movies, you shot film in either 16mm or 35mm. Projectors all over the world could show your film because they, too, came in these two gauges. Imagine, however, creating something that must play on any number of different computer and software configurations, and must be made compatible for Macs and PCs! In the early days of multimedia when we put together *Ruff's Bone*, we had to assume that consumers had much slower machines than are now routinely in use. An important step in the production that few people know about is called "QA." It involves a room full of people with every conceivable computer configuration under the sun who play and replay the CD-ROM you have created. They are looking for "bugs," and they find a lot of them! These need to be fixed. This is one of the final stages in the making of a Living Book, *(S)*.

Credits: ***(Colossal) Pictures' New Media:*** **Creative Director:** Stuart Cudlitz **Executive Producer:** George Consagra **Senior Producer:** Anne Ashbey **Producer:** George Consagra **Director:** Eli Noyes **Animation Director:** Catherine Margerin **Technical Directors:** Johnathan Levy, Dave Wise **Animator:** Beth Segal **Animation Assistants:** Ruth Daly, Karen Heathwood, Cindy Ng, Susan Tremblay **Computer Graphics Technicians:** Lee Dean, Portia DiGiovanni, Karina Jakelski, Jonathan Levy, Sara Whiteley, Dave Wise **Original Music & Lyrics:** Greg Hale Jones **Music Production:** Anderson Jones Music **Sound Designer:** James LeBrecht

Living Books: **Product Manager:** Todd Power **Sound Design:** Tim Keanini, Jane Scoleri **Technical Design:** Donna Bonifield, Rob Bell **Programmers:** Matt Siegel (Macintosh); Glen Axworthy (Windows); Misc. Tools: Dave Lucas, Mark Webster **Assistant Marketing Manager:** Laura Norman **Director of Marketing:** Susan Lee-Merrow **Creative Director:** Mark Schlicting **Executive Publisher:** John Baker **Package Design:** Ronni Valenti, Karina Jakelski **Quality Assurance Testing:** Craig Riddle (Lead QA Technician) **Testers:** Marcus Duerod, John Hamele, Lisa Irwin, Daniel Kelmenson, Erik Spencer, John Crowell (Senior QA Lead), Anne Sete

•IMPROV ANIMATION•

Written with Athomas Goldberg

17.5 Simon says: In a demonstration at SIGGRAPH 95, an interactive avatar named Sam responded to spoken statements and requests from untrained participants in a game of "Simon Says." Sam was programmed to respond to requests that were preceded by the words "Simon says." In addition, he was scripted to goof up occasionally and follow requests that weren't given to him. He would then act embarrassed at having been fooled, making him a realistic character that participants could interact with and relate to. *Courtesy NYU's Media Research Laboratory.*

If you are a young actor or a budding computer programmer, an evolving field called Improvisational Animation is where you should be heading.

Using a radical new set of computer tools, you will someday be able to create animated characters with built-in personalities that equip them to react spontaneously to situations, environments, and characters they are faced with. The software system for creating these animations is called *Improv.* It involves a combination of procedural animation and behavioral scripting techniques that are currently being developed at New York University's Media Research Laboratory.

HOW IT WORKS

To achieve this new form of animation, two different *engines* are utilized. The *animation engine* provides tools to control the body of an actor or character of an animation. By allowing the animator to layer and blend chunks of prebuilt animations, realistic motions and gestures are possible. The *behavior* or *decision engine* provides authoring tools for guiding an actor's behavioral choices. This allows the animator to determine sets of rules that govern character movement, including time information used to determine when a given animation is activated, as well as decision rules used to determine the probabilities of actions occurring under defined circumstances. For example, a character can be scripted to wave only when waved at and to be only waving with that arm at that time. If this second specification was not made, a character might try to wave and scratch his head at the same time, which would look awkward and unrealistic.

Simply put, in Improv, characters follow *scripts,* sets of author-defined rules governing their behavior, which are used to determine the appropriate animated *actions* to perform at any given time. All this comes together on the desktop of a single artist, who can create dramatic environments featuring complex interaction.

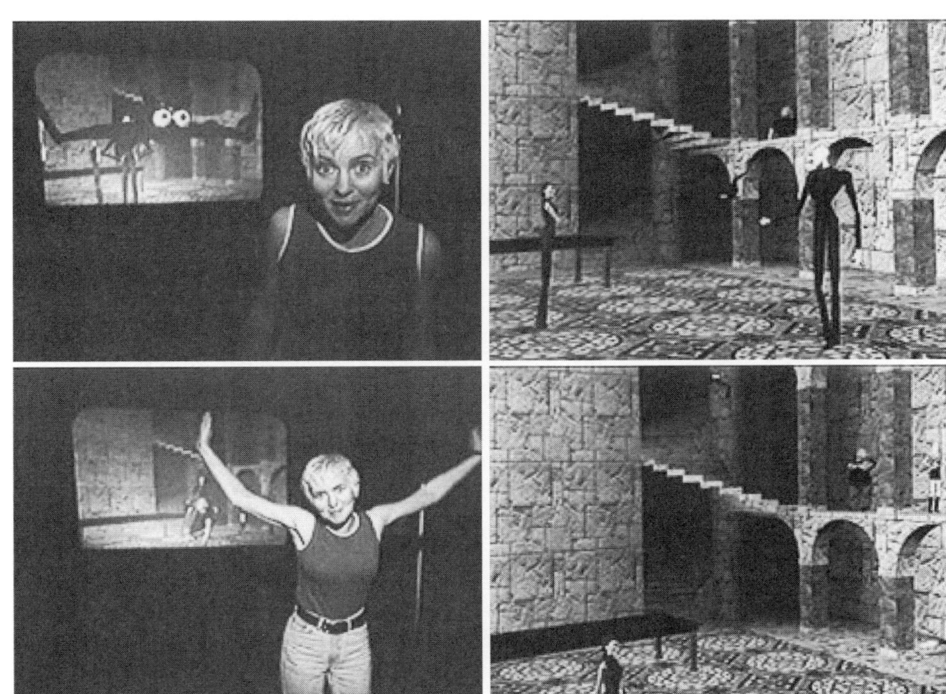

WHERE IT'S GOING

So now that you have a general idea of how Improv works, you're probably wondering how it will be used. At NYU's Center for Advanced Technology, explorations are under way to employ input sources ranging from the computer keyboard to speech to gesture recognition by computer. Figures 17.5 and 17.6 illustrate some of the simple demos that have been tried out. These examples suggest much greater possibilities in the future, including improvisational theater with 3-D characters where each performance is unique and unpredictable; computer games that are psychologically and emotionally challenging in addition to testing a player's reflexes or puzzle-solving ability; and eventually fantastic, communal worlds in which a human-controlled character (a surrogate for yourself) can interact with Improv-controlled animated characters (scripted and embedded) and with other human-directed actors. The long-term appeal of improvisational animation is that it offers ordinary individuals the chance to play on-screen roles in a variety of virtual realms—all linked to other users in real-time environments all over the world!

17.6 Flying bat: A participant appeared as a flying bat and was presented with a large rear projection of a room full of characters in conversation. Her position and simple arm gestures were tracked by an overhead video camera. As the participant walked around, the bat flew accordingly. The nearest character would stop his/her conversation and begin to play with the bat. When the participant flapped her arms, the bat would fly higher in the scene, and the camera would follow, which gave the participant a sense of soaring higher and therefore a sense of control and involvement in the scene. *Courtesy NYU's Media Research Laboratory.*

•PERFORMANCE ANIMATION•

Written with Jane White

A

B

Performance animation (AKA motion capture) is what you get when you hook up a 3-D character to a puppeteer so that, in real time, how the human moves becomes how the 3-D character moves.

This high-tech form of computer animation depends, as you might expect, on computers with large processing power. The computer's track sensors are attached to a human performer and are used to interpret the movement of the human actor into the movement of corresponding points on a character model that exists within the computer. As the actor moves, so does the animated figure.

Performance animation brings the production styles and methodologies of live action and puppeteering to all forms of 3-D animation, from low-resolution game and Internet characters to "live" performances on television to the highest-resolution special effects of animated feature films.

Because Chapter 16 covers the process of creating 3-D characters, the discussion here can focus on that critical part of performance animation wherein real movement is transformed into animated movement. *Motion capture* is a widely used alternative name for this burgeoning field within animation. That particular tag puts the emphasis right where it belongs—on a range of devices that takes on the challenge of capturing the full nuance and expressive power of human gesture.

And what a bunch of clever devices they are!

Figure 17.7 takes you into the computer lab/studio of Protozoa, a San Francisco–based outf t that is a world leader in the new medium of real-time, 3-D character animation. Using a proprietary software system called *ALIVE!,* which modifies input data to breathe life into characters, performers "work" a series of input devices including a bodysuit, pickups, face trackers, data gloves, sliders, joysticks, and even a mouse. Some performance animation systems rely upon magnetic sensors, where several wired sensors are attached to the performer, who gestures within a magnetic field. The sensors read the changes in the field and feed corresponding data into the computer. Other performance animation systems utilize optical sensors—directionally reflective markers are attached to the performer, who is then recorded with high-resolution cameras and infrared strobes. In both cases, the computer must be able to read the changes in location and map key points on the performer to corresponding points on the animated figure. There is little doubt that pioneering done by Protozoa and other performance animation shops like The Big Pixel, Mr. Film, Windlight Studios, InToons, TeleVirtual, Boss Film Studios, and Turner Entertainment will eventually make its way to desktop computer systems via off-the-shelf hardware tools and software applications.

As fascinating as the technology may be, more interesting still is the new world it opens for creative work. When motion capture first began to evolve, its palette of characters seemed anchored to the human figure. Maybe this was because people were still figuring out the technology to capture the essence of a full human, including the arms, legs, head, hands, feet, fingers, and expressions. Or maybe the computer jocks hadn't spent enough time with animators. At Protozoa, there is now just as much creative exploration as there is technical prowess. Figure 17.8 shows a bunch of stills representing different kinds of characters that have been wiggling, strutting, and floating their way into digital life.

Procedural animation is used to augment the performed animation of a character. This technique, which was first introduced in Improv earlier in this chapter, involves the use of preprogrammed mathematical expressions to create automated motions. These preprogrammed expressions can be tied to a specific time or event (e.g., an eye blink that automatically occurs every 15 seconds, or a hand's fingers opening and closing with the rotation of the wrist by the performer). Protozoa's engineers build procedural animation into a character's design and use this technique to make the character's behavior independent of the actor wearing the motion-capture suit or the puppeteer working the joysticks. In "wiring" the character into ALIVE!, special consideration is given to procedural animating techniques to give the character lifelike subtleties such as a springy tail, wobbly antennas, regular breathing, or other motions that can remain consistent regardless of the performer or the ongoing action of the scene.

Stop-motion capture animation combines traditional stop-motion and motion-control animation techniques with

C

D

17.7 Performing at Protozoa: On a barren stage at San Francisco's Protozoa studio, an actor wearing sensor devices brings Reginald, a 3-D insect who affects a Shakespearean twang, to life in real time—*(A), (B),* and *(C).* The close-up, *(D),* shows a digital animator at a Silicon Graphics computer working both a joystick (for the eyes) and sliders — input devices that are used to augment the animation of the surfaces of a performer's face, which are difficult to pick up with optical or magnetic sensors. Before performing a character, about four weeks are required for modeling and "wiring." *Photo by Niki Haynes. Courtesy Protozoa.*

17.8 MOTION CAPTURE GALLERY:

This collection of frame-grabs shows the range of both technical and aesthetic work that has been undertaken at Protozoa. Captions for the images provide a bit of detail about each one. *All pictures were provided courtesy of Protozoa, with special permission from various clients, as noted.*

A

A. A Protozoa group shot shows, in one crazy portrait, a variety of character design.

B. Soulman is a digital, 3-D representation of stand-up comedian Sinbad. *Courtesy Sinbad and Image Public Relations.*

C. A scene from *The Adventures of Worm and Fred*, a television show being developed by Protozoa. The Worm shows a refreshing departure from characters that mirror the appendages of a human actor. Fred, a Russian monkey lost in space, has procedural animation built into his eye blinks and some of his finger movements. The random movement of Fred's rocket ship also depends on procedural techniques enabled by Protozoa's proprietary software.

B

D. Dev Null, the world's first cyberhost, created by Protozoa for *The Site* on MSNBC. Dev is an example of live-to-tape performance animation. *Courtesy of ZDTV.*

E. Moxy, the first performance animation character to appear on U.S. television, is a virtual emcee with procedural animation built into many of his movements. *Copyright © 1997 The Cartoon Network, Inc. All rights reserved. The Cartoon Network name and logo and the Moxy character are trademarks of the Cartoon Network, Inc.*

C

F. Reginald, an interactive insect with a British bent, features digital springs built into his eyes to emulate real-life bug behavior. His geometric simplicity allows for excellent performance on less powerful workstations.

D

E

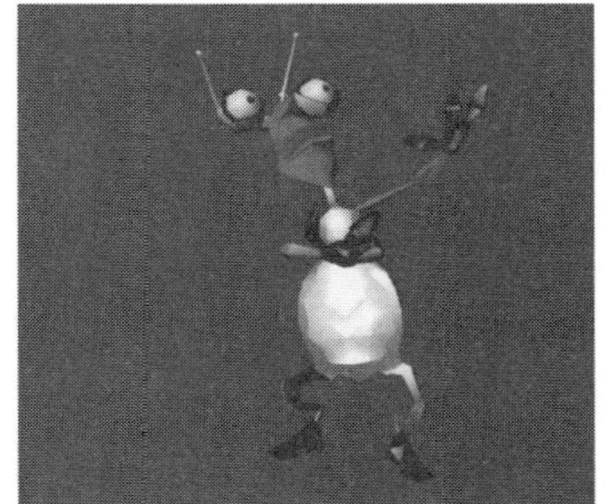

F

G

G. Saurn is a creature who speeds through the desolate world of tomorrow. This character can display a wide range of expression yet requires only the mouse to control it.

H. Dr. Finnery Dexter Klaus, a mad genetic scientist and fast-food magnate, from *Meat*, a television show that Protozoa is developing. This character is an example of high-end graphics paired with face-tracking sensors.

H

I. Frigate and Red (with pal Flit) are a feisty VRML duo appearing in regular episodes on the Web at www.protozoa.com. Low polygon-count in the characters enable them to move in real time, working within the size limitations of Web applications.

I

J. This close-up of a Squeezil shows how Protozoa has designed characters with fewer polygons to make them move faster in real-time games. The character remains highly expressive despite its simple geometry. In fact, the Squeezil has the highest score in the world on the personality-to-polygon rating scale.

K. Floops is the award-winning star of the first-ever VRML Internet cartoon. Sponsored by Silicon Graphics, the wisecracking freak of nature appears in twice-weekly episodes on the Web at www.protozoa.com.

J

K

L. The fish is a pure example of procedural animation. It is a mouse-driven character and responds with a whole range of motions when the tidbit of food is moved by dragging it with the cursor.

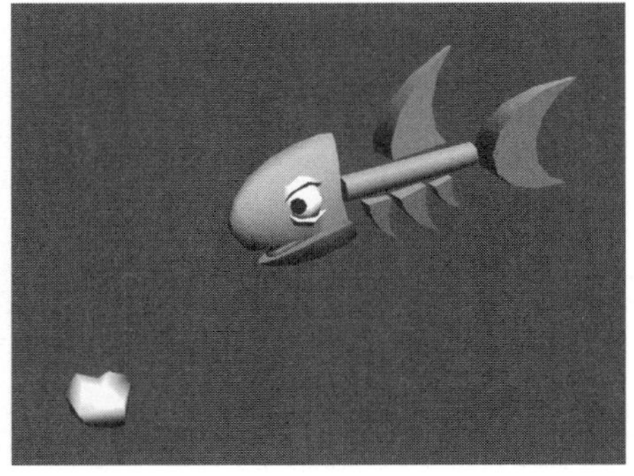

L

state-of-the-art technology. Like performance animation, a character is first modeled in the computer and then wired to magnetic sensors. In place of a bodysuit, a plastic armature similar to those used in stop-motion animation allows frame-by-frame capture. Many effects can be created this way, including flying, falling, and doing a back flip, all of which would be exceedingly difficult in a performance suit.

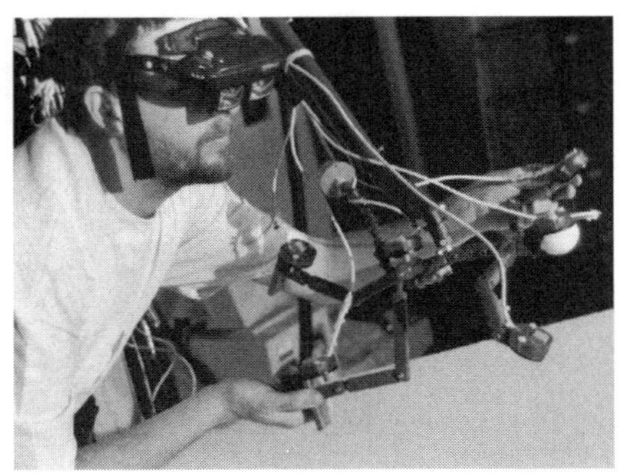

17.9 Stop-motion capture: Is this a cool way to animate, or what?!! Animator Tennessee Reid Norton, wearing goggles that let him see the animation, positions a character-shaped armature with sensors attached at critical points (hands, feet, head, tail, etc.). Each change in position generates a frame, and the resulting sequence is known as stop-motion capture animation.

VRML

Protozoa's background in performance animation has enabled them to use VRML 2.0 (the 3-D standard) to export their wacky sensibilities to the Web. The same procedures are used as discussed above, except that finished animations are written out in a standard script form that can be interpreted by a Web browser and interacted with by the user. Besides being interactive, editable, and engrossing, such animation is much more compact than video.

Something quite wonderful is emerging as the rapid production techniques of performance animation collide with the immediate publishing environment of the Web. Many new "stages" are evolving that require a fresh generation of animated "actors" to appear as VRML characters in daily comics, as host and guide characters, as avatars, and in advertising and games.

In fact, the functions and futures of animation on the Web are growing so phenomenally fast that it is almost a full-time job to track the innovations. But if that's a turf that calls out your name, read on to the next (and largest) animation frontier.

•WEB ANIMATION•

Written with Art Bell

Five years ago you sent E-mail. It seemed pretty cool: You could describe the antics of your dog, Spot. Three years ago it became way cool to send E-mail that included a picture of Spot. Two years ago you attached a sound file of Spot barking. Last year you uploaded a short video clip of Spot doing his Frisbee jump *plus* you created "Spot: The Cartoon"—a rotoscoped animation showing how your canine pal sees himself, with long flowing hair, shimmering colors, and incredible leaping panache.

Today you are working on a 3-D interactive virtual experience that will let your friends explore Spot's favorite corners in your backyard. Your personal Web site will soon contain a series of Spot animations. You are using virtual-reality technology to create a library of environments, all of which offer 360-degree views from 18 inches off the ground—Spot's eye level.

Is animation important on the Web? You bet. It grabs our attention and allows us to share our experiences. *And we ain't seen nothin' yet!* As the Web grows in capability, there is no doubt that the use of animation will continue to flourish. Right now, any piece of animation you make can be sent out via the Web—although there are limits to image size, color depth, and projection speed. The limits come partly from the technological limits of desktop computing, partly from the speed of modems, and partly because each animation file has to be created and organized so users everywhere, regardless of their type of computer, can get to and share each other's work.

ANIMATION TOOLS FOR THE WEB

As an animator who will be designing for the Web and using the Web as a distribution vehicle for your work, you have to know a bit about the tools you need today and the tools you will need in the near future.

HTML. Hypertext Markup Language is the standard for creating and viewing pages of information on the Web. HTML is a file format that defines the layout and content of a 2-D page with links to more information. *Web browsers* such

17.10 Internet speed
Average size of various types of data, uncompressed

1 page of text	1 kb	
1 second of audio	100 kb	
1 second of video animation (30fps)	1000 kb	(1 megabyte)
1 hour *compressed* video animation	500 Mb	(500 megabytes)

Average speed of data on the Internet based on the speed of your modem

MODEM SPEED	KILOBYTES/ SECOND		FILE SIZE
14.4 Baud	1kb/sec.	60kb/min.	3 Mb/hour
33.6 Baud	3kb/sec.	180kb/min.	10 Mb/hour
ISDN (pair)	12kb/sec.	720kb/min.	43 Mb/hour
T1	100kb/sec.	6000kb/min.	360 Mb/hour

Average Costs of Service and Gear

33.6 speed	ISDN	T1	
modem	$100	$1,000-2,000	Not required
avg. charges/*mo.*	$20	$60-$200	$600-$2,000

NOTES:
1. Data transmission rates via cable vary but in general fall between ISDN and T1 rates.
2. Real time (30 frames per second) *uncompressed* video operates at 27 megabytes per second.
3. "Twisted pair" and ISDN are both via your phone line. Phone lines require only special modems, while T1, T2, etc., is special wiring from the pole to your box.

as *Navigator* from Netscape, *Explorer* from Microsoft, and *America Online (AOL)* all depend on standards like HTML to access and view pages on the Internet.

TCP/IP. Each request you make on the Web requires the use of *Transmission Control Protocol/Internet Protocol.* In fact, the Internet is all the computers with access to the public telecommunications network that are using TCP/IP to communicate. The TCP manages all the packets while the IP makes sure they get to the right addresses. These packets are the technical lifestream of the Internet and regardless of the variables—type of computer, modem, telephone line, or country—it all works because of this one standard known as TCP/IP.

To understand TCP, it's helpful to use a railroad metaphor. TCP/IP packets act like train cars on a track, the track being the Internet. Different train cars head for different destinations but temporarily share a common track. Similarly, all data on the Internet move in packets. These packets are all the same size, but with faster equipment you are able to get more of them at the same time. The key here is that different packets share the Internet pipe at the *same time;* animations you are receiving, incoming E-mail, and viewing different Web pages can all occur simultaneously on your computer. This "pipe" is a combination of the speed of the computer, the speed of your modem that connects your computer to your telephone line, the speed of the company that provides your connection to the Internet (for example, AOL, Netcom, ATT, etc.), and the overall speed of the Internet on any given day. This last is referred to as the speed of the *backbone,* which fluctuates according to overall load on the Internet.

This pipe operates on the weakest-link-in-the-chain theory, where the speed of your connection is generally the speed of the weakest link in this chain, which is usually your modem. Figure 17.10 contains some information about how fast different file sizes will travel on the Web.

QuickTime. As any good animator knows, our eyes are

easily fooled, but it is far harder to trick our hearing. QuickTime, the common (standard) file type for storing and sending video and animation on the Web, is ideally suited for the Internet because it was designed to *always* play audio in real time. When QuickTime uses your computer to play audiovisual sequences, it will sometimes skip frames, although it tries hard to keep up with the picture as best as the speed of the computer will allow.

QuickTime is a cross-platform multimedia development, storage, and playback technology from Apple that can combine sound, text, animation, and video into a single file. Using a QuickTime player, you can view and control (brief) multimedia sequences. Although QuickTime files are limited by file size, not time, it's useful to know that 2 gigabytes is the maximum for one file. This is 3½ minutes at full-quality video or 12 minutes at Web quality.

File Compression Systems. There are a number of ways to compress your data, but the common methodology of all compression schemes involves the application of a mathematical formula to each frame of animation. In this process, successive frames of digital information (remember, these are just long strings of numbers) are compared and after the first frame is sent down the pipe, only the changes from the first frame to the next are sent, greatly reducing the overall size of the animation file. That's the basic idea, in any event. See Chapter 23: Computer Software for more about compression.

Creating for the Web is getting easier every day, as recent versions of all the major graphics and editing programs have added new functionality by providing a Save As choice that will create the format and most efficient file size appropriate for use on the Web. Still, there is a long way to go in compression technology, computers, and telecommunications until real-time viewing of full-size, full-color animation via the Internet is possible. The realization of this dream may well play itself out in the fusion of television with computers. The digital TV standards that have been mandated for early in the next century will undoubtedly provide rich avenues for the distribution of animation that won't need to be limited in color depth, playback speed, or frame size.

Animation and Multimedia Authoring Software. With one exception, all the major animation applications are dis-

17.11 Throbbers: These throbbers from Netscape, *(A),* and Explorer, *(B),* are pulsing animations that give immediate feedback that a user's request into the Internet queue is under way and being processed. Netscape's 4.0 throbber boasts 12 frames of animation, as opposed to only 4 available with the earlier versions.

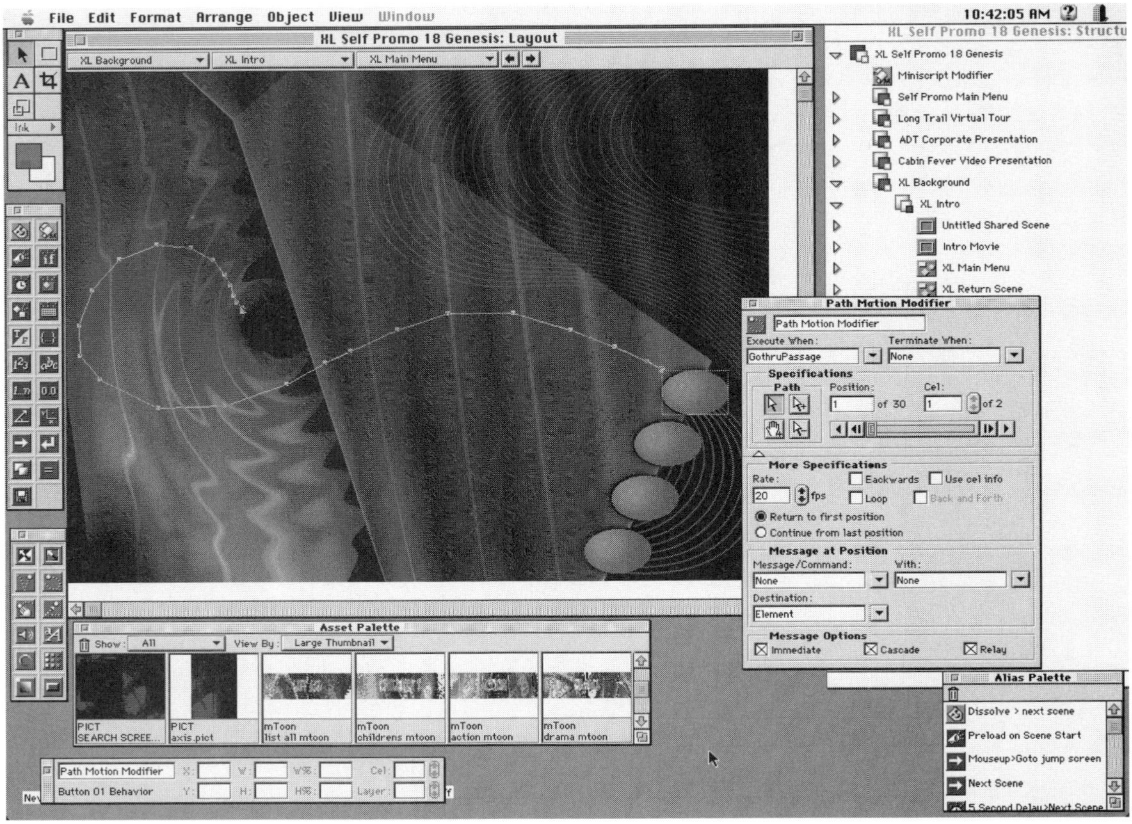

A

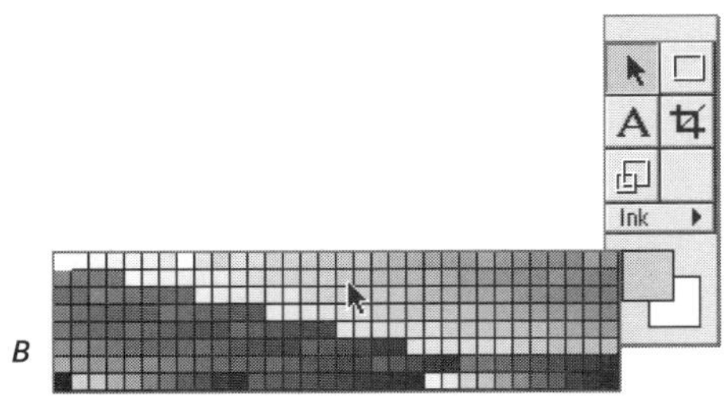

B

cussed or presented in case studies in this book, with Chapter 23 providing a comprehensive software overview. The exception is *mTropolis,* one of the just completely object-oriented applications for multimedia authoring. It was designed largely by artists frustrated with the complexity and steep learning curve of Macromedia's Director. MTropolis does seem easier for animators and artists who are not technically inclined. Figure 17.12 gives a quick intro to this nifty piece of software.

ANIMATION LEVELS ALONG THE INFOBAN

Let's move now from a look at basic tools to a look at the evolutionary pathway by which animation has come to the Web.

Animation has had a native voice on the Web since the

very beginning. The Web was born when the first browser allowed nontechnical users to navigate the Internet. Since that initial NCSA (National Center for Supercomputing Applications) *Mosaic* browser appeared in the mid-1990s, animation has played an important role that began, simply enough, with a user feedback mechanism.

Throbbers. To this day, in the upper right corner of all Web browsers there lives a small pulsing square called a *throbber,* which animates in an endless cycle to indicate that the browser is searching to find the location of the link the Web user has requested. (Folklore suggests these throbbers were possibly created as a distraction similar to mirrors in a hallway that occupy us while we await the elevator.)

Such throbbers actually had their origins closer to the beginning of computer life. They popped up as various analog throbbers to show you when your computer is "thinking," which you've undoubtedly seen.

Animated GIFs and JPEGs. A second level of animation on the Internet can be found in some form of advertisement or banner where a message, often commercial, usually in the form of logo text, has been given motion. These teasers are, of course, attention-getters that induce the Web surfer to go to another location or—more simply—to pause for a moment to recognize and simply think about a particular product or issue.

An animated *GIF (graphics interchange format)* is a single file that contains within it a set of images that are presented in a specified order. Animated GIFs can loop endlessly (and it appears as though your document never finishes arriving) or they can present one or more sequences that then stop the animation on a single image. One particularly cool piece of animation software that isn't covered elsewhere in this volume is *GIF Builder,* a great shareware program from Yves Piguet. See Figure 17.13 for a quick introduction to GIF Builder, which organizes animations in an elegantly concise, high-quality format that is widely used.

Along with the GIF file, *JPEG (joint photographic experts group)* is a file type supported by the Web protocol. As the Internet gets faster, JPEGs, which can be higher-quality than GIFs, are becoming more common. You can create a progressive JPEG that is similar to an animated GIF. Most browsers have

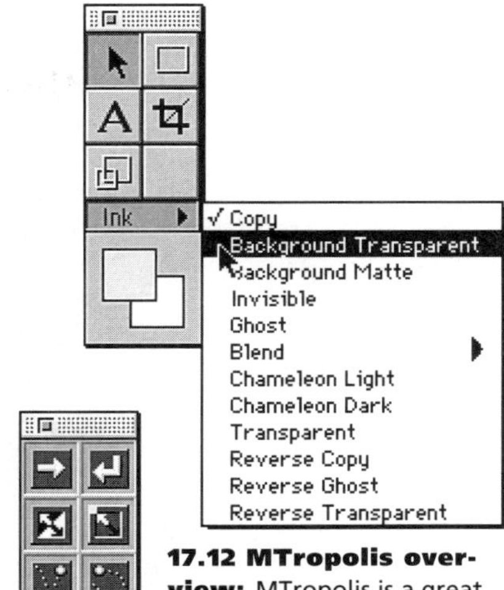

17.12 MTropolis overview: MTropolis is a great tool for giving motion to artists. If you're creating interactive animations, Web sites, and/or CD-ROMs, mTropolis takes full advantage of the latest in computing technology so you spend more time being artistic and less time being technical. mTropolis is a fully object-oriented environment, which means all objects — any image, animation, video, text, or sound — can be assigned properties and inherit behaviors from other objects. For example, if you spend time creating a fluid motion for one object, you can reuse that motion in the same or any other work, without having to know computer programming. Objects can also respond to external events and send messages to each other. New objects can be dragged and dropped into an mTropolis application and they can immediately and transparently interact with other objects.

MTropolis consists of an authoring environment (editor) and a run-time environment (player). mTropolis is essentially a collection of objects, assembled and interconnected by an author in the mTropolis editor for interaction by consumers via the player. The screen-grab, *(A),* shows a typical screen with various windows open. *(B)* shows mTropolis's color picker—which looks much like those of other programs. *(C)* provides a closer look at the menus available for different tools.

A

B

17.13 GIFs and GIF Builder: GIF Builder is a very easy-to-use program that shows you each image in sequence, its size, position, and the time you wish each frame to display. All modifications to an image's data are done directly using only these two windows. In *(A),* GIF Builder was used to set up the animation, its pacing and timing. Each individual 2-D image can be repositioned within the animation, making the eye appear to move realistically. Object-Dancer, (*B*), an example of a GIF with animated text, was made using PaceWorks' (www.paceworks.com) proprietary technology, which provides a time-based, object-oriented environment with features to facilitate the development of animation effects, sharing of animation attributes between objects, and integration of music and rhythm characteristics in the whole animation. This technology allows one to use simple tools to quickly develop lively animated multimedia for the Web, CD titles, video clips, animations, and other projects.

the ability to present GIF and JPEG files automatically and to play the animation sequence after it loads into your computer.

On aesthetic grounds, these mini-animations can be obtrusive at their worst, amusing and informative at their best. Animated GIFs are becoming so mainstream you can now purchase stock animated GIFs similar in concept to traditional stock photography.

Browser Plug-ins. The third level of Web animation requires a plug-in to be downloaded from the Internet and generally added to the Plug-in folder of your Web browser. Plug-ins are available for free and came about as Web site designers figured out how to play audio, 2-D, and 3-D sequences on a computer via drastic compression strategies. Plug-ins extend your browser's functionality, so you can simply plug-in new functionality as it is developed without having to install a new browser each time.

As of June 1998 there were about 250 plug-in applications available for Netscape and Microsoft browsers. Among the most popular of these plug-ins to download are *Acrobat* from Adobe, a document presentation and navigation program that lets you view documents just as they look in the print medium; *QTVR* from Apple Computer, for real-time navigation in QuickTime Virtual Reality scenes; and *RealAudio*'s streaming sound and video player.

Java. Java can be thought of as a cross between a plug-in and an application program that is built into your browser. Java, developed by Sun Microsystems in 1995, is a programming language expressly designed for use in the distributed environment of the Internet. Java *applets* are streamed to your computer and recognized automatically by the browser because the Java programming language is already contained in your browser. These applets are mini-applications or programs that can execute operations similar to any other application you already own. Java may be a key strategic tool for the future of the Internet, since the notion of distributed tools allows the core functionality to reside inside your browser while you download only the newest functionality for your use. Java offers greater speed, flexibility, and functionality over traditional plug-in technology. Using Java, you can view an animation similar to an animated GIF with the additional capability that the animation can be changed and updated based on new information.

17.14 Browser plug-ins: A typical folder for Netscape Navigator (Microsoft Explorer would be very similar) showing various audio, video, and QuickTime plug-ins that you download and place in your Plug-ins Folder to customize your Web browsing environment. QTVR and Acrobat could also be in this folder, but don't appear here because this particular configuration allows these applications to be used both on-line and off.

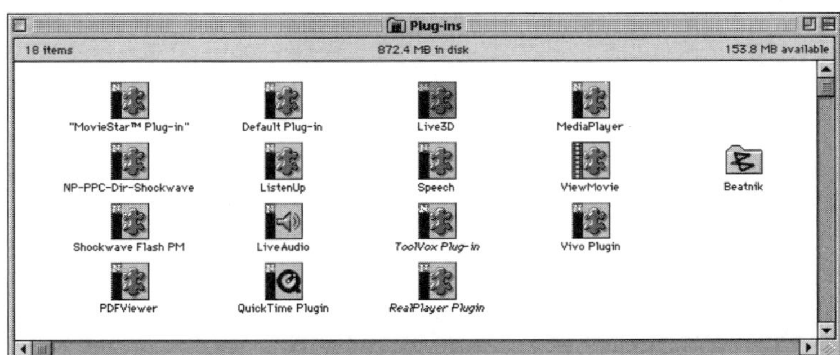

WEB AS VIDEO STORE AND DISTRIBUTION SYSTEM

As this is being written in late 1997, the quality of video and animation you can view directly on the Web, using your browser, is limited to 16 bits of color, or thousands of color choices. To your computer, 8 bits of color depth allows the computer to display a maximum of 256 colors. Note that the spectrum of color that television delivers is hundreds of thousands of colors, and for film and magazines, over 16 million colors are delivered. The limitations of the Internet result because, as you can easily imagine, more colors create larger file sizes, which slow down both the downloading and viewing of still and moving images. Thus the Web is a medium currently limited in color spectrum, although when you work on the same computer with files outside your browser, your limits are only those imposed by the size of your computer hard drive and your patience. If you have the horsepower, you can download any file of any quality and use it on your computer. The distinction here is the capability of today's browsers connected to today's modems.

Not too long from now, however, the Internet will be a true full-bandwidth delivery medium. It will turn today's video rental stores into analog antique shops. Remember that on the Internet everything is just data. Words, pictures, sounds, animations, and films are just bits we create, scan, move, share, and sell instantly without regard for location.

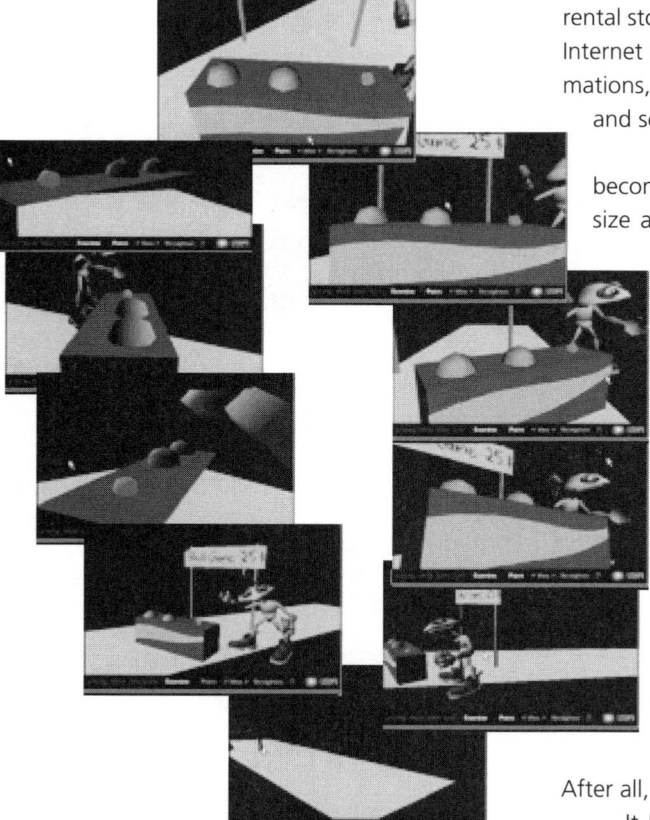

17.15 3-D Web environments: If you visit the Protozoa web site (www.protozoa.com) you can experience a VRML based 3-D environment. The circle of frame-grabs here tries to give a sense of how a visitor can "fly" himself or herself over, under, and around the simple environment. If you make the trip, you can also see the character Floops moving via real-time streaming video. *Images courtesy Protozoa.*

For those of us who enjoy animation, the Web will become a place where we can download animations of any size and color depth. Similarly, there will be no limits on what we can upload. Because of the Web, there can now be direct links between providers and consumers of electronic art, between the artists and their audiences.

But the Web can be so much more.

If animation is to flourish as an art form, it has got to have legs. It has to travel. A nifty piece that you create—a screen saver, a short movie, an animated greeting card, a cartooned gag—needs pathways by which it can reach others who will enjoy it. And when you've established interest in your work, you're on your way to paying gigs! We're talking an assignment. A sponsor. A fee. A royalty. Distribution is key in helping the animator afford to pursue his or her art. In addition to that, a robust showplace for good work lifts the art form higher. After all, we develop good taste by tasting good things.

It has always been difficult to get access to animated movies that don't have a commercial basis. When this book was first written, there was a healthy (not thriving) universe known as *nontheatrical distribution.* Prints of 16mm films and videocassettes were sold and rented to public libraries, schools, and other institutions. It wasn't a mass market, to be sure, yet there was a system in place that supported more than two dozen self-sufficient distributors who published catalogues, sold and rented animations to whoever would buy them (including television networks), and shared the revenue with the artists. But the widespread use of VCRs and video stores has largely made these distributors unnecessary.

17.16 QTVR panorama: This is half of a 360-degree panoramic view of one of the Great Barns built by the Vanderbilts in the early 1900s at their Sherburn Farms estate in northern Vermont. When encountered on the Web, the user can experience what it's like to stand in the middle of the barn and look around, up, and down. *Courtesy Art Bell.*

There's good reason to be hopeful that the Internet will mature as the best distribution system yet, far surpassing the now dead channels of nontheatrical distribution and offering a diversity of product that is much richer than what you can find in a video store or via catalogue services. The Web is naturally suited for linking creators to collectors, matching those who commission animation to those who make it, and helping quality animation find its way to the niche audiences who most value it.

VIRTUAL REALITY

Let's redefine animation. Let's expand the very core of what it means.

A simple but accurate meaning of *animate* is to bring motion to motionless objects. In the world evolving rapidly around us, *animate* can also mean to bring *ourselves* into full movement. Prepare yourself for a dive into virtual reality.

Virtual reality (VR) has been around since the 1980s, and until recently has been a very expensive and complex process both to create and to participate in. But with the development of a new level of accessible technology and the Internet, it has become possible for everyone to experience virtual-reality environments. *And* to create them.

When you enter a virtual location, your actions and destinations are not predetermined. In VR there is no beginning and no end. Users of VR tend to speak as if they have the experience of *being there,* because they were in control. You move from passive viewing to active doing because VR puts you in a driver's seat that lets you animate yourself through new worlds.

Want to be involved with this newfangled animation? The place to start is by learning how to create a place that exists only as digital bits, as a virtual-reality scene. Basically there are two types of VR environments: those that are created

17.17 QTVR Object-o-Rama: Using QTVR you can navigate 360 degrees around something in real time. This can be achieved by a series of still images, properly linked. In the future, you will be able to go to a virtual clothing store, input your height and tailoring measurements, and see how a particular piece of clothing would fit onto a model that is your body double. The shots are captured as the model rotates on a revolving platform and then these photos are scanned into the computer for processing by the QTVR software magic. *Courtesy Art Bell.*

using 3-D modeling tools and those that are made up of a series of real-world photographs.

Three-D VR. To construct 3-D computer-generated scenes you have to go through two steps. First you construct your "world"—your environment in three dimensions—using your favorite 3-D program. Second, you need to learn a scripting language called *VRML. Virtual Reality Modeling Language* is similar in concept to HTML for creating Web pages, but goes further to define the layout and content of a 3-D computer-generated world, going beyond the interlinked pages of HTML first-generation information. The difference results from next-generation tool sets that provide a standard way to distribute and a uniform way to interact, in 3-D, within the 3-D world that has been created.

VRML is a high-level scripting language that may someday replace HTML to bring a complete three-dimensional "working" environment to your computer. When you begin to think about it some, you will see that the three-dimensional world we live in every day can become the basis of new, powerful metaphors to help us better navigate electronic spaces. There's no reason to limit ourselves to the accepted 2-D interfaces just because we have gotten used to them. In VRML the two-dimensional monitor plane we normally operate in becomes 3-D, allowing the location of information itself to tell us something about that information. In other words, the data itself becomes its own interface, making much better use of visual cues and feedback. VRML obliterates the limits of computer space. While there are only so many pixels available on a 2-D surface, if you need more space in 3-D you simply move forward or turn your head. The switch to 3-D is almost alchemy: You get infinite screen real estate out of the same, finite number of pixels on your 2-D monitor.

Photograph-based VR. You can also create a Web site that has the VR experience through the use of photographs. There are two types of VR you can create using photos: *panoramas* and *objects.*

To create panorama VR scenes, the approach is easiest to understand if you imagine shooting a series of photos that form a 360-degree sphere, and then pasting them inside a giant ball. Your viewers will stand at the center of this ball and be able to turn in any direction they want.

17.18 3-D GUI prototype: These three screen shots show how three-dimensional space might be used to present various locations on the Web on different planes—near to far—to represent various locations you visit often or consider more important. The concept is meant to provide a much more physical area in which to interact with the Internet, but using conventional computer monitor technology. The idea here shares much with an "infinite zoom" interface called PAD that has been developed by Professor Ken Perlin at NYU's Center for Advanced Technology. *Courtesy Art Bell.*

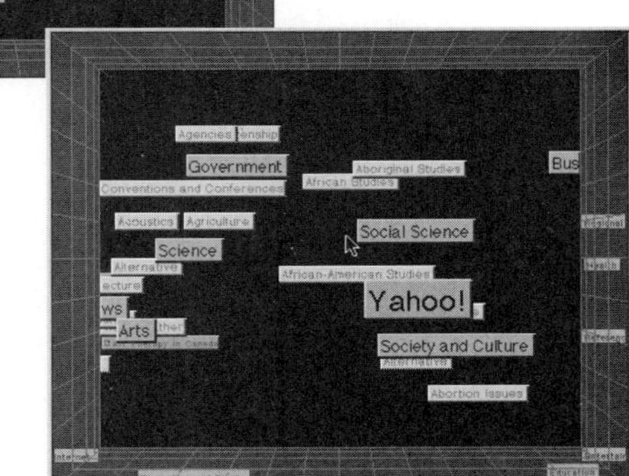

To create an object VR scene, imagine exactly the opposite: You would shoot pictures of an object from all sides, photographing it as if it were the ball and you were going to give your audience the ability to walk around it.

After gathering photos in either panorama or object format, the process of creating VR involves using an image editing program such as Photoshop or more specialized applications such as Panimation from Nodester, Spin Panorama from Picture Works, or PhotoVista from Live Picture to line up the photos into one large, strange-looking file. The final digital magic takes place when you employ Apple's QTVR standard (QuickTime Virtual Reality) software, which asks you to import this big file and decide if you are making a panorama or an object scene.

Prior to 1996, creating virtual environments required computer programming skills. QTVR allows anyone with a photographic camera and scanner or just a digital camera to produce and distribute VR scenes. Newer releases of QTVR even allow the presentation of locational sound—sometimes referred to as 3-D sound. This is a powerful addition to the illusions of virtual reality. Locational sound mirrors the real world

A

B

17.19 Shared graffiti: This is a good example of how the Internet is a great way to electronically share artistic experiences. The artist Gary Manacsa created a digital graffiti wall and lets anyone add to and re-create the piece in endless layers. The three images here represent the wall in March, *(A)*, April, *(B)*, and June 1997, *(C)*. The cat appearing in *(B)* in the lower left corner is Art Bell's cat, Walker, in her usual spot sucking heat from his monitor. Art snapped a picture of her with his digital camera, downloaded the wall, added the photo and a caption, "Cat Dreams," that referenced the Fish. After he uploaded his contribution, Art enjoyed following the next sets of collaborations. Gary's wall is at: http://members.aol.com./gmanacsa/graffiti/graffiti.html. *Courtesy Art Bell and Gary Manacsa.*

C

so that sounds become heard as you approach certain locations and fade as you move away from them.

THE FUTURE OF WEB ANIMATION

To plan for where the Internet might go, all that's important is to know that in the near future the bandwidth pipe will be wide, storage will be ubiquitous, and computation will be in real time. Better to ask what will happen to the speed of creativity.

Figures 17.19 and 17.20 provide two probes into possible deep futures of Web animation: the Graffiti Wall and Screen Saver of the Day. How soon will they happen? Will they happen at all? The tools will be there; the show is up to you.

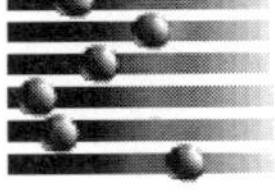

17.20 Screensavers-of-the-week club: How much would you be willing to shell out for a Web service that would use push technology to download onto your computer, once a week, a new assortment of screensavers that would amaze and delight you until the next week's batch arrived? As a club member, you'd have a gentle obligation to send in screensavers and linked images that strike your fancy and that don't infringe on anybody's copyright. So how much would it be worth to ya? *Courtesy Art Bell.*

18 ■ Production Planning

You can do a short piece of animation by the seat of your pants, but not a bigger one. What is big? What is short? My experience as a producer puts the cutoff at thirty seconds. If your project is relatively simple and if you are really, *truly* organized, then you may be able to manage in your head (and with a few notes) all the steps involved in undertaking a limited project. Try it on a larger scale and you are guaranteed to screw up.

This chapter should be a bit scary, for it seeks to remind film and digital animators just how *many* steps there are in completing a project. Lots of novice animators start out with good ideas—and the talent to back them up—but find themselves quickly stalled out and looking at a series of uncompleted projects. This doesn't happen because they are lazy or because they are stupid. It happens because they have underestimated the complexity and rigors of the overall process.

Let me offer this chapter as a producer's cram school. It will provide an overview to the different disciplines of the producer, with emphasis on areas that may not be particularly "arty" but which are absolutely essential if you are to finish the 'toons you've started.

CONCEPTUALIZATION

Without a good idea, there can't be a worthwhile project. Production planning begins with a thorough shakedown of your basic concept. Remind yourself what is at stake: You are going to be spending weeks and months breathing life and form into a single idea. So you had better be working with intellectual material that is rich enough to keep you interested and dedicated when the going gets rough and that initial infatuation has worn away.

The Tag. No matter how complicated your concept is, you ought to be able to state it in one simple sentence. Something that sounds so easy and quickly accomplished may seem like child's play before you actually try it. But you'll soon find it's not so simple. In fact, I predict you will end up telling yourself how totally unreasonable it is to squeeze the richness of your idea into one dumb little sentence. My response is tough love: Stop kidding yourself! If you can't come up with a simple tag—the kind of one-liner you read in *TV Guide* or a catalogue—you haven't yet really worked your way to the heart of your idea. It's absolutely critical for you to clearly identify and understand what your animation is about. Its eventual impact as a screened experience demands this precision.

Story Treatment. Once you've got a lock on what your project is about, it's time to study the central concept and give it dramatic and aesthetic form. Here is the place to elaborate on the larger circle of concerns and interests that the basic idea represents. Over the years there has evolved a single, basic written format, about two pages long, that summarizes what a film or tape is about, whom it is for, and how it will work. This format is called a *story treatment.* It traditionally starts with a concise statement of the project's core idea—the sort of one-liner described above. This is followed with paragraphs that provide a more thorough analysis of the film's goals: the kind of tone or feeling the film will convey; its major content requirements (information it must carry, if any); the project's function (to entertain, inform, instruct, provoke, etc.); and its audience. Of these elements, an analysis of your audience warrants additional comments.

18.1 Production map: A well-executed storyboard is the best of all planning devices. The sample board that starts here and runs in succeeding pages shows the opening action from a film titled *Why Me?* that was written by Derek Lamb and animated by Janet Perlman, when both worked at the National Film Board of Canada. The overall storyboard for this movie consisted of 128 panels. Figure 18.5 provides additional materials from this production.

(DOCTOR TAKING OUT CIGARETTE)

DOCTOR-

HUM HM...ER...MR. SPOON, I HAVE SOMETHING IMPORTANT TO...ER... TO TELL YOU..ER...I'M AFRAID IT'S GOING TO BE SOMETHING OF A SHOCK.

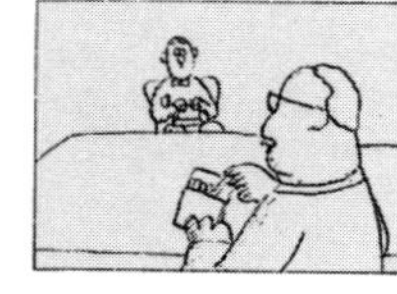

(SPOON SITS FORWARD IN SEAT)

SPOON-

I-IS SOMETHING WRONG, DOCTOR?

(DOCTOR LIGHTS CIGARETTE)

DOCTOR-

..ER... YOU'VE BEEN MY PATIENT FOR SOME TIME, MR. SPOON....

(TWO OTHER CIGARETTES BURNING)

....IT'S NEVER EASY FOR A DOCTOR TO HAVE TO TELL A PATIENT...ER... I WON'T BEAT ABOUT THE BUSH....UM. ..I...ER...

SPOON-

YOU CAN BE FRANK, DOCTOR.

(DOCTOR WIPING FOREHEAD)

DOCTOR-

MR. SPOON, IT'S LIKE THIS- YOU...ONLY HAVE A LIMITED TIME LEFT TO LIVE.

(DOCTOR DROPS ARMS, RELIEVED)

SPOON- LIMITED TIME?

...HOW- HOW LONG DO I HAVE?

DOCTOR- YOU HAVE EXACTLY. FIVE MINUTES.

(PAUSE)

If your animation is to have the impact you want, it must be carefully tailored to the specific audience you intend to reach. This may seem obvious enough. But many projects fall short of their potential impact just because the producer/creator never took the time or trouble to carefully consider who the show's audience was to be and specifically what the audience would bring to the screen in terms of taste, expectations, and prior experience.

The most important service of a story treatment is that it will force you to be certain that the animation technique you've selected is the one best suited to the basic concept you've targeted. This part of the story treatment is usually called the *approach.* Give it a good description and perhaps a sketch or two. Answer these questions for yourself: Does the style of the animation match the content? Is the length appropriate to your goals? Will the audience be able to stick with you and not get lost or bored? Are you sure you've figured out the very best way to communicate the feelings and thoughts that are bound into the project's premise?

Script and Storyboard. The treatment and tag will help you locate your next project within the world of ideas, audience, and technique. But lots of work is left before actual production is ready to begin. If your project is narrative in form, then you (as producer) will want to make sure that the script is set. No matter what kind of animation you are undertaking, it's critical that there be a full set of storyboards (review Chapter 8).

SCHEDULING

Put the concepts away. Gotta deal now with resources. Let's start with the most valuable and limited commodity. This is not money, but *time.*

Figure 18.2 crams onto one page a master list of discrete steps that you should plan for in doing your project. Don't panic at the small type or the length. Some steps will be unnecessary. As you work step by step through the chronology, I suspect you will come up with items that aren't on the list.

Force yourself to make a concrete estimate of the number of hours you will spend on each task. The schedule template also has a blank column in which, later, you should

18.2 Production scheduling template

TASK	est hrs	log hrs	Notes
Conceptualization			
Treatment			write out story idea, style, tag, content, audience; start new project's production book.
Initial Storyboard			develop storyline; show action lines; work out set-up/development/payoff (structure).
Script			write dialogue for actors to record; describe key sound effects to come later.
Character/Art Direction			do comprehensive visual studies; composite "sample" frames of key action; show all different "looks."
Animatic (film, video,digital)			shoot/scan storyboard with timing for all shots/scenes; playback to gauge pacing.
Scratch Track			audio mock-up of all tracks/music to length; goal is to fix the film's length; lay onto animatic.
total hours			be realistic about time you can put into project; spread total hours onto weekly schedule, below.
Design & Testing			
Revised Storyboard			breakout all scenes; tighten story as possible; check mix of close-ups & long shots; lock script; add titles.
Model Sheets/Models			simplify characters; draw model sheets to show front, back, side, key poses; estimate drawing time.
3D Modeling & Texturing			design, revolving & extruding to create wire frames; texturing w/ materials & image maps.
Color Tests			variations of graphic style & colorization; place characters against backgrounds; lock color choice.
Technique Run			execute a "typical" scene from initial to final phases; goal here to assure command of technique.
Visual Effects Tests			camera/software tests of all special techniques: matting, ultimatte; rear-light, Alpha channels, etc.
total hours			fight the natural impulse to jump into project thinking "I'll figure out that part later." Testing saves time.
Audio Production			
Record Music			record music independent of voice & effects; get rights to any non-original music.
Record Voice			get releases from performers; edit rough dialogue, leaving room for action scenes; put into Leica reel.
Record Sound Effects			stock effects from "cleared" library; record on location; "foley" recording in studio to emulate sfx.
Transfer Tracks			for film transfer audio tape to 16 or 35mm mag stock; for computer digitize audio into software program.
Analyze Track/Cue Sheets			locate and mark (with frame accuracy) lip sync; beats within music; location of sound effects.
Preliminary Mix			for use in Leica reel/rough assemblies and editing; keep separate copy all tracks for final mix.
Track Clean-up & Final Mix			final adjustments in editing; this step done in postproduction phase.
16mm Optical Master			film laboratory will provide from mixed master; used for striking answer & release prints
total hours			sound design, recording, editing & mixing can come either before or after production of picture elements.
Art Production & Prep.			
Layouts & Motion Analysis			work through by "scene"; draw sketch bkgs; plot action lines; breakdown timings; locate key poses.
Exposure Sheets			frame-by-frame map to guide drawing or under-camera shooting; plan moves & special effects.
Serial Drawings			separate drawings done on registered sheets of paper for subsequent cel/computer techniques.
Pencil Tests			shoot drawings via video set-up, film down shooter, or scan into computer w/ software compilation.
Leica Reel			film/video/digital "shadow" of final film uses storyboard (later pencil test) matched to track.
Backgrounds			for all layered forms (paper or digital), including cut-outs, path, line, cel & 3-D techniques.
Ink & Paint			for cel projects using either film or digital platforms.
Model Making & Puppetry			for stop-motion & object animation.
Props & Costumes			for pixilation & stop-motion animation.
Scene Construction/Lighting			for stop-motion & object animation.
Scanning Cut-outs & Objects			for cut-out, stop-motion & object animation.
Source Materials			original or acquired materials for rotoscoping.
Digitization of Video			for pixilation and rotoscoping techniques.
3D Scene Building			for 3-D animation; choreography, composition, camera setup & lighting.
total hours			
Final Shooting/Animation			
Checker & Cleaning			required for film animation before shooting cel art.
Oxberry/Downshooter			key phase in stop-motion, clay, paint-on-glass, puppet & other "direct" techniques.
Location Production			shooting outside traditional animation studio for Time-Lapse, Pixilation & Object techniques.
3D Animating			define key frames to define object's motion; inverse kinematics.
Computer Rendering			for computer projects in 3-D, where computational time required for each individual frame.
Compositing/Integration			digital merging of elements via various software programs (2-D & 3-D).
Revisions & Redos			there is *always* something. Count on it.
total hours			
Postproduction			
Assemble Dailies			one lite print off camera negative; digitize video sources; computer files -- all inserted into Leica reel.
Edit to Rough Cut			via flatbed, video editing system or computer.
Edit to Fine Cut/Lock Pix			all elements in final form, including credits.
Cut Negative			A & B roll for 16mm printing; note optical track (above) needed to strike answer and release prints.
First Answer Print			for film techniques; optical effects added in post will first be seen here. Color correction often follows.
Release Print			after close examination and notes to film lab, a timed and perfect answer print, including optical track.
Audio Mix & Lay-back			reminder that subtitle editing & re-mix of voice, music & sound effects tracks may be required.
Lay-off to Video			pieces completed in computer are transferred to video format like Beta SP or VHS formats for distribution.
total hours			
Total Estimated Hours			time estimates should come from all individuals working on a project.
Logged Hours			a running tally will show you the accuracy of your estimates.

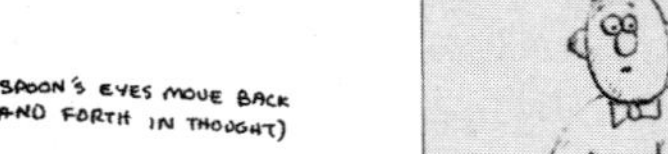

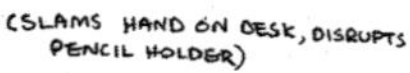

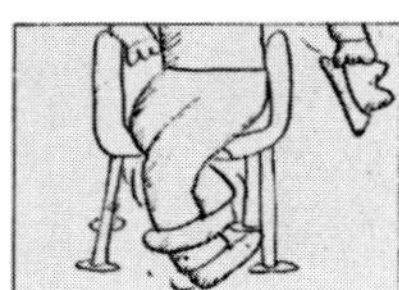

carefully record the actual time that's been spent. This second column is your early warning system. If you encounter differences between the time budgeted and the time spent, you can be sure that the project—and its budget—will not be what you intended. The larger the difference, the more you should be prepared to replan the project and, quite possibly, cut back its ambition and scale.

The schedule template is a handy production checklist (as the budget will be). Use it with flexibility. Different computer- and film-based techniques will require a different sequence of tasks. Note that I have broken out audio production as a discrete phase. This reflects my bias that no matter what the animation genre, sound is equally important as visuals. You will need to make an independent decision about when to work on the audio track(s) of your own project. Before the storyboarding starts? After the final edit has been locked?

BUDGETING

Determine at the outset exactly what the project is allowed to cost. Few things are worse than starting an animation and then being unable to finish it because you've run out of money. Figure 18.3 provides a budget template that represents the combined brainpower of myself and four colleagues with years of experience in producing animation within the filmmaking, computer, and multimedia domains. Here's how we approached the problem of creating a one-size-fits-all budgeting device.

First, we have assumed that the costs of labor will be given at the standard commercial rates. Such "book" rates will fluctuate from market to market. People working on industrial projects will often have lower fees than those working on advertising projects. What you need to know is that all rates are negotiable, especially as you sign up people or equipment for longer periods of time. Don't be scared off from making a cartoon because you take one look at the budget template and see that you can't afford it. Learn to negotiate! Every animator we know began with projects that were done very, very inexpensively.

Second, we decided there should be a default figure for each line in the budget. To come up with a ballpark guess, we had to make additional assumptions. The largest of these is that the template anticipates a finished product of between two and five minutes in length. We couldn't predict what animation technique would be used or whether production was taking place in film, digital, or hybrid platforms. So we tried to cover all of these, although none is listed in total detail.

In terms of equipment prices, we entered into the spreadsheet what we know to be reasonable rental charges. Often, of course, you will purchase a particular piece of hardware or software. In that case modify the budget and amortize the cost of gear you own. We generally space them out over three or four projects, or the same number of years, whichever comes first.

Finally, we feel obligated to point out that the budget template is either missing or setting very low default figures for some very significant categories that we, as professional producers, plug into most of our own projects. These include:

- *Payroll taxes:* Obligatory payments every employer must pay for state and local taxes, including workmen's compensation, which are about 10 percent to 20 percent of salaries. If you work out of a studio, *fringe* payments that accrue to each employee include health insurance, vacation pay, sick pay, and holidays.
- *P and W:* The pension and welfare markup that unions require for the benefit of their members—10 percent to 15 percent is common.
- *Insurance:* For liability, for faulty equipment, for errors and omissions coverage against copyright and other suits—1 percent to 2 percent.
- *Contingency of 5 percent to 10 percent:* Although it's most accurate to budget "tight" by specifying and quantifying all detail, one inevitably forgets something or runs into unpredictable delays and additional costs.
- *Marketing and press materials:* You should allocate something to the seven marketing/promotion categories listed in the template—obviously lots of money could be spent here.
- *A producer's fee or profit margin:* A pot at the end of the rainbow—some extra money that might help cover the costs of developing your next project.

18.3 Production budgeting template: Starting here you have a budget template that should be useful in projecting costs for your next animated project. Please remember that the default numbers used here are only representative. Be sure to research and verify each line item that accrues to your production. Note, too, that some expenses are "real" (involving out-of-pocket expenditures) while others are "imputed" (the costs are labor or equipment access for which you will not need to cut a check). You'll need to tally totals for each section and compile a Grand Total.

Project Personnel	**#**	**unit**	**rate**	**total**
Producer	650	day		
Director	750	day		
Designer	750	day		
Technical Director	400	day		
Visual Effects Supervisor	500	day		
Production Assistants	75	day		

Animation Crew	**#**	**unit**	**rate**	**total**
Art Director/Creative Director	500	day		
Layout/Illustrator	350	day		
Character Designer	400	day		
Storyboard Artist	350	day		
Storyboard Artist (by page)	750	page		
Art Production Supervisor	225	day		
Prop and Model Artist	400	day		
Animatic Operator (needs Avid)	300	day		
Track Reader	300	day		
Sheet Timing (per 35mm ft)	3	ft		
Animation Director		day		
Animator (3 sec/day)	350	day		
$5,000 flat for 30 sec.	5,000	30 sec		
$115/35mm ft	115	ft		
Assistant Animator	275	day		
AfterEfx Animator/Compositor	300	day		
AfterEfx (w/ workstation)	500	day		
Art Production Supervisor	275	day		
Rotoscoper	275	day		
Matte painter	350	day		
Art Assistant	175	day		
Background Painter	350	day		
Inkers and Opaquers	150	day		
Cleanup Artist	250	day		
Animation Checker	175	day		
Digital Ink & Paint Scanner	150	day		
Digital Artist	225	day		
Digital Compositing	325	day		
Animation Cameraperson	250	day		
Asst. Animation Cameraperson	150	day		
Stop Motion Animator	350	day		
Video Animation Stand Operator	175	day		
Other Crew (Specify)				

Studio Crew	**#**	**unit**	**rate**	**total**
Director of Photography	1,500	day		
Camera Operator	500	day		
Asst Cameraperson	475	day		
Motion Control Operator	400	day		
Set Designer	600	day		
Model Maker	300	day		
Prop Builder	450	day		
Gaffer/Electrician	475	day		
Grip	425	day		
Recordist	475	day		
Playback Videotape	450	day		
Makeup/Hair/Wardrobe	500	day		
Scenic Painter	400	day		
Tape Engineer	400	day		
Script/Continuity	300	day		
Other Crew (Specify)				

Digital Staff	**#**	**unit**	**rate**	**total**
Technical Dir/Lead Programmer	100	hr		
Visual Effects Supervisor	700	day		
Programmer	75	hr		
Prototyper (Lingo)	50	hr		
Digital Artist	50	hr		
Cleanup Artist	25	hr		
Quality Assurance	30	hr		
Software Support	300	day		

Talent	**#**	**unit**	**rate**	**total**
Writer	tbd	fee		
On-Camera Principals	450	day		
Voice (Character) Principals	400	session		
Puppeteer	550	day		
Audition costs	500	allow		
Casting Director	500	flat		

Postprod & Sound Staff	**#**	**unit**	**rate**	**total**
Postproduction Supervisor	350	day		
Editor	500	day		
Assistant Editor	225	day		
Optical Supervisor	250	day		
Music Director/Composer	1,000	flat		
Musicians	tbd	session		
Voice Director	1200	day		
Sound Editor/Mixer	250	day		
Studio w/ Engineer	275	hr		

Film/Video Equip. Rental	**#**	**unit**	**rate**	**total**
Camera Rental 35mm	750	day		
Camera Rental 16mm	350	day		
Single Frame Motor		30		
Sound Rental	350	day		
Lighting Rental	500	day		
Grip Rental	500	day		
VTR Rental (playback on set)	250	day		
Motion Control System Rental	2,500	day		
Single Frame Tape Recorder	175	hr		
Animation Camera w/ Operator	100	hr		
Rotoscope Camera	50	hr		
Pencil Test Camera (Video)	25	hr		

Stage and Studio Rental	#	unit	rate	total
Rental for Build & Prelight	450	day		
Rental for Strike	450	day		
Prop Rental/Purchase	250	est		
Studio Rental for Shoot Days	600	day		
Power Charges and Bulbs	225	est		
Meals for Crew and Talent	125	day		
Shop Rental	200	day		
Production Office at Studio	200	day		
Paint Hard Cyc.	300	flat		
Studio for Narration Recording	75	hr		
Studio for Music Recording	75	hr		
Foley Studio for Sound Effects	125	hr		

Digital Equipment Rental	#	unit	rate	total
Paint Workstation (PC/Mac)	350	day		
Digital Scanning Workstation	800	day		
Digital Composite Workstation	1000	day		
Video Graphics Workstation	400	day		
Storage Media Equipment	150	day		
SGI Rental	tbd	day		
Motion Capture	tbd	day		
Performance Rental Suite	tbd	day		

Art Supplies/Services	#	unit	rate	total
Paper, Paint, Brushes, etc.	250	est		
Cel Xerox	2	ea		
Stats and Kodaliths	4	ea		
Typesetting	150	allow		
Still Film/Transparencies	75	allow		
Camera Supplies	200	est		
Color Xerox	2	ea		
Ink and Paint Services	5	cel		
Pencil Test (film-to-tape)	300	allow		
Photo/Rotoscope	3.5	frame		
Dye Sublimation Prints (8x10)	10	ea		
Typesetting (esp. for credits)	250	allow		
Storage	tbd	est		

Film Materials	#	unit	rate	total
Purchase Raw Stock (16mm)	0.30	ft		
Developing	0.30	ft		
Video Dailies	0.19	ft		
Transfer to Magnetic Stock	0.10	ft		
Sync/Screen Dailies	60	hr		
Audiotape Stock (Mag, 1/4")	7	1/2 hr		

Digital Materials	#	unit	rate	total
Software Applications	500	allow		
Storage Media Software	300	allow		

Editorial: Audio	#	unit	rate	total
Sound Effects Search and Fees	250	day		
Stock Music Search and Fees	250	day		
Digital Sound Workstation	300	day		
Sound Transfers	200	est		
Mixing Studio	150	hr		
Audio Layback (to videotape)	300	est		
Optical Sound Negative	0.30	ft		
Track Reader	300	day		
Needle Drop (music licenses)	tbd	per		

Editorial: Film				
Flatbed Editing Suite Rental	50	hr		
Screening Room	25	hr		
Negative Cutting	10	cut		
Answer and Corrected Prints	0.80	ft		
Interpositive/Internegative	1.10	ft		
Optical Effects	25	ea		
Editing Supplies	150	allow		

Editorial: Videotape	#	unit	rate	total
Off-line Editing (deck-to-deck)	750	day		
Off-line Editing (non-linear)	150	hr		
Film Cleaning	125	hr		
Film-to-Tape Transfer (1 lite)	450	hr		
Film-to-Tape (w/ color correct)	800	hr		
Pin Registered Film to Tape	950	hr		
Paint Box	300	hr		
Compositing (Harry & Abekas)	850	hr		
On-line Editing (D2, Beta SP)	450	hr		
Input Camera	100	hr		
Character Generator	150	hr		
Tape Stock	300	allow		
Window Dubs	150	allow		
Masters (Dub & Stock)	250	ea		
Viewing Copies: VHS 1/2"	35	ea		
Viewing Copies: 3/4"	50	ea		
Tape-to-Film Transfer	600	hr		
Tape-to-Film (w/ color correct)	800	hr		

Editorial: Computer	#	unit	rate	total
Digital Editing Workstation	300	day		
Exabyte to Video Transfer	350	hr		
Film Scanning Setup Charge	250	per		
Film Scanning Frame Charge	5	frame		
Film Recording Setup Charge	250	per		
Film Recording Frame Charge	6	frame		
Special Effects	300	allow		

Miscellaneous	#	unit	rate	total
Legal Fees	500	allow		
Bookkeeping/Accounting	500	allow		
Location/Permits	250	allow		
Transportation	350	allow		
Messengers	150	allow		
Telephone and Federal Express	200	allow		
Working Meals	300	est		
Research Materials	150	allow		
Still Photographer/Slides	500	day		
Marketing/Promotion Materials	500	day		
Press Event	300	allow		
Talent per Diem/Air Fare	tbd	day		
Production Stills Photographer	350	day		
Wrap Party	tbd			

Overhead and Indirect Costs	#	unit	rate	total
Payroll Taxes (staff only)	0.08	%		
Pension and Welfare (staff only)	0.07	%		
Other Fringe (holiday/sick pay)		0.05		
Insurance (% of overall budget)	0.02	%		
Office Exp. (rent, elect, heat)		tbd		
Production Fee/Profit		tbd		
Contingency	0.10	%		

DESIGN AND TESTING

It doesn't matter whether you are working with a computer or with a film camera—it's always terribly painful to shoot an animated sequence and *then* find out that the equipment wasn't working right or that you didn't spend enough time refining the look for your piece.

There's no set procedure for testing your production equipment. Basically, what you want to do is identify all of the possible variables at each step and then, systematically, keep all other factors constant as you test one specific element. Take camera tests as an example. First set up the camera, its stand, and the lighting as you think they'll be used in the production. Load the exact type of film you will be using. Make up a chart that records camera model, lens, stock, lights (top and bottom, if using both), exposure, focus, shutter, motor, and artwork. Now comes the testing. You will be shooting a series of exposures (96 frames works—that's 4 seconds). For each of these, put a small pad of paper under the camera, within the shooting field, that identifies what variable you are testing. So if you were checking exposure latitudes for a specific film stock, you'd shoot (and mark) 96 frames of each of your camera lens's f-stops: f/22, f/11, f/8, f/5.6, f/4, f/2.8, f/2.0, f/1.4. The goal here is to be comprehensive. So if you are testing through the aperture range, as above, when you subsequently project your camera tests you will be able to see exactly what one stop in lighting change does to your image. Be sure to shoot a sample piece of artwork, not just a sheet of paper, which can bounce too much light back into the lens.

I hope you get the idea: Clear reasoning and patience are required for technical testing. The same basic approach (keeping all the variables constant except for one) holds for animating in any film- or computer-based technique, using any piece of hardware or software.

And what holds for gear also holds for art. You should be sure to give yourself plenty of time, and some test sequences, to work through as many variables as you can imagine for your characters, how the characters move, the range of backgrounds you will use, lighting setups, special effects done in-camera or in postproduction, and your sound tracks. Be sure to write down what you test.

PRODUCTION BOOK

I want to urge you to keep a special *production book* for every animation project you undertake. There are at least four compelling reasons for doing this. First, a production book will facilitate the process of conceptualization. It should contain your story treatment and you may want to store your storyboard here as well. Second, a production book can provide the literal framework from which you will organize the production. It should contain your schedule and your budget. It can house the letters, releases, and agreements you complete. Third, a production book can provide you with a comprehensive and detailed set of notes for use at a later time. Store your testing data here. You want to make sure that what you discover while making one project is not lost when you need that specific knowledge for a different undertaking—even after a number of years. Finally, a production book can have an important spiritual

18.4 Model sheets: One of the items to keep in a production notebook is your model sheets. Here is one for the Stick Figure Tribe that appears throughout these pages. Note that the "bible" for an animated series will have full model sheets for each character and each location.

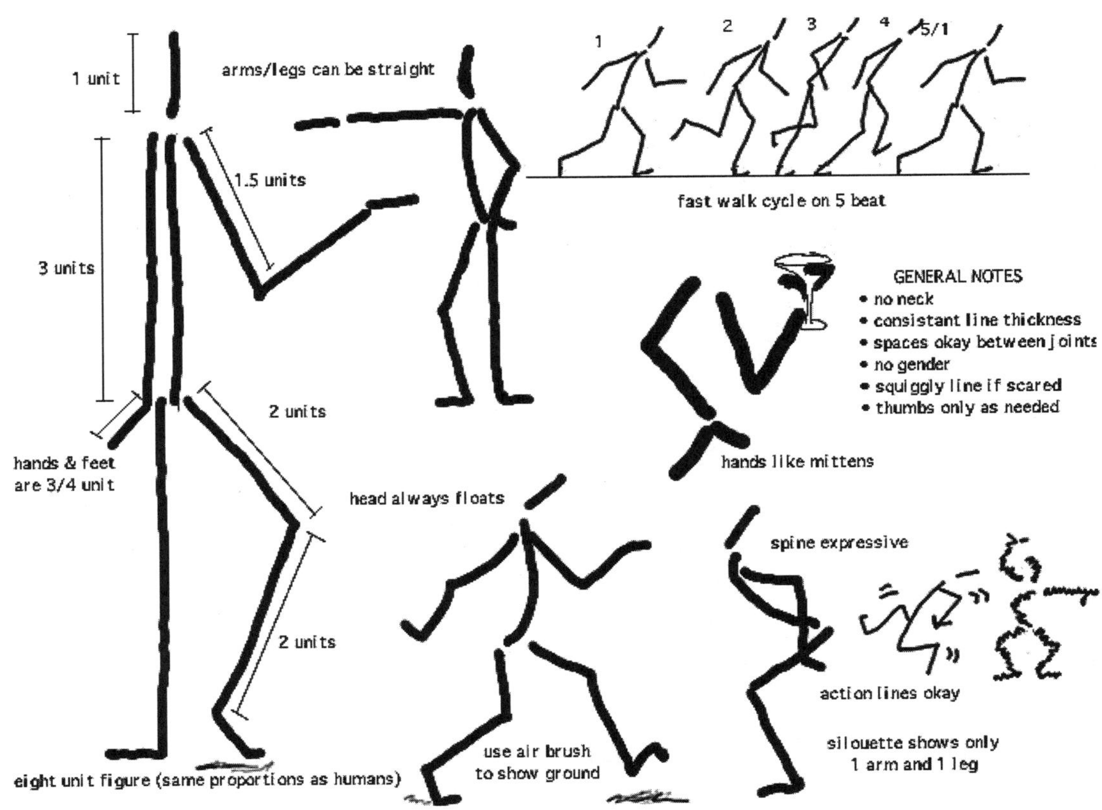

18.5 CASE STUDY: A One-Animator Approach

With careful planning, an individual animator can create an entire cartoon. The story line of *Why Me?* uses broad humor and irony to follow the progression of the emotional state experienced by someone who knows he or she is terminally ill. In the cartoon, written by Derek Lamb and animated by Janet Perlman, the clinically determined stages that accompany the acceptance of one's own death are compressed hugely, providing opportunities for character animation pyrotechnics as we follow the character "arc" of the movie's protagonist. The photos here were shot while the film was in production. Its finished length is ten minutes and the film went on to win many awards, including the Grand Prize for Animation at the New York Film Festival.

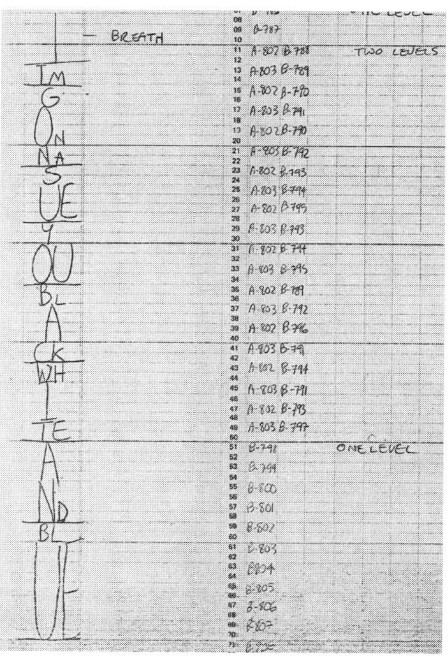

A

Production Map

Once the script had been storyboarded (see Figure 18.1) it was cast and dialogue was recorded with actor Marshall Efron as the voice of the dying man. *(A)* shows a chunk of the completed "dope sheet" in which a single line of dialogue ("I'm gonna sue you black, white and blue") is broken down in a standard exposure sheet. Note you can see how the animator has simplified elements so that there are only one or two cel levels in this particular segment. The track breakdown of dialogue against specific frames provided Janet Perlman with a production map that took her storyboards to a far more "atomic" level. Note how the analysis even shows the animator where the character draws a breath. Janet followed such exposure sheets as she methodically worked her way from the first scene to the movie's conclusion. She was the *sole animator* of the entire film!

Production

A simplified cel technique was developed for this project so no ink and paint would be required. After pencil-testing key scenes, the art production process was carefully worked out with pencil drawings and watercolors done on a heavyweight paper stock that would be able to stand up to reuse under the camera. Production design called for some of the drawings to be cut out so that up to two layers could be stacked up on a single background. This represents a clever deployment of the art creation inherent with cel techniques. *(B), (C),* and *(D)* show three levels into which one of the scenes was broken down: A small number of "rain" backgrounds, *(B),* was cycled to show the storm happening outside the doctor's office; a midground element (a single cutout showing the window frame) was used again and again, *(C);* the main character, Mr. Spoon, appears as a talking head, *(D),* placed on the top level. Thus the scene is made up of only

B

C

D

E

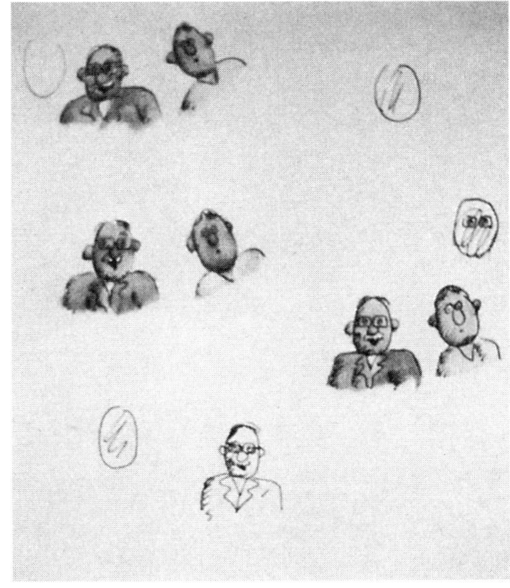

F

G

three levels which, when combined and tightly framed by the camera, *(E),* provide a surprisingly "full" look. The reduced scale of the artwork itself (working in a #6 field) saved the animator lots of time hand-coloring larger image areas. But for this to look right, character design had to mirror the simplicity of production technique.

Photograph *(F)* shows detail from preliminary character studies. In *(G),* a pile of finished drawings stands ready for filming. Here you can see that although the use of acetate cels is avoided, full animation can require a stack of cutout frames that preserve the original pegging system of registration and yet work in conjunction with other cel levels. Janet Perlman, *(H),* who single-handedly animated the film, is seen at work in her studio at the Film Board of Canada's production center outside Montreal. In *(I),* Derek Lamb confers about another project with animator Yossi Abolafia. For many years Derek was Executive Producer of English Animation at the National Film Board of Canada. *Photos by the author.*

H

I

function. Sometimes it's important to have a place where one can voice frustrations or expectations or any of the other emotions that are encountered in the course of creating an animation. A production book should be the receptacle for all such observations and comments. In a sense it can become a diary through which you stay in touch with the overall process.

PRODUCTION AND POSTPRODUCTION

One of this volume's main goals is to show that each animation technique has its own particular production requirements and creative nuances. Still, there exists a sequence of five major steps that are common to all productions and that you will hear talked about again and again:

1. Storyboarding
2. Board-o-matic/Leica reel/animatic
3. Layouts
4. Animation and shooting
5. Postproduction

The first two items will reveal the basic form of your piece. Layouts and animation (traditional and digital) go hand in hand. It is here that the piece comes to full life, scene by scene, in all its glory. Through this protracted passage you will be up to your eyeballs with details of every imaginable kind.

Postproduction is an easier and faster step for animation than it is for live-action projects. There are two big steps: editing and track work. Editing should go easily, especially if you have been slugging real footage or final digital renderings into an existing Leica reel or animatic. The ratio of created segments to final edited form should have less than 10 percent wastage. The editing process usually includes reaching a *rough cut* (when you have all your picture elements in place and try out the basic sound track) and a *fine cut* (when you tweak the pacing of scenes and "lock pix," meaning you determine that no further changes will be made).

If you have not already recorded your sound track, you do so in the postproduction phase. See Chapter 7 for a review of all the steps and elements in building and mixing a good sound track.

TROUBLESHOOTING

While the preceding methodology is comprehensive and logical, don't put too much trust in it. Don't ever assume that your planning prowess can ensure that everything will emerge in the right form at the right time. The unexpected is to be expected, and troubleshooting is the job of the producer. This is especially true if you are working as an independent animator, without a staff of specialists at your side. Each and every phase of production is yours to manage—from the first idea to the distribution of the completed project. While friends and colleagues might contribute significantly, you will be the only person who knows all the pieces. It's your job to shepherd the project, to make it your best work, and to ensure that anyone you corral into helping will feel that you have given them an opportunity to do some of *their* best work.

SCREENINGS, EVALUATION, DISTRIBUTION, AND YOUR REEL

It's a thrilling moment when you show your finished animation for the first time. The lights are lowered. The audience becomes quiet. The projector or video playback unit stands ready. Your artistic vision is about to be communicated and, believe me, it *is* thrilling. All your work must now pass the scrutiny of an audience.

Somehow everyone lives through the anxieties of a first public screening. Living through it isn't enough, however. You've got to learn through it as well. I suggest that you arrange some screenings for people whom you'll feel comfortable asking for reactions. Find out what they liked. What did they think the animation is about? Whatever reactions and feedback you can glean should be used in evaluating the film in terms of its original goals. Look back at your initial story treatment. Did the show do what you wanted it to do? What were your project's strengths and what were its weaknesses?

If you like what you've made, you should think about exhibition and distribution. One way to go about this is by entering your pride and joy in festivals that showcase work by independent or student filmmakers. Other options for distribution are reviewed in Chapter 25: Resources. At the very least, you should include the finished project in your *reel.*

Over the course of many years and many projects, all serious animators take the best of their completed projects and, working from the original or *first-generation copy,* edit it all together on a videotape (often called the *director's reel*) that can easily and inexpensively be given away. Reels are rarely longer than six or eight minutes and can contain entire projects (which is great if they are short) or chunks from completed productions. Keep your reel short and avoid the runt-of-the-litter syndrome, in which you keep a piece that has some sentimental value to you but just doesn't help the reel, which by definition is a showpiece and sales tool. Five minutes should be the maximum length of the reel if you are just starting out. A good way to show off your various projects and experiments is to create a montage of clips and great moments from different films—including some that might never have reached a final form—that are edited against a single piece of music. A reel should begin or end with the animator's name, address, phone number, and E-mail address. You want those potential fans and fellow artists to know how to reach you. Not to mention sponsors and investors!

THE NEXT ONE

In a sense, the final step in the production of one animation is the first step toward another animation. You should always keep thinking of the next project. Sometimes an idea will emerge from a test you've tried. Sometimes an idea comes from a mistake or an accidental discovery. Sometimes a particular technique of animation will spawn the next movie. Sometimes you'll learn as you work on one project that your real interest lies within a totally different approach.

There is no such thing as a valueless production or a wasted film. Every sequence that you undertake teaches you something. And this makes film production a regenerative process. One thing always leads to the next thing, although it's often impossible to know what lies ahead. Animation, like anything that matters, should be viewed as a dynamic process in which the goal is excellence, not just one excellent piece of work. The end of one project marks the beginning of the next.

18.6 CASE STUDY: Creative Process Step-by-Step

A

B

C

In the early 1990s, MTV created a television show called *Liquid Television,* a half-hour show made up entirely of short animations. The show ran for three seasons and succeeded in fulfilling its mission as a development laboratory for new kinds of animation and new approaches to storytelling. Two MTV series that got their start in *Liquid Television* were *Beavis and Butt-head* (by Mike Judge) and *Aeon Flux* (by Peter Chung). The contributors to *Liquid Television* included some of the most innovative directing, writing, art direction, and animation talent working in the early 1990s.

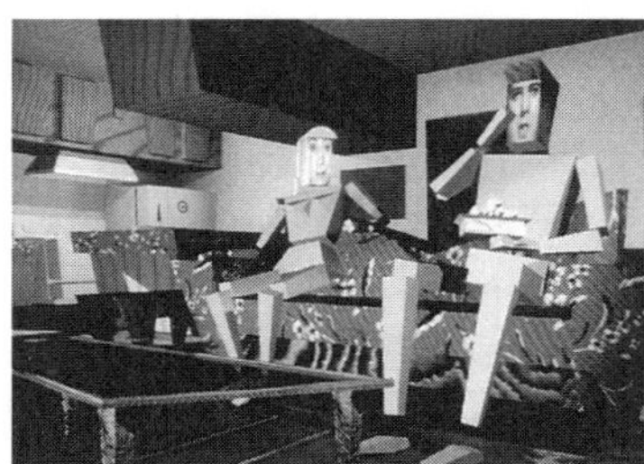

D

This case study takes a look at *The Blockheads*, a series of three, two-minute segments that was developed at (Colossal) and directed by Eli Noyes, who provides the following commentary. It focuses on the forces that drive innovation—not the least of which was a tightly constrained budget. When one is starting out in animation, innovation is *always* a major component of production planning. Hence the appropriateness of this particular case study in this chapter.

E

Along with everybody else, I was fascinated when Brad deGraf set up some SGI machines with proprietary motion capture capabilities at Colossal. We all wanted to see if we could create some computer graphics characters, put them in a story, and then have actors use motion capture devices to breathe life and character into their bodies. Japhet Asher, executive producer and originator of the *Blockheads* idea, found cartoonist Peter Bagge from Seattle, and commissioned him to write several short scripts featuring two characters in scenes typical of Peter's cartoons — the middle class caught in urban angst.

Brad and his crew wanted to keep costs and computing power low by making the characters out of simple, computer-friendly polygons: *blocks.* And so we started designing blocky characters who would need to move, express emotions, drive blocky cars, live in blocky rooms, and most of all, talk and relate to each other. Frames *A* through *E* sample what emerged.

The Blockheads Episode 1: The TV — Page 2

Blaine pauses, stops chewing for a few seconds.

BLAINE:
... What is the point of learning something if yo[u] don't do anything with that knowledge? I nev[er] saw you cooking anythi[ng] French ...

:28

Blanche fumes, grabs remote control and changes channel. Sounds of a wrestling match come from T.V.

BLAINE (shocked):
W-Why did you chang[e] the channel?! I thoug[ht] you hated wrestling!

Blanche points remote at T.V. again, starts changing channels at a fast pace.

BLANCHE:
What difference does i[t] make? I can't hear th[e] T.V. anyway! Thanks t[o] your chewing!

:39

Blaine grabs one end of the remote, as Blanche tugs at the other end,

BLAINE:
Hey! Whatareya --
Gimmie that thing!

Truck past them with remote. Follow to fish bowl

F

Script and Storyboards

We were severely limited by the crude state of motion capture in 1993. The system we used could only record the motions of one actor at a time — therefore, scenes where we had two characters had to be carefully storyboarded so we knew where everybody was at every moment, *(F)*.

G

Character Design

Character design was a delightful challenge. We hired artist Dave Gordon to help us with the characters and their world. He helped us sculpt a female and a male body from blocks, *(G)* and *(H)*, and to think creatively about block cars, block furniture, block buildings, and even block clouds in the sky.

I

A particular challenge was the question of what the two characters' faces looked like. We knew we could "texture map" a face on the front of the character's block head, but were unsure about what kind of face it should be. After considering animated eyes and mouths, we discarded that idea for cost reasons and started to think there might be a way to shoot video of actors' faces and map them onto the front of the blocks. Some experiments we did involved cutting out eyes separately from the nose and the mouth, *(I)*. We also experimented with different Photoshop filters to see if we wanted scan lines, various kinds of distortions, etc., *(J)*. In the end we did the simplest: mapping of the face directly onto the front of the block, *(K)*. The sides and top of each head were made from freeze-frames of ears and hair texture. This worked well, much to our collective delight.

H

J

K

CGI/Background World

Dave Gordon also helped to design the world our Blockheads lived in — made from blocks, of course. We did a layout of what we thought the living room "set" should look like *(L)*. We were mostly concerned with furniture placement, size of the room, where the "actors" should sit, etc. This was important to work out in advance, since we wouldn't have the luxury of a real set in which we could move the furniture and block out actions on the spot. Dave then took the living room and added his unique style to it, *(M)*.

L

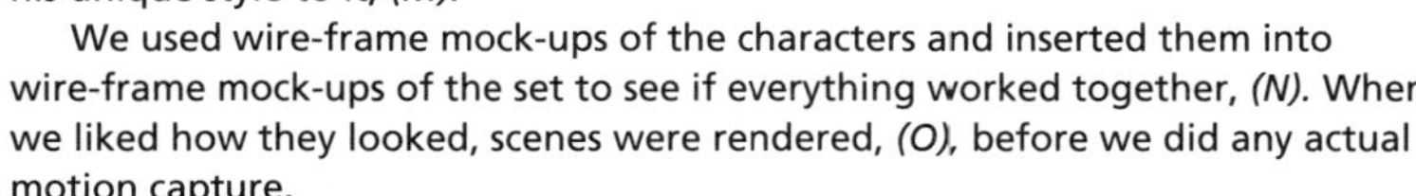

We used wire-frame mock-ups of the characters and inserted them into wire-frame mock-ups of the set to see if everything worked together, *(N)*. When we liked how they looked, scenes were rendered, *(O)*, before we did any actual motion capture.

The final product was a delight. Mark Swain, our CGI wrangler, spent many hours importing motion capture data to our Blockhead models, and bringing in video frames of our actors' faces through the Macintosh for texture-mapping on the moving characters. He also built all the models and lit them for final rendering.

M

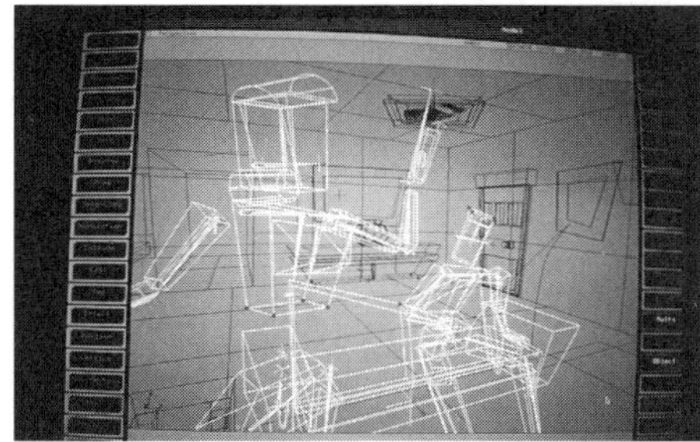

N

O

Motion Capture

Although we used an optical motion capture system, images, *(P), (Q),* and *(R)* show a more advanced motion capture system that Brad deGraf's group (now incorporated as Protozoa) is currently using. The operator wears a bodysuit with magnetic sensors on it. These sensors are connected by an umbilical cord to a central computer that can track as many as sixteen different points on his body simultaneously and use that data to drive a CG character with similar control points on its body.

P

The actor must wear head-mounted TV screens so he can see a representation of himself at all times, even if he is facing away from the camera. This system, like the one we used, can only capture the motions of one actor at a time, making scenes with multiple players in them a challenge to organize and perform. Since the virtual character has different proportions than the actor who performs it, the live actor may find he must put his hands and body into strange positions to get the motions he wants. The sign shown in *(S)* provides a droll reminder to the performance animation actors.

Q

The quick advance of animation technology will surely eliminate most of the limits we faced in creating *The Blockheads*. But those same limits forced us to explore things in ways that yielded unexpected discoveries.

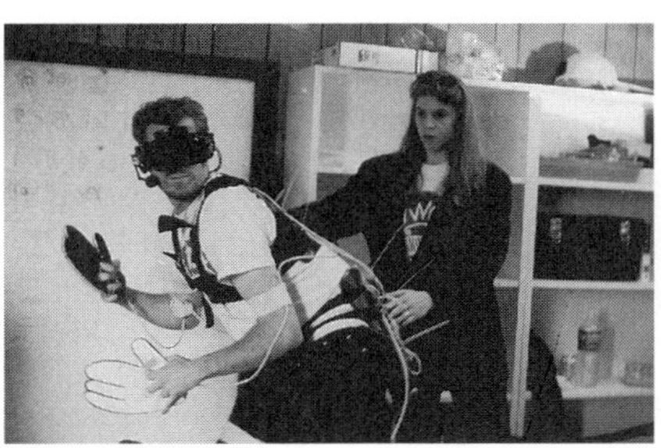

R

S

Credits: Photos courtesy Eli Noyes, Protozoa, (Colossal) Pictures, and MTV. Like many projects done on tight schedules and tight budgets, no one seems to have saved the credits list, so they cannot be cited here. High-level kudos are possible: to Abby Terkuhle, who honchoed *Liquid Television* through its three seasons at MTV; to Debby Beece, who green-lit the initial development deal; and to Judy McGrath, MTV's president, who has had a career-long love affair with independent animation.

III · TOOLS

The next seven chapters comprise a catalogue of animation tools. Collectively they provide a list of every type of tool you will need to work in film (a discussion of both super 8mm and 16mm formats) and in the digital domain (reviews of computer hardware and software). Whether you are a motion-picture animator or work digitally, the sampling will familiarize you with tools appropriate to different production levels—from the most sophisticated to the simplest. Throughout, technical prose has been boiled down so that you can readily distinguish the major features and basic utilization of each tool.

These chapters do not reveal the Absolute Truth. When one "talks tech," it can be like entering a free-fall zone. The deeper one goes, the more questions, choices, and details one encounters. The array of unknowns seems to expand and accelerate. One can tumble forever with no bottom in sight.

Because of this free-fall phenomenon, it's prudent to introduce Part III: Tools with five statutes of limitation:

Comprehensiveness. You will not find complete listings of every last piece of equipment used in animation. Tools that are highly specialized—or really expensive—are not discussed in nearly as much detail as the more accessible items. This revised edition has required a cutback on what had been a rather encyclopedic coverage of film gear. Some of the least essential tools have been cut entirely.

Depth. Detailed technical specifications are left for each manufacturer's operating manual and/or user's guide, or for that library of specialized books that can provide step-by-step, feature-by-feature discussions of various tools and techniques. For example, there is only a superficial treatment of film and video editing here. As you get into this area, you'll probably want to consult one of the excellent books that are available. The same holds for computer programming, an orientation to microphones, and other corners in the world of animation.

Timeliness. While filmmaking technology is relatively stable, computer development evolves so quickly that the entire spectrum of graphics tools migrates through a new generation every eighteen months—or less! The information in this book is sure to be dated and incomplete, since the research stopped months before publication. It turns out, however, that this matter of cold reality is a freeing element rather than a confining one, for it has focused the writing of these pages on the most fundamental and immutable elements of computer hardware and software. The two chapters on digital tools have been designed to give you the right questions to ask when it comes time to consider products that haven't even been dreamed of at the time of this writing.

Cost. Although the information here is also sure to become dated, I feel it can be really valuable to see the *relative costs* of key animation tools. The following policy applies to prices in this volume:

- Items costing under $50 are rounded off to the nearest $5.
- Items costing between $50 and $300 are rounded off to the nearest $10.
- Items costing over $300 are rounded off to the nearest $50.

The value of putting prices in this volume is less what the actual figures are (which march ever upward following inflation) and more about making relative comparisons to get an overall ballpark of potential expenditures.

Recommendations. It would be way out of bounds for this book to offer hard-and-fast recommendations for specific configurations of filming or computer hardware. There is a wealth of periodicals that are much better equipped for this task, offering well-researched, up-to-date, and empirical bench tests on comparative equipment. It's slightly different for computer software. The leading animation and graphics applications have become institutions in themselves. As the technology continually evolves, the leading programs tend to change with them, always adding new options and greater control. The case studies in previous chapters—plus a few more to come—single out venerable software programs there is no risk in recommending.

Chapter 25: Resources is designed to give you recommendations on where to go for more information. So when you come across a problem or you have a question that falls outside the parameters of these pages, check the last chapter of the book.

A CAUTION FOR TECHNO FREAKS

It is very, *very* easy to become enamored with the tools of animation. Component parts and features, the beauty of a well-made tool, curiosity about special effects, the pride of technical competency—each of these honest responses can become a seductive mania that draws the animator further into hardware and technique and further away from creativity and meaning.

I'd like to issue a warning: Check the impulse in yourself toward becoming a techno freak. It may help to remember that no matter what tools you begin with, you will eventually outgrow your equipment. The desire to purchase the next level is never satiated. There's *always* another level. Besides, if you begin with the fanciest and most versatile tool set, you may find you never use some of the capabilities you've paid the most for.

So in the beginning, try to get along without the complicated gizmos. You'll surprise yourself by finding that a rich source of creative energy is released when one works with limited hardware and is forced to explore the capabilities of relatively simple tools.

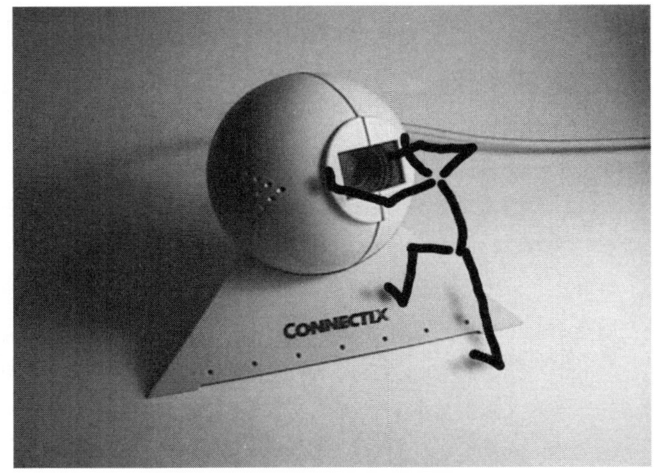

As your interest in animation advances and matures, you'll most likely find yourself shopping for new tools and making tough choices about which techniques and formats are essential to the kinds of animation that fascinate you. Building one's equipment base can be one of the most rewarding aspects of being an independent animator. And this is most true when the acquisition is done patiently, carefully, and with a sense of creative as well as financial ecology.

Here are some general tips for tooling up.

New Versus Used Equipment. Serious animators take good care of their equipment. For this reason it is often very wise to consider purchasing used equipment. Local stores that sell filmmaking equipment will often know where you can find used equipment. As noted above, computer technology is changing so fast that what some people have traded in will be more than ample for your needs. Look for reconditioned computer gear from manufacturers.

Borrow or Rent. The place to sink one's real investment is into the time that is required to learn the craft of animation. Don't overextend yourself so far in acquiring tools that you're left with no resources to do the creative work itself.

Get Warranties. Don't purchase any piece of equipment without some kind of warranty that allows you to return the item to the dealer if you discover a problem with it soon after purchase. Most computer purchases can be returned after a few days if one isn't satisfied. So once you make a purchase, unwrap it, check it over, and dive right in!

Enough of warnings, limitations, and tips. The next six chapters ought to thoroughly orient you to the animator's world of tools and processes, gadgets and supplies, and hardware and software.

19 ■ Film Cameras and Accessories

The most important tool in film animation is the camera. Its mechanical precision and its range of image-making capabilities set the parameters within which the animator must work.

Literally hundreds of super 8mm and 16mm cameras are capable of animation. These vary greatly in their relative costs, sophistication, and quality of workmanship. Yet there are certain basic elements common to all movie cameras, and there are special features that operate much the same regardless of camera size, format, or cost. Because of manufacturing standardization, it is possible to talk in general terms about cameras and accessories. In identifying and describing the elements that make a camera work, I've placed emphasis on those features that are especially critical in animated filmmaking. This chapter also presents representative models of super 8mm cameras and then 16mm cameras. My selections represent simple to sophisticated models, inexpensive to costly ones, and models that can be used for all filmmaking needs as well as those designed only for animation.

If you can combine the data catalogued here with your own hands-on inspection of different cameras, you'll quickly forge a working knowledge that should get you quite far along. When it's useful, information becomes data. Otherwise it's Dada. Eventually, you are going to encounter a problem, question, reference, or demand that forces you to expand and refine your understanding of animation cameras. When that happens, return to these pages.

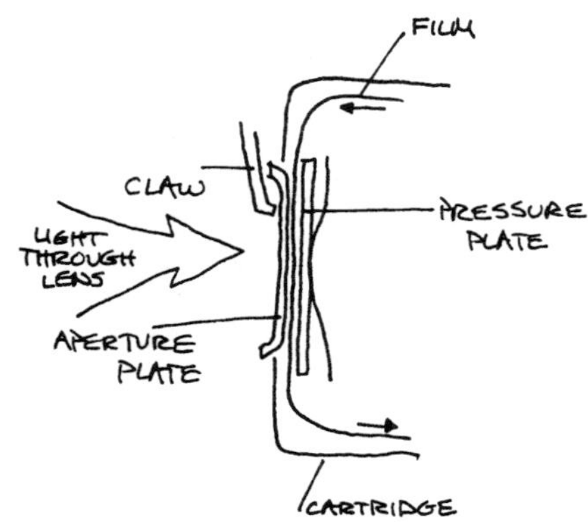

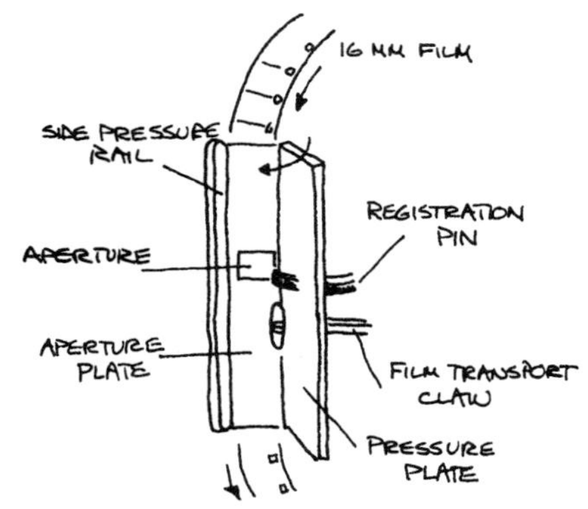

19.1 Camera registration: Drawing *(A)* represents the cross section of a super 8mm camera. It shows the four elements that provide registration: the *aperture plate,* a channel through which film travels; the *claw,* which advances the film by pulling down on successive sprocket holes; the *pressure plate,* built into every super 8 cassette; and the *film stock* itself, bearing a precise configuration of sprocket holes. Drawing *(B)* shows the registration system of a 16mm motion picture camera.

GENERAL FEATURES: CAMERAS

The movie camera is a lot like a still camera except that it takes its pictures in rapid order instead of one at a time. But the movie cameras *can* expose individual frames one at a time. In fact, that's what makes animation possible.

The mechanics of the camera are relatively simple. A *claw* device interacts with the *perforation holes* on the film surface to pull successive frames into the *camera gate,* where the frame is exposed. All this happens quickly—twenty-four times in one second when the camera is running at sound speed.

In order to use a motion picture camera for animation, the *registration system* that brings the film into place during the exposure must be very accurate. In sophisticated and specially designed animation cameras there is a set of *registration pins* that holds the film precisely in place during exposures. In any movie camera, however, the claw device combines with both the aperture plate and the sprocket holes to produce a fairly accurate registration system.

All movie cameras require a source of power to advance the film. A power system is often required for other mechanical tasks—to operate a power zoom lens, for example. In most super 8mm cameras, electric current is used to power the camera. Its source is a set of alkaline batteries. Many 16mm cameras are also powered by batteries of one form or another, sometimes mounted in the camera and sometimes carried independently and attached to the camera by a cord. Some 16mm cameras, however, are powered by *spring-wind motors.* Either power system will work, yet both have liabilities. Animation drains batteries very quickly and there is a tendency for spring motors to vary somewhat in the length of exposure time they provide for each frame of film. For these reasons and others, animation is often done with *auxiliary motors* that are attached to the camera and are themselves powered by electric current from a normal wall outlet. Auxiliary motors are discussed later.

In both 16mm and super 8mm formats, the execution of an exposure of a single frame of film is often achieved by use of a *cable release*—a cloth or plastic tube that has a metal cable running through it and a plunger device at one end. The plunger activates a metal pin in the other end of the cable. This pin is inserted into the camera body and it triggers the release of a single frame of film. The cable release usually screws into a threaded receptacle that is located either within the camera's trigger or at some adjacent location on the camera body.

Some cameras employ an alternative method of executing single-frame exposures. Many cameras have a *single-frame position* marked on the camera's variable speed control. With this design, the trigger is pressed by hand (as in normal filming), but the action releases just a single frame. Because even the lightest touch of a trigger can cause a change in the relationship of camera to subject, this mode of single framing can mess up the stable registration required in animating. Note, however, that many of today's super 8mm cameras have *remote contact mechanisms* that allow operation without applying any pressure to the camera body.

Movie cameras have *standard filming speeds.* For 16mm this is 24 fps and for super 8mm it is 18 fps. Actually, the super 8mm format has two standard speeds. Eighteen fps is the silent speed and, increasingly, 24 fps is becoming the standard for super 8mm sound filming. At the present time both 18 and 24 fps are considered standard and many cameras have settings for both. Depending on the particular 16mm or super 8mm camera, the filmmaker may be offered additional speed options. The fastest speeds are generally 64 fps and the slowest 6 or 8 fps.

Obviously it is of absolute importance for the camera to expose the film to just the right amount of light to get the proper image. Achieving the *correct exposure* requires the control of two factors: f-stop and shutter.

The *f-stop* or *aperture* is a function of the lens, which permits the filmmaker to control how much light will pass through the camera onto the film. This is done by means of an adjustable *iris* or *diaphragm* that is built into every lens. In almost all super 8mm cameras there is an *automatic metering system* that measures the light reflected from the object at which the camera is pointing. These metering systems compute and mechanically select the f-stop setting that will give the correct exposure. In doing this, the camera automatically makes an adjustment for whatever *film stock* is being used. Different stocks require different amounts of light.

If possible, animators always try to use a camera that has a *manual override* in addition to an automatic light-metering system. Animation techniques such as line or paint-on-glass (or

19.2 Variable shutter: The drawings represent four positions of the variable shutter on a 16mm Bolex camera. Pictured to the right in each drawing is the *shutter blade* and *aperture hole.* Drawing *(A)* shows the shutter in its normal position (open) and in drawing *(D)* it is seen in its fully closed position. To the left in each of the four drawings is the corresponding position of the *variable shutter control,* which is located on the outside body of the camera.

anything with backlighting) require the camera operator to film relatively thin lines and small markings against very bright backgrounds. In this situation, an automatic exposure meter tends to let in too much light. Most 16mm cameras used for animation do not have automatic metering systems and require manual light readings and manual setting of the camera's f-stop.

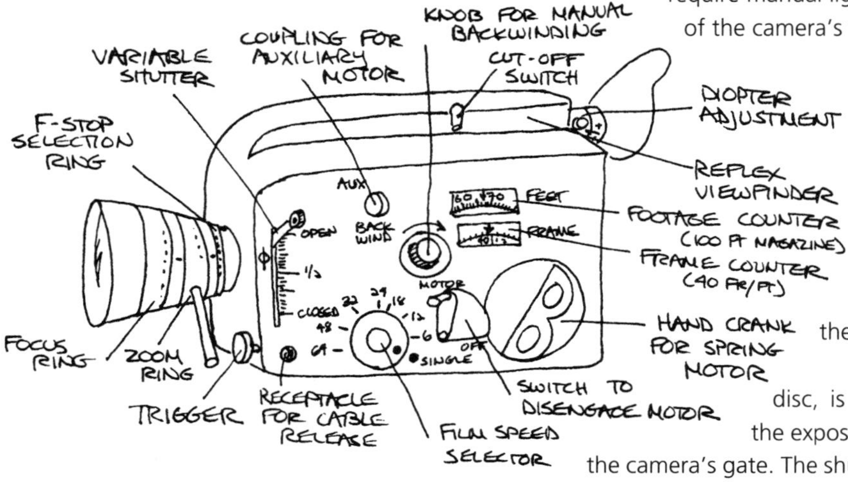

19.3 Universal 16mm camera: Following loosely the design of the Bolex 16mm camera, this drawing identifies all the standard features of 16mm cameras. This model contains elements from many different camera makes, and for convenience all its features are located on the visible side of the camera body.

In animation, the f-stop setting is an extremely important variable because it controls *depth of field,* the amount of distance in front of the camera lens in which things are sharply focused. More about depth of field in the next chapter.

The *shutter,* a rotating metal disc, is the second factor controlling the exposure of each frame of film within the camera's gate. The shutter is mechanically linked with the claw mechanism to ensure that light passes onto the film only at that moment when the film is being held motionless in the camera's gate.

A *variable shutter* allows the animator to use a number of *in-camera effects.* A *fade-out* is accomplished by closing the variable shutter gradually during a successive number of single-frame exposures until there is no light hitting the film as it passes through the gate. The reverse procedure accomplishes a *fade-in* effect (Figure 19.2).

Superimposition also requires a variable shutter—a feature available on many but not all super 8mm and 16mm cameras. To create a superimposition, the shutter is set so that roughly one half the normal amount of light passes onto the film during a first "pass" through the camera. The film is then rewound within the camera. To get the superimposition, the film is advanced through the camera a second time. Half the normal exposure is again used in this second pass. Many of the more recently manufactured super 8mm cameras offer the special feature of an *automatic dissolve control.* This device enables the animator to superimpose, fade, and dissolve using the standard super 8mm film cassette. The best dissolve controls also allow manual control of the shutter. Check your instruction booklet to find out if your camera has this feature.

A close relative to the superimposition or "super" is the *dissolve.* Here a fade-out is precisely superimposed over an equally long fade-in. If the dissolve covers a long period of time —say, 2 seconds (48 frames) or more—it is called a *lap dissolve.*

Doing supers and dissolves requires two additional camera features. *Backwinding,* as the name suggests, refers to the ability to run film through the camera in reverse. In order to achieve accurate fades and dissolves the camera must be equipped with a *frame counter,* a device that lets the animator keep an accurate count of the frames as they are exposed or backwound. *Footage indicators* are common to all cameras in all formats.

And every camera must have a *lens,* a combination of glass elements that reduces the image in front of the camera to that small, exact replica that is recorded on the emulsion of the film stock. A more detailed discussion of various lenses follows. It bears noting in this context, however, that some cameras are equipped with a built-in *zoom lens,* some cameras have a *fixed lens,* and some cameras accept *interchangeable lenses.* An important requirement for most animation techniques is for the lens to focus on objects very close to it. This is accomplished by means of a built-in *macrofocusing* capability and the use of a special *close-up* lens or *diopter,* which can be attached to the front of a normal lens.

The close proximity of camera to artwork makes it particularly critical in animation to have precise *framing* and *focusing.* A *reflex viewing system* is a built-in feature of most super 8mm cameras. It allows the camera operator to view the image exactly as it will fall on the film stock. Sharp focusing and accurate framing are made possible with this ability to view through the same lens that will expose the film. However, there are a number of excellent 16mm animation cameras that don't have reflex or through-the-lens viewing. Instead, such cameras have a *viewfinder system.* This kind of design presents the animator with a problem called *parallax.* What the lens records on the film is not precisely what the viewfinder shows to the camera operator. Many such nonreflex cameras have some sort of a calibrated system to assist the filmmaker in compensating for parallax.

Even with reflex systems there can often be a slight error between what is seen on the camera's viewing screen and

19.4 Super 8mm cameras: When the camera is running at the standard live-action speed, the actual *exposure time* for a single frame is approximately one fiftieth of a second. For various mechanical reasons, however, the exposure time for single framing is always somewhat longer. Usually the shutter is open for one thirtieth of a second per frame. Check the operating manual of your camera for specific calibration data.

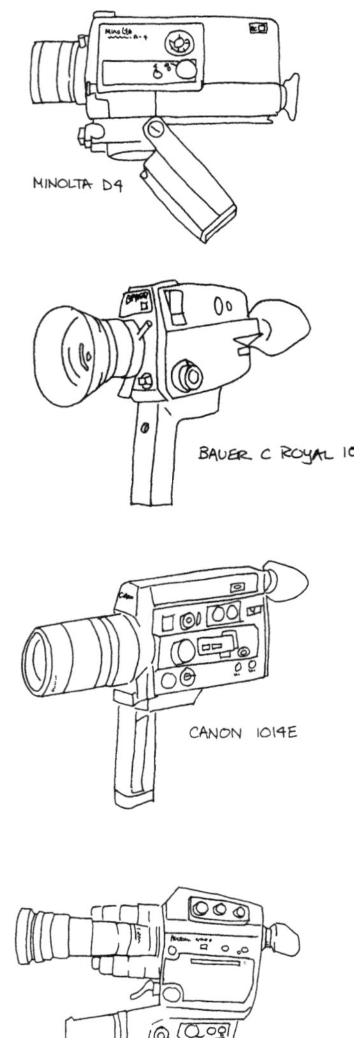

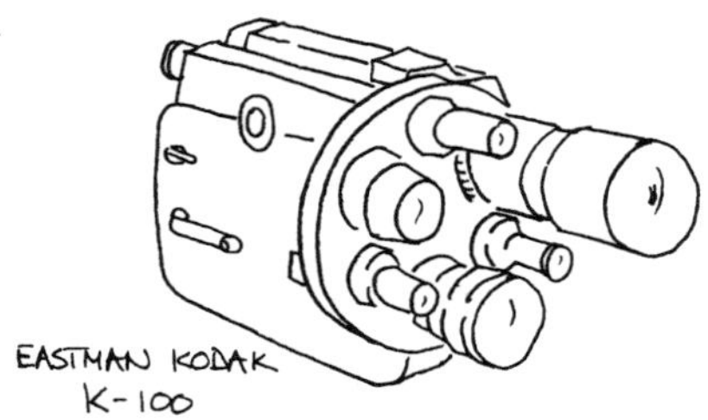

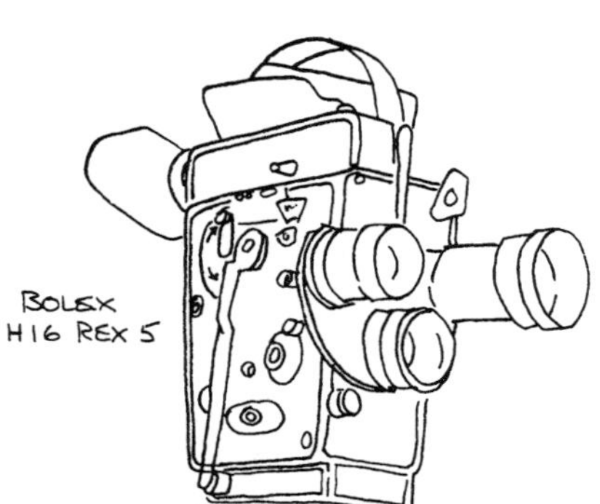

19.5 16mm cameras: These illustrations give you a quick introduction to specific 16mm cameras. As with the sampling of super 8 cameras, the list here goes from simple to complex, from older models to newer ones, from less expensive to more expensive. Independent animators often get into 16mm production through used equipment. Although prices vary wildly according to the demand, original cost, and condition of used cameras, it should be possible to get a good used Bolex or Kodak K-100 for under $500. New equipment costs more, of course. A new Bolex Rex 5 is around $1,000. Special process cameras such as the Oxberry and Forox models are sold with their stands.

what is actually being recorded on film. Animators should always test the framing and focusing mechanisms of their cameras. Another note: Reflex viewing systems require a *cutoff mechanism* that prohibits light from entering the camera body via the viewfinder when film is being exposed and the filmmaker's eye is not pressed to the viewfinder as it normally is in live-action filming.

LENSES

Imagine that you've painstakingly designed and executed a five-minute film using, say, paint-on-glass technique. Everything has been done under the camera—no artwork exists after your labors. Okay. You rush home with your processed film and thread it through the projector. The lights go off and the image appears on the fabled silver screen. Just as you are about to grin with pride, you suddenly notice that the focus is soft. Part of your image is fuzzy. No matter how hard you work the focusing knob, the image remains blurred. The film is ruined.

This little horror story could be prophetic. It's happened to me, and if you are not careful, it can happen to you. I recount the tragedy as a reminder that optical precision is essential in animation. If a lens is even slightly out of alignment, the image will be sharp in places and out of focus in others. Indeed, any imperfection in a movie camera's lens system will show up most clearly when you are screening animation.

Most of today's super 8mm cameras have built-in lenses. When purchasing or using this kind of camera, you should conduct a thorough test to verify the optical quality of the lens. But whether you are using a built-in lens or a removable lens, the comments that follow are particularly important. They pertain to five facets of lens control and lens modification.

CLOSE-UP FILMING

Many of the techniques described in this book require a focusing distance that is *less* than what standard lenses will accommodate. The majority of super 8mm cameras, for example, cannot be focused on a subject that is closer than 4 or 5 feet. As an animator, you will constantly be required to operate inside this minimum focal range. The process of close-up filming, also called *macrofilming,* is achieved through one of four procedures or attachments.

Built-in Macrofocusing. Many of today's super 8mm cameras have built-in zoom lenses that can be operated in a macro range. A manual adjustment is made on the lens or camera body that allows the variable focal mechanism of the zoom to act as a focusing device in ultra-close-up cinematography. A standard zooming motion is not possible when such a macrofeature is in operation.

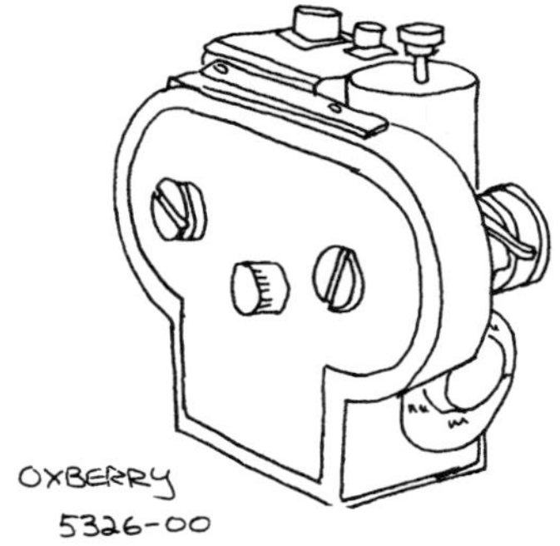

OXBERRY 5326-00

Plus Diopter Lenses. By placing an auxiliary lens in front of the regular camera lens, one can simply and effectively achieve close-up filming. *Plus diopters* is the technical name of such close-up lenses. These accessories are generally sold in sets of three, and they are sized according to the threading within the metal housing on the front edge of the normal camera lens. Measurements for the diameter of the diopter, given in millimeters, must match that of your specific camera lens.

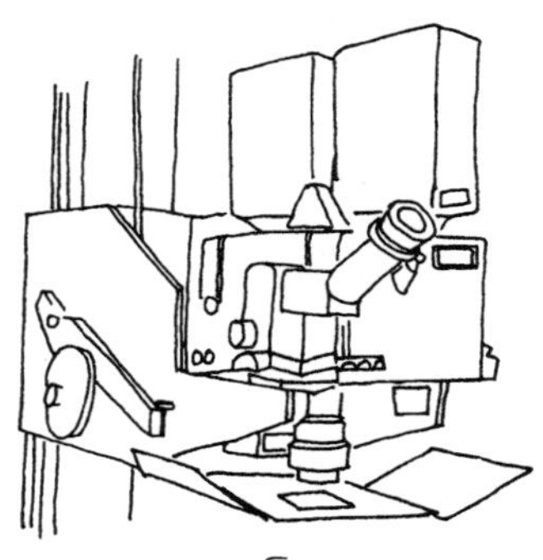

FOROX MODEL SSA

Diopters are rated according to their magnification power: +1, +2, and +3. The threading on each lens holder permits the filmmaker to stack two or more lenses simultaneously. This achieves various multiples in magnifying power. When more than one diopter is being used, the highest-power diopter lens is positioned closest to the normal camera lens. Note that the use of too many diopters can have deleterious effects, including soft edges and "barreling" of the field of vision.

Although simple in construction and use, plus diopters must be ground and mounted with extraordinary precision. For this reason, quality diopters are expensive. A set of three diopters for super 8mm cameras will cost around $25; 16mm diopters are usually sold separately and, because they are bigger, cost more. Prices range from $20 to nearly $200. The manufacturers of diopters provide charts to help you determine how a particular close-up lens will affect the effective minimum focal length of the lens it is attached to.

THE ZOOM LENS

A *variable-focus lens,* usually called a *zoom lens,* can be one of the most technique-extending tools an animator can acquire

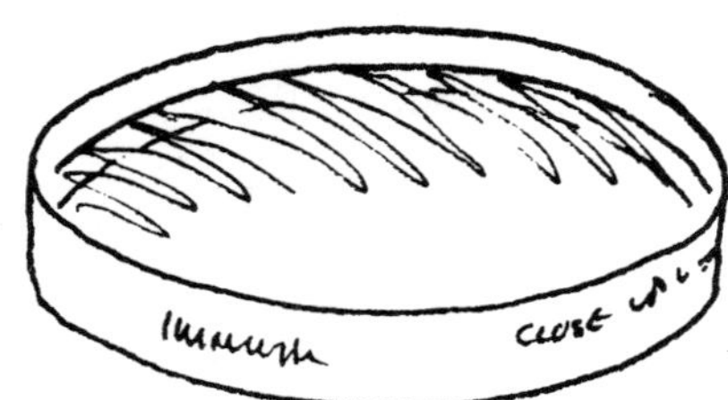

19.6 Close-up lens: A set of close-up lenses is available for virtually all super 8mm and 16mm cameras.

for his or her camera. The popularity and usefulness of a zoom lens is shown quite graphically by the fact that it is almost impossible to find a super 8mm camera that is without one. Because 16mm cameras are usually sold without a lens of any kind and because these cameras generally accept different lenses (unlike super 8mm cameras), most animators will try to acquire a quality zoom lens.

The zoom lens permits the animator to quickly adjust the size of the field being filmed beneath the camera. A zoom lens also permits an effective camera "movement" toward or away from the subject. A special homemade device can be easily attached to almost any zoom lens in order to provide the animator with precise control of the lens during frame-by-frame filming (Figure 19.7).

A zoom and a set of diopters combine to produce the most versatile lens system for animation. It is possible, for example, to add enough magnification so that one can zoom within a field the size of a standard 35mm slide. Obviously, the "stronger" the zoom the more flexible it will be. Zoom lenses are measured by the ratio of their minimum to maximum focal lengths—a 12mm-to-120mm zoom lens has the ratio of 1:10.

Zoom lenses for 16mm cameras come with different types of camera mounts (Arri Standard Mount, Arriflex Bayonet Mount, Bolex Rex Mount, Bolex Bayonet Mount, "C" Mount, Cinema-Products [CP] Mount, and more) and in different sizes (9.5mm–57mm all the way to 12mm–240mm). Zoom lenses can be acquired with built-in metering systems and they come with reflex viewfinder attachments.

LENS FEATURES AND PROBLEMS

Depth of Field. As noted earlier, depth of field refers to the amount of physical distance in front of the camera that will appear in focus. Depth of field is a complicated phenomenon —a function of a number of elements, including lighting, film speed, distance from subject to lens, focal length of the lens, and f-stop. The f-stop is the easiest variable to control—the higher the f-stop (say, f/16 or, better, f/22), the greater the distance in front of the camera that is in focus.

As you can clearly see, depth of field is a critical factor in many animation techniques. For example, in clay, puppet, or time-lapse techniques, the animator may want to blur the background by using a low f-stop and less lighting, thereby reducing depth of field. Sometimes it may be important to have as much territory in focus as possible, with a high f-stop and lots of lighting. More times than not, animators try to shoot at a high f-stop in order to maximize the distance in which everything will be in sharp focus.

Other important factors affecting depth of field include the focal length of the lens (wide-angle lenses give greater depth of field than telephoto lenses); the distance from the art to the camera (the farther away, the greater the depth of field); and the film speed (the slower the film stock, the greater the depth of field, if lighting is constant). Note that the shutter speed is not a factor in depth-of-field control because, in animation as in all live-action filmmaking, the shutter speed is fixed at roughly 1/50 second (live action) or 1/30 second (single-framing).

Aperture. A final and very critical quality of every lens has to do with the range of its *aperture settings.* For most moviemakers, the "faster" the lens the better. The lowest possible f-stop opening is desired in order to extend the camera's range when shooting under minimal lighting conditions. For animators, however, it's the other end of the scale that is most significant. Because of the artificial lights under which animation is filmed, it is easy to get bright, even illumination. Therefore the animator selects a lens because of its *highest f-stop*—f/16, f/22, or even f/64. It's nice to celebrate pleasant serendipities when you find them. It's much cheaper to buy a lens that does not have a low f-stop, and for animators, low f-stops aren't really needed.

Filters. There are many different filters that operate as camera attachments. Filters provide effects such as stars, multi-images, and just plain weird colors. An introduction to filters will be found among the comments about lighting tools for animated filmmaking (see Chapter 22).

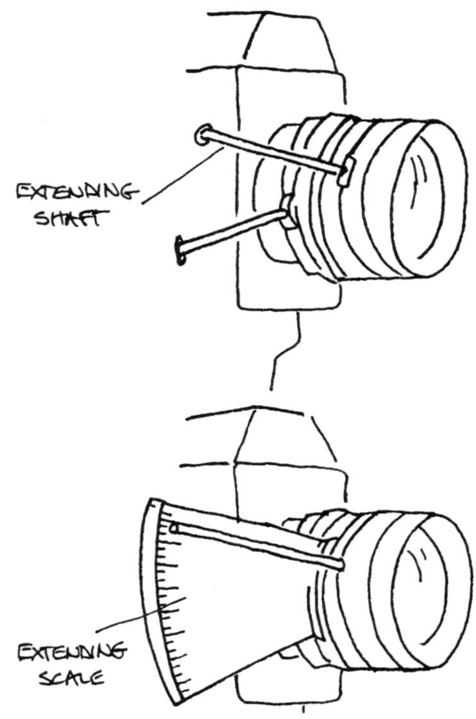

19.7 Zoom extenders: Most zoom lenses are equipped with short metal shafts that stick out of the lens and are used as levers in adjusting the lens before or during filming (many super 8 cameras have automatic zooms, a feature of little use to the animator). By replacing the short shaft with a longer one of metal or wood, the animator gains greater precision in causing the incremental changes that are required in executing a frame-by-frame zoom. Registration of the zooming action is further extended by attaching a reference scale to a fixed part of the lens shaft or the camera body. The scale allows the animator to control precisely a zoom movement and it also helps in planning ahead—for example, in estimating what change is required for a zoom that covers the entire zoom ratio in exactly five sets of two-frame exposures.

CABLE RELEASES

To most live-action filmmakers, single-framing means nothing more than the technique of exposing one frame at a time. To the animator, however, single-framing is a matter of tremendous concern. Animators perceive single-framing as an act filled with many hazards and nuances.

The registration between the camera and the artwork has to be perfect in animation. This means that the process of exposing a frame must be done with a feather touch. The duration of the exposure must be precise too—down to a hundredth of a second. All frames in an animated sequence must have an identical exposure time as well as a perfect registration.

In order to minimize the pressure on the camera during filming, a *cable release* can be used to trigger the advance of single frames. The cable release provides a sheath in which there is a metal pin and a spring. One end of the release threads into a single-frame release socket usually found in the trigger of the camera or nearby. The other end of the cable release has a plunger. Manual activation of the plunger causes the metal pin to trip a frame release mechanism inside the camera. The spring returns the pin quickly to prevent the exposure of multiple frames.

Cable releases come in different qualities and different lengths. Some 16mm cameras such as Bolex require a special form of cable release. A useful feature on some cable-release systems is a foot pedal or air bulb that allows the animator to operate the release by foot. The hands are then free for work under the camera.

In recent years, the more sophisticated super 8mm cameras have been designed with electronic rather than mechanical single-framing devices. These electromagnetic systems expose single frames by means of a remote cable attachment that bypasses the trigger and permits the animator to advance the film without touching the camera at all. This is a preferred method, because while cable releases push the camera to some degree, electromagnetic operation is accomplished without the application of any physical force to the camera body.

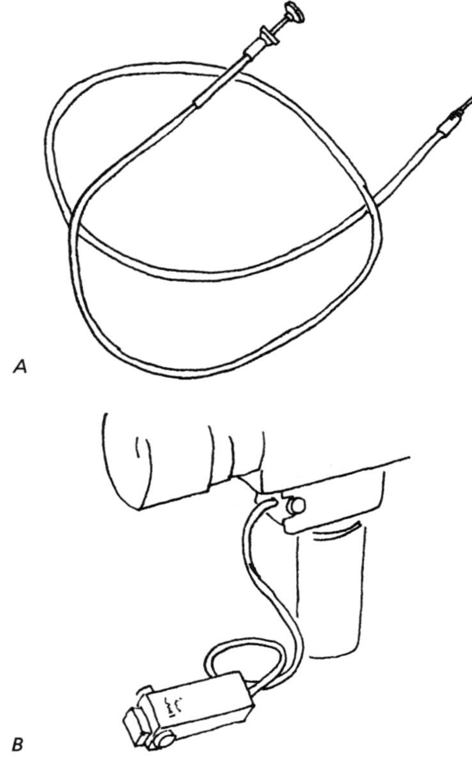

19.8 Cable release: The familiar cable release, *(A)*, can cause the camera to jiggle. the newer electromagnetic release systems, *(B)*, do not require any external pressure to trigger the camera. Pictured here is a Minolta remote release. The Minolta super 8mm camera system is particularly well designed for animation and its price is relatively low.

ANIMATION MOTORS

Many 16mm cameras, and a few super 8mm cameras, will accept an *auxiliary animation motor.* When such a motor is used, the built-in power system is disengaged. These auxiliary motors, powered with electric current, are linked with the camera body by means of a coupling device. The drive wheel of the animation motor connects with the frame advance mechanism within the camera.

Auxiliary motors are designed to fit particular cameras and there is no standardization of design for the coupling system, the exposure time, the triggering device, or other specific features.

The J-K Camera Engineering Company manufactures a number of motors for animation and optical printing. The most often used is the J-K Animation Motor (Model #A-60B), which is designed for use with Bolex cameras and costs approximately $300. It features a 60-frames-per-minute synchronous motor, and all the operating controls are located in a remote-control box that is connected to the motor by an 8-foot cord. J-K has six other motors, including a more expensive model with an electronic preset memory, bright digital display, a self-contained intervalometer, and other features.

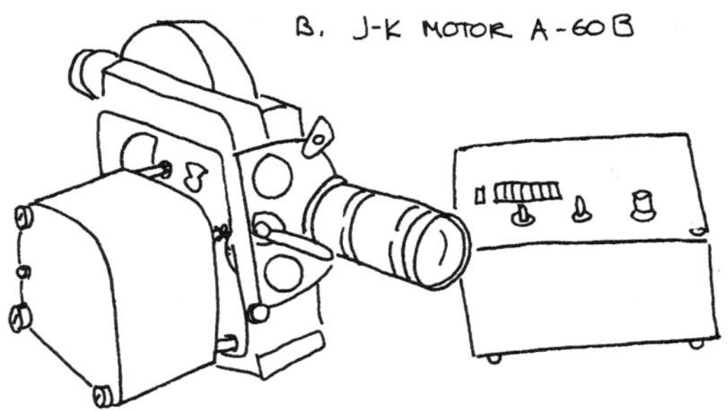

19.9 Animation motor: The illustration shows the component parts for the J-K Engineering auxiliary motor and control box, shown with a Bolex.

20 ▪ Animation Stands

Animation stands come in all shapes and sizes. You can build your own rig or you can purchase a professionally designed and commercially manufactured system. The cost of an animation stand can vary from a few dollars for 2-by-4's and plywood to something well over $100,000 for a top-of-the-line, fully computerized animation stand. But whether you are in the market for a sophisticated tool or just planning to make do with whatever is available, a familiarity with the general features of animation stands will extend your competencies as an independent animator.

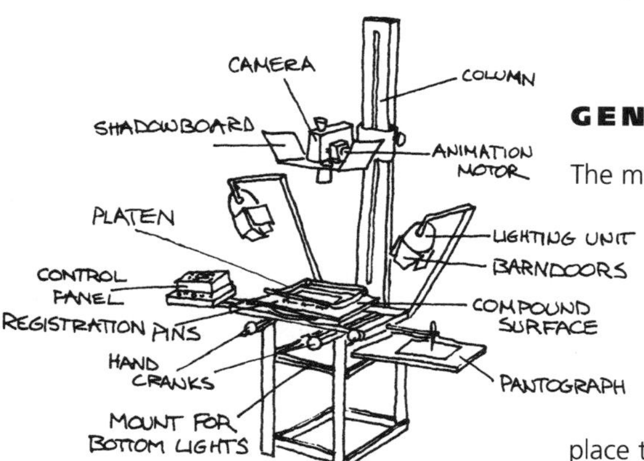

20.1 Universal stand: Consisting of elements from many different commercially produced animation stands, this single sample exhibits all the basic features of an animation stand.

GENERAL FEATURES

The main feature of any animation stand is its *structural stability.* It must hold the weight of the camera in an absolutely steady position in relation to the artwork being filmed.

Requirements of stability extend beyond the stand itself. Check the room in which you are working for vibrations that can come from other sources. If you can, place the stand in some out-of-the-way spot to avoid accidental bumps or jarring. Once the stand is in position and the camera is mounted, try to avoid moving the stand or having to take it apart. The location you select ought to have good security.

Obviously, every animation stand must have a *camera mount.* Every super 8mm and 16mm camera has a threaded receptacle called a *bush* that receives the standardized screw of the tripod or other camera-mounting hardware. Naturally, any stand must be able to mount the camera you will be using. But because of variations in design, not all stands accept all cameras. Sometimes a stand's manufacturer will sell adapters especially designed to fit a particular camera to a particular stand.

An animation stand will often be designed for a specific camera. Whether you are building a stand or buying one, check carefully to determine what kind of camera is recommended, or required, for the stand you are considering. The preceding chapter should help you to identify those features you'll need to meet your own working objectives. Items to check include the camera type, film selection, viewfinder system, lens, filters, film advance, a single-frame device, film speed, film rewind, footage and frame counters, shutter, aperture and metering system, and available lens and camera accessories.

All stands hold the animation camera over a flat surface on which the artwork is placed during filming. Some stands can also be turned on their sides to present a horizontal puppet stage. An important thing to check for is the ease with which you can sight through the camera's viewfinder during filming. Also, make sure that the stand's height permits easy access to important camera controls, such as a fade/dissolve lever.

Many stands feature a vertical column that is engineered to allow the entire camera mounting arm to be adjusted to various heights above the table. Usually there is a single column (the larger stands have double columns), which carries inside it a ball-bearing system and counterweights. Such *column movement* permits the operator to adjust the camera's height with ease and accuracy.

Vertical movement of the camera toward or away from the artwork produces a *zoom effect,* sometimes also called a *vertical truck.* The precise height from focal plane to film surface is often calibrated right on the column and there must be a locking device to hold the camera 100 percent steady at any position along the column. Some vertical movement mechanisms are motorized. The fancy expensive stands also have an automatic focusing device that keeps the lens sharply in focus at any height.

All animation stands require a registration system, a way to ensure that the camera holds a constant relationship to the surface below it and that artwork can be precisely aligned.

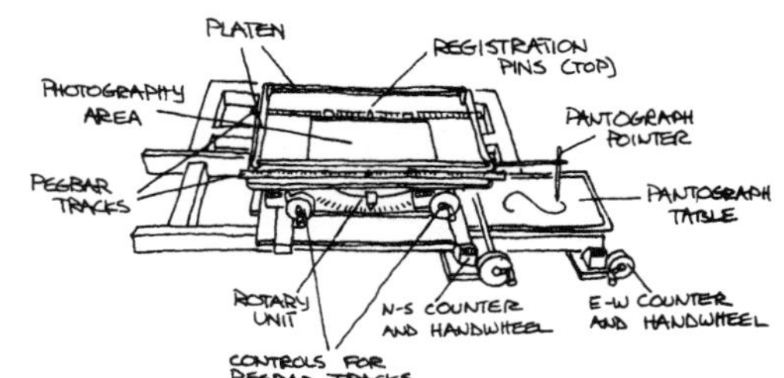

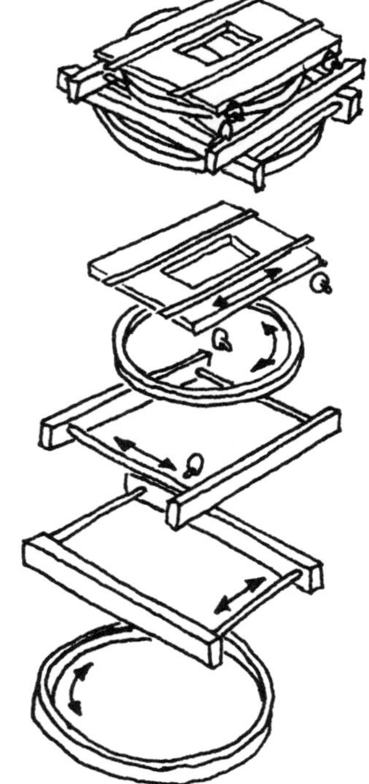

20.2 The compound: Illustration *(A)* shows the individual elements of a professional-caliber animation stand. Illustration *(B)* indicates five component pieces and the direction of movement each facilitates. The latter drawing is based on an illustration of a rostrum compound in *The Animation Stand* by Zoran Perisic, Focal Press Ltd., London. Drawing *(A)* is based on an illustration in *Animated Film* by Roy Madsen, Interland Publishing, New York.

Registration on both these matters is generally attained with an Oxberry or Acme pegging system. These industry standards are discussed in detail in Chapter 21. It is possible and sometimes necessary to use alternate registration systems. But whatever the system, it must be used consistently in preparation of artwork as well as for filming.

On commercially manufactured stands, the surface bearing the registration device may have two sets of pegs—one on the top and one on the bottom. In some stands, there are as many as four sets of pegs—including moving peg bars, calibrated in twentieths or even hundredths of an inch.

The surface that holds the artwork is called the *compound.* In many stands the compound can be moved laterally in any direction. Compass coordinates are used to indicate horizontal directions: north, south, east, and west. Such N/S and E/W movements are usually done by cranking mechanisms, often calibrated in hundredths of an inch. While compound movements are usually done with hand cranks, sometimes the drive mechanism is electrically powered. A locking system is always present.

As noted in Chapter 9, the movement of artwork beneath the camera can be facilitated by a *viscous-damped compound.* This device makes possible ease-ins, ease-outs, and constant-speed panning at real-time filmmaking speeds. The viscous-damped system yields an omnidirectional compound with a grease-based braking action. It is tremendously useful in composing the frame over artwork.

Many compounds also provide a rotational movement through the full 360 degrees. The *rotating disc* always has a lock mechanism, and its spin movement is calibrated in degrees.

The *platen* consists of a metal carriage that holds an optically pure sheet of glass. It is used to force artwork flat on the compound table. Because of differing thicknesses created by cels, cutouts, and other artwork, the platen is designed to "float." The glass is mounted on a pivoting center and hinged on spring-loaded brackets. Platens are operated manually, even on the most expensive stands, and they are generally removable for oversized artwork. A standard size for platens is a #12 field. There are #10 and #16 field platens as well. A commercially manufactured platen will cost between $100 and $300. Note that an ordinary sheet of glass can be easily rigged to form a platen.

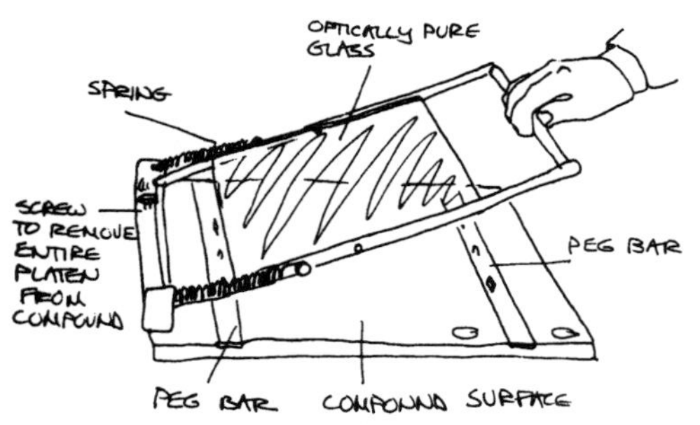

20.3 Platen: The platen illustrated here bears features common to all platens: pure glass that floats in a center mounted framework; spring mounting; a handle for manual operation; and a removable mount on the compound surface.

Animation stands must have toplighting. The lights themselves are color corrected and mounted at a 45-degree angle to the plane of the camera. This prevents reflection of the light directly into the camera. A useful feature of animation stands can be a capability for bottomlighting or backlighting. More information is provided in Chapter 22.

The *pantograph* is a device that allows the animator to execute complicated camera movements. In using a pantograph, the movement is first planned out on a separate sheet of paper. Often this pantograph grid is somewhat smaller in total size than the full compound field. However, the planning graph, like the pantograph table on which it sits, is accurately aligned to the camera tracking mechanism of the compound. The physical movement the artwork will make underneath the fixed camera is represented by a thin line that the animator draws on the pantograph. Often the positions at which the animation camera will take exposures are represented by crosshatching on the camera direction line. During the actual filming, a pointer or a crosshair device follows the line of movement.

Depending on the camera and the sophistication of the stand itself, there will be a correspondingly complicated *control system.* The simplest control is a cable release (to expose a single frame) and a footage indicator (to measure the amount of film that has been shot). Some super 8mm cameras carry both features a bit further by means of a foot pedal for the cable release and a frame counter as well as a footage counter. Additional camera controls such as forward/reverse, lighting, focusing, and exposure switches are required when using a specially designed animation camera or when using an auxiliary animation motor. In many of the more expensive stands, all the camera controls and stand controls for automatic column and compound movement are presented on a single control panel, which also provides an accurate frame count.

The animator respects an animation stand for its precision. But what makes a stand lovable is its idiosyncrasies. Almost every stand, manufactured yourself or by a commercial company, also has its own unique features, which represent, really, the personal biases and passions of its designers. You'll have to check instruction booklets or consult each builder of homemade stands for what passes as special features or plain quirks.

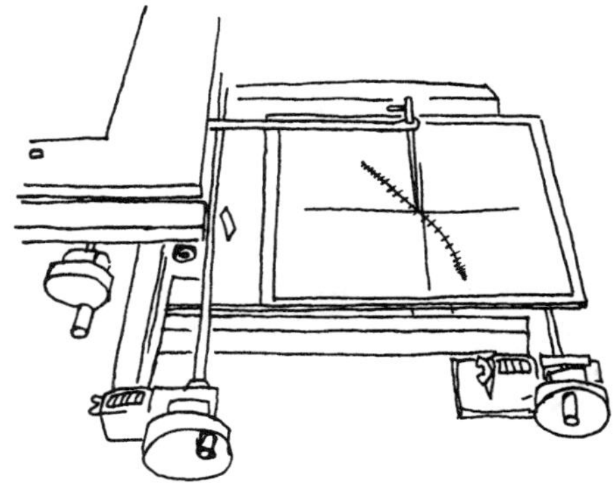

20.4 Pantograph: The pointer on the pantograph is attached to the movable compound surface. The pantograph surface never moves itself but, rather, carries plotting lines that the camera operator will follow in filming a complicated sequence of moves during an animated scene.

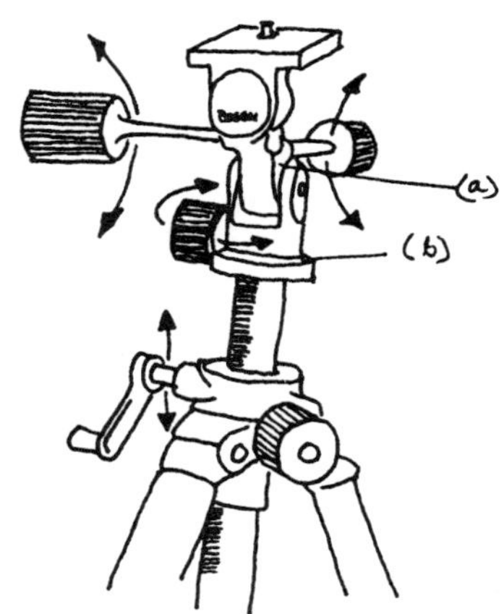

20.5 Tripod head: The arrows indicate directions that different elements of the tripod allow (or limit) movement. To calibrate a particular movement, measurement lines can be taped onto or etched into a fixed piece of the hardware at a place where it meets the corresponding area of movement. The model pictured here is a Bogen 3030, costing around $100. It is manufactured with calibrations at points *(a)* and *(b)*, as indicated.

20.6 Tripod positions used in animation

TRIPODS

The tripod is an accessible and highly versatile tool. It constitutes a relatively primitive yet effective sort of animation stand. Note, however, that all tripods are not alike. The ones that can be used for animation must have these three qualities. They must be strong enough to hold the weight of the camera and its accessories. They must be sturdy enough to keep the camera absolutely steady when shooting. Tripods must also adjust easily but hold various positions firmly.

The following animation techniques *require* a tripod: time-lapse, pixilation, puppet, clay, and small-object animation.

Often the technique will call for *controlled movement* of the tripod through pans, tilts, and even dollies. Such movements can be plotted out most accurately if the animator improvises a marking system right on the tripod itself.

For close-up photography, tripods can be "tied down" with *gaffer's tape* against a raised surface. (In case you are unfamiliar with it, gaffer's tape is a standard supply in filmmaking. It is a super-sticky and super-strong tape that is cloth-backed, usually gray in color, and comes in widths up to 2 inches. It is sometimes called duct tape.) With some tripods, it is possible to reverse the head of the model entirely. A collection of tripod positions is illustrated in Figure 20.6

The least expensive tripod you can use with a super 8mm camera will cost in the neighborhood of $75 to $150. Brand names to investigate include Testrite, Bogen, and Husky.

Top-of-the-line 16mm tripods are purchased in two parts. There are the legs, or the tripod, usually made of wood with metal-fortified joints; and there is the head, made of highly machined metal. In live-action filmmaking a fluid head is desired for professional 16mm work. However, this viscous-cushioned design is not required for animation. A high-quality friction head is recommended. Costs for legs are $150 to $500 and a tripod head runs from $200 to $600. Note that tripods and tripod heads are not standardized. Be sure the units you get are compatible.

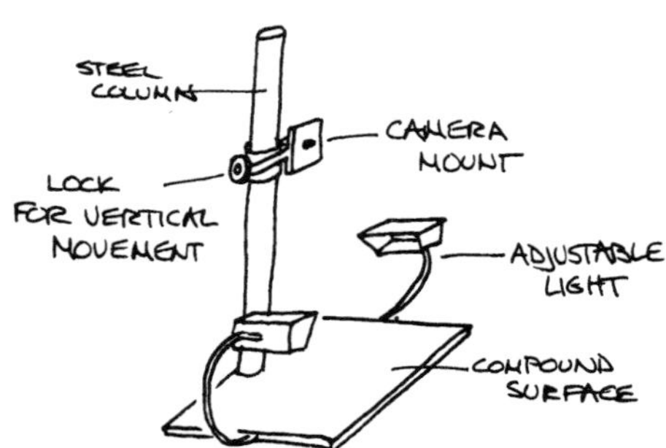

20.7 Copy stand: Ranging in price from less than $50 to many hundreds, copy stands must be strong enough to hold firmly your particular motion picture camera.

COPY STANDS

Although originally designed for still photography photo copying work, commercially manufactured copy stands can be used for animated filmmaking too (see Figure 20.7). Their structure is simple. A metal column is firmly mounted on a solid baseboard. A bracket allows the mounting head to be moved vertically up and down the column. Copy stands generally come with a pair of lighting fixtures that are mounted at 45-degree angles to the camera plane. A good copy stand, like a good tripod, is one that is strong, steady, and adjustable. The least expensive copy stand is priced around $50. Top-of-the-line copy stands will cost hundreds of dollars.

I should point out here that unless it's been specially modified, using a copy stand for animation requires one to work upside down because of the camera mount. This can be a real inconvenience with those techniques that require manipulation of materials directly under the camera.

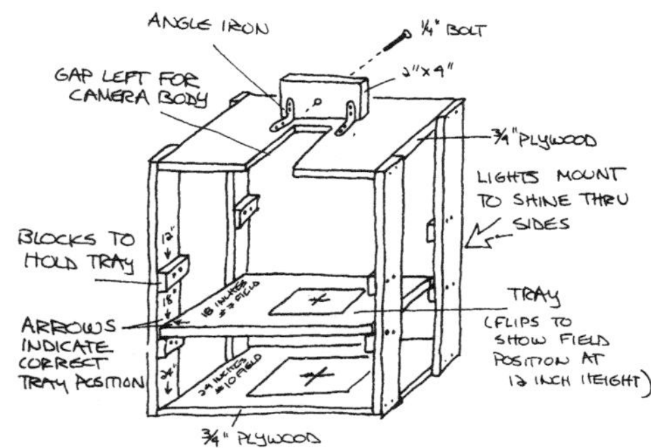

20.8 Laybourne's box stand: Selection of heights for the removable tray is determined by the minimum focal length as marked on the lens of the particular camera being used. If this is a nonreflex camera, remember that a test roll must be shot to determine the precise location and size of the fields at the selected heights. In such a situation, once the camera is mounted it is removed only when absolutely necessary. Modifications may be required so that the camera can be loaded and adjusted without dismounting.

HOMEMADE STANDS

If there were ever an international contest to search out the best examples of human cunning—of the inventiveness, tenacity, and resourcefulness of the human mind—there's no doubt that an animator would get the grand prize. Nowhere, it seems, is there a greater manifestation of sheer inventiveness than in the design and development of homemade animation stands. Here are three very simple examples:

Laybourne's Box Stand. A box design is particularly well suited for a nonreflex camera. The frame holds the camera in a fixed position, but allows it to be loaded, wound, and single-framed. The heights of a movable shelf are determined by the focal length of the camera's particular lens. Because the camera we are discussing has no through-the-lens focusing, there is no way to determine the precise focus at close distances without relying on the distance indicator marks on the lens casing. In order to locate the precise boundaries of the fields at the preselected heights, a camera test must be shot. In this test a field guide is placed on each level and film is shot of it. Only when this film returns from the lab and is screened can the exact frame demarcation be determined. This is marked on the surface for future reference.

NFB Super 8mm Stand. I encountered another simple and inexpensive stand at an animation workshop conducted by the National Film Board of Canada. In constructing your own

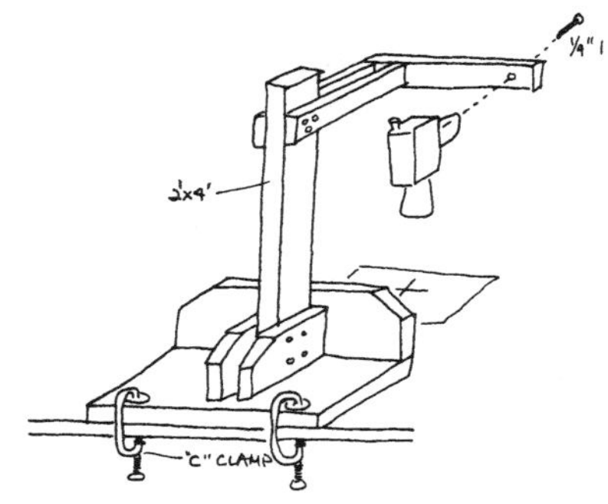

20.9 NFB super 8mm stand: This drawing is based on a sketch of a wooden stand specially designed for super 8mm cameras that was used in a workshop coordinated by Co Hoedeman of the National Film Board of Canada. You'll have to experiment with heights and other measurements so that the stand will suit your camera and your animation plans. A nice feature is that the stand "reverses" the camera so that the field is not inverted as it would be if a super 8 camera were mounted on an ordinary copy stand.

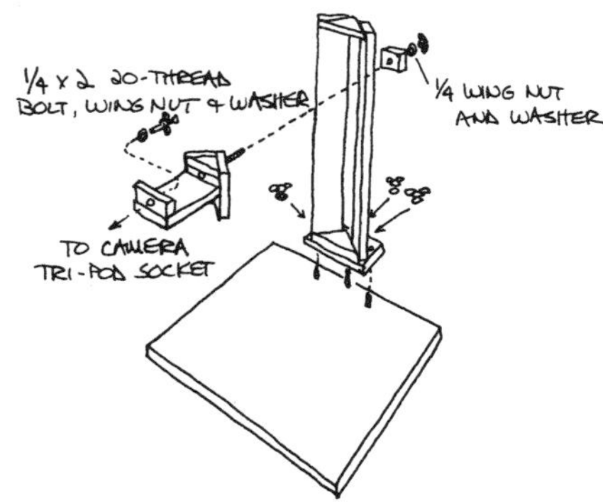

20.10 Kodak copy stand: For approximately $15 you can build a copy stand with an adjustable column. Pamphlet T-43 from the Eastman Kodak Company (Rochester, New York 14650) provides further specifications for the stand illustrated here. *Reproduced with permission of Kodak.*

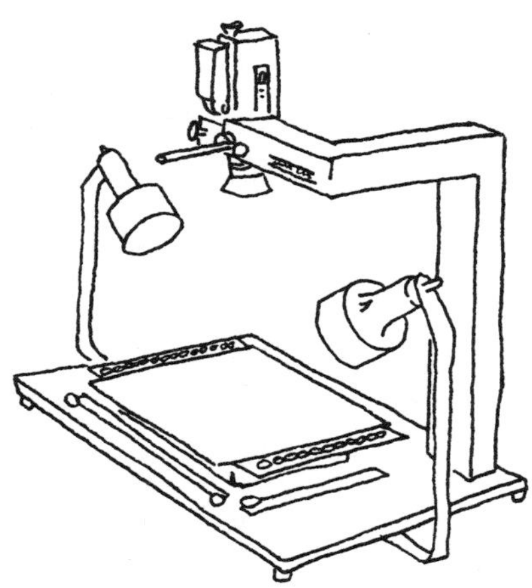

20.11 Oxberry Animator-8: This low-end stand was specially designed for super 8mm filmmaking. The model features a viscous-damped compound consisting of a special traylike device that uses thick grease to smooth the manipulation of the compound. This model features a spring-mounted platen, 12 inches square, that is removable. It has toplighting but not bottomlighting. The original cost, without a camera, was under $700.

version, determine the vertical height of the column after you've played around with your own super 8mm camera. You'll need to select a height that affords the largest range of field sizes, using the zoom at a wide-angle setting all the way to the telephoto position with diopters or macrofocusing. Like the box stand, this design is portable. The reflex camera can be easily removed for live-action filmmaking or for those kinds of animation that require a tripod.

Kodak Copy Stand. The Eastman Kodak Company circulates information on how to build a simple wooden copy stand that is sturdy and features a vertically adjustable column. Figure 20.10 provides a detailed breakdown of the Kodak stand.

Other Designs and Commercial Components. There are endless variations. Depending on the range of animation techniques that you want to do and on your camera, various building materials can be combined in any wondrous concoction.

Slotted steel angle irons can be used to create a strong and adjustable framework. It's like a giant Erector Set. Modified photographic enlargers, lathe beds, X-ray machines—all have been used to create successful animation stands.

As you begin to plan your own stand, you may want to consider integrating into its design some of the commercially produced elements that have been introduced in this book, such as backlighting units, an auxiliary motor, a zoom-lens extender arm, and so on. For example, the J-K Engineering Company offers a compound table for do-it-yourself animation-stand builders. The compound mounts on two metal pads and its features include a black anodized aluminum tabletop, 360-degree rotation, traveling peg-bars, #12 or #16 field platens, rear lighting, graduated hand wheels for 20 inches of east/west movement and 12 inches of north/south travel, and mechanical counters. The unit costs around $2,000.

COMMERCIAL STANDS

When this volume was first published in 1979, the term *animation* connoted the film medium. True, there was animation on TV, but all of that was generated by motion picture technology and then transferred to video. Back in 1979, computer animation was an exotic frontier with tiny dribs of work coming out of Bell Labs and a handful of other places.

20.12 Oxberry Media Pro: The Media Pro stand was designed to work with a Bolex H-16 SBM 16mm camera attached to a single frame motor. The stand features a heavy-duty aluminum column that can be adjusted vertically by a spring-balanced ball-bearing camera carriage and lock. The compound movement features the same sort of viscous-damped movement as the Animator-8, with a #12 field, spring-loaded glass platen to hold artwork flat, plus a #8 field pantograph. A terrific feature of this medium-level animation stand is its bottomlighting. The Media Pro takes a 17-inch cast-aluminum disc with two peg tracks cut to match the movable peg-bars included with the compound table. There is a #12-field opal-glass insert over a #12 field 3200° K, cold cathode backlight. The backlighting stage is on legs that raise it above the compound during filming. When the stage is removed and the legs are detached, it can be used as a standard drawing disk for preparation of art, tracing, inking, and opaquing. When this model #2500 was being manufactured, it cost $2,300 without a camera and $6,300 with the Bolex and motor.

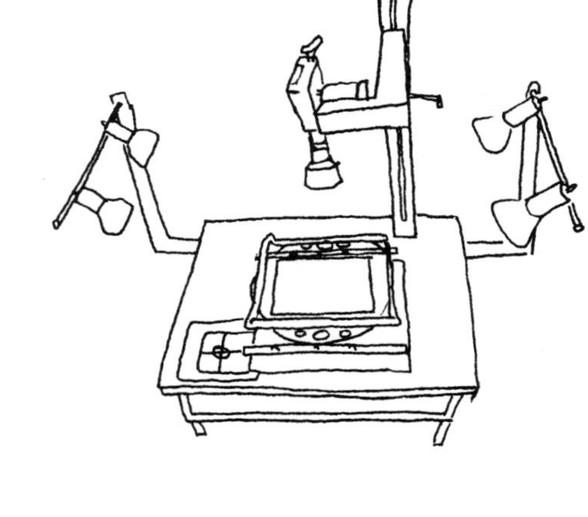

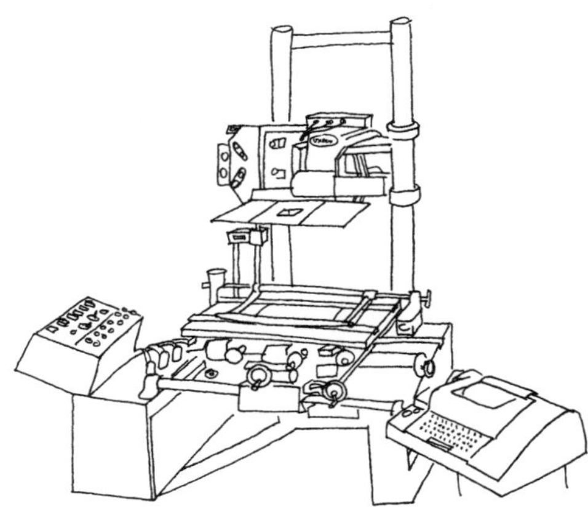

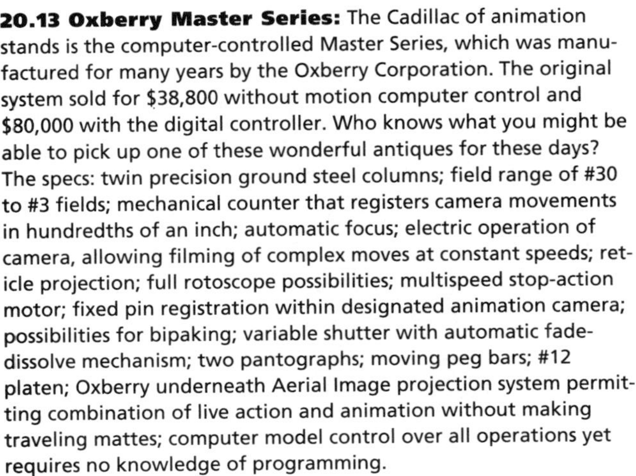

20.13 Oxberry Master Series: The Cadillac of animation stands is the computer-controlled Master Series, which was manufactured for many years by the Oxberry Corporation. The original system sold for $38,800 without motion computer control and $80,000 with the digital controller. Who knows what you might be able to pick up one of these wonderful antiques for these days? The specs: twin precision ground steel columns; field range of #30 to #3 fields; mechanical counter that registers camera movements in hundredths of an inch; automatic focus; electric operation of camera, allowing filming of complex moves at constant speeds; reticle projection; full rotoscope possibilities; multispeed stop-action motor; fixed pin registration within designated animation camera; possibilities for bipaking; variable shutter with automatic fade-dissolve mechanism; two pantographs; moving peg bars; #12 platen; Oxberry underneath Aerial Image projection system permitting combination of live action and animation without making traveling mattes; computer model control over all operations yet requires no knowledge of programming.

20.14 Oxberry Filmmaker animation stand: Another step up in expense brings you to a dedicated animation stand with a 16mm Oxberry (brand) process camera, automatic focus mechanism, self-supporting structure, motorized tracking, automatic remote-controlled reticle projection, shadowboard with hinged wings, quartz-halogen top- and bottomlighting, compound with four peg tracks, compound movement with hand wheels and digit counters, rotation units, pantograph, full control consul, and rotoscope attachment. These babies used to sell, new, at just under $15,000.

One way to measure the degree to which digital tools have overtaken the mechanical ones is by looking at manufacturers. Twenty years ago there was a robust marketplace in animation stands with model lines from companies like Oxberry, Animation Sciences, Forox Corporation, Fax Company, JK Camera Engineering, and Ox Products, Inc. Today, only two of these companies are making animation stands: Fax Company and Oxberry (a Division of Richmark Camera Service). Furthermore, the kinds of stands they make are geared more for photocopy work than for filmmaking.

All is not bleak. During the fifty years when filmmaking technology ruled supreme, many thousands of animation stands were sold. These are rugged beasts. And animation stands are not subject to the same wear and tear of, say, a live-action camera. Hence, the enterprising animator who wants to work in film should be encouraged to use his or her ingenuity to locate used stands that run just as well as they did the day they were manufactured. Oxberry—who was the largest maker of animation cameras and stands—still stocks and sells all parts for their full line of animation stands, which are sampled in the accompanying line drawings. The company has a special warm spot for independent animators and can be contacted for product information at Oxberry, 180 Broad Street, Carlstadt, New Jersey 07072.

20.15 Walking: A frame from Ryan Larkin's *Walking*. *Courtesy of the National Film Board of Canada.*

21 ■ Registration Devices

Registration is the animator's mania. Throughout the long hours of preparing artwork, the animator must be focused—physically and mentally—on keeping things lined up, keeping them registered. This awareness is intensified during the process of filming—the artwork must stay perfectly aligned under the camera.

Almost every piece of animation equipment exhibits a technical preoccupation with registration. Cameras must advance single frames with perfect registration; zoom lenses must be modified if the animator wants a fully registered movement; the stand itself is nothing more than a mechanical system that provides fully controlled registration of all filming variables. Registration is also a big issue in digital animation.

The animator's choice of different *pegging systems* has been discussed fully in Chapter 14. Improvised registration systems for individual sheets of paper or acetate include the use of: (1) a standard ¼-inch ring punch, the type employed with loose-leaf notebooks; (2) 90-degree corners on sheets of paper and acetate cels to provide a simple way of lining up successive images; (3) index file cards, a variation of the corner method; and (4) some kind of "bug," a continually redrawn mark that is consistently employed.

The specifications of commercially produced pegging systems have also been discussed earlier. The two major peg systems bear the names of the companies that first developed them: *Oxberry* and *Acme.*

Animating with either paper sheets or acetate cels

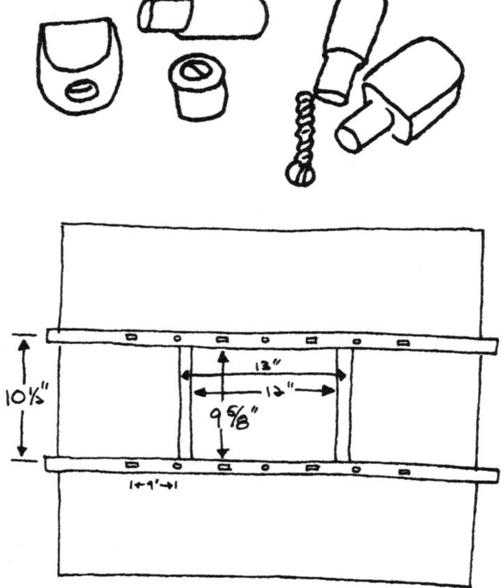

21.1 Loose pegs: Metal pegs can be purchased unmounted. These loose pegs are equipped with screws so that you can mount them onto a drawing or filming surface. Whether Oxberry or Acme pegs are used, the distance between both round and flat pegs is always exactly 4 inches (center to center). Also illustrated is a conventional arrangement of pegs on a compound surface. Note that a number of peg sets can be mounted either above or below the drawing surface. This eliminates much of the need for sliding pegs.

requires the artist to make continual comparisons between the position of a drawing on one sheet or cel and the positions of that drawing's variations on other sheets or cels. In order to work easily, an animator needs to combine a pegging system with a lighted drawing surface. Hence, the evolution of special drawing discs and other underlighting devices, which will be discussed shortly. All these registered drawing aids also, of course, provide a standardization and registration of specific areas that will be filmed by means of a field guide. As described earlier, the field guide is used in conjunction with the pegs and drawing surface (see Figures 14.2 through 14.6).

A catalogue of these tools, along with other items that help an animator plan and plot and polish his or her work, follows.

PEGGING DEVICES

Loose Pegs. Sets of three loose, removable, screw-in pegs (Figure 21.1) are available for Oxberry or Acme systems. Each set contains two oblong and one round peg. A set of three: $20.

Peg Bars. The standard peg bar is made of aluminum and is 3⁄16 by ¾ by 10½ inches. Available with either Acme or Oxberry pegs, the bar is mounted on a wooden board or light box. Note that a channel for accepting the bar must be routed out of the surface. A bar with nonremovable pegs: $20. A bar with removable pegs: $35.

Extra-long black anodized-aluminum peg bars are available in Oxberry or Acme systems. These long bars carry three round and two flat pegs and are scaled in twentieths of an inch. They can be glued or taped to a surface. Dimensions: ⅛ by 2 by 18 inches. Price: around $50.

Peg Plates. A particularly flexible pegging device is an aluminum tape-on peg plate bearing either Oxberry or Acme pegs. It can be easily mounted on a light board or under the camera by simply taping the thin plate to the surface. The plate, which is also easily removed, is 1⁄32 by 2 by 10½ inches. Price: $30.

Cel Punches. High-precision cel punches can be purchased that allow the animator to punch Oxberry or Acme registration systems (holes) into paper or acetate sheets. The design of the most expensive punch is "progressive"—the system can be used with extra-long backgrounds, pan papers, and cels. Oxberry's punch is no longer manufactured, but prepunched paper is available.

Quarter-inch Wooden Dowel System. Lumber and hardware stores sell round wooden dowels that are ¼ inch in diameter. These can be used to create an inexpensive, homemade registration system that is accurate enough for most work. Pegs can be cut from the dowel and inserted into drilled holes on a drawing or shooting surface. Similarly, short pieces of doweling can be cut and glued on heavy posterboard or matte board. In either case, the exposed portion of the dowel/peg must be sanded carefully to prevent the scratching or tearing of paper and acetate sheets.

The distance between holes on such homemade peg plates and peg bars should be carefully selected so that it matches one of the standard two-ring or three-ring spacing distances that are used in standard school loose-leaf notebooks. I recommend the three-ring system in which ¼-inch-diameter holes are exactly 4¼ inches apart. If the pegs are placed correctly, the animator ought to be able to use rotary-cut and punched loose-leaf papers.

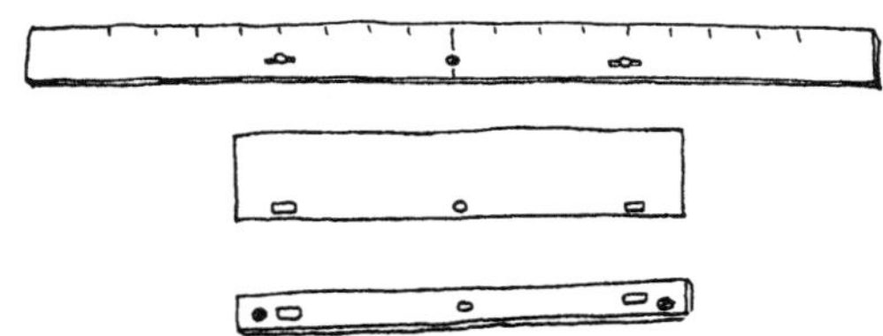

21.2 Peg bars and plates

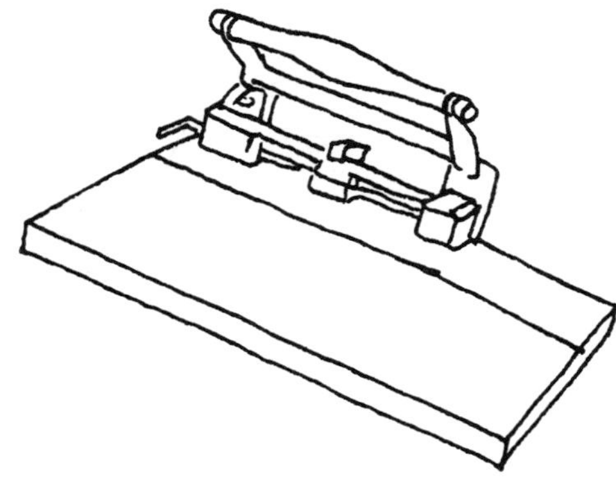

21.3 Commercial cel punch

DRAWING DEVICES

Animation Disc. The Oxberry Corporation still manufactures and sells a high-quality, black anodized animation disc that features a #12-field translucent glass window mounted within a machined cast-aluminum disc. These discs can be clamped to a baseboard for under-camera cel registration and background moves. It is better, however, to mount the disc in a 16⅜-inch-diameter hole in a drawing table or surface to be placed beneath the camera. When so mounted, the disc can be rotated for drawing, tracing, inking, opaquing, or for 360-degree rotation moves under the camera. The disc's outside dimension is 17⅜ inches.

On such drawing discs, sets of pegs are mounted top and bottom, 10½ inches apart, and accommodate up to a #12 field. An Oxberry brand animation disc with movable pegs (bearing either Oxberry or Acme pegs) costs $415.

Ink and Paint Boards. A number of suppliers have simple boards for use in drawing, tracing, inking, and opaquing.

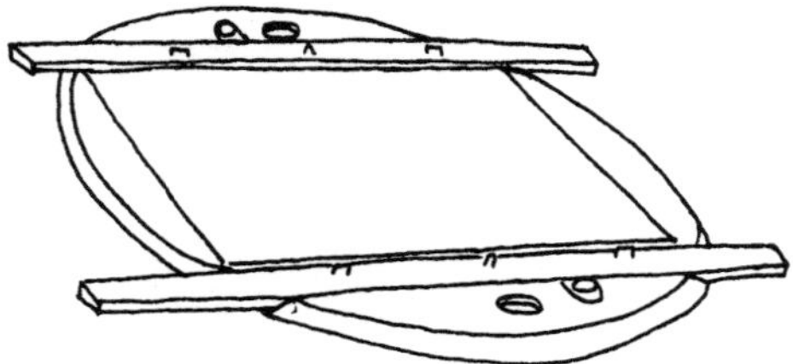

21.4 Animation discs: This drawing is of an Oxberry disc.

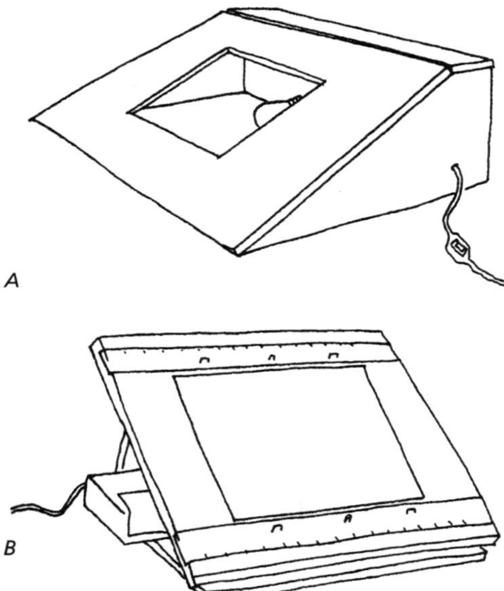

21.5 Light tables: Ordinary plywood can be used to make a simple light table such as that in *(A)*. The square hole is sized to accommodate a #12-field sheet of white Plexiglas. Standard wooden molding keeps the drawing surface in place and mounted flush to the wooden surface. A porcelain socket is mounted for use with a 60-watt incandescent bulb. A peg plate can be taped to the wood or to the Plexiglas and then repositioned beneath the camera where the registered sheets or cels are subsequently filmed. Illustration *(B)* shows a portable light table from Heath Productions. It has two 18" pan peg bars and a removable fluorescent lighting unit.

Choices include size (#12 field or #16 field), material (clear Plexiglas, aluminum, or white opaque plastic), and top pegs only or top and bottom pegs. They are generally available with either Oxberry or Acme pegging systems. Prices range from $25 to $75.

Oxberry manufactures a thick black anodized-aluminum inking peg board—11 by 16 by ⅜ inches—with three permanently mounted Oxberry or Acme pegs. Price: $50.

Light Tables. Oxberry has a light table that mounts any Oxberry aluminum drawing disc. There is an instant-start, fluorescent backlight. The table is recommended for preparation and registration of paper sheets and acetate cels (Figure 21.5). Price, without disc: $130.

For $75 to $95, Cartoon Colour Company has a range of light boxes made from Plexiglas.

Another variation of a light table is the portable steel tracing box. Its 16- by 20-inch glass surface has a light diffuser and either two or three 15-watt bulbs with a three-way switch. The metal stand allows the top surface to be tilted. This kind of tracing box has to be used with a tape-on peg plate. Price: around $75.

Field Guides. The standard reference for the area beneath the animation camera is the Oxberry field guide. Field guides that are printed on .02-inch plastic sheets provide #1 through #12 fields. They are available punched to Oxberry or Acme standards. Price: $100.

Field guides that indicate TV cutoff areas—that familiar rounding of the rectangular shape of the television screen, which can mean a loss of up to 20 percent of the screen's area—are also available. These clear plastic sheets are like a field area chart, but they show the precise cutoff line that exists when a film is projected on television. Such TV charts show a #12 field through a #14 field and are available punched or unpunched for $20 (Figure 21.6).

OTHER AIDS

Exposure Sheets. Should you not wish to detach and duplicate the exposure sheet provided later in Chapter 25, you can purchase a pad with 50 sheets at a cost of $8 from Cartoon Colour.

Footage and Timing Charts. A 16mm–35mm footage

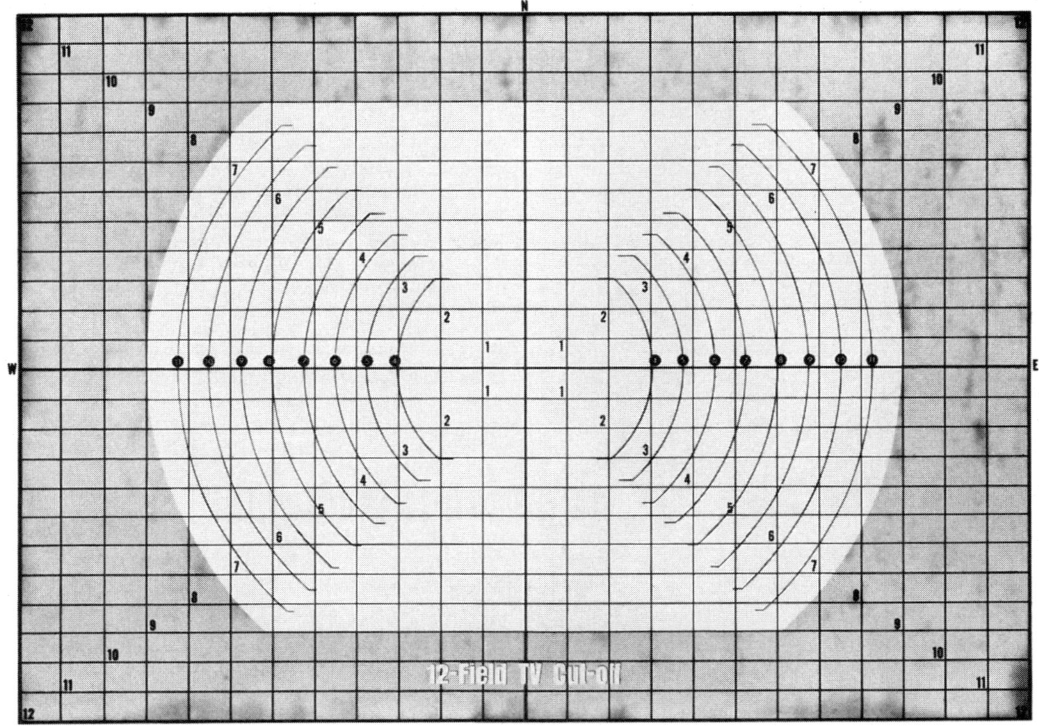

21.6 Field guide with TV cutoff (not to scale).

and timing chart, printed on thick plastic, can be purchased from the Cartoon Colour Company. Price: $2.

Stopwatches. A stopwatch is often useful for determining the precise duration (and subsequent number of frames, number of drawings) of a particular movement or portion of a movement. A standard stopwatch will work, but for those with *everything,* Cartoon Colour will sell you a stopwatch with time-footage scales for 16mm, 35mm, and super 8mm animation. Price: about $150.

Animation Platen. If you are building your own stand, you ought to know that you can purchase an Oxberry #12 field platen that is designed to hold cels and other artwork of variable heights flat on the compound table. The glass sheet is ¼ inch thick, float-mounted on a pivoting center, and hinged on a spring-loaded platen bracket. The glass itself is optically pure. Since the entire assembly fastens to the surface with knobs, it is easy to mount and then remove completely when filming large art. Price: under $300. Platens are critical for registration in many cutout and collage techniques.

22 ▪ Filmmaking Gear

Although this chapter is most useful for animators working in film technology, a number of elements within it should be of value to digital animators as well. For instance, the discussion of lighting setups for animation stands is also of direct use to digital animators who will mount a video camera pointing at a flat surface that must be lit (or underlit). The section about art supplies is equally useful for all animators working with registered sheets of paper and traditional drawing and painting media.

The seven topics that Chapter 22 covers are:

Lighting
Audio
Editing
Projection
Film Stocks
Art Supplies
Lab Services

GENERAL LIGHTING REQUIREMENTS

Because animated films are almost always shot indoors and under controlled situations, the element of lighting is less problematic for the animator than for the live-action filmmaker. If you set up the right lighting system once, you may never need to think about it again.

The human eye and brain are always making compensations to allow us to perceive colors as constant under various lighting conditions. It is different with film. The same colored surface—say, a white piece of paper—will film with very different hues depending on the source of light and the characteristics of the kind of film stock that is being used.

Motion picture film is manufactured with different chemical balances. These produce different measurements of *color temperature*—the sensitivity of the emulsion to whatever light falls on it. *Daylight* film stocks are structured so that a white sheet of paper will appear white when filmed under normal daylight—the light provided by the sun. *Tungsten* film stocks are designed to be used with artificial lighting. As far as color balance is concerned, there are two variations of tungsten stocks (also called *indoor* stocks). Type A film is balanced to one specific measurement on a temperature scale (3200°K), and Type B film is balanced to a slightly different reading on the same scale (3400°K). Daylight films, by the way, have color temperature of 5500°K.

You may remember from your science classes that light, at least in part, behaves like a wave, and that these waves occur at different frequencies within the electromagnetic spectrum. For the filmmaker, the practical result of this phenomenon is that different kinds of light sources—the sun, an incandescent bulb, fluorescent lights—will produce different colors even if the same object is being filmed on the same film stock.

So how do you get the "right" film for the "right" light source? Or how do you compensate so that an indoor film can be used outdoors?

The *Kelvin color temperature scale* is the measurement system used by filmmakers to make sure that an apple will look like an apple and that your favorite movie star will have the "correct" flesh color when he or she appears on screen. The Kelvin scale is a scientific measurement scale that is used in film and photography fields to measure degrees of color temperature. Sunlight is rated at 5400°K. The 100-watt household bulb is rated at 2860°K. A candle flame is 1850°K. Almost all artificial lighting systems used in moviemaking are manufactured so that their rating will be either 3200°K or 3400°K. Obviously the

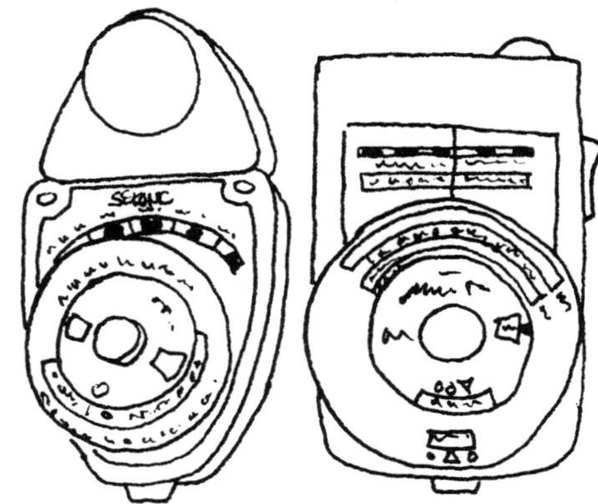

22.1 Light meters: At left is an incident light meter (Sekonic model #L280), which measures radiated light directly, whether it comes from the sun or a flood lamp. On the right is a reflectant meter (Gossen Luna-Pro) that can also take incident readings. A reflectant meter measures light that bounces off a surface.

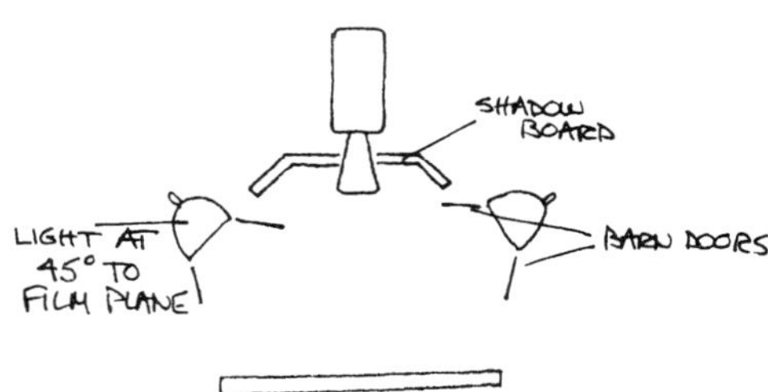

22.2 Avoiding reflection: The 45-degree angle of the lights should mean that reflections will never bounce back into the camera's lens. That assumes, of course, that the surface being filmed is flat. This is one reason why a glass platen is recommended for most kinds of animation. The shadow-board's black matte finish absorbs light and helps shield the lens from direct flaring from a nearby light source or reflecting surface.

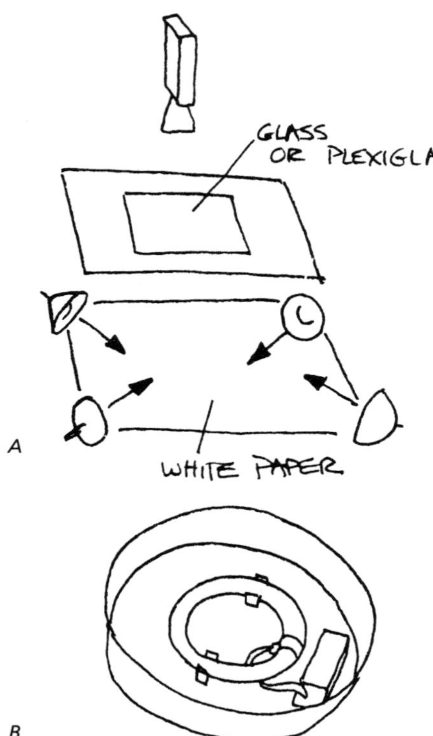

22.3 Backlighting: Self-rigged backlighting, *(A)*, can be done by mounting two or (better) four lights under the compound surface. These must be arranged so that they provide even illumination through the translucent surface that is mounted into the compound. Commercially manufactured backlights must also provide even illumination. A Fax backlight is pictured in *(B)* with a fluorescent bulb.

two ratings are very close. Film stocks designed to be used with artificial lighting are also rated 3200°K or 3400°K.

The quality control of commercially manufactured film stocks and lighting equipment virtually ensures that if you match the Kelvin scales of both film stock and lighting tools, what you see with the naked eye will be what you see when the exposed and processed film is projected on a movie screen. If problems occur, it is usually because there are mixed sources of light, say a fluorescent light or ambient sunlight in addition to the proper artificial lighting. In case you ever have a weird shooting setup that keeps giving you problems, you may want to use a special *color temperature light meter.* These specialized devices will let you measure the precise temperature value of any lighting situation.

LIGHTING PROBLEMS

Metering and Exposure. Many super 8mm cameras have built-in light-metering systems that will automatically set the camera's diaphragm at the proper f-stop opening for the given light intensity and for the particular sensitivity of the film stock being used. As noted earlier, there is a tendency in animated filmmaking for these built-in, automatic meters to overexpose the film's subject when a brightly colored background is being used. For this reason, a manual override capability is strongly recommended.

Or better, the animator should take manual light readings by using a standard photographic light meter (Figure 22.1). Light meters measure *reflected light* (bouncing off the surface) or *incident light* (light that is falling on the surface being filmed). The utilization and price of meters won't be covered in these pages. Your local camera shop will explain the tool or you can check a volume on the basics of filmmaking. What ought to be noted here, however, is that the animator should always shoot test footage in order to *judge by eye* the proper setting for a specific kind of artwork under a specific lighting setup. The process of testing exposure is called *bracketing,* and it is accomplished by filming the same surface under the same lighting but at aperture variations of ½ an f-stop. By studying the exposed film, the perfect setting can be selected.

When measuring light with either an automatic or a manual metering system, you should place a *gray card* under the camera. These special pieces of paper or cardboard of grayish hue have a value that is exactly halfway between white and black; the card's purpose is to help the meter estimate the f-stop that will yield an exposure that best captures the midrange of color and brightness intensities. The gray card will compensate for the tendency of meters to underexpose the film. But if you're unsure about a reading and can't do a test, follow the time-proven rule of still photographers—expose for shadows. When in doubt, underexpose rather than overexpose.

Even Illumination. The use of artificial lights can make it difficult to evenly illuminate the surface underneath the camera. Because the lights are mounted so close to the compound surface, there is a tendency for a *hot spot* to develop. This is an area that is more brightly lighted than the rest of the surface. In order to get a balanced intensity of lighting, animators will usually mount the same size light sources in matching positions on either side of the artwork.

Reflection and Glare. Artificial lights tend to reflect off almost any surface, even one that has a matte finish. The woeful result can be glare or flashing in what is otherwise a perfectly exposed animation sequence. To fight this problem, animation stands are designed so that lights are mounted at a 45-degree angle to the plane of the camera (or compound).

The use of a *platen* to hold artwork flat under the camera is also important in defeating reflection and glare. Finally, many stands feature *shadowboards* with hinged wings. These devices are mounted close under the camera, with a hole cut for the lens. The shadowboard also reduces reflections caused by the camera body and the lens itself.

Heat. Artificial lights get hot. Just how hot they get will be appreciated the first time you work under them for any length of time. Animators are always trying to escape the heat. This can be accomplished by the use of *low-wattage* bulbs and lamps when possible; *barn doors* on the lighting units in order to focus brightness on the table as much as possible; *heat-absorbing glass filters* in front of the bulbs; and the use of *fans* to cool the lighting units (or an air conditioner to cool the camera operator!).

Bottomlighting. A *backlighting* unit is required when filming layers of paper sheets or with techniques such as sand and ink-on-glass animation. Naturally, bottomlighting requires

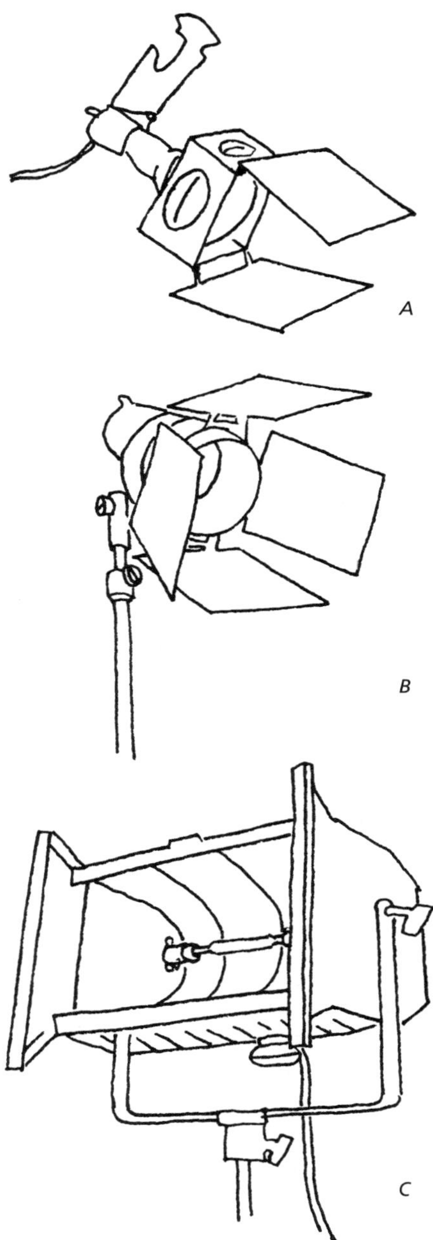

22.4 Lighting systems: The Lowel-Light with barn doors, *(A);* the Colortran Mini-Pro with barn doors, *(B);* and the Colortran Multi-Broad, *(C).* Both Colortran models use quartz lighting.

QUARTZ LIGHTS

Among many, many styles and designs of quartz lighting systems, here are three possibilities, from lowest in price to highest.

- ***Smith Victor Model 770 Quartz Broad Hi-Fi Studio Light*** provides an even, broad lighting, which makes it specially suited to animation. Of all-aluminum construction for cooler operation, the light mounts to a standard ⅜-inch stand. Barn doors must be used to hold scrims and various filters. Price: with accessories, about $75; lamps, $20 each (75 hours).
- ***Berkey Colortran Mini-Pro Lights*** have a recessed power receptacle permitting use of a detachable 120-volt or 30-volt power cable. Bulbs are interchangeable. The lamp has a 3:1 variable focus and weighs 30 ounces. Attachments facilitate mounting in almost any situation. Price: a complete unit, about $150; bulbs, under $25.
- ***Colortran 750-Watt Soft-Lite,*** a relatively small and lightweight studio lamp, provides a smooth, diffuse, shadowless light. Its dimensions are 13½ by 16 by 6½ inches. Price: $250; bulbs, around $20.

a transparent or translucent surface mounted under the camera. Here are three options, in recommended order: (1) a sheet of white, opal, translucent, and shatterproof Plexiglas (Plexiglas 2447), (2) a sheet of frosted plate glass, or (3) a sheet of clear glass that is covered with a thin piece of tracing paper in order to diffuse the light.

Professionally manufactured bottomlighting units are recommended. Specific models include a 3200°K cathode lamp mounted in a metal tray, produced by Oxberry. A 110-volt backlight costs $900 and a 220-volt backlight costs $950. This backlight provides even illumination beneath the drawing disc and it can be used under the camera for pencil tests and other shooting. It covers a full #12 field (10½ by 12 inches) and can be mounted under a drawing disc to provide art preparation as well as photographic backlighting.

LIGHTING SYSTEMS

Although *movie lights* are often used in amateur live-action filmmaking, they are *not* recommended for animation. Color-balanced at 3400°K, they have a very high wattage (650 watts to 1,100 watts) that produces an intense and very, very hot light output, which is uneven and harsh. They are also expensive.

Photoflood bulbs and *reflector scoops* are one of the least expensive ways to light for animation. The bulbs fit standard sockets and some of them, called EAL bulbs, have a built-in reflector, which eliminates the need for a reflector scoop. Photofloods are available in most camera stores in either 3200°K or 3400°K color temperatures.

Photofloods come in various power ratings, from 150 watts to 1000 watts. Animators generally prefer the lower-wattage versions. They are cooler to work near, yet they produce enough brightness for midrange f-stop settings with even the slowest stocks. Photofloods are relatively cheap (less than $2), but they don't last a long time. The average life of a photoflood is six hours and the color temperature will shift with age, caused by the carbonization of the bulb.

Low Kelvin (2750°K) lamps have a long life—up to 2,000 hours. The unacceptable color can be compensated for with an 82C filter on the camera lens. Ask your local hardware store for 75-watt *reflector floods.* The low wattage is a plus factor in cutting heat on the artwork. Using four bulbs also helps to achieve a more even coverage at #12 field. These lamps fit into any incandescent socket, have built-in reflectors, and cost under $5. They are also called *flood lamps.*

Quartz lamps, which are specially manufactured for film production, including halogen and iodine cycles, come in a variety of forms rated at 3200°K or 3400°K, with wattages from 250 to 2000. Through the use of quartz, a high-temperature glass, these lighting systems are able to maintain the same color temperature their entire lives by eliminating carbonization.

FILTERS

Camera manufacturers and filter supply companies offer filters of all kinds for all cameras. Here is a primer that applies to both 16mm and super 8mm formats. How one actually mounts a filter will vary from camera to camera—check an instruction booklet. Most super 8mm cameras take glass filters, which are screwed into the front of the lens casing. While many 16mm cameras take the same kind of front-mounting glass lenses, others often use a filter gel, either directly in front of the camera gate or in front of the lens via a matte box or similar mounting device.

GENERAL AUDIO REQUIREMENTS

There is an especially intimate relationship in animation between the sound track and the moving image. Many of the techniques for working creatively with sound have been discussed in Chapter 7. The comments that follow here deal primarily with audio tools. First there's a brief recapitulation of the sound production systems available within today's 16mm and super 8mm technologies. After that you'll find a fast run-through of recording tools and audio editing processes. While brief, this information should at least prepare you to approach audio with the respect it deserves.

Once a film has been completed, it's easy to create a *wild track,* an audiotape recording that provides (and sometimes combines) music, voice, and sound effects. Wild tracks can be created for both super 8mm and 16mm films. The lia-

FILTERS

- ***Color Correction.*** The 80A filter corrects daylight-balanced films to 3200°K (tungsten) lighting conditions, while the 80B filter corrects daylight stocks to 3400°K. The 82A filter compensates Type A film (most super 8mm color stocks, which are valued at 3400°K) by cooling the light to match the 3200°K output of most quartz lamps. Filming with a Type A film without this filter will produce a slightly reddish or "warm" tone. The 85 filter, which is built into all super 8mm cameras, converts 5500°K rated daylight film so that it can be shot under tungsten lighting. An FLB filter is designed for use with fluorescent lighting to compensate for the green-blue cast that fluorescent lighting produces on tungsten film.
- ***Damping Light Intensity.*** Sometimes during animation filming you will discover that more light is present than can be used. In these situations a *neutral-density filter* can be employed to cut down on the amount of light entering the camera. Such filters have no effect on color. Tiffen, a producer of still and motion picture filters, manufactures fifteen different neutral-density filters, which are measured according to light transmission percentage and f/stop increase. Here are a few samples:

ND 0.1	80%	½ stop
ND 0.3	50%	1 stop
ND 0.6	25%	2 stops
ND 0.9	13%	3 stops

- ***Polarizing Filters.*** Polarized filters are useful in animation because they help block out and reduce reflections from glass, water, and dust. The use of polarizing filters is quite complicated and you are advised to check a thorough production manual for more detail on how they work and what they will do.

bilities of such audio accompaniment is that it cannot be matched very accurately to the progression of the film.

A more satisfactory system is made possible by the super 8mm *sound-on-film* technology (Figure 22.7). After an animated film is shot and edited, the film can be *striped* with a thin surface of audio recording tape that is applied directly to one edge of the film stock. Striped super 8mm raw stock is also available. An audio track can now be recorded directly on the film. This is done with a *super 8mm sound projector.* These projectors are used both to *record* the sound on the film and then to *play back* the sound with the film images during normal projection.

Such a system permits the animator to create a fairly well matched track, for the audio can be recorded and rerecorded on the completed and striped film until it is "just right." Thereafter, the track will be synced properly with the movie's visuals. A similar system exists for striping 16mm film, but it has become so infrequently used that you're advised not to try it.

Sixteen-millimeter film technology permits a more detailed analysis of audio materials. This is accomplished by transferring recorded sound from a reel-to-reel or cassette tape recording to *16mm magnetic recording film,* called *mag stock.* Various tools permit frame-by-frame *sound reading* and *track analysis.* The most important tool is a motorized *film editing console* and/or a *gang synchronizer* with a *soundhead* and an *amplifier.* These tools permit a very, very precise matching of frames of picture to corresponding frames in the 16mm sound track. In fact, most sophisticated films begin with an analysis of the track. The images are then designed to match the track.

When a 16mm film and a set of 16mm audio tracks are fully edited and synced, the tracks are *mixed* to form a *master track.* This is then taken to a lab and turned into an *optical master,* which is subsequently printed along with the original film on a single strip of 16mm film. Voilà! You have a completed 16mm *optical print,* which can be distributed and screened wherever there is a standard 16mm sound projector.

SOUND RECORDING EQUIPMENT

Cassette Tape Recorders. There are hundreds and hundreds of cassette recorders. Many come with built-in microphones.

Some accept "line" input from other recorders or from phonographs and some take auxiliary mikes. A major drawback to these recorders is that it is almost impossible to do sharp editing with them.

Reel-to-Reel Tape Recorders. Because of quality and ease in manual editing, reel-to-reel recorders are generally preferred for filmmaking. Decent monaural recorders begin at about $150 and are manufactured by all major audio firms. Synchronous monaural tape recorders made by Nagra and Stellavox cost at least $3,000. They are *not* required for animation. Stereo reel-to-reel recorders can be useful for the amateur and independent animator. They enable one to record two different tracks—one on each channel—and then to "mix" these down to one track via a "y cable" that transfers the stereo onto a single track machine or onto the edge of the film itself. The mixing of 16mm tracks is usually done at a professional sound studio, which provides more control.

Microphones. You can spend just about anything you want for a mike. The cheapest is about $15; the most expensive, around $1,000. There are totally different mikes that will handle different recording jobs. The list includes omnidirectional, cardiloid, and shotguns. There are condenser mikes, dynamic mikes, and radio mikes. Along with these are mixers and amplifiers and filters and all manner of auxiliary attachments, and a universe of plugs.

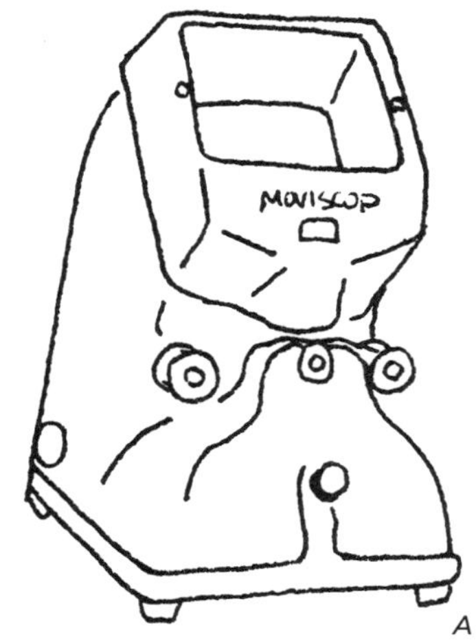

22.5 Viewers: The Zeiss Moviscope is pictured in *(A)*. A Bolex super 8mm viewer, with rewinds, is shown in *(B)*.

AUDIO TAPE EDITING

By connecting two audiotape recorders with a patch cord, it is possible to selectively transfer audio material from one machine to a fresh roll of audiotape on the second machine. In this way, a simple editing of sound tracks can be quickly, easily, and inexpensively accomplished.

A more precise way to edit involves actually splicing the recording tape. The tape is run forward and backward past the recording head until the exact point for a cut is selected and identified. A soft-lead pencil is used to mark the spot. When cut diagonally with a razor blade (a grooved metal splicing block is used to position and hold the tape), the end of one segment can be spliced together with the start of another segment. A special audio splicing tape is used. This procedure is

22.6 Editing consoles: The Movieola 6-plate "flatbed" is illustrated in *(A).* A super 8mm editing machine is shown in *(B),* the MKM model 824. In illustration *(C)* the surface of a Steenbeck 6 plate is shown. The top two plates carry the picture (16mm work print) and the bottom two sets of plates carry two different sound tracks (16mm magnetic recording stock).

fully explained in numerous source volumes or in the directions accompanying an audio splicing kit, which can be purchased for under $20.

OVERVIEW: EDITING AND SPLICING

Editing matters less in animation.

This is so for a number of reasons: animators preplan their films with greater precision than live-action filmmakers; there is greater control during the shooting process itself; animated films are generally shorter than live-action films; certainly the ratio of shot film to used film is far, far lower in animation than in any other kind of film production.

Depending, of course, on which technique is being used, the animator approaches the task of editing with two objectives. First, there is a need to place the exposed sequences into the proper order—an order that has usually been carefully storyboarded or scripted in advance of the actual production. Second, the animator edits the film so that it conforms to a prerecorded sound track. This involves a series of *fine cuts*—minor alterations in the picture and sound track so that they run together synchronously. If a wild track is being used, the editing process is simplified further.

In contrast, the live-action film editor must constantly make selections between possible "takes" and also determine the precise flow and order of the edited footage (a 15:1 shooting ratio is common). In live-action filmmaking there can be up to three or four different sound tracks that must be interrelated and cut to go with the visuals.

The job of editing is generally broken down into a number of sequential tasks. Each editor will have his or her modifications to the process—each will handle the choice-making process in a different way and a different order. For what it's worth, however, here is my own approach to editing, briefly reviewed.

Processed footage (the *rushes*) is screened again and again until the filmmaker/editor is certain of what footage is to be used and in roughly what order it will appear in the finished movie. No cuts are made in the footage until at least a first *rough-cut* version is determined.

The film is next run through a *viewer* or an *editing con-*

sole while the editor marks the film at the precise frame at which cuts will be made. The film is then literally cut, with *scissors,* and hung up in an *editing bin* or improvised system that labels each individual strip of film. This is an important stage. The editor has to be certain that the "head" end of each strip of film can be identified and that it will be possible to quickly locate various strips of film in the order they are required for the *rough-cut* assemblage. Note that no footage is ever thrown away. Those pieces that are not included in the final film are called *outtakes* and they should be carefully saved.

Using *rewinds* and a *splicer* and beginning with a strip of *white leader,* the editor now begins to splice pieces of film together—the "head" of one piece to the "tail" of the preceding piece. Splices in both super 8mm and 16mm formats are made in either of two ways: a *butt-end splice* or an *overlapping cement splice.* (Splicing tools are described later in this chapter.)

The rough cut is spliced to a piece of tail leader and can now be projected with either an editing console or with a projector. At this stage the animator generally makes refinements, which lead to what is called the *fine cut.* In 16mm film editing the picture is edited in relationship to the 16mm magnetic stock that bears the sound track(s). Often, of course, the track will be cut and rearranged so that it fits the film's images. When both sound and picture are set, the completed film can be previewed by projecting in *double system* with a special kind of film projector or by using a motorized editing console. If all is as the animator wants it, the *camera-original film* (a *work print* having been used in editing) is conformed to the fine cut, using a *gang synchronizer,* a tool that advances individual tracks and pictures together. A gang synchronizer can be equipped with a *soundhead* and *amplifier.* This gives the editor another stage at which to analyze the film to make certain that the picture and sound are aligned as they should be. As noted earlier, the synchronizer is used commonly in animation in order to preanalyze a recorded sound track before the actual filming takes place.

A few paragraphs are not adequate to introduce all the stages, techniques, and tools that are used in film editing. Entire volumes have been written on this topic, and you'll need to consult one if you are beginning to learn editing. But to familiarize you a bit further with the editing and splicing

22.7 Super 8 sound motorized editing bench: The illustration shows, from left to right, a rewind crank (note long shaft), a modified cassette tape recorder, a super 8mm viewer with soundhead mounted directly in front of the viewing gate, a two-gang synchronizer with a motor drive assembly (the rear section), and a forward wind crank that carries two up-reels.

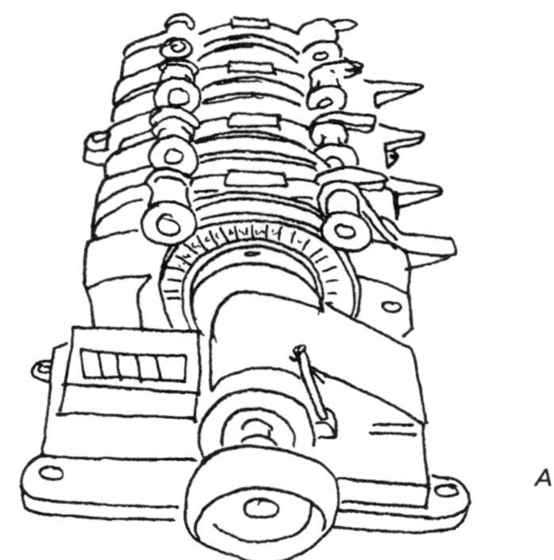

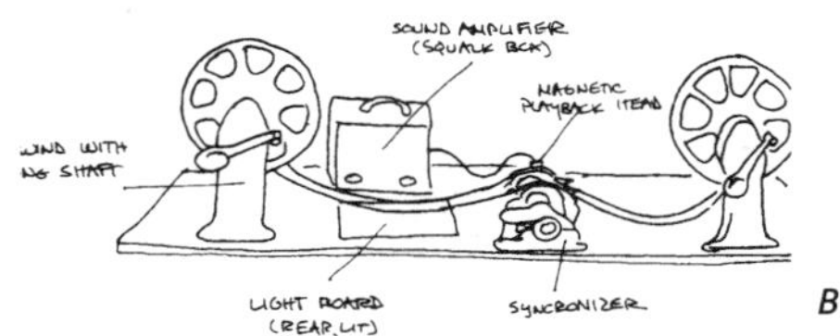

22.8 Synchronizer: A four-gang synchronizer is pictured in drawing *(A).* The synchronizer's position on an editing bench is indicated in drawing *(B).*

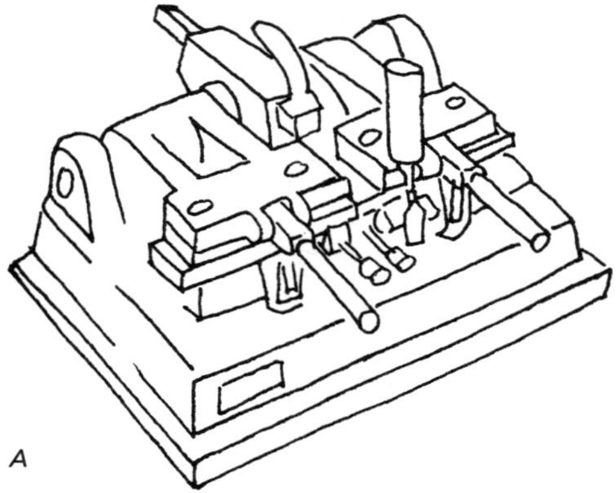

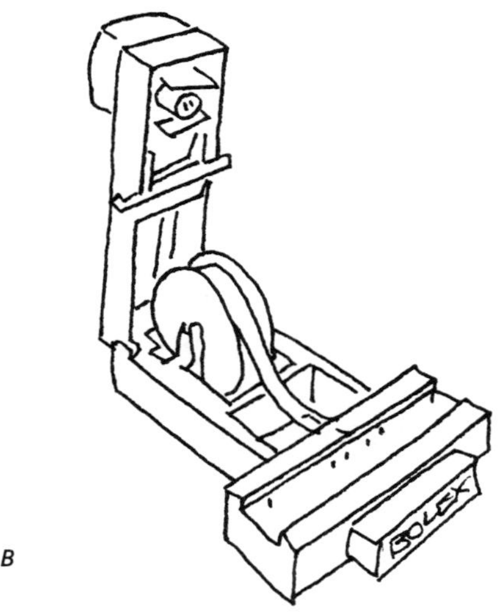

22.9 Splicers: The Maier-Hancock portable hot splicer is shown in *(A)*. Bolex's super 8 Guillotine splicer is shown in *(B)*. A Rivas 16mm tape splicer is shown in *(C)* and an inexpensive splicing bar that uses Kodak Press-Tapes is shown in *(D)*. There are many other brands and styles of cement and tape splicers for both super 8mm and 16mm editing.

process, and to help you determine what tools you'll need to borrow, rent, or purchase, here is a catalogue of editing and splicing equipment.

EDITING EQUIPMENT

Viewers. Available in a wide variety of super 8mm and 16mm models, viewers are operated by passing the processed film from left to right through a precision transport mechanism that ensures smooth film movement without scratching (Figure 22.5). The viewer has a rotating shutter with a reflecting prism that casts a projected image past polished condenser lenses and onto a ground-glass viewing screen. The sizes of the viewing screens and the intensity of the image vary. Viewers are used in conjunction with hand-operated rewinds. Prices run from under $50 for super 8mm viewers, to around $300 for a 16mm Zeiss Moviescope.

Editing Machines. Modern *flatbed editing consoles* provide speed, accuracy, and convenience through precision engineering (Figure 22.6). These machines are designed in modular units with a large-screen viewing surface, variable speed movement in forward and reverse modes, quality soundhead(s), footage/frame counters, instant start/stop, and other features. Flatbed machines hold and transport both film and sound in a horizontal position. The capabilities of such machines are measured by the number of *plates,* the loading and take-up positions that hold *cores* on which film is reeled. A four-plate console has one picture head (for the film and usually designed with a 26-frame advanced optical reader) and one magnetic soundhead (for the mag stock). There are eight- and even ten-plate models.

Rewinds. Sixteen-millimeter rewinds have gears with a cranking ratio of one handle revolution to four reel revolutions. They are available with friction drags, long shafts, elbow breaks, spacers, tite-winds, and spring clamps. A basic long-arm rewind costs $50. Super 8mm rewinds cost about the same. Many super 8mm viewers have attached plastic rewinds that operate in a one-to-one cranking ratio.

Synchronizers. These essential tools are designed to keep multiple picture and sound tracks locked in absolute mechanical sync. Film and sound sprockets are held by rollers and hinged pressure arms. These roller arms are often drilled to mount a magnetic head, which rests gently on the sound film and is connected by wire to a sound amplifier/speaker called a *squawk box.* The rollers are referred to as *gangs.* Synchronizers cost in the $200 to $300 range and are available in one- to six-gang models. All have footage counters. The rollers themselves mark the frame count (Figure 22.8).

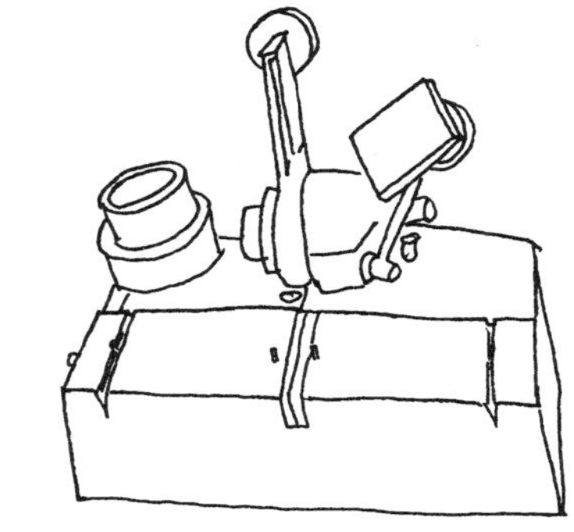

C

D

Splicers. Cement splicers are preferred (often required) for printing 16mm films. The technique of *wet splicing* involves scraping off the emulsion from one of the ends being joined. These ends are then slightly overlapped and fused together by a *cement* that is specially designed for the purpose and must be fresh or else the splices won't properly bond. An inexpensive 16mm cement splicer such as the Griswold HM6 costs $40; Bolex makes a super 8mm cement splicer at $75.

Hot splicers are the professional standard. They are precision tools with a carbide scraper and a thermostatically heated blade, which makes a quick and perfect weld. Cement is also used. Maier-Hancock Model 816S will take 16mm and super 8mm films. Splicers can be rented for a few dollars a day.

In what's called *butt-end splicing,* the two ends are cut on the frame line and then taped together without overlapping. *Tape splicers* are preferred in rough editing because they can easily be taken apart without losing even one frame. A tape splice can be done far more quickly than any cement splice. The tape is specially fabricated to withstand the rigors of projection and handling and to remain pliable and colorless.

Guillotine splicers are tape splicers that have machined die-set posts that guide the cutting blade on the frame line. Guillotine splicers also have a diagonal blade for use in editing magnetic stock. When two ends are placed end to end a Mylar splicing tape is stretched across it. With a single stroke of the guillotine handle, the tape is perforated and cut to conform with the film. Guillotine model M.2-16mm-2T costs $225. A professional super 8mm version (Model S1) costs $275. Guillotine also manufactures a plastic model for super 8mm filmmaking. It's called the Semi-Pro and costs $20. Rivas 16mm tape splicers perform the same function as Guillotine splicers but use a perforated splicing tape. The Rivas model is preferred by some editors; cost is about $200.

EDITING SUPPLIES

- ***Trim bins*** are used to hold 16mm film that has been cut and awaits splicing.
- ***Reels*** come in sizes of 400 feet, 800 feet, and 1,200 feet, and are made of either steel or plastic, although steel is recommended.
- ***Plastic cores*** are used to edit on flatbed machines. Get a *split reel* for projection, which will fit around a core, holding the film in place while it is being projected.
- ***Film leader*** for super 8mm and 16mm formats should be black-and-white. Sixteen millimeter leader comes in single or double-perforated forms. The *single perf* is recommended. You may also want to purchase *academy leader,* which has that 3-2-1 countdown you've seen all your life. Don't forget to purchase some clear leader if you are going to try drawing onto film—see Chapter 2.
- ***Grease pencils*** come in white, red, and yellow and are used for putting marks onto the celluloid to indicate cut points.
- ***Sharpie fine-point felt-tip pens*** are permanent markers used to label white leader.
- ***Editing gloves*** are worn when cutting negative and doing Michael Jackson impressions.
- ***Lupe magnifier*** is a model with 8X magnification that is useful for eyeballing the surface and frames of both 16mm and super 8mm footage.
- ***Splicing tape*** is something you can never have enough of. Be sure to get extra roles for whichever splicing system you use.

One of the simplest tools in filmmaking is a 16mm *splicing block.* Ediquip's model, which costs $20, is used with a single-edged razor blade and with perforated splicing tape or precut perforated tabs such as Kodak Press-Tapes. The latter, while expensive, are fast and easy ways to make a good butt-end splice with 16mm or super 8mm gauge.

GENERAL PROJECTION REQUIREMENTS

Film projectors are much alike. They project and rewind movie film from metal or plastic reels. Many of them feature *automatic threading systems.* In 16mm sound projectors there is an *excitor lamp* and an *amplification/speaker system.* In super 8 sound projectors there is a *record/playback magnetic head,* plus amplifier, speaker, and input receptacles.

An important feature of any projector is the loving care it provides to film. All projectors claim to be safe, but some aren't—particularly the self-threading mechanism. *Beware the auto-shreds.* Ask around before you purchase a projector.

Zoom lenses. These are only occasionally helpful. It's a feature built into many super 8mm projectors. With 16mm projectors you select a specific lens to provide the width of *throw* you'll require.

Variable projection speed. This is a very, very useful feature for animators. Slower-than-normal speeds allow the study of animated movements in detail, as if the film had been shot at six or seven frames at one time instead of the usual two.

Sound. As discussed elsewhere, many of today's super 8 projectors have the capability to record and play back an audio track that is placed on one edge of the super 8 film. Other projectors have outlets that allow them to be synced up with external audio recording and playback devices. There are many, many special features and options and specifications to consider when purchasing a sound projector. Consult a dealer or a technical manual for more discussion and guidance.

FILM STOCKS

Although many, many different film stocks are manufactured for 16mm and super 8mm filmmaking, only a few of these are used regularly for animation. This section introduces them to you. As with everything else in animated filmmaking, there's no end to what you can know about the nature and qualities of different motion picture films—or about the processes of developing and printing.

Reversal film is developed so that the *camera original* provides a series of *positive* images that can be projected. All super 8mm film is reversal stock—what goes through the camera is what comes back from the laboratory, ready to be screened. In still photography, a familiar example of reversal film is the ordinary color slide. The film stock that enters the camera unexposed is the same object that is subsequently projected.

Negative film is developed into a negative image in terms of the object that has been photographed—light areas appear darkest on the negative film, etc. When developed, negative film *must be printed* onto another negative film in order to create a projectable series of images. Thus, with negative film the animator always has a "negative"—an unprojectable camera original that has to be reprinted in order to be seen. (Almost all black-and-white still photography involves use of negative stocks that are developed and then printed onto negative papers.)

The overall visual characteristics of reversal films are generally considered to be better than negative films. Prints can be made easily from either kind of camera original. In the printing process, however, negative film yields a greater ability to change the exposure levels. But it is easier to handle and print reversal stocks and, on balance, many animators prefer reversal to negative.

There are, of course, *color* and *black-and-white* stocks. The characteristics of emulsion that differentiate films in either category are classified under these headings:

Contrast refers to the separation of lightness and darkness (called tones) in a film or a print. Adjectives like "soft" and "hard" or "flat" and "contrasty" are used to describe variations of contrast. Note that there are special *high-contrast* stocks that polarize the image toward the extremes of black and white in

22.10 Projectors: The illustrations show the Kodak S-8 Ektagraphic projector, one of the best ever made, *(A)*; the Kodak Instamatic M-100A super 8 projector, *(B)*; the Bolex SM80 Electronic super 8 sound projector, *(C)*. All the manufacturers of 16mm and super 8mm cameras also produce projectors. Used projectors are a good value.

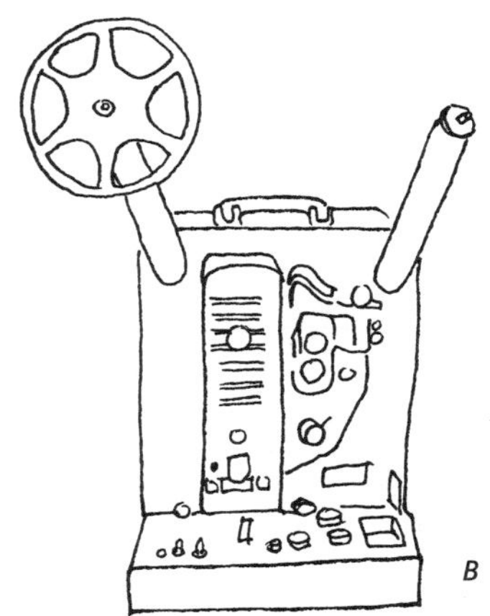

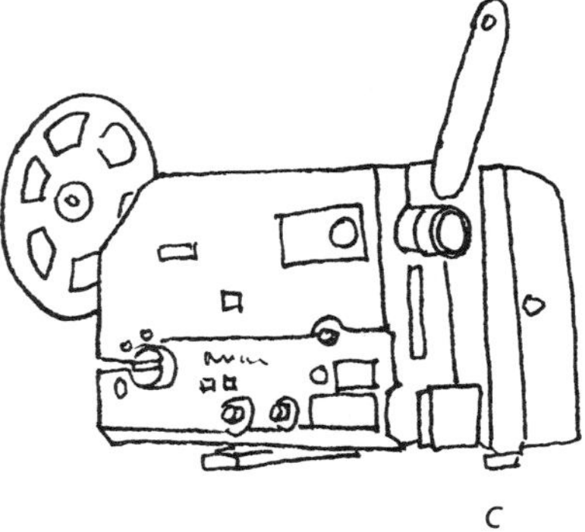

ART SUPPLIES

- ***Animation bond paper***—10¼ by 12½ inches (#12 field size), punched for either Oxberry or Acme pegging: 500 sheets, $30; 5,000 sheets, about $225.
- ***Animation bond paper***—10¼ by 25 inches, 250 sheets, $25.
- ***Animation bond paper***—10¼ by 12½ inches, unpunched, 500 sheets, $25; 5,000 sheets, $250.
- ***Background paper***—2-ply weight, 10¼ by 14 inches, white vellum finished, punched for Oxberry or Acme, 50 cents each.
- ***Tracing vellum***—10¼ by 12½ inches, punched for either Oxberry or Acme, 100 sheets, $10.
- ***Flip-books***—white pads, 5 by 7 inches, 50 pages. Less than 50 cents each, the cost varies with the number purchased.
- ***Color-Aid papers***—These special papers are used because of their adaptability and quality. Each is silk-screened with flat matte color that doesn't bleed. Color-Aid papers come in 210 coordinated colors (24 basic hues with 5 tints and 3 shades each, plus 16 grays and black and white). Single sheets, 18 by 24 inches, 50 cents; 24 by 36 inches, $1. Color-Aid swatch book (210 colors), 5 by 7 inches, $15. *Note:* Quality colored papers are also manufactured by Color Vu, Pantone, Tru-Tone, etc.
- ***Construction paper***—inexpensive, assorted colors, 50-sheet packages; 12 by 18 inches and 18 by 24 inches, under $5.
- ***Day-Glo papers***—come in assorted colors and sizes.
- ***Background color cels***—#12 field size, punched, pastel colors and black; 50, $15.
- ***Background bristol paper***—#12 field, punched; 65 cents each.

the tonal range. These high-contrast stocks (available in reversal or negative 16mm film) are useful to animators in filming white paper sheets and many bottomlighting techniques.

Speed of a film stock refers to the inherent sensitivity of an emulsion under specified conditions of exposure and development. Film stocks are constantly being developed that have higher speeds—that is, they require less light to achieve a good image. Such stocks, while very important to live-action filmmakers (especially cinema verité documentarists), are relatively unimportant to animators, who have great control over lighting conditions. A more significant factor for the animator is the sharpness and granularity that is related to film speed. The slowest color emulsions are the sharpest and have the finest grain.

Graininess or *granularity* is a quality of the image's physical/chemical structure. An individual frame of film seen via projection or through magnification will possess a grainy structure. The effect of grain—its perceived value—is largely an aesthetic choice. But granularity is directly related to *photographic sharpness*—the ability of an emulsion to record fine detail distinguishably. Most animators prefer a sharp, highly defined, low-grained image.

Exposure index is a scale that provides a number that is used with photoelectric meters to help determine the correct exposure of a particular stock. Index ratings are given in *ASA numbers*—that appear on every film container. There are two ratings for each stock—one for use of the film outdoors (daylight) and one for use under artificial lights (tungsten). Animators prefer a low index number for overall emulsion quality.

Color saturation and *spectral sensitivity* are terms that refer to color reproduction characteristics. Different stocks will yield slightly different colors—some are "warmer" (emphasize the red/yellow end of the spectrum) and some are "cooler" (blue/green are emphasized). The reason for this is that color dyes (cyan, magenta, and yellow), which make up the emulsion, will respond differently to variations of light wavelengths. Also, different manufacturers of film stocks make different judgments as to what constitutes a pure shade of, say, red. Color specification, it turns out, is a highly subjective judgment. And so also are variations of "richness," of reproduced colors vis-à-vis the original colors being photographed. This latter quality is called *saturation*. Underexposure tends to encourage "rich" saturation.

THE MOTION PICTURE LABORATORY

Listen to your mother: Be nice to people who can help you. Don't forget to say thank you. I'm speaking from experience. You never know when you'll need help. Don't worry but be careful. Remember to wear a nice smile. And remember, many a live wire would be a dead end without its connections. Believe me. I'm your mother and I know.

Place the staff of your local film lab at the top of your Maintain Friendly Relations list. The expertise, cooperation, and pride of craft of the film lab combines to form a silent partner for the animator in any creative production.

Although the standard lab services (described below) ought to be enough to convince you that Mother's advice is sound, it is only when you encounter a real problem—a crisis—that you'll appreciate the invaluable counsel and irreplaceable resources that a first-rate lab provides. Next to close relatives, you can count on lab people.

Along with the "normal" services catalogued here, the lab can deal a bit of wizardry that you should know about. The lab can push or flash your raw stock. It can color-correct. It can blow up, reduce, and otherwise ready a finished film for sale and distribution. The lab can repair damaged footage (the process is called rejuvenation). Labs can help you hide cel scratches, clean dirty film, and in other ways help your movie look as good as possible.

Best of all, the lab's highly trained and thoroughly professional staff can give you the advice, encouragement, and direct support that comes from years of experience with every imaginable film production problem. The lab can troubleshoot for you, direct you to specialists and specialized services, save you time and money and heartache. So, as with a cup of hot chicken soup, equate a good lab with nourishment, friendliness, and survival.

SUPER 8MM LAB SERVICES

Although super 8mm animators generally tend to deal with a local camera shop for purchasing and processing super 8mm film cartridges, it's important to know that you can go directly to labs that process the film. You just need to find out

CELS

- ***Clear acetate animation cels***—punched for either Oxberry or Acme, .005 mil. thick, 10¼ by 12½ inches (for a #12 field); 100 sheets, $30; 500 sheets, $150.
- ***Clear acetate animation cels***—unpunched, 10¼ by 12½ inches, 250 sheets, $75; 1,000 sheets, $270. Wider sheets are also available.

IMPLEMENTS AND COLORS

- ***Animation paints.*** Professional cel paints are specially fabricated for use on acetate cels, vinyl, foil, glass, illustration board, watercolor paper, and Color-Aid paper. They can be used with an airbrush, are thinned with water, and will not chip, crack, or flake. There are approximately fifty standard colors plus thirty shades of gray. All these colors are available with cel-level compensation.
- ***Black cel ink.*** This ink is opaque, waterproof, free-flowing, and acetate-adhering; 2 ounces, $3.50; 4 ounces, $6.
- ***Inking pens.*** Traditional inking pens are composed of a simple hardwood holder and steel pen points. The holders cost $1.50 each and the pen points (medium, fine, and flexible), 75 cents each. Some animators like to use Rapidograph pens that have interchangeable points. Pens, $20; replacement points, approximately $13 each.
- ***Strip chart.*** A set of acetate sheets showing 44 basic colors, 32 grays, and more than 600 compensating tint colors. A strip chart is used in selecting and ordering animation cel paints. Price: $30.
- ***Opaquing brushes.*** A crow-quill brush costs under $1. Sable watercolor brushes with round points are available in 16 sizes and are priced from $4 to $50.
- ***Poster (Tempera) colors.*** These opaque paints come in jars, are fully intermixable, and dry to a flat, brilliant surface. They can be used with brush, pen, or airbrush and are available in boxed assortments of colors or in individual bottles.

what labs there are, what services they provide, and at what cost.

Here are some of the special services you may want to find out about:

- ***Push processing***—footage shot under very low lighting conditions can be specially processed as if it had a higher ASA setting
- ***Enlargement***—to 16mm format
- ***Contact printing***—to strike a print from the camera original
- ***Opticals***—see list under 16mm services, below
- ***Sound striping***—after an 8mm film has been edited

16MM LAB SERVICES

Filmmakers working in 16mm use their local lab for many different activities and services.

- ***Forced developing***—the same as push processing, but costs a few extra cents per foot
- ***Reduction***—if you have 35mm and need to reduce it to 16mm

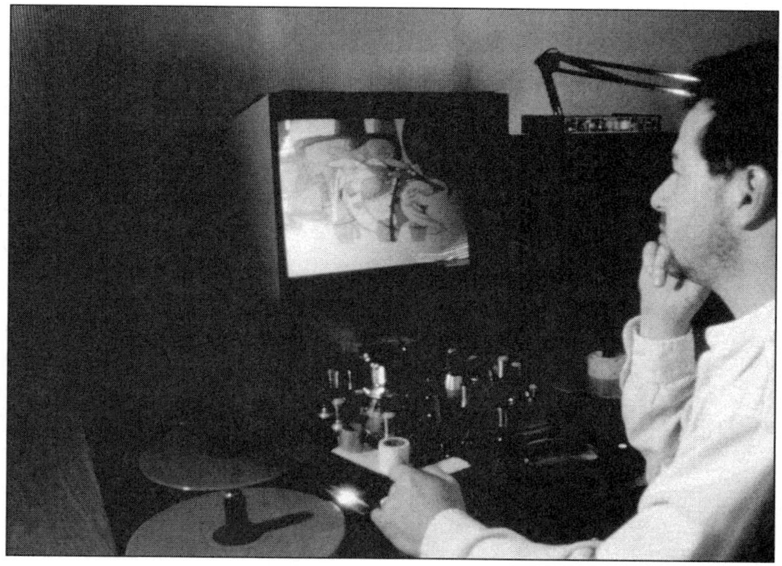

- ***Contact printing***—to get ***work prints*** and, at the very end, ***release prints***
- ***Printing of A-and-B rolls***—after the animation has been edited, the filmmaker carefully cuts the camera original, which is laid out in a special way so that, in the printing process, the splices are masked and don't show in the final film. This special layout is called A-and-B rolling. Consult a filmmaking book to see how it is done. Your lab will print from the A-and-B roll to get an . . .
- ***Answer print***—the movie's first version, which often requires reprinting so that color and density of various shots will all match together. When the film is just right, the lab will make you one or more . . .
- ***Release prints***—if you are doing a number of these, you will want the lab to make you an ***internegative.***
- ***Opticals***—done during the A-and-B roll phase. Some require special work on the ***optical bench.*** Costs are charged by the foot. The most common optical effects include cropping (frame line or image repositioning), freeze-frame, reverse action, matte wipes, lap dissolve, superimposition, and step printing.
- ***Sound transfer***—many labs provide the service of transferring audio from cassette, DAT, or quarter-inch format to 16mm magnetic recording stock.
- ***Optical sound master***—similarly, labs will take your mixed 16mm mag stock master—mixed at a recording house—and turn it into an optical printing master that is used with the A-and-B rolls or the internegative to create the final 16mm films, ready to be played on any projector in the known universe.

22.11 Six Plate Flat Bed: That's how this Steinbeck 16mm film editing machine is known in filmmaking parlance. *Photo courtesy Nickelodeon.*

Miscellaneous Materials

- ***Storyboard pads.*** Storyboard layout pads are specially designed as "work-up" aids in visualizing a sequence or entire movie. Each frame (drawn within the format of a television screen) has space beneath it for audio information (dialogue/music/effects) or for script notations. Storyboard layout pads come in a variety of different sizes and sets of frames. Two options: 8- by 18-inch pad with 4 sections per sheet (75 sheets, $5) and 19- by 21-inch pad with 12 sections per sheet (50 sheets, $10).
- ***Reinforcements.*** Double-ply bristol reinforcements are used to strengthen damaged peg holes or to protect heavily used cels, backgrounds, or papers. These are available in strips or in self-sticking single peg-hole reinforcements. The strips cost about a dime. (Specify Oxberry or Acme when you order.)
- ***Peg hole strips.*** Bristol paper strips (.009 inch thick), 12½ by 18 inches, are available in Acme and Oxberry punch systems. These are taped onto undersized cels, paper, or background paintings. Also used for changing animation, overlays, or backgrounds and to convert any artwork to standard pegging registration. Price: 100-peg strips, around $6.
- ***Cotton gloves.*** White cotton gloves are worn in cel animation in order to protect cels from the oil of hands. Ink and paint won't adhere to an oily surface. Price: small and large sizes, 80 cents a pair, $8 a dozen.
- ***Additional art materials*** that are often required for animation include: masking tape, pushpins, scissors, matte knife, single-edged razor blades, rubber cement with applicator and thinner, Wite-out, white tape, Scotch tape, gaffer's tape, pencils, sharpeners, kneaded and pink-Pearl-type erasers, and grease pencils. Experiment with other materials, for through them you can discover a new technique well suited to your drawing style.

23 ■ Computer Software

WRITTEN WITH GEORGE ESCOBAR

Whether your software is store-bought, downloaded, or hand-delivered by FedEx, you act quickly when you get it home. There's something about opening new computer software that fills you with anticipation, be it floppy disks, a shiny CD-ROM, or a big file that magically appears on your computer desktop. A manual accompanies it, which after a year remains begging to be read. And there's a registration card with serial numbers, and sometimes a coupon for software add-ons. But you ignore this paraphernalia, staying focused on the promise of amazing new tools.

Double-click on an unfamiliar yet promising icon, and a splashy screen heralds the successful launch of the newest "Great Software." A new blank page appears with scroll bars and a neat row of buttons. You're *in*! A new interface looks at you. You stare back at it, catching your breath as you hope that *this* software is *the one*. You fantasize about a dazzling array of graphic gadgets and fresh techniques within its bits and bytes that will unleash your creative powers. If you've

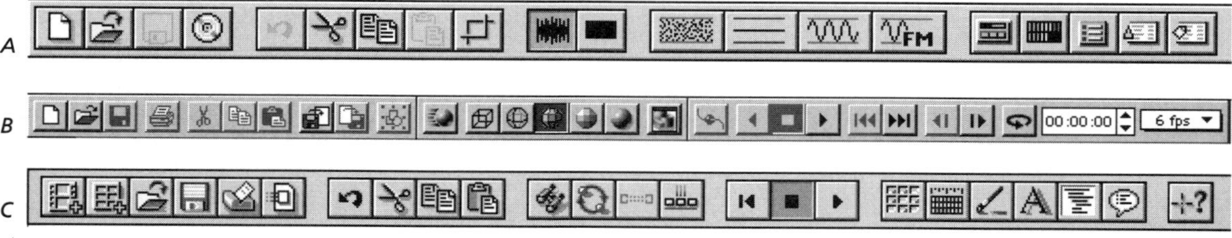

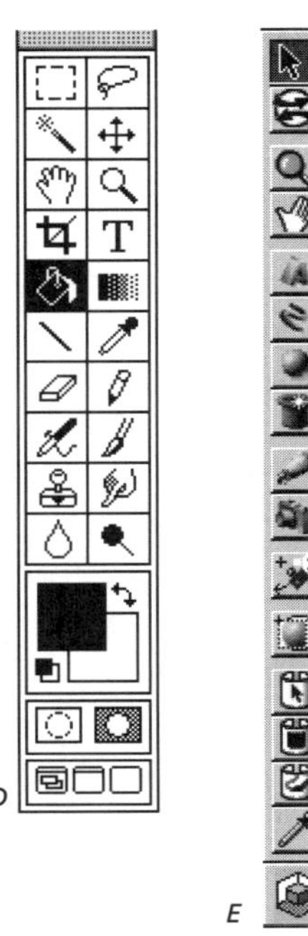

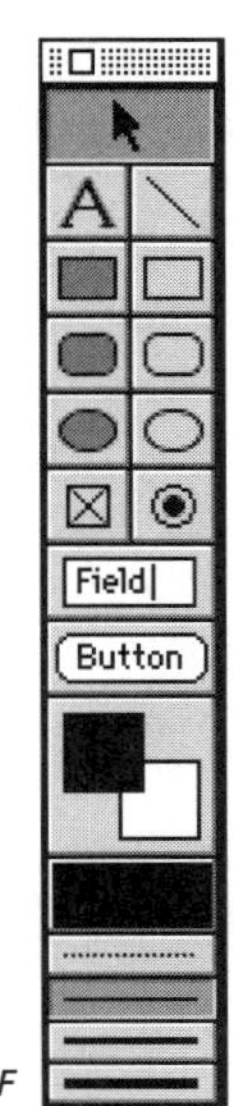

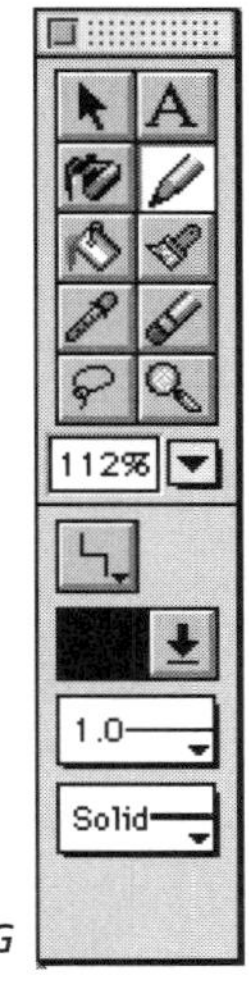

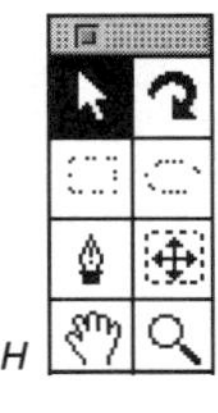

23.1 Interface anatomy: Software programs use many standard icons to perform similar functions: an arrow to point and click; a pencil for freehand drawing; the letter *A* or *T* for entering text; a lasso for selecting odd-shaped objects; a magnifying glass for zooming in or out; a hand to grab and reposition objects; a paint bucket to pour color; a folder to open files; and squares or circles to draw basic shapes. Unique icons are also used for proprietary functions such as manipulating sound waves or reshaping 3-D objects. Sometimes, as with Premiere, words are used to evoke the commands. Here are some representative toolbars and toolboxes from different, widely used programs. Together they are known as GUIs, or Graphic User Interfaces. *Courtesy George Escobar.*

A. SoundEdit 16 (Audio)
B. Ray Dream Studio (3-D)
C. Director (Interactive/Multimedia)
D. Photoshop (Image Processing)
E. Ray Dream Studio (3-D)
F. Director (Interactive/Multimedia)
G. Flash (2-D)
H. After Effects (Compositing)
I. Premiere (Video Editing)
J. Painter (Image Making)

already been doing digital animation, you're eager for potent cures for chronic software "bugs" and a tenfold productivity increase as you use this new software to become a world-class animation factory. If you're just starting out in digital animation, the excitement is even more intense!

Such is the promise of new computer software. Does it deliver? Sometimes. More often than not, the perfect animation software remains just beyond reach and into the next release. Sometimes a new version of a familiar application does indeed have new features, but you find you don't really need or use them after all. So what's going on? It is a game called perpetual upgrade, designed to extract the maximum amount of (your) dollars for a minimal supply of (their) features.

THE RULES OF THE GAME

It's a fun game. And it's one that you don't even want to win because the software that collects inside your computer, like the creativity and expertise that collects inside your brain, should be an ever-expanding universe of possibilities.

But. But. But. As everybody knows, if you are going to

From this:

Commands
Import Multiple... F5
Import Folder... F6
New Title F9
Movie Capture F10
Batch Capture F11
Movie Analysis F13
Add Folder...
Audio Snap
Ripple Delete
Edit Line Insert F7
Edit Line Overlay F8
Mark In F14
Mark Out F15

. . . an "interface editor" creates a two-column menu along with a key command now in gray.

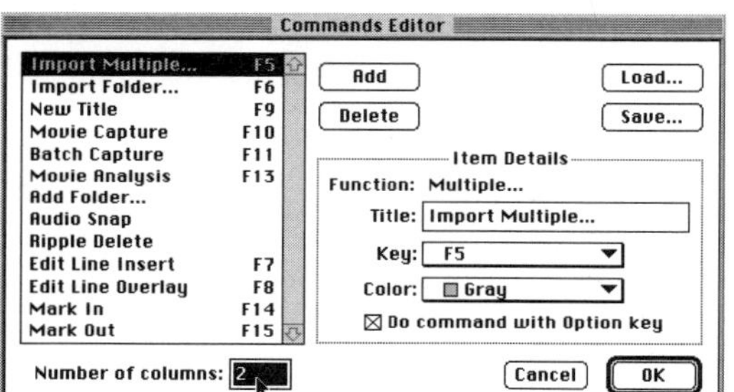

Commands
Import Multiple... F5 | Audio Snap
Import Folder... F6 | Ripple Delete
New Title F9 | Edit Line Insert F7
Movie Capture F10 | Edit Line Overlay F8
Batch Capture F11 | Mark In F14
Movie Analysis F13 | Mark Out F15
Add Folder...

23.2 Custom interface: Some software applications allow you to easily change or edit the tool set or set of commands to suit your style of working or the needs of a particular project. These settings can be saved and chosen again later, making it easier to use the software and leaving the screen less cluttered. For popular programs like Premiere, settings files others have found useful can even be found in the developer's Web site, user groups, or in how-to books. *Courtesy George Escobar.*

play the game, you gotta know the rules. That is what this chapter is all about. It takes on the nearly impossible task of explaining what every animator needs to know about computer software. As you will see, the focus holds tight on information that is relevant to animation—and only animation. If you are looking for a wide orientation to computers in general, you will need to buy one of many good books that talk about the different platforms and operating systems. If you are looking for a narrower focus on a particular piece of software, you will need to work your way through the manual and tutorials that come with the particular *application* (or *program,* or *package*—the words are used interchangeably) or the array of books on the market devoted to each application. If you are looking for close comparisons of one program to another, then you will need to consult one of the excellent magazines that serve the Windows PC and Macintosh platforms for a bench test that breaks out each and every feature and function.

But if you are an animator who wants to know a little something about the full spectrum of digital software tools, you're in the right place. Read on.

WHAT SOFTWARE IS

What is software, anyway? Regardless of what type of computer you use, for the most part you live and interact with what's on the screen, and that's the software. The components of software are fairly straightforward. It is the way these qualities are mixed and served up that determines the software's usefulness and makes one product better than its competitors. If you can acquire a fundamental understanding of seven essential ingredients, you will have a good foundation for understanding what is really central to an application and what is less important—that is, bells and whistles.

User Interface. There are two ways to look at this. *User interface* is what you see on the computer screen (buttons, text, menus, operational devices, etc.) and how the computer requires you to manipulate these things. Once you've learned the conventions of an interface, the exchange between what you see and how you interact becomes almost automatic. The other way to think of a user interface is more poetic: It is the dance in which your hands and mind are partnered with a computer's computational power. Sometimes you lead and sometimes you follow. The interface is where you hold your partner. The level of innovation and creativity that comes into your work is a measure of how well you dance together.

Either way, all software programs operate via an *interaction model,* which develops as a result of how these on-screen elements are packaged and organized. Together the user interface and its interaction model enable users to access the software's functionality. Over the past few years, a set of user-interface conventions has been adopted by numerous software developers. These include commonly shared icons that represent tools, actions, and functions. It also includes standard formats and navigational controls.

Adoption of common conventions has been both good and bad for users. It is good to have a familiar set of tools that behave exactly the same way regardless of the software product. Imagine the nightmare of having to drive a car if the accelerator and brakes functioned differently in each make of auto. The bad side is that at some point the interface of every software behaves in an awful, convoluted way in order to adhere to convention.

Functionality. Functions (also called *features*) define the software's usefulness and provide a checklist of why you might wish to buy it. The overall feature set and user interface become the basis for how the software is marketed and to whom.

One function that has become a favorite among artists, designers, videographers, and animators is the ability to customize the user interface. This *personalization* of software will be a growing trend as software engineers provide better and better methods for "opening up" their software by making the programs extensible.

Another set of functions common to graphics and multimedia software are *file translators.* These allow a person working in one software program to work on artwork or movies created by another software application. Such translators are ways for different programs to both import and export different types of file formats that can then be used somewhere else in the

23.3 Toon tech specs: This box lists minimum Mac and PC requirements for using all the software applications discussed in this chapter. *Courtesy George Escobar.*

	MAC	PC
CPU	603e - 200 MHz	Pentium 200 MHz
OS	Mac OS 7 or 8	Windows 95 or NT
RAM	32 MB	32 MB
Hard Drive	2 GB	2 GB
Extensions	QT, QuickDraw 3D	QT, QuickDraw 3D

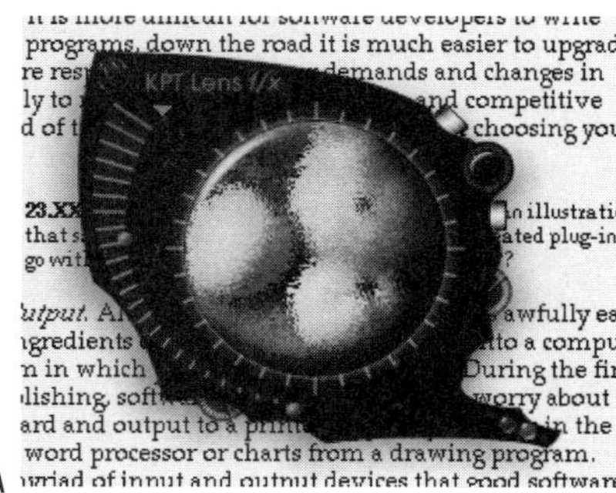

A

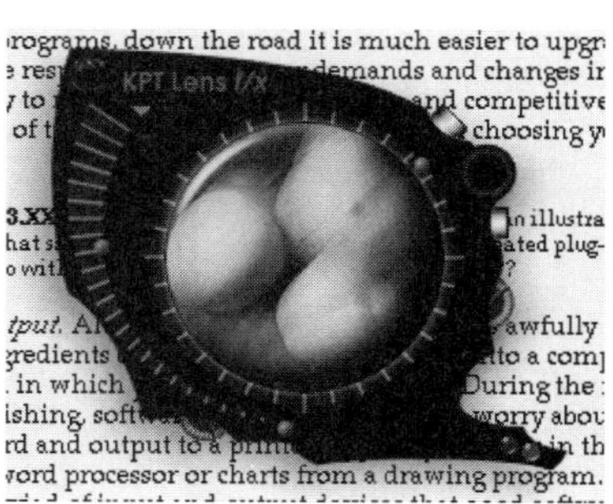

B

23.4 Kai the plug-in king: It's easy to see why Kai Krause, the inventor and genius behind *Kai's PowerTools,* has made a significant mark in the software landscape. His easy-to-use, intuitive, and powerful plug-in tools make it downright fun to manipulate images. Almost any serious artist and certainly every professional one uses Kai's tools. With them, relatively "staid" applications such as Photoshop or Painter are given power and functionality heretofore unimagined by the original developers. By using plug-ins, you can keep your applications strong and fresh without having to upgrade them to the next version. These are some samples of how images can be altered using this software. *(A)* and *(B)* give the illusion of using a physical lens to modify an object. *(C)* curls a page corner to simulate a page turn. *(D)* adds a "perspective plane" to flat services. *(E)* wraps images into a sphere and lets you modify it, and *(F)* distorts the image by creating an artistic twirl. *Courtesy George Escobar.*

creative process and are transparent to users. Do you remember the "universal translator" that the characters in *Star Trek* used to communicate with aliens? It effortlessly translated the intergalactic languages into English and vice versa. In the world of computer data, this cool piece of science fiction has indeed become fact.

An old typographical function that's becoming more popular in illustration software is the ability to convert text shapes into artwork. This is done by "tracing" the outlines of the text and adding control points that pass through spline curves so that the converted text can be manipulated as if it were a piece of drawn artwork. This function has helped to spread the use of text as a key element in animation, especially in sophisticated television commercials where word forms and logos become unique characters, not just labels.

Extensibility. Increasingly, software upgrades in the form of plug-ins or extensions to the original program are being produced by both the original manufacturer and third parties. You add plug-ins to gain new functionality, therefore extending the usefulness of the software. *Extensible software* includes those applications with built-in facilities allowing other pieces of software code (the plug-ins) to "hook" into the main software product and usually to show up right inside your program. In effect, the software adopts the new code and smoothly integrates it into its own. While it is more difficult for software developers to write extensible software programs, down the road it will be much easier to upgrade them so that they are responsive to both user demands and changes in hardware, thus making them more productive and competitive over a longer period of time. Look for this capability when choosing your software.

Input and Output. Although it may seem obvious, it's awfully easy to forget the dual ingredients of how you get information into a computer (input) and the form in which you can get it out (output). During the first days of desktop publishing, software engineers only had to worry about input from a keyboard and output to a printer. Typically this was in the form of text from a word processor or charts from a drawing program. Today there are a myriad of input and output devices that good software must support, especially for animation. For *input,* these include scanners, digital pens and tablets, digital still and video cameras, infrared remote controls, faxes, modems, and microphones. For *output,* the printer is still paramount, but close behind are VCRs and digital encoders; more distant are recordable CD-ROMs, film recorders, and DAT machines. The next chapter discusses these hardware tools.

Cross-platform Support. Learning to use multiple computer platforms (Mac or PC) is hard enough. Imagine how confusing it would be if the same software (Photoshop, for example) behaved differently for each type of machine it ran on. Fortunately, sanity has prevailed in the software industry instead of expediency. Almost all major software applications made today are *cross-platform.* Try to stick with these. The data or files that were created in an application using the Mac platform will run on the version of the application created for

C

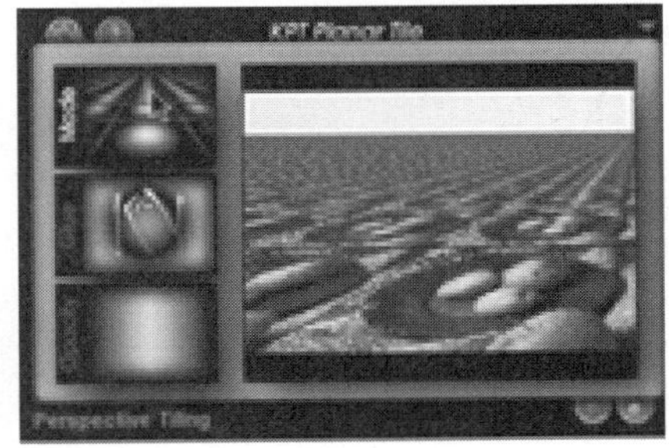

D

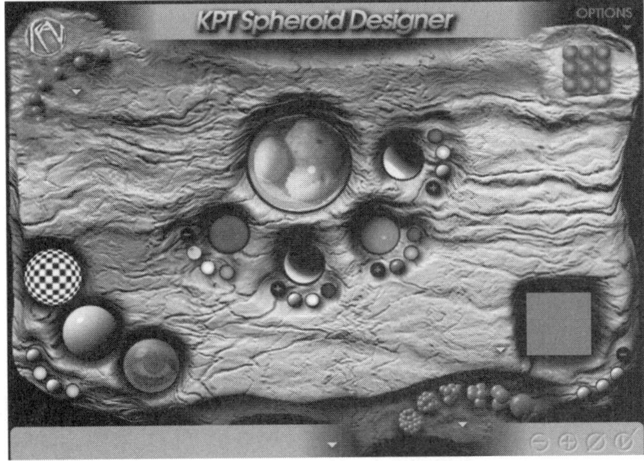

E

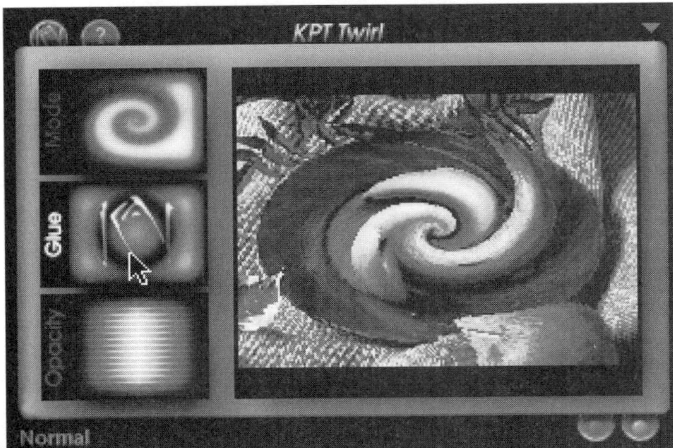

F

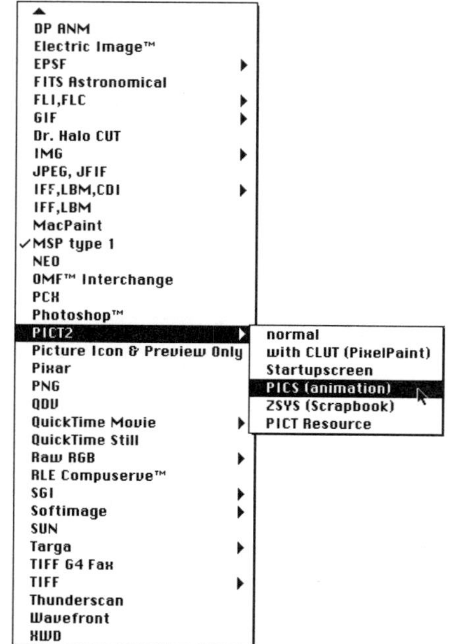

23.5 DeBabelizer: If Kai is the king of plug-ins, then DeBabelizer is the queen of file formats. Like Will Rogers, DeBabelizer "never met a format it didn't like." In fact, every format ever invented—whether it's arcane (like FITS Astronomical) or wildly popular (like GIF)—is supported by DeBabelizer. *(A)* shows the list of supported formats that comes with the software, and other formats not included can be purchased as separate options. With DeBabelizer you can open a file or a set of files in its native format, change its dimensions, resolution, color, palette, and convert it to another file format. Using its automatic batch-mode setting, *(B)*, DeBabelizer can take the tedium out of converting dozens, hundreds, or thousands of files by allowing you to select a list of images to convert and process in a batch. *(C)* shows the program's toolbar. The only bad news is DeBabelizer's horribly confusing user interface. Even with the manual, using DeBabelizer is more or less a trial-and-error affair. *Courtesy George Escobar.*

the PC platform. Note that the software program itself is not cross-platform. If you purchase a Windows 95 version, don't expect it to work on a friend's Macintosh. For an introductory discussion about the two platforms that comprise the desktop animation world and the one platform that is used for professional-level tools, see the next chapter.

Efficiency. For users there is no way to test the efficiency of software except by actually working it or by reading magazine reviews. If it feels sluggish or slow (compared to similar software or its predecessor version) then you know the application is bloated—not lean and mean. If you experience frequent crashes, screen freezes, or system errors, then you know the software is "buggy." Sometimes running the software on a fast computer (with more CPU horsepower) alleviates the sluggishness of the software, but that's an expensive cure.

Minimum Hardware Requirements. All of today's animation software has a minimum set of requirements for it to run. You will see these appearing as a list on the outside of the software package or listed inside the user's guide and manual. Be sure that the desktop computer system you are operating at least matches the minimum conditions requested by a particular application. And know that "minimum" does not always mean it will run well in everyday, real-world situations.

Figure 23.3 provides what might be called "'toon tech specs"—a generic chart of minimum hardware specifications for digital animation. We've checked the current versions of all the software packages discussed in this chapter and kluged one set of requirements that will fit all the software. Don't despair if the configuration of your hardware setup is less than listed here. Most of the applications will operate on less powerful and less memory-enriched computers. You just have to be more resourceful and more patient.

FILE FORMATS

In the early days of personal computing, software developers for the PC and Macintosh sought to gain advantage over their competitors by using proprietary file formats. In effect, a file created in one application would not work in another. This is similar to the Beta-versus-VHS videotape format wars. The hardware vendors also played their part in promoting their favorite

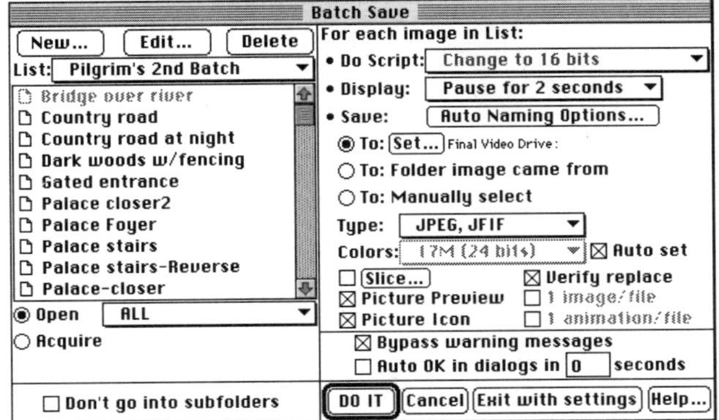

B

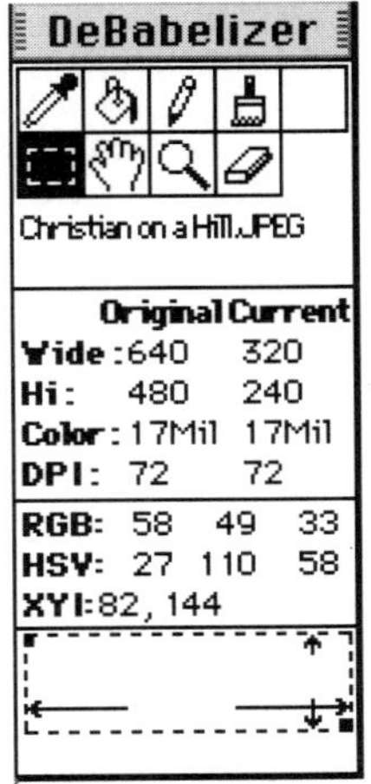

C

file format. Fortunately this interplatform warfare did not last long, because users demanded compatibility. Since no one wanted to concede defeat, the developers decided to "support" other file formats besides their own. What has emerged are numerous basic file formats that animators need to care about. They are PICT, TIFF, GIF, BMP, EPS, JPEG, MoV, M-JPEG, AVI, and DXF. The first six deal mostly with still images, while MoV, AVI, and M-JPEG handle digital video, and DXF is for 3-D models. For audio, the common file formats include AIFF, WAV, SND, MDI, and AU. Each computer platform has its own preferred file format, and so do different image and audio editing programs. Happily, most programs can read different file formats and convert them to the one you need.

New formats continue to emerge, especially for the Internet. These include HTML, VRML, RA, MPEG, Shockwave, and QuickTimeVR.

The function of a file format is twofold: first, it is to fully exploit the capabilities of the application software, allowing the artwork or digital video to retain its exact form when it is saved and subsequently reopened, converted for use in another program, or shared with other users. Second, some file formats serve to *optimize* the saved file by reducing its disk size and allowing it to open more quickly in memory.

Check any software you are considering to make sure it supports as many standard file formats as possible. As a final resort, there are dedicated file format programs, such as *DeBabelizer* from Equilibrium, which convert one file format into another. For most artists and animators DeBabelizer has become an essential tool because of its flexibility and powerful

23.6 Color maps: Imagine looking at a color cartoon using only fifteen colors for its palette. The cartoon is of an egg about to hatch in the middle of a farmyard. The egg is colored white. There is black outlining all of the artwork. There's a mother hen, also white, but with an orange beak and red crown on her head. The nest uses four shades of brown as it rests on three shades of green grass. Behind the scene is a blue sky with a yellow sun peeking out behind white and gray clouds.

Now let's see how the computer looks at this scene in the raw, without compression. First, the cartoon's size is 640 x 480 pixels for a total of 307,200 individual picture elements. Because we are using a 4-bit palette (to support up to 16 colors), we multiply 307,200 x 4 to get 1,228,800 bits of information, or 153K, which the computer must manipulate to display this one picture. That's a hefty load for a very simple picture.

When compression is used, the computer segments this scene according to color, dividing the scene into a color map (a *bit map*). Instead of counting each pixel individually, it can group them, assigning only a few color bits to describe the blue sky. For example, it looks at the starting position of the blue color, then at its ending position. These are called coordinates. Then it simply records this information: "From here to there, is the color blue." It repeats this process for all the other colors and now it has a color map of the cartoon. It's similar to making a map of your neighborhood, recording only the relevant information to show the major streets and landmarks. So instead of a million bits of information, the cartoon is compressed or reduced perhaps tenfold, to only 122,000 bits, or 15K. This simple type of compression is called "run-length encoding." There are many other types of compression techniques that use different and more complex mathematical formulas known as *algorithms*. These algorithms were developed to solve a multitude of large-file imaging challenges. *Drawing by Chris Allard.*

features, including batch conversions and a simple method for automating frequently used functions (see Figure 23.5).

MULTIMEDIA/INTERACTIVE MEDIA/NEW MEDIA

The term *multimedia* was coined in the mid-eighties to refer to a computer-based world where text, animation, audio, and video were all colliding in new formations. Of course, this convergence of media began way before the 1980s, but the word didn't catch on with users and the news media until more powerful and affordable personal computers made it easy to use mixed media. Later, as it became clear that the user was at the center of these new configurations of visual information, this world became known as *interactive media.* But that name implies limits to this burgeoning, interconnected sphere. Hence a new title has evolved for the whole shebang: *new media.*

Whatever it's called, the explosion of digital forms has made huge demands upon the basic data-crunching power of personal computers. Machines have needed to get faster, as you will see in the next chapter. The momentum of digital forms has also forced all forms of software to squeeze the raw data tighter and tighter. Hence your introduction to software must look at compression technology and then at the distribution formats that allow the animation files one has created to be easily duplicated and distributed.

COMPRESSION

The large size of digital images—both still and motion—makes compression a necessity for artists, videographers, and animators. By using compression technology, the fidelity of the original image file can be largely retained while reducing the need for expensive computer resources required to store and play back such files. In other words, the goal of compression technology is to re-create the original image as quickly and economically as possible.

Compression algorithms can be characterized as being either "lossy" or "lossless." Lossless algorithms preserve all of the original data and are generally preferred, particularly for still images. Lossy algorithms do not preserve the data exactly. Such compression formulas purposely lose some image data, which can never be recovered after the compression algorithm is applied. This latter approach attempts to compress the data as much as possible without decreasing the image quality in a *noticeable* way. Because of the nature of persistence of vision (our eyes tend to smush together individual frames into one fluid, comprehensible pattern), such lossy algorithms are acceptable in situations where the data is moving. The viewer's perception is tricked and doesn't notice the missing image data.

Software applications such as *Premiere* from Adobe (for video editing) support multiple *codecs* (compression/decompression algorithms), allowing you to choose the compressor best suited to your project. Here are some of Premiere's standard compression utilities:

Animation. The animation compression scheme keeps the best sound and image quality possible for a clip. However, the final file size is large, taking up lots of disk space. Consistent playback of files using this codec is not always guaranteed on every computer since it uses more data. If, however, the final destination is videotape or the image quality is important, then use this compressor.

Video. This compressor is designed to play back video footage at real-time rates if possible. However, the quality suffers if the image has large areas of continuous tone, like most cel animated films.

Graphics. The graphics compressor is limited to 256 colors and supplies a greater amount of compression than the

Original size at low compression

A

Image is clear and quite acceptable

Enlarged size at low compression

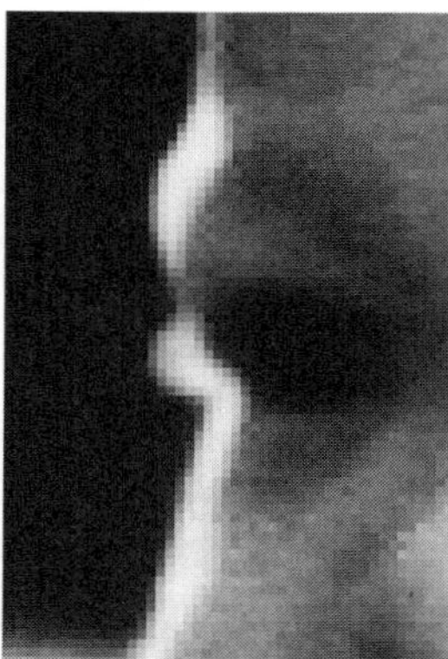

B

Gets blocky at edges

Enlarged size at medium compression

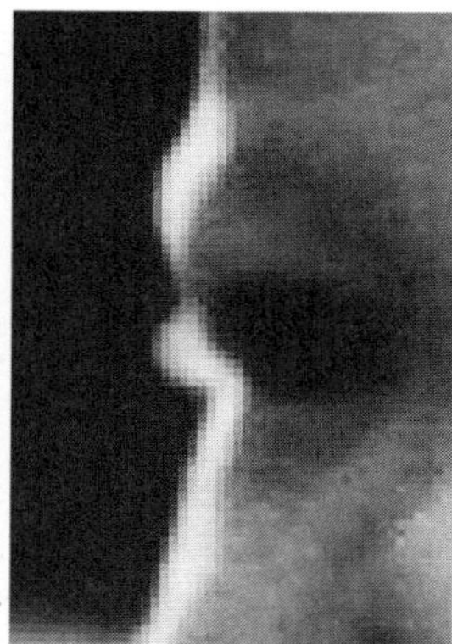

C

More blockiness

Enlarged size at high compression

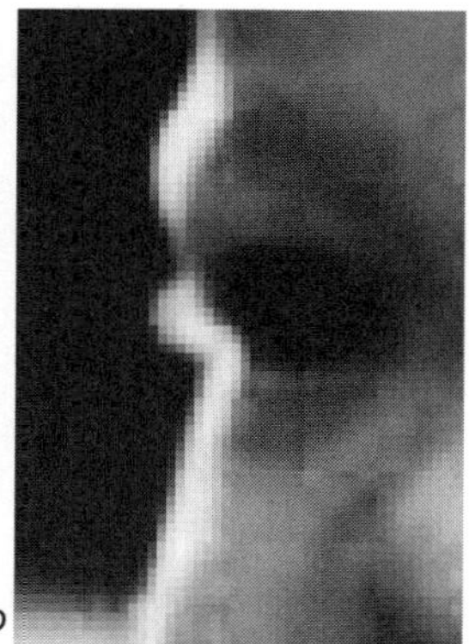

D

Blocky and loss of detail

23.7 Codecs at work: It's a good thing that compressed images are hard to detect when the image is small or normal-sized, *(A)*. When they are blown up, *(B)* and *(C)*, artifacts such as pixilation and stair-stepping (or jaggies) are clearly visible. By the time JPEG "high compression," *(D)*, is applied to this image, which was originally scanned at 72 dpi, the resulting picture can be noticeably blocky and undesirable. *Photo courtesy George Escobar.*

animation codec. Unfortunately, the playback isn't as fast, sometimes resulting in skipped frames.

JPEG. This is the international standard for compressing images. It can reduce files greatly, but playback is poor since it's designed for stills and not moving images.

M-JPEG. To accommodate moving images, JPEG was modified with a "motion" algorithm, hence the name *Motion-JPEG* or *M-JPEG.* Many video hardware boards use M-JPEG as its algorithm engine, but it needs powerful processors to deal with video. The advantage of M-JPEG is that it facilitates the editing of individual still images from a set of images that comprises a video sequence. It's also extremely fast and can compress and playback video in real time. The disadvantage is that it uses a relatively large amount of hard disk storage despite its use of compression.

MPEG. To deal with the disk storage problem of video, a new standard called MPEG was created MPEG works by discarding large amounts of redundant data from one image to another. Only the visual information between key frames are kept. It takes a lot more processing power to compress video into the MPEG format, but because the resulting movie file is so efficient, it can be played back using only software. Since the human eye cannot detect MPEG's removal of visual data, it has become the low-cost standard for the new DVD (digital videodisks) as well as the new generation of 18-inch digital satellite dishes.

Cinepak. This codec is often used for video clips targeted for CD-ROMs or distributed over the Internet. It provides smooth motion that plays back on almost any computer. The image quality is usually acceptable, though not the best. It provides a high amount of compression to save disk space, hence the reason it also takes longer to compress a clip.

23.8 QuickTime parts: QuickTime is a software add-on or extension to the operating system that provides powerful multimedia capabilities such as recording and playback of video. The other parts of QuickTime, including the PowerPlug, Musical Instruments, and MPEG are themselves extensions to augment the initial QuickTime component. This ability to plug-and-play new functionality to QuickTime has made QuickTime the industry standard for audio and video. *Courtesy George Escobar.*

QUICKTIME

QuickTime is an extension to a computer's operating system, making it possible to view and edit video, animation, music, and even text. It was invented by Apple to create an "architecture" for multimedia that includes compression, user-interface, and file-format standards. This unifying architecture has enabled multimedia to grow because it provides developers, producers, and audiences with an easy-to-use technology that is completely cross-platform—it works on PCs and Macs.

Today all computers come equipped with QuickTime extensions. It is the de facto standard for multimedia and might well be an element in your basic operating system. Should you find you don't have a QuickTime program, you can get one free of charge by downloading it from the Apple Web site (see Chapter 25: Resources). Most graphics programs and CD-ROMs will also provide you with a free copy of the latest QuickTime version, which is automatically mounted into your Systems Folder (on a Mac) or Resources (on a PC) when you launch the new software for the first time.

Technically, QuickTime supports two basic types of files: pictures and movies, which may also contain audio information. Picture files contain still images, and movie files support multiple images that are linked to time-based data. Within a movie, QuickTime supports various media files, including video, sprites, audio, music, text, QuickDraw 3-D, and time code.

A QuickTime movie can contain any or all of these media types, putting them into "tracks" so they can be individually controlled. This is what makes QuickTime invaluable. It keeps these "tracks" automatically synchronized so that a movie's images, sounds, and text are played and displayed at the proper time. Moreover, the QuickTime architecture is flexible enough to support new media types as they emerge, such as

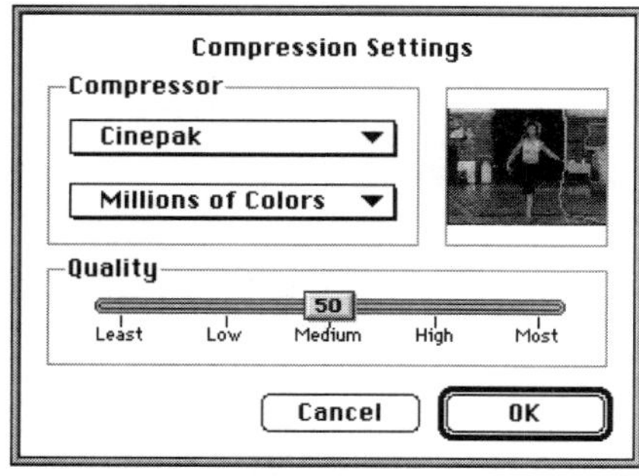

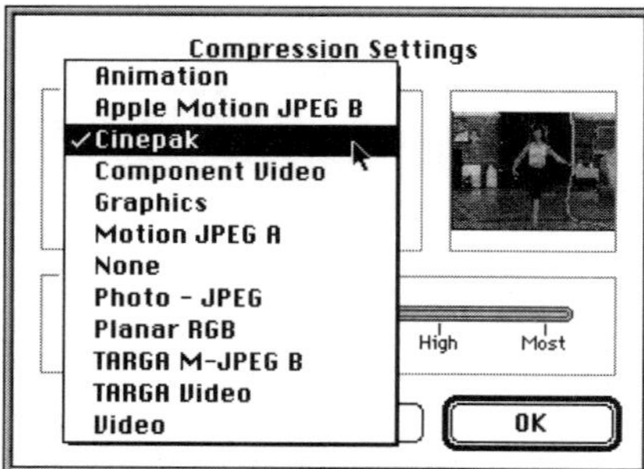

23.9 QuickTime interface: As an operating system extension, QuickTime provides additional functionality such as compression (shown above). When using QuickTime compressors, software developers do not need to invent their own interface in order for a user to set, for example, "Compression Settings." Instead, developers can use the standard interface provided by QuickTime. This way users need to learn only one interface method. *Courtesy George Escobar.*

virtual reality (see Chapter 17). With this flexibility, software developers can create new custom media components to supplement the generic ones originally supplied by Apple.

PAINT AND DRAW PROGRAMS

Until fairly recently, artists had to make difficult choices about what kind of artistic software tool to buy. There were three separate types of programs with many similar capabilities. Eventually one would need all three. So for most computer artists that meant purchasing *SuperPaint* (paint) and *Photoshop* (image editing) from Adobe, and *FreeHand* (draw) from Macromedia, which also meant paying triple the money for many of the same functions. What makes these programs distinct yet similar—and necessary?

A *paint program* distinguishes itself from a draw program because it more closely mimics traditional methods: The artist takes a brush, puts paint on it, and applies the paint onto a canvas. If the paint is applied correctly, the artist continues on. If there is a mistake, the artist scrapes it off or paints over it. Computer paint software operates similarly but adds features that are delightfully familiar and useful to artists trained in traditional graphics materials. For instance, with today's robust paint programs, the artist has unlimited ability to undo, a function that permits the removal of previous steps one at a time. Also, because computers are painting with electrons, you can have an unlimited choice of surfaces (canvas or different paper finishes), and even a choice of artistic brush stroke—from impressionistic to medieval Gothic—as well as the size, width, and density of your brush.

A *draw program,* on the other hand, allows artists to easily draw straight lines, smooth curves, polygons, and basic shapes (e.g., squares and circles). These *primitives* can then be manipulated using *control points,* without having to erase or redraw the original image. By using these control points, artists actually change mathematical equations (or vector coordinates) that make up the image.

For a number of years there was quite a heated argument about which approach to computer graphics was better: painting programs or drawing programs. The right answer depended on your background and on what kind of images you were making, and for what purpose. While the traditional applications continue to flourish, the question has been put to rest by the evolution of programs that combine both draw and paint approaches.

Foremost among these is Fractal Corporation's program *Painter.* It is the only software on the market that can do seemingly everything with great style and elegance. Moreover, it has room to grow because the basic program is designed to receive plug-in modules that extend its functionality. For animators, Painter has added value because it supports onion-skinning, drawing tablets, and pressure pens, has a rotating disc, and can import, play, and make QuickTime movies. Painter uses "natural media" tools that simulate art tools and materials like pencils, charcoals, watercolors, oil paints, bristly brushes, and over a hundred different paper textures. You can also import digital video frame by frame, allowing you to apply natural-media tools and features to QuickTime movies. There is also a function called Net Painting, where multiple artists can log onto a network for a collaborative painting session. In fact, art lessons using Painter are available on the Internet.

23.10 Bit maps vs. vectors: The image on the left was hand-drawn, then scanned into the computer as a "bit map" image. This means that the original image was broken up into thousands of little dots (which become "bits" in computer talk). These bits can be manipulated individually, which is sometimes advantageous or downright tedious, depending on the objective. The image on the right was drawn directly on the computer as a set of vectors using Flash from Macromedia. Vectors consist of location points on the computer screen. Each point contains its horizontal and vertical location, which become control points for the image. When points are joined or connected, they create a line between them. This line can be pulled, stretched, and resized at the control points. Qualitatively, you can see that vectors create more uniform lines, while bit maps can be rougher, more fluid. Being able to use both styles is a boon for artists and animators. *Drawings by Chris Allard and George Escobar.*

IMAGE EDITING PROGRAMS

An *image editing program* is a different breed of software from its draw and paint cousins. For starters, it was conceived as an application to work with imported images—mostly photographs.

The mother of all image editing programs shows this origin. It's called *Photoshop* and has been refined over many years by the Adobe Corporation.

Photoshop provides familiar digital darkroom techniques such as enlarging, cropping, dodging, and burning, as well as a host of new filters and special effects like bursts, waves, ripples, and texture maps. Collectively, these imaging tools facilitate modes of artistic expression that are impossible with traditional media. Photoshop gives the digital artist powerful new options, such as image layering, to create composite images that incorporate individual image transparency values, so the artist is able to combine multiple images and text, then highlight different layers or parts of those layers.

Anyone working professionally in computer graphics knows how vital Photoshop is to the graphics and imaging industry. Perhaps no other software has such wide support and praise, and because of its popularity, an almost unlimited number of plug-ins and extensions are made for it. If you can imagine an image in your head, Photoshop can help you execute it. Such power comes at a price, however, both monetarily (at $600) and in its complexity. Don't expect to become a Photoshop guru overnight—it will take months or years!

A tip. While there are many fine draw and paint programs on the market, including *Illustrator* from Adobe, *FreeHand* from Macromedia, *CorelDraw* from Corel, and *Canvas* from Deneba, only Painter from Fractal and Photoshop from Adobe offer a complete draw, paint, and image editing package that can support all of the techniques described in this book. Both of these are mature applications that excel in all seven essential categories that make for superior software. There are currently over six dozen books and tutorials available for using Photoshop and Painter. Before spending hundreds of dollars buying the software, you might want to get one of these books to get a close-up, real-world look at their power and see how closely they match your artistic style and meet your creative goals.

3-D ANIMATION PROGRAMS

It's a bit surprising that the first of the affordable animation programs for the Mac and PC evolved in the domain of 3-D rather than the world of established 2-D character animation. Yet the simple truth is that the software pioneers with the greatest impact in the world of animation were those who created ways of providing logo fly-throughs within three-dimensional space. Animation programs soon expanded from there.

Three-D programs are still used primarily for animating hard objects such as cars, planes, type fonts, and spaceships. But with the increase of computer processing speed and memory, 3-D programs are becoming more and more proficient at 3-D character animation. Chapter 16 provides an overview of the 3-D process. There is also a case study that shows a strong 3-D application called *3D Animation Pro* from Martin Hash.

Perhaps not as well known as other 3-D applications, 3D Animation Pro has a starter version available for $200 and a professional version for $600. This software was designed from the ground up to do character animation. It can create spaceships, robots, furniture, cars, and other "hard" objects. But what really makes it a great package is that it can help you create nifty characters. The biggest drawback of this software is its poor documentation. But if you are patient and willing to undergo a steep learning curve, then 3-D Animation Pro is the most affordable character-based tool available.

Another 3-D program we can recommend is *Ray Dream Studio* from Meta Creations, an integrated software package that includes a Bézier-based spline modeler, a ray tracing renderer, shaders and a shader editor, and an event-based animation system. You can animate everything in a scene, including all lights, cameras, shaders, and objects. There are also deformer tools to twist, bend, stretch, and squash for special effects. Ray Dream uses a unique interface, utilizing a perspective window view instead of the more traditional top, left, right, and 45-degree window views carried over from computer-aided design (CAD) programs. With a perspective window you can immediately see the height, depth, and width of your work. Ray Dream Studio also includes high-end features such as inverse kinematics to ensure accurate and realistic motion when you create animations.

For low-end to midrange PC and Macintosh computers, there are over a dozen major developers of 3-D programs, including *Extreme 3D* from Macromedia, *StudioPro* from Strata, *Infini-D* from Meta Creations, and *CorelDraw* from Corel. These programs are feature-rich, providing many control points, lighting effects, and special effects previously available only from high-end programs. The drawbacks of these programs are that none of them is very easy to use and they take a long time to master. Nevertheless, they are fertile training grounds before making the leap to the higher-end 3-D programs, costing between $2,000 and $7,000, from companies such as Kinetix, Electric Image, New Tek, and Microsoft SoftImage. *Forms-Z* from Autodessys is also very important in this list, especially since it is the only "solids" modeler on a PC—which in the future will be the way all 3-D modeling will be done. Those working at the highest end of PC modeling/animation

23.11 High-end digital animation systems

TIC-TAC-TOON	For SGI and Windows NT on DEC
Toon Boom Technologies	$25,000 per seat
818-954-8666	$2500–$7500 for modules

Preproduction: Includes exposure sheet, storyboard, and automatic lip assignment. This module links and registers all parts of the production.
Scanning: A single automatic process that scans, cleans, aligns, and vectorizes drawings in only 35 seconds per image.
Animation: A virtual animation disk for design, sheet timing, layout, animation, effects, cleanup, and in-betweening.
Backgrounds and Effects: Complex backgrounds and effects can be created using a brush editor.
Paint: Resolution-independent image painting with auto-gap closing, color styling, inking, painting, and checking.
Virtual Editor: Capability for compositing, special effects, and rough cut.
System Management: Integrated database for artwork management, routing, archiving, reports, import/export, and file conversion.

SOFTIMAGE TOONZ	For SGI and Windows NT
Microsoft	$13,000
818-365-1359	

Xsheet: Emulates look and feel of traditional exposure sheets, plus real-time previews and full-color display. Unlimited number of Xsheet levels; infinite number of pegs. Includes special effects like transparency, backlight, motion blur, glows, shades to modify your artwork.
Scan and Cleanup: Scan drawings then filter them for cleanup.
Palette Edit: Create, edit, and save color palettes for color modeling and markup.
Ink and Paint: Create color drawings that conform to specified colors.
Batches: Run processes such as compositing in the background.
Other Features: *Camera Stand* feature to enlarge, reduce, and rotate image, or create complex movements. *Pencil Test* feature to view animation quickly, with onion-skinning, just like a flip-book. *Audio* feature to load sound and synchronize it with animation.

are probably using *Forms-Z* or *Lightwave* to model, and *Electric Image* to animate.

For high-end workstations, namely those from *Silicon Graphics Inc. (SGI),* there is a plethora of very powerful, sophisticated, 3-D animation packages that generally cost $10,000 and above. The development of this category of software was driven by the entertainment industry, led by the top special-effects powerhouses: Industrial Light and Magic (ILM), Pixar, Pacific Data Images, Digital Domain, Robert Greenberg & Associates (RGA), and others.

High-end 3-D programs require top-of-the-line hardware in order to make their expense worthwhile. What you're paying for is speed, finer controls, more elaborate (and therefore more complicated) features, and special effects. For example, features such as *footstop-driven key-frame animation, motion mapping,* and *advanced inverse kinematics* from Kinetix's *3-D Studio Max* make animating two-legged characters almost automatic. High-end programs also offer development kit options that allow you to modify or add custom functionality to the software (if you know C programming). The problem with high-end equipment (besides cost) is its complexity. It takes as much computer programming knowl-

AXA TEAM 2D	For Windows 3.1
AXA Corporation	$1,495 Personal Edition
714-560-8800	$2,995 Professional Edition

Pencil Test: Incorporates editable exposure sheet for real-time playback with synchronized audio. Also includes auto antialiasing, 100 levels of animation plus shadows and highlights. Preview multiple scenes with sound and camera moves (using optional Camera fx module). Includes *Capture Module* to input drawings using a scanner or video camera. Capture automatically cleans up and positions drawing into the exposure sheet.
Ink and Paint: Uses color models for painting and auto repainting and re-inking. Also has specialized tools to create automatic shadows, backlighting, and highlights, with soft edges.
Camera fx: Modeled after an Oxberry and includes multiplane capability, including rack-focus. Up to 100 peg bars can be panned, zoomed, and blurred independently, plus truck in or out while blurring images and rotating the virtual camera.
Options: *AutoScan* for using automatic scan feeder; *AutoRecord* to record composited scenes directly to an output device; *AutoMatte* for generating tone mattes; *Telecine* to convert animation drawn at 24 fps into 30 fps for NTSC. All included in Professional Edition.

THE ANIMATION STAND	For Mac, Windows NT, and SGI
Linker Systems	$5,000 Mac/NT
714-522-6985	$9,000 SGI

Ink and Paint	$1,495 Mac/NT; $3,000 SGI
Pencil Test	$1,495 Mac/NT; SGI (N/A)
ScanLink	$995 Mac/NT; SGI (included)
Art-Director	$400 Mac/NT; SGI (included)

ScanLink: Supports batch or manual scanning, cleans up cels, fixes lines, aligns, builds exposure sheet.
Animation: Includes exposure sheet, sound synchronization, field controls, transitions, multiplane camera controls, motion paths, pencil test, frame-accurate output, antialiasing, automatic drop shadows, also output and device controls with SMPTE, including real-time output.
Sound: Display waveforms, beat marking, squash/stretch sound, frequency shift, also supports multiple concurrent sound tracks.
Ink and Paint: Complete drawing and painting tools—up to 64 bits of color, controls for lighting, color, airbrush patterns; NTSC legal colors, CMYK separations, paint with textures, and 32-bit masking.
Optical and Special Effects: Unlimited effects in one pass, multiple exposures, mattes, transparencies, rack focus, gels, glow, backlight, and support for bluescreen/greenscreen.
Art Director: Generates markup sheets from colored cels, automatically updates color names.
Programming: Although not required to use The Animation Stand, there is a built-in programming language so you can build your own custom paint tools or create proprietary special effects.

edge as artistic acumen to really take advantage of the software. Within many shops that use high-end programs, an artist/animator is typically matched with an engineer/operator to perform really tough maneuvers.

The good news is that the processing power of PC computers and that of high-end workstations are converging, which may take us into a new golden age of computer animation software. The first sign of this new age comes from Microsoft, which acquired *SoftImage* in late 1994. Within a year, Microsoft released desktop versions of SoftImage's core products for the PC, dropping their cost from $25,000 to $15,000 at launch, then later down to $7,000 and soon to less than $1,000. Other 3-D animation companies have had to react to Microsoft and now many offer products for the PC at substantial savings from their SGI workstation siblings.

2-D ANIMATION PROGRAMS

For the longest time it seemed that software developers were approaching the design of animation tools without talking with any traditional character animators.

Early programs worked off a conceptual model in which "stacks of cards" represented individual actions and became stacks of frames. *HyperCard* from Apple was and is an example of this. Early programs developed facility in prescribing pathways for rigid images but didn't accommodate the stretch-and-squash world of cartoons. HyperCard-based tools (including parallel capabilities in Photoshop, Illustrator, and Premiere) allowed for "onion-skinning" that would show one, two, or even three previous image layers, yet there was no on-screen way to select two different key frames and then extrapolate the in-between drawing. Also, for a long time there was no 2-D tool that mimicked the animator's light table and provided the simple yet elegant control of selecting two completed drawings and then referring to both while making a third, new drawing.

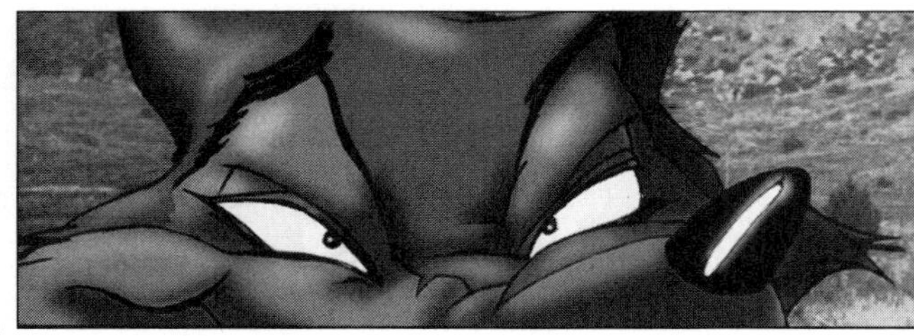

A

B

23.12 Digital ink and paint: Although these four black-and-white images don't do justice to the subtlety of their full color versions, you can get a sense here of how the new digital coloring can bring form and roundness to cartoon characters. Using a higher-end animation package called U.S. Animation by Toon Boom Technologies, animation director John Hays and his colleagues at Wild Brain in San Francisco rendered corresponding eye frames for a coyote *(A)*, a cat *(B)*, a buzzard *(C)*, and a dog *(D)*. Note the highlights and shadows. With digital ink and paint, the "lines" of a drawing can be given any color once the drawing has been scanned into the computer. With care and patience, the same quality of coloring can be achieved using Photoshop or a graphics program like Painter. *These images are from a project in development by Kit Laybourne with Nickelodeon.*

Even after almost a decade of multimedia, software specially conceived and designed for two-dimensional desktop animation is still at an immature state.

Fortunately, a new day is dawning, thanks to some innovative companies. One such company is FutureWave, which developed *Splash,* a $200 program designed for animators by a team that included an animator. It was the first vector-based animation software designed especially for distributing animations on the Internet. It also has interactivity built in so that clicking on the animation triggers other events or animated segments.

Perhaps what makes Splash most special is its interface. It takes familiar tools found in paint and draw programs and makes them better by adding animation elements such as filling a specified area with paint, even if there are gaps in the outer lines of the drawing. It also includes an intuitive timeline for making scenes and sequences of key-frame animation. The "onion-skinning" tool in Splash makes the task of drawing multiple images much easier. On top of that, Splash can automatically render any graphic or text on the fly as an antialiased image, instead of waiting until the very end. *Antialiasing* gets rid of the jaggies, letting you see your work with crisp edges and no banding, as it would appear on television. Splash quickly proved itself to be so powerful and innovative that multimedia powerhouse Macromedia purchased the company and its product and renamed it Flash. To learn more about this program, see the case study in Figure 23.13.

C

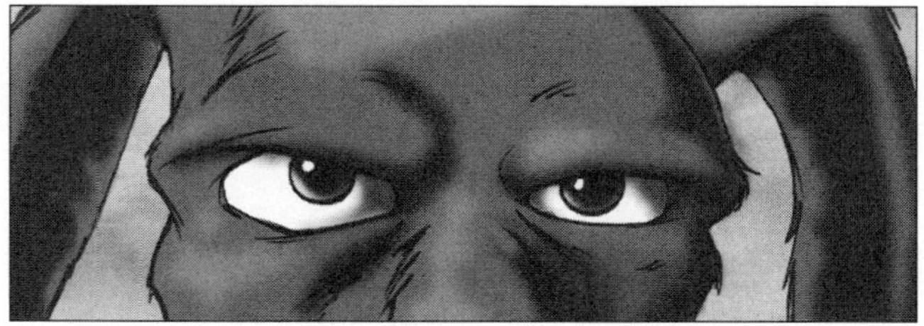
D

HIGH-END INK AND PAINT PROGRAMS

At the high end, 2-D animation programs are evolving rapidly and offer automatic mouth assignments and lip-syncing, database management tools, and auto pin registration of scanned art. The companies that are developing such 2-D animation applications are specialized and much smaller than their 3-D counterparts. Enterprises like Toon Boom Technologies and Linker Systems are not as well known to the PC industry, but they are setting the pace for developing artistically friendly interfaces with professional products such as *Tic-Tac-Toon* and *Animation Stand,* respectively, each selling for around $4,000 per copy.

VIDEO EDITING

Desktop video had been a long time in coming. For many years the software tools were so expensive and hard to use that most consumers steered clear of the process of editing digitized videotapes. This was in contrast to the world of desktop publishing, which was immediately embraced by enterprising writers and by the publishing and print design industries. Until recently, only motion graphics professionals and semipro videophiles had the bucks and the stamina to pursue desktop video.

All of that has changed with the introduction of powerful 200 MHz Pentium and PowerPC processors running on PC or Macintosh consumer models (see the next chapter on computer

23.13 CASE STUDY: Using Macromedia's Flash

A

Pilgrim's Progress is an animated musical feature film based on John Bunyan's classic allegory, published in 1678, about a man who becomes converted to Christianity and leaves his home to seek the Celestial Kingdom. The screen adaptation was written and is being produced by George Escobar from his home-based animation studio in Encino, California. During the design phase of *Pilgrim's Progress,* George has taken hand-drawn sketches and turned them into vector-based drawings using Macromedia's Flash software. Later during production, he will convert computer-generated 3-D characters to look like cel-animation using Ray Dream Studio or a similar software. Finally, he will composite the characters into multilayered backgrounds using Adobe After Effects.

As George Escobar describes the design of *Pilgrim's Progress*, take note of how Flash—a software tool made primarily for the Internet—can be successfully adapted for feature animation production.

Writing and Music

In feature animation you have ninety pages or minutes to tell your story. Yet typical of most adaptations, the source material is usually two or three times more voluminous. Bunyan's classic is no exception.

As an allegory, Bunyan's original text is filled with dozens and dozens of characters whose names, dialogue, and actions represented a particular archetype, fact, or belief. The first writing challenge was to select and recombine characters to less than a dozen key players. The next was to capture the essence of their two- to six-page discourses with five to ten lines of dialogue.

B

I decided early on that *Pilgrim's Progress* would make a great musical because the ideas expressed by the characters are so emotional and spiritual—key ingredients for powerful songs. Another reason is that *Pilgrim's Progress* is quite a "preachy" story. I realized that music and song would allow me to exploit the emotional arc of our central character, a woodcutter named Christian. I gave my song and lyric ideas to Leslie Burkart, a songwriter who has written over 400 songs. In less than three weeks she gave me back eight songs with Broadway-quality melodies. With real songs in the script, I could now begin the visual design of the movie.

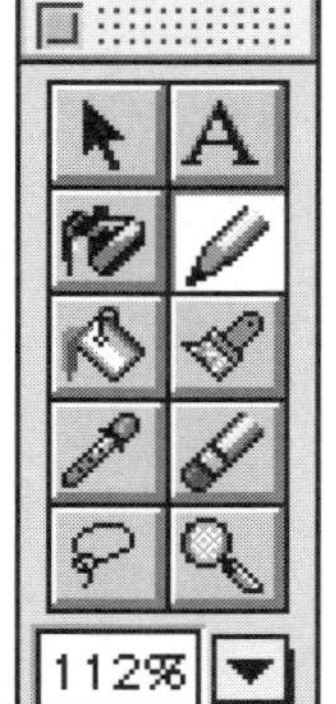

C

Design and Character Development

To find my artist for character design and storyboarding, I used the Internet, logging onto the Animation World Network Web site at http://www.awn.com. Using the site's Career Connection area, I selected half a dozen artists to interview by phone. Five were nearby in southern California, but the one I hired, Chris Allard, was in

D

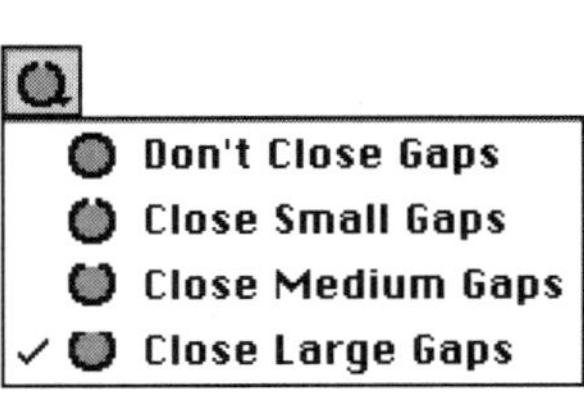

E

Massachusetts. Despite the cross-country distance, I chose him because of his "can-do" attitude. Chris immediately read the script and within a week sent two dozen character sketches, *(A).*

We spent the next four weeks, via fax, phone, and America Online, reworking and refining the characters—the toughest one was our dragonlike character, Apollyon, because he is described as having the head of a lion, claws of a bear, wings like a bat, and fire and darts coming out of his belly. Our easiest one was Young Ignorance. We just filled him with haughtiness, *(B).*

In reworking these characters, I began using Macromedia's Flash, not because of its auto-trace tool, which is indeed powerful, but because of its line-drawing tools, *(C).*

Creating vector-based versions of our hand-drawn sketches was easy because of the program's "smoothing function" and excellent control points. The reason for converting drawings to vectors is that it gives you great flexibility in resizing the images without any loss in resolution or detail, *(D).* Because vectors are mathematical formulas, rather than pixels, there are no "jaggies" even when the image is enlarged by 800 percent.

Flash also makes inking and painting much easier, because it lets you paint within the lines even if the lines have gaps, where in other standard paint programs, the paint "leaks" into unwanted areas. In *(E),* the large gaps in the face are filled in, stopping just short of the character's left eye even though it has a small gap. In other paint programs, this small gap would result in the eye being "painted" as well.

Another advantage of working with Flash is the ability to send your drawings and animations efficiently over the Internet. Your animation will play back on any computer, exactly as you designed it.

Layouts and Animatic

Flash is ideal for laying out the movement of characters within one or more backgrounds—especially during the design phase. Because Flash has multilayering capabilities, we could easily break up the elements of our backgrounds—for example, separating the trees in the scene so that trees farther back could move more slowly than foreground trees when panning across the scene. We can also have the character walk behind the tree. Doing these simple layout tests using traditional animation tools would have been prohibitively expensive during the design phase. By using Flash we have been able to get the rough timing we needed to start budgeting the screen time for each major sequence. It also helps for creating the breakdown list of the elements we will later use for final compositing in Adobe After Effects, *(F).*

F

While nearly 1,500 panels for the storyboard were being completed, the dialogue and songs were recorded. I used Adobe Premiere to combine sound and picture elements, yielding a ninety-minute animatic, complete with camera movements and simple effects such as fades and dissolves. As the production proceeds, these storyboard panels in their animatic form will be replaced with pencil tests, which themselves finally become replaced with color animation. The only thing left to do now is get the $4.8 million needed to complete the project!

Credits: **Producer and Screenwriter:** George Escobar **Music and Songs:** Leslie Burkart **Music Arrangement and Orchestration:** Jonathan Neal **Storyboard Artist:** Chris Allard **Character Design:** Chris Allard and George Escobar **Voice Consultant:** Pat Fraley **Sound Production Consultant:** Michael May hardware). Moreover, the cost of full-motion video cards has plummeted from $3,500 down to $500. The good news for animators is that video editing applications are now at broadcast-quality, professional levels, but priced for regular consumers.

A

23.14 Digital compositing: When images become digital, they can be "recomposed" or "composited" in endless ways. Photographic stills can be combined with cartoon drawings. Those can be combined with live-action footage. To give you a sample of what's possible, here is a frame, *(A),* in which two coyotes are circling a feast of carrion. Five elements are composited in this one image. There is the shot of a dead tree, *(B).* Note that some of the limbs were eliminated as well as the background. The mesa background, *(C),* replaces the shrubbery background. The cartoon coyotes *(D)* and *(E),* were resized as they got placed into the image. Finally the carrion, *(F),* appears as a foreground element. By placing all these elements on different layers in After Effects and then "panning" the layers at different speeds (less in the rear, more toward the front), the animators were able to achieve the illusion of movement with the scene traveling past the viewer's eye. The effects achieved here with simple computer software are much like those of the Multiplane camera devised by Disney almost fifty years ago. And it would be a simple step to add true live-action footage—say, a time-lapse sky above the mesa landscape and stalks of grass in the foreground, being blown by a western wind. *This deconstruct designed by Wild Brain, Inc. Courtesy Kit Laybourne and Nickelodeon.*

Corel's *Lumiere* for Windows 95 and Windows NT is a powerful editing tool with motion controls, allowing a video clip to follow a prescribed path. Lumiere can also scale, rotate, and distort the clip. It has over sixty transition effects and sixty video filters for creating special effects and transparency controls for superimposing images (in effect, compositing). The cost? Amazingly, only $69.

The industry mainstay, however, is *Premiere* from Adobe, available for the PC and Macintosh. It comes in two versions: Premiere LE, which is less than $99 and is often available for free as part of a bundle with a video card; and the full version of Premiere, which usually retails for $499 and is more suited for broadcast applications, plus an incredible array of plug-ins and filters to extend its functionality, like camera 3-D controls, lens distortion, strobe, and mosaic filters. The advantage with Premiere is that it's based on QuickTime, making its movies transportable across computer platforms and applications, and able to play over the Internet.

For animators, it's important to realize that video editing software like Premiere provides a great tool for creating storyboards and animatics.

COMPOSITING

After Effects from Adobe is perhaps the only software from the multimedia world to emerge as a powerful tool yet retain an elegant user interface. After Effects' primary specialty is motion-based *effects* that involve multiple image layers: moving text, images, and video clips superimposed over full-screen video or patterned backgrounds. For animators, After Effects is great for cutout animation and title sequences or for assembling mixed-media animation, including 2-D and 3-D images.

There are a number of case studies in Part II: Techniques that cite the use of After Effects.

The basic operating style of After Effects involves using two main windows: the *composition window* and a *time layout window.* The composition window is a large pasteboard on which you can layer and position assets such as video clips, text, and still images. The time layout window lets you control the position and characteristics of assets as they change over time. All together, this set of assets and their associated motion and effects settings is called a *composition.*

A significant feature of After Effects is that it treats time and motion as two distinct attributes. This enables you to specify an asset's motion path independently of its speed along the path. Another advantage is the unlimited number of layers of movies and stills that can be composited. Additionally, a project can have numerous composition windows where you can nest one composition within another to simplify working with complex projects or to reuse parts of a project.

Another significant strength of After Effects is its ability to apply up to thirty-two effects to a single layer. The *effects settings window* lets you control the settings of each effect and the order in which to apply them. As for masking, After Effects replaces polygon masks such as oval, rectangle, and polygon shapes with Bézier masks. Béziers provide Illustrator-like control points that enable you to create precise masks on any shape.

The only weakness that After Effects has is its audio support. You can vary the volume of an audio track over time, but there are no audio effects or sophisticated processing options. Audio is still best handled by dedicated sound editors or plug-ins such as those from Waves.

After Effects comes in two versions, costing about $650 through mail order for the base version and $1,900 for the production bundle from Adobe *VARs,* or *value-added resellers,* who tend to offer turnkey systems along with very specific and often very high-performance add-on hardware, training, and support. The production bundle includes additional effects and functions via plug-ins that appeal to video professionals. These plug-ins include a set of key-frame assistants, so you can draw complex motion paths in real time and calculate motion paths using a built-in scripting language. Another plug-in is a *motion stabilize command* that removes

- ***Beginners.*** *Audioshop* from Opcode is a $150 product for two-track (left and right) recording. Audioshop's interface looks similar to the audio CD-player utilities that come with most CD-ROM drives. But unlike those utilities, Audioshop lets you convert audio CD tracks into digital audio files. Audioshop can also record audio from a source connected to the computer's sound-input jack. For a low-end product, Audioshop provides strong editing features. It has a waveform window for selecting, deleting, and rearranging clips of a recording; a tool palette for applying a variety of special effects, and a mixer command for combining two recordings and controlling their relative volumes.
- ***Semiprofessional.*** *SoundEdit 16* plus *Deck II* from Macromedia is a powerful two-program combination package for $400. The two programs let you create audio for nearly everything—from multimedia to music to the Internet. Like Audioshop, SoundEdit 16 has built-in effects and a basic mixing function that lets you combine tracks. However, SoundEdit 16 isn't limited to just two tracks; you can create as many tracks as your disk space and memory allow. It also can import tracks from audio compact discs, converting them to any audio file format and/or compressing them to reduce disk space and memory requirements.

  While SoundEdit 16 supports numerous tracks, its mixing and recording features can be cumbersome.
- ***Professional.*** This is where *Deck II* comes in. It currently supports up to thirty-two tracks of real-time audio effects. It also has an array of nondestructive editing tools, with which you can manipulate original audio files without permanently altering them. Unlike SoundEdit 16, Deck II has a scrub feature, letting you use a mouse to play a recording forward and backward over a waveform to find an exact point, like the start of a sentence. This is essential for editing voice soundtracks. Deck II also has strong support for MIDI as well as time codes for film and video.

handheld-camera jitter, while a *motion tracker command* lets you synchronize the location of one element with an exact point on another element.

The more expensive production bundle also includes significant enhancements to After Effects' keying features, nine additional distortion filters, and plug-ins for controlling high-end digital disk recorders from Abekas and Accom. However, it is not necessary to have this production bundle to produce outstanding results. The base version of After Effects can sustain almost every creative ambition of novices and professionals, but you can upgrade the base version to the production bundle for an additional thousand dollars at any time.

With After Effects, an animator on a Macintosh or a PC can add world-class, broadcast-standard animated effects without compromise. The only requirements are talent and output to a single-frame video or digital recorder to record your animation sequences.

SOUND EDITING AND MIXING

Sound editing programs for Macs are superior to those found on PCs. This is because the Macintosh, from the ground up, was designed with built-in sound support, microphone, and speakers, so it doesn't need additional sound cards. On the other hand, there are many more sound editing products on the PC, most of which support the *Sound Blaster* standard. Regardless of platform, if you are an animator wanting to add sound to your film or animation, there is a choice of software-based audio products for you to consider.

Investing in high-end audio hardware is no longer necessary to achieve professional results. Today's digital audio software products can record and play back near-CD-quality material that sounds great and provides editing precision and mixing flexibility that analog systems can't match. Since it's done with software, rearranging sections of a recording or mixing tracks is as easy as cutting and pasting audio files, with no degradation of quality.

There are currently hundreds of audio software products, ranging from $10 shareware to $10,000 developer bundles. Described in the margin are three products available for those with a tight budget.

23.15 THE SOFTWARE SHRINK			
Type of Software Tool	**Low End $300 or less**	**Midrange $300–$600**	**High End $1,000 or more**
Drawing	Integrated packages such as **Microsoft Works** or **Claris Works** provide both drawing and painting tools that can be exported later to more expensive programs.	**Macromedia's FreeHand** and **Adobe Illustrator** are the mainstays in this category, although **Deneba's Canvas** and **CorelDraw** both provide more functionality for the money, including painting.	Can add plug-ins to the base price of **Photoshop, FreeHand, QuarkExpress,** and **Illustrator.**
Painting	Numerous shareware programs.	**Fractal Painter** practically owns the natural painting category. There are numerous midrange general purpose 2-D painting programs.	**Parallex Matador** is used in matte paintings for films such as *Speed.* It specializes in wire removals and other retouching effects.
Image Editing	**Macromedia's Xres,** when included in a bundle, is a good deal for the artist. Similar to **LivePicture. Xres** uses fractals to create "proxy" images of the original. This permits faster performance.	**Adobe Photoshop** edges out **LivePicture, Xres,** and **Fractal Painter. Photoshop** has the widest selection of plug-ins and by far has the largest number of users. A new entrant, **Illuminaire Paint,** is designed for video broadcast–quality work.	**Photoshop** is available in a **Unix** version for SGI's. It has the same functionality as its PC cousins, but runs much faster for twice the cost. On the extreme high end is **Kodak's Cineon** used in special effects.
3-D Animation	**KPT Byce 2** is a powerful, specialized tool for creating realistic, as well as surrealistic, landscapes, and backgrounds. The best in its class, it has no competition. **Fractal Poser** is also a specialized program for creating the human form. Both are designed to work with other 3-D packages.	There is a slugfest for 3-D dominance in this area. **Ray Dream Studio** is the overall best choice because of features and its $399 price. The others, including **Extreme 3D, Infini-D,** and **Strata StudioPro** are equally as powerful but more expensive. For character animation there is only one package—**Hash's 3D Animation Pro.**	For the PC, the high end means using **Lightwave 3D** or **3D Studio Max.** For SGI's, the trio of **SoftImage 3D, Alias PowerAnimator,** and **Wavefront Explore** are competitors. It takes a lot of skill and horsepower to use these incredible tools that have become film and TV industry standard tool kits.
2-D Animation	**Macromedia Flash, Simpletoon,** and **PaceWorks Dancer** are 2-D animation tools designed for the internet. They can also be used for standard animation since they can output QuickTime movies.		**ToonBoom, Animo,** and **Animation Stand** are the best known commercial production-quality animation systems available. Studios like Disney and Pixar have invented their own proprietary systems.
Video Editing and Compositing	**Streta VideoShop,** formerly from **Avid,** is an easy to use video editor that has suffered market share against **Premiere.** It's a solid product and now quite inexpensive. **Corel's Lumiere** is also a good buy. It has motion controls, transparency, and a title window.	**Adobe After Effects** and **Premiere** are the industry standards. (New entrants such as **Illuminaire** are trying to break in but **After Effects** is so strong and compelling that it continues to dominate. **Premiere,** however, is vulnerable to **Macromedia's Key Grip** video editor, created by **Premiere**'s original programmer, **Randy Ubillos.** It should be a **Premiere**-killer.)	**AVID** and **Media 100 Proprietary** video and compositing systems. Essentially you buy the hardware and it includes the software. The exceptions are SGI workstations. You still have to buy **Flint, Flame,** or **Wavefront Composer** software to edit or compose.
Sound Editing	**SoundForge** is the standard for **PC Windows.** It has a plug-in architecture, making it extremely flexible and powerful.	The combination of **SoundEdit** 16 and **Deck II** make this offering from **Macromedia** an unbeatable combination. Together they generate professional work.	**Avid's Digidesign** product line occupies the high end for sound editing.
Multimedia	**SuperCard, Toolbook,** and **HyperCard** remain extremely powerful multimedia authoring tools. If you don't need a time line interface for your project, any of these three tools will be good.	**Macromedia Director** reigns supreme in this category. More titles are created with **Director,** hence more plug-ins. Lingo code and authoring engineers are available. A good second choice is **mFactory** from **mTropolis.**	

23.16 CASE STUDY: Finding Heart

A

Since the late 1980s, Derek Lamb and Kai Pindal have been making animated films for the poorest kids on earth. *(A)* and *(B)* show preliminary sketches for *Karate Kids, AIDS,* a twenty-minute adventure cartoon to teach children about AIDS. *(C)* shows two of the characters on a cel and *(D)* has two character sketches drawn from life. The photograph, *(E),* shows Derek with kids who live at the city dump in Guatemala. *(F)* is a storyboard sketch from *Goldtooth,* a film about prostitution. The series also includes a twenty-minute film about drugs, *The Big Fire.* All three films are available from Street Kids International, 398 Adelaide Street West, Suite 1000, Toronto, Canada, M5V 1S7; Ph: 416-504-8954; Fax: 416-504-8977; or check them out on the Internet at: www.streetkids.org. Take a moment to read Derek's own account of how he has approached the challenge of making movies that really matter.

B

C

When Peter Dalglish, founder of Street Kids International, Toronto, asked me in 1989 to help him create films for the hundred million or more homeless children around the world, kids who live hand to mouth, marginalized, vulnerable to every kind of danger and abuse the world can offer, I was apprehensive. These were to be films that would talk to kids about AIDS, drugs, pedophilia, behavior of all kinds — not a world I knew about. But I supposed Peter had confidence in my so-called abilities to write stories that communicate to kids, and he understood the universal appeal of animation. Within forty-eight hours, my animation partner, Kai Pindal, and I were on our way with him to Central and South America.

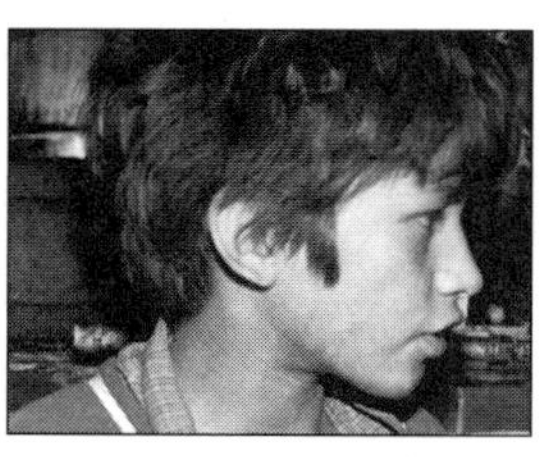

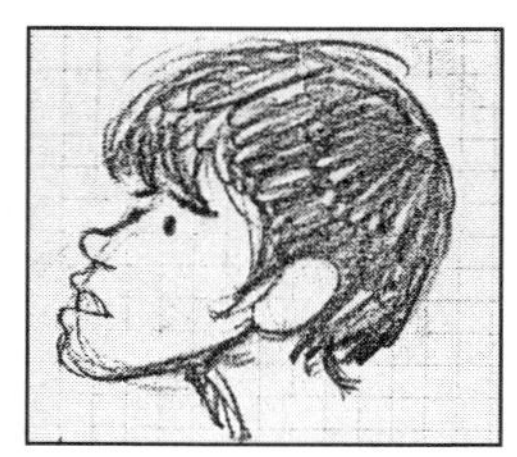

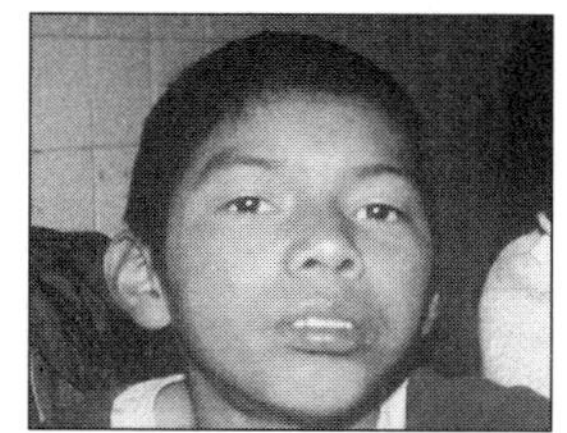

D

I needed to ask myself some questions:

- Who is the audience?
- How do street children live day to day?
- How do I start to think in their terms, be in their shoes, and be honest to their experiences?

I believe these questions are key to creating most kinds of scripts. Over time, I've learned that writing stories exclusively out of my own head is a bad idea. It's much better to generate characters and ideas in collaboration with my audience, especially this audience.

In Mexico City, Pindal and I mixed with hundreds of street children: We were in street clinics, workhouses, prisons — a world of Charles Dickens, where kids beg on the streets, juggle, sing for money. In Guatemala we met children living on garbage dumps and competing with vultures for bits of food. We met kids who sniff shoe glue to get high and forget their

E

pain and fear. We took photographs and made a lot of drawings — they became our kids.

At what point do a script and storyboard begin?

For me I always look for one simple, clear image — an inspired moment that encapsulates the overall theme. My films invariably start out that way. In the case of *Karate Kids, AIDS,* one day we saw two boys performing at the traffic lights, one of them juggling while sitting on the other's shoulders. I said to Kai, "Eureka! That's it, if one of those boys contracts AIDS the team is in trouble." *That* was the metaphor, a seed image for the film. A storyboard could now be developed around that central idea, about the adventures of those two boys and how AIDS came to affect their lives.

Why do I take so much time to test my storyboards in the field, show Leica reels, often in several countries, before starting to animate?

Animators love to start animating. That's what we like to do best, but often way too early in the process. I say count to ten before starting, get out of the studio, get your stories watertight before the expensive process of animation begins; it can pay off in spades.

At Street Kids International, we often test stories by having an actor perform them live, like a play, to watch the children's reactions and listen carefully to their comments. Kids are your best and most honest critics. They've told us how to better tell the story, introduced new characters, switched genders, and killed ideas — a ton of things for the better!

Karate Kids, AIDS, is now in forty languages in more than a hundred countries. It's seen rear-projected on the sides of trucks in Brazil, in cassettes carried by every health worker in Thailand, dubbed into languages we've never heard of, pirated everywhere, which is great!

F

Kids love to watch this series of films again and again. A key reason is they see themselves in it: It's their humor, their street smarts, their lives shown in a nonjudgmental way; a validation of themselves and the messages.

I would say the way Kai and I work is in the tradition we learned at the National Film Board of Canada, making entertaining films in the service of public information.

A final thought: If you aren't already, study improvisational theater techniques; try performing it, too. You can find it an amazing tool for character and story development, and a heck of a lot of fun.

Materials courtesy Derek Lamb and Kai Pindal.

MULTIMEDIA AUTHORING

Authoring a multimedia title, either for a CD-ROM or the Internet, is akin to orchestrating a production studio on your desktop. It is a complex process, demanding mastery of multiple skill sets, ranging from drawing to design to sound to animation to programming. As daunting as this may seem, it is not unlike creating a short animated film. The main difference is you do not have to limit your thinking to being linear, as in "The audience will view my work from A to Z, start to end." In multimedia you have the exciting opportunity to design a world where your audience will have choices—"Here's a place for some users to go one way and others to find a different path." In multimedia you must think about interactive buttons by which users will navigate within your CD-ROM or across Web sites. In multimedia, the audience is free to go where and when they want to—and you have to allow for this. It's no wonder that a leading authoring tool, *Director* from Macromedia, incorporates a frame-based animation timeline metaphor for its user interface.

Director is a mature and robust system with tools for image creation, painting, editing, animation, and a sophisticated interactive programming language called *Lingo*. Multimedia authors typically rely on other software packages (such as those mentioned in this chapter) to create their media "assets." Director's own asset-creation tools are quite good, but they are not nearly as complete as stand-alone packages. Hence, Director is primarily used for arranging and sequencing media assets in a purposeful way so that end users are entertained, engaged, and informed.

CHOOSING THE RIGHT SOFTWARE

The Software Shrink chart in Figure 23.15 is designed to assist you in selecting the right mix of software for your project. Note that blank boxes mean there are no recommendations or products in the category.

GET STARTED NOW!

To understand why software is like it is (confusing, delighting, frustrating, magical), it is helpful to consider where it comes from and who makes it. Almost all software is designed from the ground up by a single individual or a small team that has a distinct point of view—usually a technical one. And hardly ever does a software creator bother to state his or her beliefs and biases.

The end users, you and I, are typically the last folks to have input about the software. Don't laugh. Up until about ten years ago, the American auto industry operated similarly and was creamed by the Japanese. The software industry, not having great incentive to change, still "codes" software rather than "crafts" it to end-user specifications, especially for creative endeavors. However, a few companies are starting to change and the future is exciting. There will be further simplification and integration. There will be greater focus on building tools to better serve different animation techniques like 3-D, character animation, and cutouts. And doubtless there will be new forms of animation that bubble up through the ever-expanding tool base from which the independent animator can do his or her thing.

24 ■ Computer Hardware

WRITTEN WITH GEORGE ESCOBAR

There are two basic computer platforms for consumer-level digital animation: the Macintosh and the Windows PC. Over the past few years, both the Mac and the PC have been evolving so quickly that they threaten to overtake more costly and complex UNIX workstations that have produced those dazzling effects we all associate with the computer era of animation in TV and feature films.

And the pace of innovation keeps accelerating! A new generation of Macs and PCs are now appearing every six months as manufacturers seek to outdo each other in competition for buyers. Five years ago, this cycle was a bit more manageable at eighteen to twenty-four months. The quicker introduction of new products is a recipe for chaos and confusion.

To make your way through the turbulence, this chapter offers a high-level introduction to the various tools that are used in digital animation. By "high-level" we mean that our discussions about the functions of each kind of hardware should be so universal, so fundamental, that they can be matched against any computer that you can get your hands on—or that may appear in the foreseeable future. (Note that "high-level" is not the same thing as "high-end"—which means expensive.)

At the conclusion of this basic orientation to computer hardware there is a self-diagnostic grid that should help digital animators determine what cost levels and peripheral gear they will need when working in eleven basic different tasks involved in computer animation.

THE PRICE IS (ALWAYS) RIGHT

An interesting thing about the cost and performance of computers is that, over the years, the price has remained pretty constant. The *low-end* cost of a desktop computer (plus related hardware) is always somewhere between $2,000 and $4,000. The *midprice* level is from $4,000 to $8,000, and anything above the $8,000 mark is considered *high end.* Generally, software adds another $1,000 to $4,000 to the cost of doing animation on your own system.

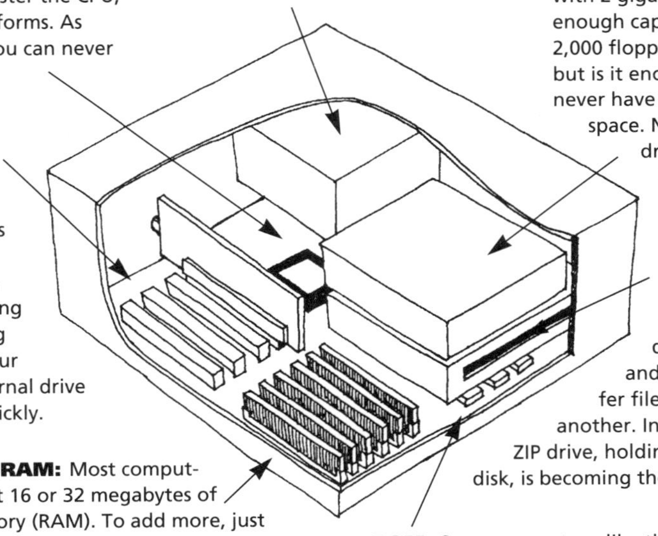

24.1 Generic computer cutaway: Every personal computer uses the same component parts, which are described here. What makes computer models different from each other is the sum of their individual parts, consisting of capacity, performance, and quality. *Drawing by George Escobar.*

The CPU: This is the brain of the computer, processing instructions from applications (stored in the hard drive), and orchestrating its executions among the different parts of the PC. The faster the CPU, the zippier the PC performs. As with the hard drive, you can never have enough speed.

Power Supply: This unit connects the computer to the electrical outlet, giving power to the PC components. If you plan on having many add-on cards to the PC, get a large-capacity power supply—something over 300 watts is good.

Hard Drive: How much information you can store and the number of applications you can run are determined by your hard drive capacity. Most computers come with 2 gigabyte drives. That is enough capacity to store about 2,000 floppy disks—which is a lot, but is it enough? No. You can never have too much hard drive space. No matter how large a drive you get, you will fill it up in no time.

Expansion Slots: These slots allow you to add capabilities to your computer, such as accelerating 2-D/3-D graphic computations, capturing and displaying video, and transferring large files between your computer and an external drive or a network more quickly.

Floppy Drive: This drive reads and writes to diskettes that hold about 1.4 megabytes of data. It is still a common and popular way to transfer files from one computer to another. Increasingly, however, the ZIP drive, holding 100 megabytes per disk, is becoming the standard.

Memory Slots for RAM: Most computers come with at least 16 or 32 megabytes of Random Access Memory (RAM). To add more, just insert them in these slots. The number of open slots determines how much you can expand your memory capacity. RAM memory is temporary. It holds its data only as long as the computer has power. When the PC is turned off, RAM memory is erased.

ROM: Some computers, like the Macintosh, store commonly used computer instructions in "Read Only Memory" (ROM). The data is stored permanently and is "read" as necessary by the computer in conjunction with the operating system.

A

B

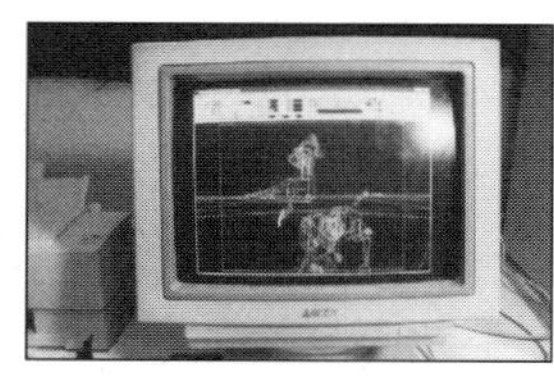

C

D

24.2 Digital finds its own voice. This example of 3-D computer animation shows a standout effort by independent animator Doug Aberle. While much 3-D work is highly representational, Doug's storytelling takes a more abstract approach.

Here are two frame-grabs, *(A)* and *(B)*, from *Fluffy,* a four-minute film produced at Doug's home in Battle Ground, Washington. Production still *(C)* shows a wire-frame rendering of the dog character. Note how the character design takes advantage of geometric "primitives" that are used in buildings 3-D segments. Three independent elements make up the dog's head, and these are not attached to each other. In motion, however, the creature comes alive and exhibits a terrific range of expressions. *(D)* shows the animator with his digital tools — a consumer-level rig. *Copyright © 1995 Aberle Films.*

While prices have remained fixed (after all, people have consistent boundaries for what they can afford to spend), hardware performance has improved dramatically, mostly because of all those new bells and whistles that attract new buyers. The computer manufacturers know that enhanced and expanded performance forces current computer owners to trade up. Those semiannual computer shows—MacWorld, PC Expo, COMDEX, and others—are showroom circuses designed to produce *digitalis wishlhadits*—also known as computer envy.

Make no mistake about it: There is an ongoing battle for your imagination and your computer dollars. The manufacturers' tactics all aim at making it impossible for you to resist the urge to splurge. Your defense against the attack should be to buy with an upgrade strategy in mind and to keep a map handy that reminds you exactly what creative destination you (not they) have chosen. The aim of this chapter is to give you such defenses. But keep this in mind: Year in and year out the cost of computing power drops about $100 a month. If you wait six months, the same raw power of your dream machine will be $600 less than it is today, *but you will have learned nothing.* The cost of knowledge is the same as $100 per month. Are you worth it?

THE PLATFORM

Despite what you might have read or heard, *Macintosh* and *Windows* PCs (both manufactured by a range of different companies) are two fundamentally different computer platforms that yield similar results because of software. Strictly from the hardware perspective, the Macintosh is far easier to set up, maintain, and upgrade than a Windows PC—even with the latter's "plug-and-play" architecture.

With a Macintosh there are no DIP switches, jumpers, and arcane software patches to mess around with if you want to add a new input device such as a scanner, making the Mac much easier to use. However, what you give up for this convenience is choice. Macintosh parts and devices are generally more expensive because there are few of them made and there are fewer manufacturers to make them. In the Windows PC world (which at the time of this writing dominates the personal computer industry, with 90 percent versus 10 percent for the

G

H

I

J

K

L

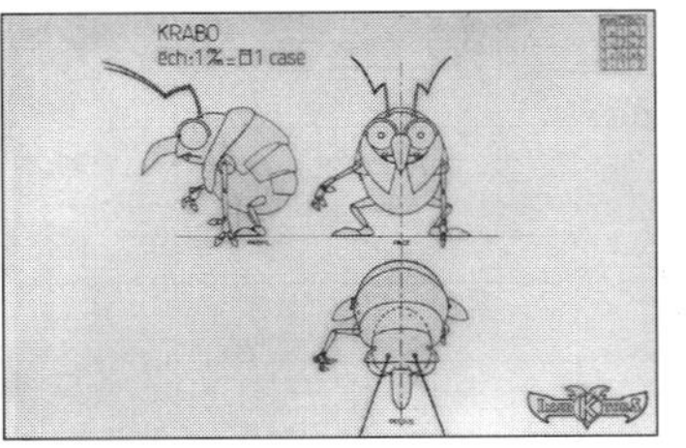

M

24.3 Character design in 3-D animation: *Insektors* is a 3-D computer animation series created and directed by the French artists Georges Lacroix, who heads the studio of Fantome Animation, and Renato. This series exemplifies what a talented group can do with state-of-the-art hardware. Photo *(A)* shows Georges Lacroix and project manager Jean François Schneider reviewing drawings for Insektors III.

The stories center on the fantastic saga of warring insects—the dark and sinister Yucks are always trying to invade and destroy the joyful universe of the Joyces. Frame enlargements *(B)* and *(C)* suggest the richness of the *Insektors* digital world, where the imagination and charm of the animators makes even the darkest Yuck, the evil Lord Krabo, extremely humorous. The evolution of the Krabo character is shown step-by-step.

Even in the cutting-edge world of 3-D, you begin with simple drawings, *(D),* which evolve into tighter renderings, *(E).* These in turn lead to the step of modeling in which character specs, *(F),* become a piece of sculpture. *(G)* shows design and sculpture artist Yves Vidal sculpting Krabo. A grid is drawn and the surface of the sculpture and the XYZ coordinates of each intersection on the grid are entered into the computer with a data pen. *(H)* shows Fantome's manager of modelization, Franck Clement Larosiere, modeling the character Spoty. The character of Krabo now lives inside the computer, appearing as a wire frame, *(I).* Animation takes place using a simplified (usually a monochrome) version of the character, *(J),* that the animator directs by entering principal positions (or key frames). The computer will subsequently calculate the "in-between" positions. The character doesn't take its finished form until the computer "renders" the 3-D figure with full color and texture, *(K).* The last two frames, *(L)* and *(M),* show Krabo in full action.

Insektors *is a production of Fantome Animation and a co-production of Ellipse Programme—France 3—Medialab—RTBF Television Belge—Neurones—Finatoon—Club d'Investissement Media (Programme Media de l'Union Europeene) in association with SOFICA VALOR 4, with the participation of Canal +, and with the support of Centre National de la Cinemathographie, and Cartoon (Programme Media de l'Union Europeene). All images provided courtesy Fantome Animation, Copyright © 1996.*

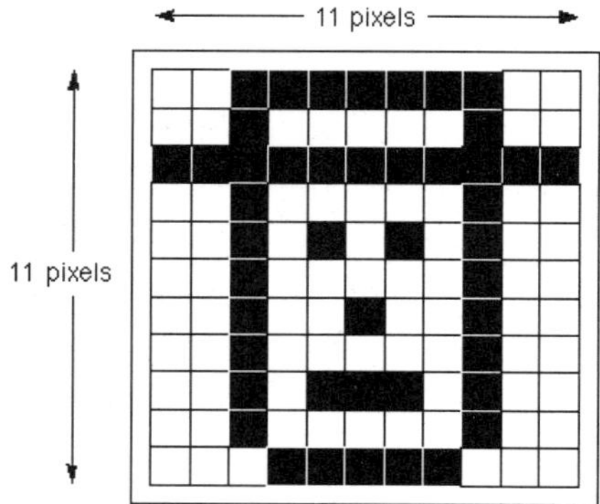

24.4 What a bit looks like: A good way to understand bits is by using a "bit-mapped" image. For this type of picture, the graphics program needs only two pieces of information: the location of a pixel and whether to turn the pixel on (black) or off (white). The man in the hat consists of 121 pixels. Since in this example 1 pixel is equal to one bit, there are 121 bits used by the program (or around 16 bytes because 8 bits equals 1 byte). *Courtesy George Escobar.*

Mac platform), prices for parts and devices are lower, the selection is abundant, and product improvements are more frequent. But this point about ease of use is not a trivial one, especially if you think you'll be adding new gear, drives, scanners, monitors, and so on. Frustration issues extend through the life of ownership. Also, there are several PC operating systems: DOS, OS/2, Windows 3.1, Windows 95, and Windows NT. Today there are more Windows 3.1–based machines than there are computers with Windows 95, and this incompatibility will continue through the year 2000.

If you want the best of both worlds, Apple's Macintosh can support two processors: One is the *PowerPC* microprocessor for Macintosh operations and the other is a *Pentium* processor for Windows operations. The Pentium processor comes in an "add-on" card that goes into a PCI slot. Orange Micro sells them for only a few hundred dollars more, but you get two computers in one.

At the high end of the spectrum, Silicon Graphics and Sun Microsystems dominate the workstation market, which uses the *UNIX* operating system. These workstations come in several flavors: from funny-looking, toasterlike boxes to mainframe-style behemoths requiring controlled air-conditioning. The lowest end of this high-end category typically costs just slightly more than their lower-end siblings—around $6,000 to $10,000—yet they pack a wallop in terms of raw speed. These "entry-level" workstations are designed to compete with a top-of-the-line Macintosh or PC. What makes them expensive overall is the cost of their UNIX-based software, which is often three to twenty times more expensive than identical software running on a Macintosh or PC. Why the steep markup? Because soft-

24.5 CPU Chronology: Brief Chronology of CPUs and their Speeds (MHz)

IBM Compatible/PC/Windows/Intel

Debut:	(1984)		(1987)	(1989)	(1991)	(1993)	(1995)	(1997)	(1999)
CPU:	8088	8086	80286	80386	80486	Pentium	Pentium Pro	Pentium II	Merced
MHz:	4–6	8	16	33	66	133–200	200–250	300–400	500+

Macintosh

Debut:	(1984)	(1987)	(1989)	(1990)	(1994)	(1995)	(1996)	(1997)	(1999)
CPU:	68000	68020	68030	68040	PowerPC: 601	603e	604e	G3	G4
MHz:	8	20	40	60	66–100	160–300	180–350	233–300	400+

ware developers in the UNIX market have a significantly smaller number of buyers when compared to the mainstream computer market. Another disadvantage is the high maintenance cost of UNIX software in terms of training and complexity. Don't expect to learn the software on your own. It's more than likely that you'll need professional training and ongoing software support because UNIX software is a lot less bulletproof than off-the-shelf software. In fact, animation and special-effects companies using Silicon Graphics workstations (SGIs) often employ a team of UNIX engineers to keep their system running.

24.6 Homage to the motherboard: Typical desktop motherboard with Intel® Pentium® II processor. *Courtesy Intel Corporation.*

If you are a current computer owner and can't even consider spending money on new hardware, let alone an SGI-class workstation, you shouldn't feel stuck—even if your Mac or PC platform is old and you suspect it is a clunker. Today even the low-end computers have the capacity for robust upgrading. This was not true until recently. And it means that you can get into the animation game right now, with just about any computer you can get your hands on. You'll be wasting precious time if you think that next month's or next year's model will be the perfect computer for you. It does not exist. You will always lust after next year's model!

THE CPU, THE OS, AND PROCESSOR SPEED

The *central processing unit (CPU),* or *processor,* together with the *operating system (OS),* constitutes the heart of the computer. The processor literally has a heartbeat, expressed in *megahertz* (MHz), that measures how fast the computer can run its operations. The OS, defined as a set of "low-level" instructions, determines how software applications are written to the processor and how everything in the computer appears to the user.

Processors Circa 1998. When this edition of *The Animation Book* is published, the Mac and PC platforms should still be operating with the *PowerPC* and *Pentium* CPUs. The former was developed by Apple and IBM working together, while the

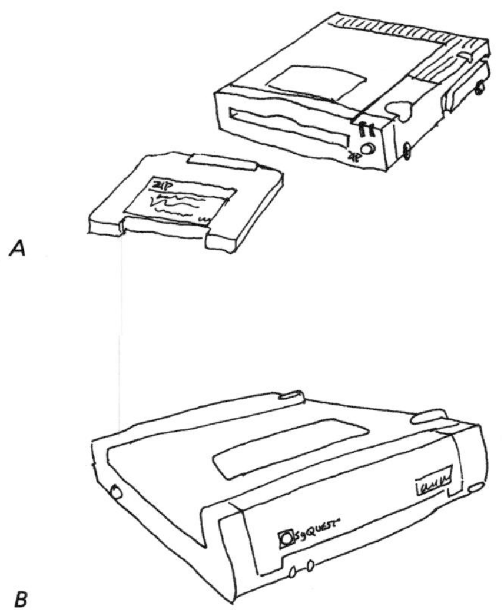

24.7 Portable storage devices: The cost of storage devices ranges from $150 or less for a 100 MB Zip drive, $250 for a 230 MB SyQuest EZFlyer drive, or up to $400 for the 2 GB Jaz or $300 for the 1.5 GB SyJet drive. The cost of the portable storage media used with these drives ranges from $20 for a 100 MB Zip cassette to $100 for a 1 GB cartridge used in a SyQuest drive. DAT tapes come in packs of five for $25 or ten for $50.

(A) Iomega Zip Drive $150 (drive)/$14 (cartridges); 100 MB storage; 5.25": Magnetic cartridge that is a floppy disk replacement

(B) Syquest EZFlyer $230 (drive)/$30 (cartridges); 230 MB storage; 3.5": Not in widespread use, but can serve as a super floppy disk.

(C) Iomega Jaz Drive $200 (drive)/$125 (cartridges); 1 GB storage; 5.25," $400 2 GB: Like a near–hard disk.

(D) SyQuest SyJet $300 (drive)/$125 (cartridge); 1.5 GB storage; 3.5": Market acceptance uncertain, but has near–hard disk capabilities.

(E) DAT (Dynatex, Sony, etc.) $1065 (drive)/$15 (cartridges); 2 GB storage; 3.5": Relatively slow, sequential nature of tape limits its usefulness.

latter was developed by Intel Corporation. Both processors are manufactured with different speeds—ranging from 133 to 500 MHz.

The combination of processor speed (CPU), memory (RAM), and the data transfer rate (bus or *backplane,* the place where the motherboard plugs into the CPU) determines how fast the software and its files will run. For functions such as drawing, painting, or inking, most animation software will perform adequately using the standard, consumer-level machines with PowerPC-based or Pentium-based *motherboards* operating at the 200 MHz level. But for more demanding digital functions such as editing sound, video-image capturing, playback of full-screen animation sequences, or rendering, a faster version of these processors is required—and these will work only when coupled with a lot of memory.

Processor speed is crucial in animation because a higher speed can dramatically reduce the time it takes to create an animated scene and because processor speed affects how well an animation file will play back on the computer once it is completed.

Let's take a moment to look at the playback phase within 3-D animation, which makes huge demands upon a computer's computational strength. As you've seen in Chapter 16: 3-D Animation, the computer must calculate the amount of sequential change in character or background between one frame and the next. The CPU has to complete this calculation at a fast enough rate of new frames per second so that the illusion of movement is sustained. The more complex the makeup of a given frame, the more calculations the processor must perform, making processing speed very important.

Take a moment to look at the creative phase, which places even larger demands upon the CPU. If a piece of digital animation is to run at 24 frames per second and the sequence is 20 seconds long, then the computer must render 480 frames in order to complete the file that will become the animation itself (usually a QuickTime movie). The animator working on a slow computer (say, 133 MHz) will discover it can take an hour to render each frame the first time. Rendering this piece would take 480 hours, which is twenty days of computer time! On a fast PowerPC or Pentium computer running at a processing speed of 200 MHz, the very same frame might take only 10 minutes to render. The same 20-second sequence now requires a mere 80 hours of computer time—only three and a half days, if you get it right the first time!

This is precisely the scenario you will encounter when working in 3-D, where the rendering process must calculate and display each model's height, width, depth, overlaying or "mapping" textures, applying illumination from various light sources, and compositing with background and foreground elements. Because this is so much work to do, it's easy to understand why a powerful processor is essential.

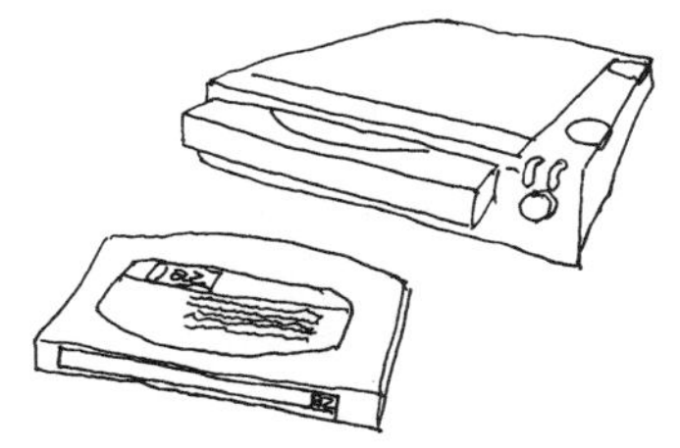

C

A performance tactic worth knowing about, yet often overlooked, is to "soup up" a relatively slow computer with a fast 2-D and 3-D graphics and video accelerator card. What this card does is take over from the main processor the functions of rendering, drawing, and displaying an image. (More about accelerators is in the Peripherals section of this chapter. Be sure to read it. It could save you hundreds—maybe thousands!—of bucks.)

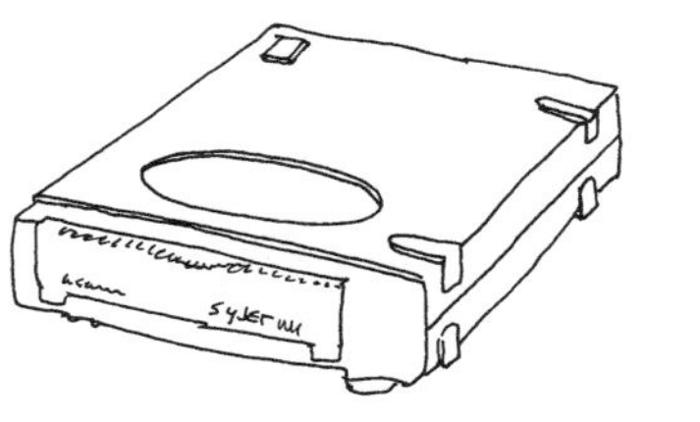

D

PowerPCs and Pentiums are the latest-generation processors, and they operate from two to ten times faster than their immediate predecessors—the 68040 CPU for a Mac and the 486 CPU for a PC. These older CPUs still power many computers at home, school, and work. Retailers also sell them at a discount, typically under $400 for a complete system with a monitor. Should you buy them? Probably not, especially if you think you will seriously use computers for animation.

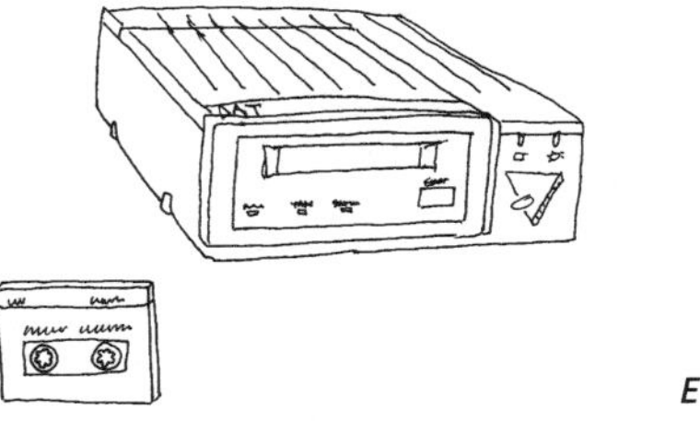

E

If you already own one of these older computers, the good news is that you can upgrade it by adding an accelerator board, replacing the CPU itself, or entirely swapping out the motherboard. Your exact upgrade strategy depends on the computer model you own, how much you can spend, and the type of animation you intend to do or grow into. Upgrades are not inexpensive. They typically cost between $400 and $1,200. You may be better off purchasing a new computer and selling your old one as a down payment.

How fast a processor should you buy? The faster the better! Any processor over 200 MHz for the Mac and 233 MHz for a PC should be enough for most cel animation projects and even for simple 3-D animations. If you are pushing beyond that—say, complex 3-D projects involving characters who talk and interact (as opposed to space vessel fly-throughs)—then processors ranging from 266 MHz to 400 MHz are necessary.

Top-of-the-line Macintosh and PC systems, priced

24.8 Megabyte scales: The formula for figuring out the storage requirement or file size of an uncompressed image is as follows:

$$\frac{\text{(Pixel image size) X (The bit depth)}}{\text{divided by eight}} = \text{image size}$$

For example:

$$\frac{(640 \times 480) \text{ X } (8\text{-bits})}{8} = 307\text{K}$$

Sample Chart:

File Size 640 x 480	1-Bit	8-Bit	16-Bit	32-Bit
	39K	307K	614K	921K

Units of measure for memory or storage size:

1 bit is the smallest unit
8 bits = 1 byte
1,000 bytes = 1 Kilobyte
1,000 Kilobytes = 1 Megabyte
1,000 Megabytes = 1 Gigabyte
1,000 Gigabytes = 1 Teraflop

around $6,000 to $10,000, now use more than one processor to go even faster. Special software is required to take advantage of this capability. Fortunately most software developers recognize this growing trend and are rewriting their software to support multiple-processor computers.

Right now the speed game is being played out in a marketing war between Intel and Motorola, the makers of Pentium and PowerPC, respectively. Each side is declaring itself king of the hill. So who really has the fastest processor? An independent study by *BYTE* magazine says the hands-down winner is the PowerPC, beating the fastest Intel Pentium chip by up to 100 to 200 percent based on raw processor performance. This can be misleading; real-world use of these processors finds them operating almost evenly in many applications since the PC uses a faster system bus and better, more flexible memory schemes. Nevertheless, the edge in real-world use still goes to the PowerPC and may in fact increase as the Macintosh switches to the same motherboard architecture as the PC, negating any system bus or memory advantage currently enjoyed by the PC.

The future for both the Pentium and PowerPC is very bright. Top processing speed will go from the current 400 MHz (in 1998) to well over 800 MHz and perhaps even 1 GHz (gigahertz) by the year 2000. Keep your seat belts on; it's going to be a fast and bumpy ride.

MEMORY

Digitized information—the little zeros and ones that make up computer code—can fill up a lot more space than you might imagine. You'd better get used to the idea, right now, that if you are going to do digital animation, you will surely become a memory and storage hog.

As you are about to see, computer memory comes in different forms and even comes in different pieces of equipment. Generally for multimedia use, memory increases the speed at which your computer works, and ample storage allows you to deal with the various components of your work in a single process. Here's a primer.

Hard Drive. Most home computers come with an average of 2 *gigabyte (GB),* or 2,000 megabytes (*MB*—a *megabyte* is one million of those zeros and ones) of hard drive, the built-in storage device that holds all those files and stores the software applications. A 2 GB hard drive is roughly equal to 2,000 3½-inch diskettes. That may seem like a vast digital pool, but it's not. With the growth of multimedia and on-line services, many off-the-shelf computer models feature 4 GB hard drives—what it takes to handle heavy loads of sound, video, and animation. The problem is (no surprise here), the bigger the hard drive, the bigger the price tag. If you are buying a new computer, the smallest hard drive you should consider is 2 GB. If you are adding an external hard drive to an existing computer system (which is easy to do), you can expect to pay $150 for 1 GB or $250 for 2 GB of hard disk storage.

A single color image created in an application like Photoshop can easily be 30 MB or larger. Since you sometimes need hundreds of such images to create a single piece of animation, not to mention the sound and video files you'll need to work with, it shouldn't be a surprise that hard disk space will disappear quickly.

A small but essential point of speed for animators and video editors is *throughput,* or *sustained speed,* of hard drives. These don't matter much to most folks because 3 MB/sec—the standard throughput for regular hard drives—is great for text. However, Photoshop and video capture/edit systems run much better with *ultra* (or at least *fast and wide*) stripped and arrayed drives running at 7,200 RPM for a minimum throughput of 7 to 21 MB/sec. Many animators and digital editors pay serious attention to this.

RAM. Random Access Memory is one of the most important things to consider before purchasing a computer. RAM helps determine how fast a computer runs and how many different pieces of software can be used at the same time. The standard amount of RAM for a home computer hovers around 32 MB, just enough for an animator to pull off smooth animation. Ideally you will want to add at least another 32 MB (for around $50) or 64 MB (for $100).

Because of the complexity and size of operating systems

24.9 CASE STUDY: 3-D/2-D Fusion

A

B

C

As digital tools become more accessible, they become more easily and more widely explored. In the process, the relative strengths of different animation techniques start to become more apparent and their boundaries less limiting. All of which leads, importantly, to the breakdown of those unperceived compartments we seem to create for ourselves as each new wave of innovation offers its approaches to animation.

D

As a case in point, let's look at character animation and 3-D animation—two forms that seem so very different. Thanks to the power of desktop computer systems and to the ever-increasing sophistication of software products, the well-mapped world of traditional character animation is rapidly finding ways to merge with the more recently discovered domains of 3-D image manipulation. The result could be called *Animation Fusion*.

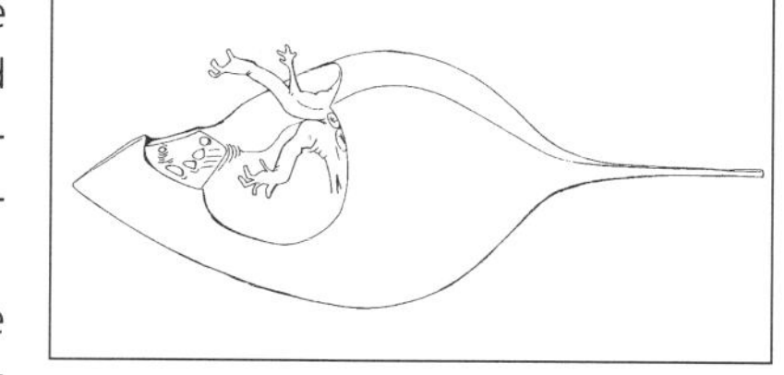

F

The figures that accompany the analysis here provide a quick look at full animation sequences created by San Francisco's Wild Brain studio for an interactive CD-ROM game titled *Flying Saucer*, a stand-alone product developed and published by Postlinear Entertainment. Jason Porter, Wild Brain's Technical Director, points out some dimensions of the project that suggest yet broader innovations that will come as digital tools and increased bandwidth become creative realities during the next few years.

E

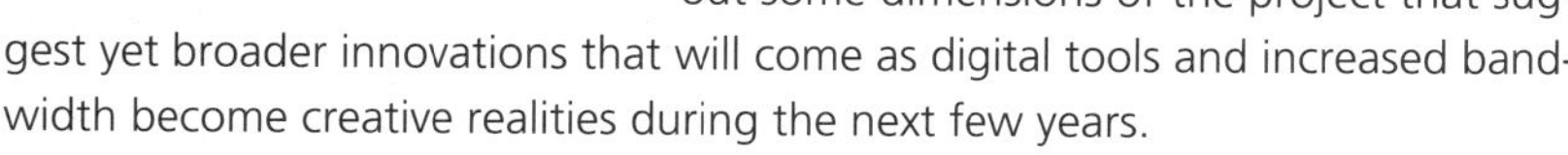

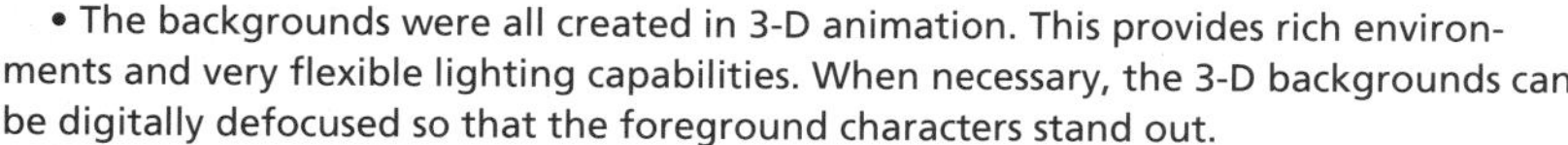

- The backgrounds were all created in 3-D animation. This provides rich environments and very flexible lighting capabilities. When necessary, the 3-D backgrounds can be digitally defocused so that the foreground characters stand out.

G

H

I

J

• The character animation is done in traditional form: hand-drawn images that are scanned into the computer. This allows animators the full control required for expressive and subtle movement and lip sync.

• Digital ink and paint allows for a soft-edged, "airbrush"-styled coloring. The look intentionally mirrors the rendering techniques found in Japanese Anime—a high-octane style of representational animation that is having a large impact upon comic and animation aesthetics.

• The world of computer gaming is itself a fusion combining narrative or "story" elements with interactive sequences in which the viewer becomes a participant (for the most part, jockeying a flying saucer through 3-D environments).

• The marriage of 2-D and 3-D elements into a seamless composition is not easy. In the *Flying Saucer* cinematics, the compositing was achieved primarily using Adobe After Effects run on a Macintosh 8500. Many other compositing applications are available or in development that allow extreme flexibility in mixing 2-D and 3-D, and working in a 3-D workspace. 3-D applications are also becoming more capable in the handling of 2-D objects and 3-D rendering treatments are allowing 3-D objects to look like 2-D, or even traditional abstract or surreal art styles.

Frames *(A), (B),* and *(C)* sample the rich fantasy landscape that appears in the game. Note how the backgrounds use 3-D techniques to build complex environments for the action. Character design was done in a quasirepresentational style that would permit drawn cel characters to fit into the 3-D world. The four poses in *(D)* show head details of one of the protagonists, Emily. The pair of drawings in *(E)* show two "turnaround" poses of a character called Grey Elder. Figure *(F)* shows a podlike creature called Signer.

K

The hero of the piece (in addition, of course, to the game viewer/player!) is named Boone. *(G)* is a color chart for Boone. Even here in this black-and-white reproduction, you can make out the soft-edged, airbrush-shading techniques employed in the digital ink and paint phase.

The fusion of a single scene is represented in the remaining images: *(H)* shows the animator's rough drawing for Boone, as he looks out a window. *(I)* is the cleanup that was scanned to make a digital file. *(J)* is the colored version of the same element. *(K)* shows the window and background (rendered from a 3-D model). *(L)* shows one of the Alpha channels that was used to matte and composite the different source images together. Image *(M)* is the final composite image of this scene.

Photos are provided courtesy of Wild Brain Studios and Postlinear Entertainment. The Flying Saucer *(© 1997 by Postlinear Entertainment, Inc.) game is due to be released in 1998.*

L

M

24.10 Monitors are not TV screens: But they sure do look like TV screens before you turn them on! The latest monitor designs feature adjustable pedestals, as shown here. Well-respected manufacturers include Sony, Samsung, NEC, Mitsubishi, ViewSonic, and Apple.

and other systems software, you need to have at least 10 to 20 MB of free RAM to run most of the motion graphics packages featured in this book, along with the 10 MB needed to run the system. Production domains that require working in full-motion video, like editing, can require even more. Programs like Premiere and Photoshop save multiple copies of images in the memory of your computer, enabling you to perform those convenient "undo" and "revert" functions, which takes a lot of RAM. Premiere recommends a whopping 128 MB for video editing. *Always, always, always* buy as much RAM as possible. If your budget is slim, purchase a decent amount now and more later. Your local computer store will be glad to show you what to purchase and help you install the extra memory. It's not difficult to pop open your computer and insert a memory card into a slot that awaits it. But it's easiest to get a pro or an experienced friend to help you make such a memory upgrade the first time.

Supplemental Storage Devices. Sooner or later you are going to find it extremely useful to have one of the external storage devices that are available for desktop computing. Inevitably, your hard disk's storage space will dwindle, no matter how much you compress files or how often you clean it out. So look into acquiring an *external drive* or *removable storage device* such as SyQuest, Zip, Jaz, or DAT drives. Besides allowing more room for your work, they're perfect for backing up files, archiving large chunks of data, and swapping files with others with whom you will be working. Ask around your creative community to see what format people are using. Currently, the two most popular types are made by SyQuest and Iomega (Zip and Jaz).

MONITORS

Almost all computers in use today feature color monitors that can display millions of colors, provided there is enough video memory in the computer to display the colors. *Video memory* or *video RAM (VRAM)* is similar to application RAM, except that it stores images instead of documents or applications. VRAM serves as a temporary storage space for an image just before it is displayed on the monitor's screen. If there are any changes to the image, even a single pixel (which is the smallest unit of a screen image), the VRAM is where that change is stored and from where it is fetched for screen display. Most computers come with 2 MB of VRAM, but they can be upgraded to hold up to 8 MB of VRAM. Why so much? Because of two factors: screen resolution and color depth.

Screen Resolution. Most monitors, regardless of their actual physical size, support multiple screen resolutions (these are often called *multires* or *multisync* monitors). They allow a screen to display 640 by 480 pixels, 834 by 624 pixels, or even up to 1,024 by 768 pixels. In a standard 15-inch monitor, using the 640 x 480 resolution mode provides an ideal image size that's not too small or too large and displays an image at approximately the same size as it would print. If in the same 15-inch space you then set the display resolution within the control panel to the maximum 1,024 by 768 pixels, everything will suddenly look twice as small. With a 17-inch or larger monitor the "smallness" effect is not as pronounced. For a 17-inch monitor, 834 by 624 resolution approximates a 1:1 ratio of image display on the screen and the actual image size that will be printed. For a 21-inch monitor, 1,024 by 768 is the ideal.

What happens when you increase the resolution is that you double the amount of pixels to display within the same screen size, allowing more screen objects to appear and providing more screen real estate to maneuver. It's like zooming in and out with a camera telephoto lens. But this flexibility does not come for free.

Color Depth. By increasing the amount of pixels to display, you *may* reduce the number of colors or the color depth of the image because there isn't enough video memory to hold all of the image information. For example, 1 MB of VRAM can yield from 256 colors up to 32,000 colors for pixel sizes of 640 by 480 and 834 by 624. If the image size is 1,024 by 768 pixels, then there is only enough video memory to show 256 colors. But by adding another megabyte of VRAM, up to 16 million colors can be seen at a resolution of 640 by 480 and 834 by 624 pixel images, and up to 32,000 colors for 1,024 by 768 resolution.

If you are going to be working in animation, it's a good idea to get at least 4 MB of video memory.

Large Screens = Luxurious Animating. The size of the monitor can also be important. Creating animation will often involve using many software applications at once, and each

24.11 Geometric fables: Frame from a series of fifty five-minute pieces created by Fantome's Renato and Georges Lacroix for international television. © *1996 Fantome Animation.*

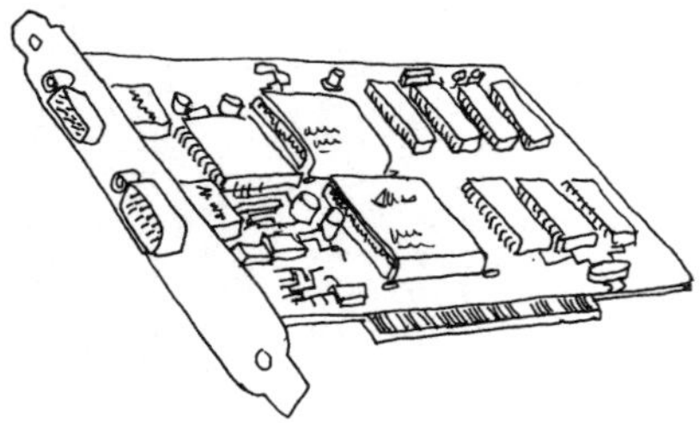

24.12 Accelerators: This is an example of a 3D add-on accelerator card.

application can use many different windows and tool palettes. A standard 14- to 15-inch monitor will work fine but doesn't allow the extra room often needed to navigate through layers of windows, folders, and files. A 21-inch monitor provides a great screen size to work on, since it offers plenty of room to view the image file or animation and still manages to provide screen space for the tool sets of multiple applications. It's worth noting, however, that 21-inch monitors are huge and will hog the top of your (literal) desktop.

Some people set up two monitors (using a second video card) that can operate off the same Macintosh computer. In the Mac dual-monitor setup, a user can move the mouse and all (virtual) desktop objects seamlessly from one screen to the next. The typical configuration is to use a large monitor to show the animation field being worked on while the second monitor, sometimes smaller in size and of a lower grade, is used to handle the windows and files of the various graphic software applications. Up to seven monitors can be driven by one Macintosh, provided there is a video card for each monitor.

Buying a Monitor. The rule: Never buy a monitor you haven't seen, regardless of price, because if you don't like what you see, you won't be happy in the long run. Once you've seen the actual monitor in operation, you can then buy it from a retailer or by direct mail. Because monitor technology is continuously improving, read the latest computer magazines with reviews of monitors before making a purchase. Today, many monitors include built-in speakers, headphone jacks, tilt-and-swivel bases, and front-panel controls to adjust brightness, contrast, resolution, size, and so on. The array of features can be intimidating, and some add to the price tag, so stick to the fundamentals.

One of the most important things to consider is *dot pitch* or how finely the pixels are displayed. Monitors are made using a mask that focuses and filters the image before it reaches the physical screen that we see. The finer the focus and filter (its dot pitch), the sharper the image. Get only .28 dot pitch or less. Cheap monitors sometimes have .39 dot pitch, giving you a fuzzy picture with grainy detail, with which you may get killer headaches and think you are going blind. Avoid them. Also, make sure the *screen refresh rate* is no less than 60 Hz, preferably 75 Hz or more, otherwise you'll notice a flicker (especially in your peripheral vision), resulting in migraines.

Don't get too enamored of the physical size of the monitor. Check instead the actual *viewable image size.* It should be no less than 13.9 inches for a 15-inch monitor, 15.6 inches for a 17-inch one, and 18.7 inches for 20-inch monitors. As far as prices, expect to pay from $150 to $250 for 15-inch, $300 to $700 for 17-inch, and $1,300 to $2,000 for 20- to 21-inch monitors.

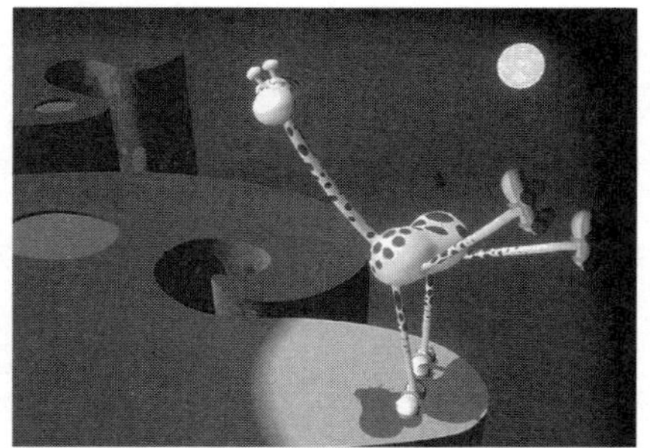

24.13 Les Girafes: Two frames from a nonverbal series of fifty one-minute pieces developed for international television and adapted from Guillermo Mordillo's comic *Les Girafes © 1996 Fantome Animation.*

PERIPHERALS IN GENERAL

In the world of graphic computing, the word *peripherals* covers a lot of ground. Following are short descriptions and accompanying line drawings that will introduce you to all the basic tools you can add to your computer system.

Keep in mind that fledgling animators—well, even experienced ones—do not need to own each and every component of the digital animation studio. Lots of times you can find a business or educational organization that will feel fine about lending you access to costly peripherals. And there are places called *service bureaus*—small businesses that provide specialized computer services on a charge-per-service basis. There are good and bad service bureaus, of course, and you just have to find one that will give you the great technical support and help you need, plus access to first-rate equipment, and fair prices.

Small-business operators and the IS (information services) folks at large corporations often seem to have a soft spot for artists. Don't be shy about introducing yourself as an independent animator. Tell people about the project you are working on. Ask their advice. If you are passionate enough, sometimes the business or institution that owns and/or controls the peripheral you need will work out a special deal.

GRAPHIC ACCELERATORS

An *accelerator* is the single most important addition to your computer animation system. For around $100 you can increase the drawing, rendering, and playback speed of your animation from between 100 percent to 400 percent. How's that for a return on investment? Spending more will give you even better performance. The midrange of accelerator cards run around

$300 to $600, which is still way, way less than the cost of upgrading to a new computer system. Top-end cards cost $1,000 to $2,000.

There are five things you need to be smart about once you are smart enough to look into accelerators.

First, the accelerator comes in the form of a plastic and silicon card that fits in the *expansion slot* of your computer. So read up on the physical size requirements of the card and its connector type (such as PCI, ISA, or NuBus) in your operator's manual.

Second, be aware that the accelerator has to support the *graphics library* of the software and operating system you are using. Graphics libraries are stored collections of graphic "primitives" or basic shapes such as squares, circles, cubes, and triangles that all animation software uses to create everything from simple images to complex ones. Think of the graphics library as the "alphabet" of computer visualization that works analogously to a writer using text alphabets to compose poems or stories. Your graphics library will most likely be one of the following: OpenGL, Rendermorphics, Heidi, or QuickDraw 3-D. Check the software manuals of your favorite programs to learn what libraries they use.

Third, the accelerator card should support *different shading routines* such as Phong and Gourad as well as other features like *texture-mapping, QuickTime,* and *MPEG* (see Chapter 23: Computer Software).

Fourth, the card should be *upgradeable* either by software driver updates or by plugging in a new chipset on the card. (Chipset plug-in capability is usually for high-end cards only.)

Fifth, some accelerators now include a TV tuner so you can watch regular TV programs in a window on your computer monitor. If you buy an accelerator with this feature, you can also use the card to capture still or video images (although you may be faced with many compromises in picture size, frame rate, and image quality).

24.14 Drawing tablets: Shown here is the full line of Wacom drawing tablets. The smallest has a 4 x 5 inch screen ($150) and the largest is 18 x 25 inches ($2,000). Accessories include DuoSwitch Erasing Pens; four to six button pucks. There are a few manufacturers, some with lower prices.

VIDEO EDITING CARDS

To "capture" video on your computer and then play it back at the proper frame rate, you will most probably need a special piece of hardware called a *video editing card.* Like graphic accelerators, a video editing card offloads processing work from the main processor, allowing you to capture video at nearly broadcast levels. These cards capture and play back video at a rate of 30 frames per second (60 fields per second of alternate scan lines), using video compression technologies to make the resulting video file manageable on a hard disk (see Chapter 23: Computer Software for more about compression).

Video editing cards are very complicated and sophisticated peripherals. While they are getting cheaper (around $500) and are now much easier to use, in order to get the best results they require very precise operating parameters that are matched to your specific computer system. If you acquire a video editing card, you will need more hard disk space to go with it, probably between 2 GB and 8 GB. Another option is to use 1 GB removable cartridges to expand the memory. Regardless of which storage media, video gobbles tons of memory—around 100 MB per minute of video at the low end. Overall, expect to spend somewhere between $1,000 for VHS-quality video capture (consumer standard) and $15,000 for broadcast-quality video capture and playback.

Video cards lead the animator toward a burgeoning area known as *desktop video.* Recent advances in computer speed and storage are now allowing people who work in their homes or at small businesses to operate with the same tool set that would have cost hundreds of thousands of dollars just a few years ago. Digital cameras, multicamera switchers and effects generators, nonlinear editing systems, and multitrack mixing have all made their way to desktop video.

You will need to investigate much deeper than these pages should you get into serious desktop video production—especially editing. There is lots of information available via books and on-line services, but perhaps the best approach is to talk about your interests with a professional video editor or with one of the people who run a video postproduction shop. Have them show you the interface of a professional-level nonlinear editor. Take a look at the configuration of monitors, control tools, and storage media that it takes to operate professional-level video suites. Or you can find expertise (and support) from companies with digital studios, cable companies, trade shows, or—perhaps best of all—from the guys who do wedding videos. They'll know firsthand, usually through seri-

24.15 Scanners: For around just $100 you can get a good color scanner that gets the job done right. Minimally it should capture 24-bit color (16 million colors) at up to 300 dots per inch (dpi) of resolution. Interpolation software that comes with the scanner can increase the resolution to anywhere from 600 to 1200 dpi. The drawing here shows a Nikon Flatbed Scanner AX-110 with optional document scanner ($400). It has 24-bit color with 300 x 600 dpi optical resolution. Scanners range from the low end — 30-bit color depth ($100 to $300) — and from the midrange — 36-bit color depth ($500 to $900). In animation, there isn't a practical value for high-end scanners because the additional resolution they digitize can't be discerned.

If you are going to use your scanner for lots of pencil testing, an optional feature to add is a multiple sheet feeder that will automatically scan in a stack of drawings. This isn't essential, but it can be handy when an animation contains hundreds of drawings, and scanning them all in one by one can become tiresome.

ous trial and error, what really works. Another great (and often untapped) source of free information, training, and access to video suites is your local cable company, who is required by law to have a community video access program.

DRAWING TABLETS

For the traditional animator making the transition to a computer, a *drawing tablet* is an essential investment. In tandem with the stylus and graphics software that come packaged with it, a tablet allows you to draw directly into the computer, much as you would with a normal pen or marker, and then manipulate the image.

Most tablets today are pressure-sensitive, allowing you to change the darkness and width of lines by applying more pressure to the tablet surface with a *pressure pen.* This is particularly valuable for people used to working with traditional drawing tools since it mimics their look and feel. Digital drawing tablets come in a range of sizes, with some large enough to cover half a desk and others that can neatly fit in the palm of your hand.

SCANNERS

If you are going to marry the precision of traditional character animation techniques with the speed and facility of digital ink and paint, then you are going to be spending lots and lots of time hunched over a scanner. This specialized piece of hardware can take a photograph or drawing and *digitize* or copy it into the computer. Chapter 15: Digital Ink and Paint shows the techniques of scanning drawings into the computer. Remember, too, that scanners can be used to generate images from many sources other than registered paper sheets—see the case study in Chapter 5: Cutouts.

Scanning at resolution rates above 24 or 30 bits is not necessary for most animation projects because the additional image fidelity cannot be noticed on a computer monitor or a TV screen. Besides, high-resolution scans take up a lot of hard disk space and will slow down the playback of your animation

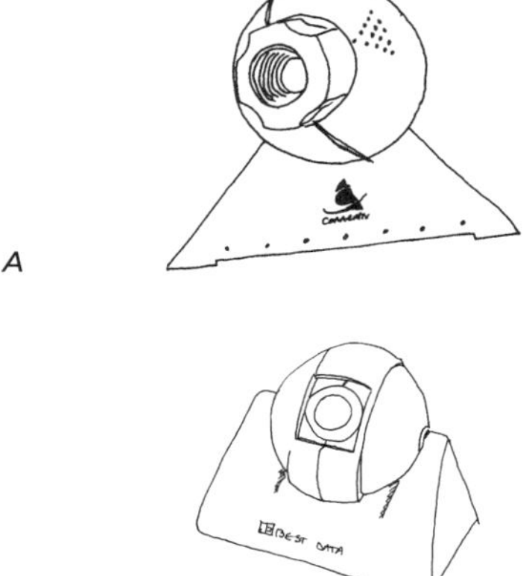

24.16 Desktop cameras: Connectix QuickCam, *(A),* at about $230, and Best Data's SmartOne, *(B),* at $200, both get high marks.

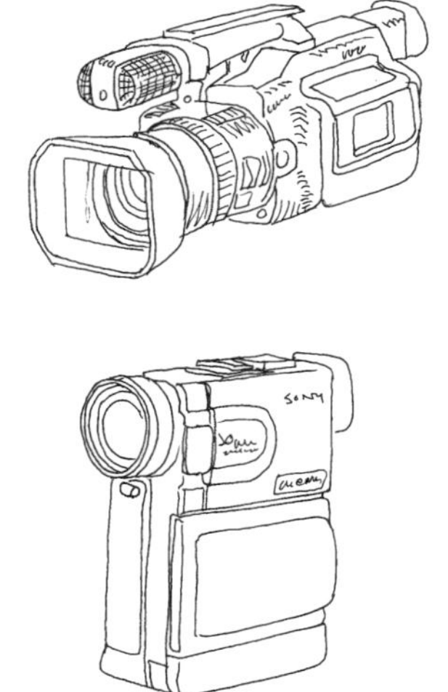

24.17 Video cameras: There are hundreds of consumer video cameras from dozens of manufacturers. Pictured here are two all-digital cameras from Sony — one from the low end and one from the high end. Prices are changing so fast, you will need to use the Web or talk with a good dealer to check out options for your budget.

because the computer has to carry the extra baggage of high image definition, even though it can't be seen.

There are no hard-and-fast rules as to how you should scan your artwork. It depends on the scope of the project, your computer system, the devices used to play back the animation (TV, computer, video, or film), and the overall look and feel you are trying to achieve. By owning your own scanner, however, you will be free to experiment until you get the best results.

DIGITAL CAMERAS

Sometimes a whole corner of computer technology seems to make a quantum jump overnight. It feels as though this is happening in 1998 within the realm of *digital cameras.* Here are three different kinds of cameras that have been around for a while but which promise to undergo a simultaneous burst of technological improvement and cost reduction. There is no doubt that computer animators will soon have a variety of digital cameras in their hands that will expand the creative process while also speeding it up and making it easier.

Desktop Minicams. Connectix Quickcam pioneered the capture of "postage-stamp" videos from a computer desktop, allowing you to send them via E-mail or embedding them in reports or newsletters—all for $99. Now there are over two dozen manufacturers of these tiny video gizmos offering black-and-white and color models, ranging in price from $49 to $300. Some computer manufacturers include them as a standard feature similar to CD-ROM and modems. This is in response to the explosive growth of the Internet and its use for videoconferencing. Can you use these tiny video cameras for animation? Absolutely. They make a terrific tool for storyboarding or for conceptualizing your animations because they capture not only moving images but still images. Another good thing about minicams is that they don't need video capture cards to input the video to the computer. Instead, they plug into a computer's standard serial or parallel port and use their special software to process the input signal. If you can't afford to get a scanner, then get a minicam.

Portable Digital Still Cameras. If you can wait a week to get your regular 35mm film and prints back, then you can use any camera as a digital camera. Most photography stores offer

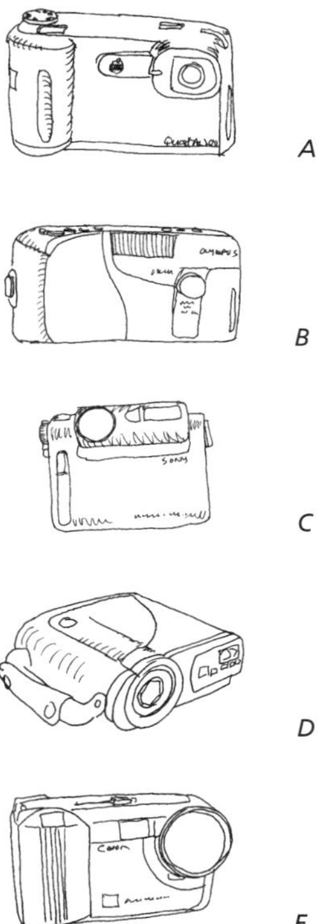

24.18 Portable digital cameras: These tools are changing so fast. We are just seeing the end of the 640 x 480 resolution limit, which was originally thought to be the standard because VRAM chips were cheap and because camera makers thought TV would be the main viewing medium. We are just seeing the introduction of the third generation of these cameras, where 1,000 x 800 is available for $900 and less. These represent the recommended choices for portable digital cameras.

(A) Apple QuickTake 200 ($600): 2 MB RAM memory; 640 x 480 resolution limit

(B) Olympus America D-300L ($900): 6 MB RAM; 1,024 x 768 res

(C) Sony DSC-F1 ($850): 4 MB memory; 640 x 480 res

(D) Kodak DC40 ($600): 4 MB memory; 756 x 504 res

(E) Canon PowerShot 600 ($950): 1 MB memory; 832 x 608 res

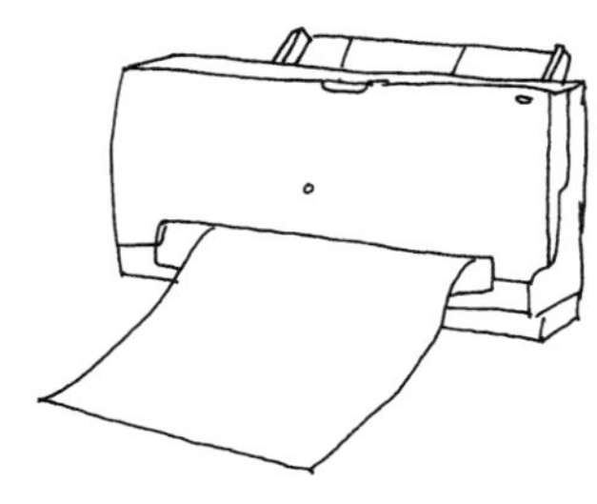

A

B

24.19 Ink-jet color printers: Here are two typical models: *(A),* an Apple StyleWriter ($200), and *(B),* a Hewlett-Packard Ink Jet 870Cse ($400).

two choices when you bring in a canister of exposed 35mm film: You can have it processed in the old-fashioned way and get either prints or slides, depending on the kind of film you shot, or you can have it developed and digitally scanned and get a photo CD or a diskette of your photos, ready to slip into the computer and view.

If you can't wait a week, then expect to spend between $350 and $2,500 for a digital camera that can connect directly to your computer, download its digital picture files from its memory banks, erase the old pictures, and allow you to take new ones.

VCRs and Video Cameras. Your home VCR and video camera are wonderful tools for animation. A VCR can play back for your study and analysis most of the best animated films ever. And if you can input the video signal into your computer using a graphics accelerator or video editing card, then you study and trace over, frame by frame, the animated segments you've captured (see Chapter 13: Rotoscoping).

The standard consumer video cameras can work the same way, except you get to create your own original images and not just work with footage that comes from an existing movie or TV show. Take a video of a baby crawling, go for a spin on a skateboard or bike (don't crash and ruin the camera), tape yourself and friends involved in a favorite sport, or simply use your video camera to get head shots of friends and family. Input such images into the computer and use them as source material for your next animation project.

If you are thinking about the purchase of a video camera, we recommend that you opt for the desktop minicam solution simply because a minicam uses the serial or parallel port of the computer to input the video and its accompanying software does the rest.

PRINTERS

Printing text and graphics is essential to the life of an animator. It's the only way to do storyboarding, scripting, budgeting, and production planning. Animators need to print out exposure sheets, character studies, model sheets, and layouts so that they can pin them to a bulletin board and have various members of the animation team study them. In short, a printer is essential just about everywhere. Because of this necessity, printers are commodity items and there are literally hundreds of models to choose from, ranging in price from $199 for an ink-jet color printer up to $10,000 for a desktop commercial laser color printer.

Before going further, let's take a moment to look at how printers work. Printers generally come in two types: ink-jet or laser. Ink-jets work by spraying ink through superfine apertures (pinpoint holes) onto the paper surface, moving side to side across the page according to a pattern set by the computer. In effect, ink-jets are literally painting a picture on the page line by line, not unlike spray-painting the side of a house. Although the inks dry very quickly, touching them soon after printing may cause a smudge. Laser printers, on the other hand, use dry dustlike toner particles that adhere to a highly polished metal drum that has been magnetized according to the image it's trying to print. As the paper goes through the revolving drum, the dust is transferred to the paper, while a heating unit bakes the dust to the paper to permanently adhere it.

The recommendation that follows will cover only color ink-jet printers. This is because laser printers, regardless of brand and price, are universally good. The variances in print quality between inexpensive models and expensive ones are microscopic. Most brands of laser printers compete in superfluous bells and whistles and the amount of memory the printer has. For home users interested in basic laser printing, get the most affordable laser printer that supports PostScript and it will be just fine.

What should you look for in a color ink-jet? First, high resolution, around 600 dpi, should be the minimum resolution for printing both color and black-and-white images. Second, it should support dual ink cartridges—one for the black ink, the other for the color inks. Some models support only one cartridge, typically for the color inks. To print "black," the primary colors are mixed together, but the result is not true black—it's a grayish black. This also uses up the color inks more quickly, especially if you are printing mostly black-and-white text, so avoid single-cartridge ink-jet printers. Third, the printing speed should be at least 2 pages per minute when printing color documents and 5 to 7 pages per minute for black-and-white. And finally, upgradeability, especially for PostScript support, is important. (*PostScript* is the universal language of printers, giv-

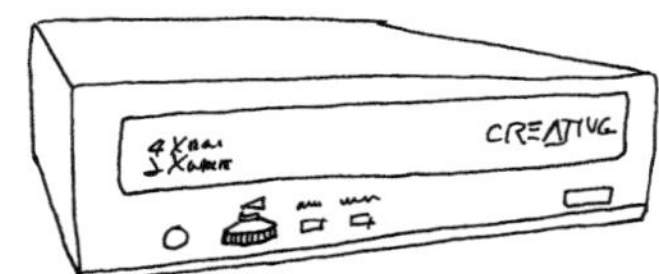

24.20 CD-ROM and friends: CD-ROMs get faster and faster—now reaching 24-speed and soon beyond. CD-Recordables are finally affordable and easy to use. The drawing here is a Creative Labs Blaster CD-R 4210, which doubles as a quad speed CD-ROM drive. The cost is $500. Soon DVD units will be routinely packaged inside desktop computers.

24.21 George's piano: In the world of serious cooking, a chef's stove is called his "piano." Here is the home animation setup of contributing author George Escobar, with his two production assistants, Gregory and Stephen. *Courtesy George Escobar.*

ing you better printing control and flexibility to enlarge and reduce images than the generic print drivers that come standard with the printer.)

CD-ROM AND DVD

The CD-ROM has become a standard playback device and comes packed inside most of today's computers. One CD-ROM holds as much data as 450 3½-inch diskettes. But if you think that is something, wait until you see the next generation of dual-sided DVDs (digital videodisks), which hold as much as ten CD-ROMs. If you get into making long digital animation or multimedia productions, look into acquiring your own CD-ROM recordable or rewritable drive (CDR). The recordable drive allows you to imprint a CD once; a rewritable can be imprinted repeatedly, but these recordable drives cost about twice as much.

MODEMS

Solitary animators locked to their drawing tables, cameras, or computer monitors cry out in search of a kindred spirit, who understands that a passion for animation can lead to a profound isolation. But they are not alone!

Salvation resides in a few circuit boards that link your computer to a phone line. We're talking about *modems,* of course. This important component of your modest digital animation studio connects you with the major animation studios (Disney, Pixar, and Warner Bros., for example), animation software companies (Adobe, Electric Image, SoftImage, Strata, Alias/Wavefront), animation schools, animation magazines, and best of all, other animators. There's a big, vibrant, and growing animation community out there. So make sure you learn to use your modem. Since many computers now are shipped with a built-in modem, there is no excuse not to connect. (To learn more about the Internet, read Chapter 25: Resources.)

If you are going to use modems as a *distribution* vehicle, you need to know something about how they work. Modems are basically interpreters. They convert analog signals into digital ones that a computer can understand and vice versa. If you've ever picked up the phone handset and listened to a modem or fax connection on the other end, you have undoubtedly heard the actual voice or screech of a modem. That unintelligible noise is generated by a modem chip that has converted digital ones and zeros into various chirps and beeps that can be carried by standard phone lines. Because the chirps and beeps are so packed and condensed, being created from typically up to 28,000 digital bits of ones and zeros per second, the actual analog signal sounds out of this world. On the receiving end, another modem converts the analog signal back to digital ones and zeros and then the computer takes over from there, turning these digital bits into pictures, words, or music. (Fax machines work the same way too!)

Connecting a modem is fairly straightforward. Take a standard phone wire and connect one plug into the modem's phone jack. Then take the other end and plug it into the phone jack on the wall. And finally, take the interface plug from the modem and plug this into the modem or serial port on the back of your computer. Launch the communications software or just use an *Internet service provider* (such as America Online, which has an easy built-in procedure for signing on), enter the required phone numbers, names, and passwords, and you're connected. If you're a new member, you don't even have to pay for the first month of service. So be sure to try it out. The other advantage of modems is that you can talk directly to other modems to send files to service bureaus, clients, or your friends. It's really a great way to stay in touch and to learn the latest in the world of animation. For some specific information on speed and operating costs, see the section on Web animation in Chapter 17: Animation Frontiers.

WHAT DO I NEED? WHICH SHALL I BUY?

This entire chapter can be seen as a primer for that big moment when you've decided to either purchase your own computer system or upgrade an existing one. What do you need? Which will you buy? Those are the questions that this chapter has been anticipating.

Choosing the right computer tools, both hardware and software, starts with understanding what you want to accomplish, how quickly you want it done, and of course how much you can afford. But before deciding on the exact computer model, you must also know what to expect.

The Right Tool(s) for the Task(s). To help you determine the computing power you need, Figure 24.22 provides a table to give you a self-diagnostic matrix that seeks to combine five disparate elements:

1. *Computer Power* (emphasis on CPU/processing speed)
2. *Peripherals* (all described in this chapter)
3. *Cost* (low end, midrange, and high end)
4. *Animation Tasks* (eleven of them are broken out)
5. *Animation Techniques* (by the categorization in Part III: Techniques)

24.22 The Escobar Analysis Grid:

THE ESCOBAR ANALYSIS GRID			
Production Tasks	**Low End: $2–5K** CPU = 75–100 MHz	**Midrange: $5–10K** CPU = 100–166 MHz	**High End: $10–20K** CPU = 200 MHz plus
Writing/Publishing	More than adequate for outlines, scripts, and sophisticated desktop publishing.	Blazingly fast for this task alone.	Complete overkill.
Storyboarding	Can handle most "slideshow-based" storyboarding tasks in black and white and color. Adding audio and transitional effects will start to slow playback of storyboards from the computer—but you can print them just fine.	On the upper end of this range, the storyboard can use any media you want to add such as, audio, photo, digitized video, and animation. This "multimedia-based" storyboard will probably require more up-front processing time, but it should play back nicely.	Much faster processing to prepare "multimedia-based" storyboards. Playback should be zippy.
Recording	Expect to record FM radio to nearly CD-quality audio, but you may need to get a sound board, especially with PCs, for reliable CD-quality audio.	Software-based audio solutions are possible. But most models at this level include built-in audio processors to easily record CD-quality audio.	Can handle multitrack recording and playback like a professional studio, depending on the sophistication of your audio software. No additional sound board should be necessary.
Scanning and Image Processing	Most of the initial heavy lifting is done by the scanner, however the computer still does the post-processing of the scanned image. Depending on the resolution (size of the image) and bit-depth (color intensity) of the scan, the computer may struggle in this task. It should handle black-and-white scans in a breeze.	Can do much better than the low end, but can still go slowly with large, colorful, complicated images. You can add special boards to improve image-processing task so it's closer to real-time performance.	You can post-process the scanned image at will without any hiccups or in most cases see it done in real time. For example, resize, rotate, flip, skew, remap, texturize, rotoscope, or do almost anything you can think of and this level of computing can handle it.

Production Tasks	Low End: $2–5K CPU = 75–100 MHz	Midrange: $5–10K CPU = 100–166 MHz	High End: $10–20K CPU = 200 MHz plus
Video Capture, Playback, and Transfer (to a video tape recorder)	Will capture and play small-size videos at between 10 to 15 frames per second (fps). Can be upgraded with video processing board to record and play back full-screen videos at 30 frames per second. Most low-end computers do not have video-out (transfer) capabilities. A video board solves this.	Can capture, play back, and transfer larger videos at faster frame rates, but not at full-screen size. Video processing board still necessary for Super VHS or Beta SP quality.	Capture and playback at 30 fps, full-screen videos, but still not as good as can be obtained using a video processing board. Can also perform MPEG video encoding (formatting) and decoding (playback) all in software. Encoding function will be slow, decoding will be fast.
Drawing and Painting	Speed is generally not the issue in this task. Low-end computers can handle painting and drawing programs just fine. It is when the finished drawing has to undergo further image processing (see above) that additional processor muscle is required.	More than adequate for this task and performs reasonably well for most image-processing assignments.	Exceptionally capable for drawing and painting tasks, including any subsequent image processing.
Modeling	Similar to drawing; should be no problem for most modeling jobs. Complicated objects or multiple models will likely be slow, but this computer can be enhanced with 3-D processing board.	Can handle more complex models before needing assistance from a 3-D processing board.	Complex modeling is this computer's lifeblood. Go ahead, make your model.
Animating	Pre-rendered animation can be done, but post-rendered animation playback (without compression) will be slow or necessitate resizing to a smaller animation image.	Can do simple pre- and post-rendered animation equally well, but complicated animation sequences could benefit from a 3-D processing board.	Unless this computer has multiple processors, even it can't guarantee superior post-rendered animation support. Better get that 3-D processing board.
Rendering	Don't expect 3-D rendering to occur in real time. Depending on the model, lighting, and animation settings, this could take a few seconds, minutes, hours, or days. Generally it will be long minutes and hours. Rendering in days is not unusual.	Some rendering can be done in a few minutes; most is done in several minutes; and more than occasionally in hours. Rendering in days can be expected.	Some rendering can be done in real time, often in minutes, sometimes in hours, and infrequently in days.
Printing	Similar to scanning, the heavylifting is done by the device (in this case the printer) and not the computer. But if your printer is not intelligent, lacks a processor or enough memory, then the computer has to do the work. The low end should perform well, but it can appear slow because it is waiting for the printer to render a line of image.	More than adequate.	Overkill.
Editing	This function taxes the computer in a similar fashion as rendering and image processing. The low end should be adequate for small, rough cuts of your animation. Without a video board it can't serve as your final output, unless it is for the Internet or CD-ROM.	Larger animation can be edited with this computer, but a full-screen size at full-motion video rates may be a struggle without the video board.	Full-screen, full-motion videos at 30fps are doable, but rendering of effects and fancy transitions will remain relatively slow.

CPU AS PERFORMANCE GAUGE (NOT PRICE)

Notice in Figure 24.22 that the speed of the processor is shown in megahertz (MHz). While this is not the only factor in the overall speed of a computer, you can use it as a handy gauge for deciding the level of computer that meets your requirements. It can also be used as your price-point gauge. A year or two from now the processor in the midrange column will be in the low end, while the high end shifts to the middle, and a new high end will emerge. The prices are likely to stay about the same for even more power in each category. That is the reality of computing.

Knowing this, do not be dissuaded from buying today. The opportunity cost for waiting another year or two will be greater than any perceived savings you may fantasize about. "Opportunity cost" is a business term for the income that is lost (or not earned) by doing nothing when an alternative (or opportunity) to earn income presented itself.

Admit it. You are thinking, "But if I wait to buy, I'll be able to do so much more!" Well, that is absolutely true. The problem is that the expectations of the audience rise proportionally so that what was acceptable as a low-end (or even a high-end) product no longer measures up. In the meantime, the skill level you need to climb to deliver audience-acceptable animation becomes even steeper. Those who are doing low-end work today can more easily handle the high-end work tomorrow.

A computer can help in many creative and production areas. But it cannot do one very important thing: It does not create the visual and emotional impact that comes with every piece of successful animation. *You* have to supply those critical ingredients. The computer-generated movie *Toy Story* succeeded because it was a *good, well-told story*—not because it was done using a computer. The computer is just a tool, like a pencil or a light box. If it is put to good use with enough skill and power, then your story can be well told too.

25 ▪ Resources

WRITTEN WITH TRISTA GLADDEN

The information in this chapter is bound to be quickly dated. The field of animation has seen tremendous growth since the first edition of this volume, and that growth continues, faster and faster.

The resources that fuel animation will change too. But the strategies and pathways for finding out what you need to know will most likely remain constant. At least that's the premise upon which we offer what we hope you will find a good starting point for learning more.

This chapter explores the categories of Books, Periodicals, the Internet, Organizations, Schools, Suppliers, Distribution, and Festivals related to animation. These lists are not meant to be comprehensive. They are quite selective. We've chosen only what we know to be accurate and durable. In compiling these resources, we leaned heavily on three sources: (a) New York University (NYU) Animation Station at http://www.nyu.edu/tisch/filmtv/animation/station.html, (b) the University Film and Video Association (UFVA) at http://www.temple.edu/ufva/index.html, and (c) the Animation World Network (AWN) at http://www.awn.com. Checking out these sites will provide the most up-to-date information on resources available in animation. Also, contributing writer Jan Cox of Manga Entertainment supplied much of the information on distribution.

At the very end of this chapter, there are two full-page reproductions that you may want to duplicate for use in your own animation: an animation field guide and an exposure sheet. There are also some special treats to lure you through these pages—miscellaneous cool pictures and photo spreads that make good reading as well as good looking.

25.1 Animation Year One: The history of animation begins with a 1906 film by James S. Blackton entitled *Funny Faces.* Here is a frame from it. *Courtesy John Canemaker.*

BOOKS

The bibliography that follows has been organized into three categories: Books on Making Animated Films; Books on the History of Animation, Studios, Animators; and Books on Topics Related to Animation. The last group is a hodgepodge of titles that, for one reason or another, seem to deserve mention within these pages. This is by no means a comprehensive list; it serves only as a sampling of the immense library available on animation in all its forms.

You'll note that a few titles in each list have been given longer annotations than others. In our judgment, these works constitute a "core collection" on animation. We've read them and liked them. This is not to say that other titles aren't valuable. But to be honest, we've only skimmed many of the books and some are included because they come highly recommended by several sources. You will, of course, have to do your own browsing to find works that serve your interests and your needs. Further recommendations can be found on the Internet at animation sites, as well as at www.amazon.com and other booksellers on the Web. They can often find out-of-print books as well.

There are a few references you should know about: the *Animation Industry Directory,* available from *Animation Magazine* for $75; the Animation World Network Web site at www.awn.com, which contains a monthly magazine as well as a slew of up-to-date resources; and the LA or NY Work Book, which lists artists and illustrators of all kinds.

BOOKS ON MAKING ANIMATED FILMS

Andersen, Yvonne. *Make Your Own Animated Movies: Yellow Ball Workshop Film Techniques.* Boston: Little, Brown, 1970. Sixteen millimeter animation for children. Specific techniques are described for flip cards, clay, drawing on film, cutouts. Filmography. Updated in 1991 and published as *Make Your Own Animated Movies and Videotapes,* with expanded focus on techniques for making flat and three-dimensional animation, exploring special effects, computer animation, editing, sound tracks, and where to buy equipment.

_____. ***Teaching Film Animation to Children.*** New York: Van Nostrand Reinhold, 1970. The director of the Yellow Ball workshop describes simplified animation techniques and the equipment needed for animation workshops.

Blair, Preston. *Animated Cartoons for Beginners.* Laguna Beach, Calif.: Foster, n.d. A beginning set of approaches for classic cartoon-type characters.

_____. ***Animation: Learn How to Draw Animated Cartoons.*** Laguna Beach, Calif.: Foster, 1949. Part of the "How to Draw" series. Published also as Advanced Animation. This large-format publication features Blair's own drawings along with a simplified methodology of using geometric shapes as structural foundation in designing and drawing your own animated characters. Although the book doesn't reach beyond the classic American or Disney style of character animation, it covers that art thoroughly and with such abundance of examples that the book is a classic.

_____. ***Cartoon Animation.*** Tustin, Calif.: Walter Foster Publishing, 1994.

Cawley, John and Jim Korkis. *How to Create Animation.* Las Vegas, NV: Pioneer Press, 1990.

Culhane, Shamus. *Animation From Script to Screen.* New York: St. Martin's Press, 1988. "How-to" book on animating and the creative process in general from a veteran of Disney, Fleischer, Lantz & Television commercials.

Education of a Computer Animator. SIGGRAPH '91, Course 4. Try a University library for SIGGRAPH materials, which are no longer published, but are floating around. You can also try www.siggraph.org for more recent SIGGRAPH publications.

Foster, Walter. *Cartoon Animation: Basic Skills.* Tustin, Calif.: Walter Foster Publishing, 1989. How to Draw and Paint Series, No. 25.

_____. ***Cartoon Animation Kit.*** Tustin, Calif.: Walter Foster Publishing, 1996.

Godfrey, Bob. *Animation as a Hobby.* BBC Books. It's out of print, but if you can find a copy, it's a good book for beginners.

_____. ***The Do-It-Yourself Film Animation Book.*** BBC Books. Out of print, but still around.

Heath, Bob. *Animation in 12 Hard Lessons.* West Islip, New York: Robert P. Heath Productions, 1972. Step-by-step guidance for the self-taught animator, with a good chapter on camera mechanics. Points to lots of good resource material.

Lasseter, John. *Principles of Traditional Animation Applied to 3D Computer Animation.* SIGGRAPH '87, Computer Graphics, Vol. 21, No. 4. Lasseter is a leading computer-animation director who worked on *Toy Story.*

Madsen, Roy P. *Animated Film: Concepts, Methods, Uses.* New York: Interland, 1969. This is one of the best volumes on animated filmmaking. Madsen's book is thoughtfully designed and presented. The volume's unique strength is the detail it provides on professional production techniques and tools used in full-cel animation. Additional areas of value include chapters on The Filmograph (Kinestasis), Planning and Drawing Cartoon Animation, Sound Recording and Bar Sheets, and Exposure Techniques. An excellent glossary.

Morrison, Mike. *Becoming a Computer Animator.* Sams, 1994. Covers computer animation history and technology, and has a slew of resources. The accompanying CD-ROM contains sample animation.

O'Rourke, Michael. *Principles of Three-Dimensional Computer Animation.* W. W. Norton & Co, 1995. Thorough look into concepts of 3-D computer animation, regardless of computer software or hardware.

Whitaker, Harold and John Halas. *Timing for Animation.* London, New York: Focal Press, 1981. Good for understanding the mechanics of timing and drawing.

White, Tony. *The Animator's Workbook.* New York: Watson-Guptill Publications, 1986. Essential cel-animation reference with extensive coverage of marking breakdowns for assistants and in-betweeners. Has an explanation of the shift-and-trace in-between system.

BOOKS ON THE HISTORY OF ANIMATION, STUDIOS, ANIMATORS

Adamson, Joe. *Tex Avery: King of Cartoons.* Da Capo, 1985. General information on the art of animation and on Avery's contributions to the field.

Canemaker, John. *The Animated Raggedy Ann & Andy: The Story Behind the Movie.* New York: Bobbs-Merrill, 1977. This book has value beyond the context of its development, which was the making of Richard Williams's feature-length animated film *Raggedy Ann and Andy.*

_____. *Before the Animation Begins: The Art and Lives of Disney Inspirational Sketch Artists.* New York: Hyperion, 1996.

_____. *Felix: The Twisted Tale of the World's Most Famous Cat.* Pantheon 1993/Da Capo 1996.

_____. *Storytelling in Animation: The Art of the Animated Image.* Los Angeles: AFI, 1988.

_____. *Tex Avery: The MGM Years, 1942–1955.* Turner, 1996. Probably hard to find, but large with full-color pictures.

_____. *Windsor McCay: His Life and Art.* New York: Abbeville Press, 1987.

Culhane, Shamus. *Talking Animals and Other People.* New York: St. Martin's Press, 1986. Autobiography that covers most of the major studios.

Finch, Christopher. *The Art of Walt Disney: From Mickey Mouse to the Magic Kingdoms.* New York: H. N. Abrams, 1973, 1995. Historical examination of the Walt Disney studio and productions. Information on Disneyworld and Disneyland. Bibliography.

Jones, Chuck. *Chuck Amuck: The Life and Times of an Animated Cartoonist.* New York: Farrar, Straus, Giroux, 1989. Autobiography with a little history and a lot of anecdotes, philosophy, and great art.

Koenig, David. *Mouse Under Glass: Secrets of Disney Animation.* Bonaventure Press, 1997. Well-researched behind-the-scenes details about Disney films and theme parks from *Snow White* to *The Hunchback of Notre Dame.*

Maltin, Leonard. *Of Mice and Magic: A History of American Animated Cartoons.* New York: New American Library/Plume, 1980. Best objective history of Hollywood animation.

Thomas, Frank and Ollie Johnston. *Disney Animation: The Illusion of Life.* New York: Abbeville, 1981. Out of print, but arguably the best book on drawn animation ever published.

BOOKS ON TOPICS RELATED TO ANIMATION

Blacker, Irwin R. *The Elements of Screenwriting: A Guide for Film and Television Writers.* Macmillan Publishing Co., 1996. Similar to Strunk & White's *Elements of Style* for English. Essential for film and television of all types.

Campbell, Joseph. *Hero With a Thousand Faces.* Princeton, NJ: Princeton University Press, 1968. ©1949. Also Fine Communications, 1996. This is a classic for screenwriters and other storytellers. Campbell's analysis of the world's great myths and epic stories provide insight into the kinds of worlds and characters that animation is so uniquely qualified to develop.

Edwards, Betty. *Drawing on the Right Side of the Brain.* Los Angeles, Calif.: Jeremy P. Tarcher, Inc., 1979, 1989, 1992. Teaches how to draw what you see and how your brain interprets visual images.

25.2 Windsor McCay: Animation historian and writer John Canemaker holds an original drawing from McCay's film *Gertie the Dinosaur* (1914). *Courtesy John Canemaker.*

Egri, Lajos. ***The Art of Creative Writing.*** New York: Citadel Press, 1965; reissued 1995.

_____. ***The Art of Dramatic Writing.*** Simon & Schuster, 1977. The bible for telling a structured and thoughtful story for any medium.

Gray, Milt. ***Cartoon Animation: Introduction to a Career.*** Northridge, Calif.: Lion's Den Publications, 1991. Points to lots of other good resource material.

Hoffer, Thomas W. ***Animation: A Reference Guide.*** Westport, Conn.: Greenwood Press, 1991. Lists books, articles, periodicals, research centers, and has a short history.

Muybridge, Eadweard. ***Animals in Motion*** and ***The Human Figure in Motion.*** New York: Dover, 1955 and 1956, respectively. The time-motion studies by Muybridge are a continuing source of fascination, appreciation, and instruction to animators. If we caught your interest with our book's cover, treat yourself to these volumes.

Nicolaides, Kimon. ***The Natural Way to Draw.*** Boston, MA: Houghton Mifflin Co., 1975, ©1941. Famous "how-to-draw" book recommended by people like Culhane and Jones.

Pintoff, Ernest. ***The Complete Guide to American Film Schools.*** Viking Books, 1994.

_____. ***The Complete Guide to Animation and Computer Graphics.*** Watson-Guptill Publications, 1995.

Walter, Richard. ***Screenwriting: The Art, Craft, and Business of Film and Television Writing.*** New York: New American Library/Plume, 1988. Enjoyable to read, just for fun.

PERIODICALS

The following periodicals often contain information of interest to those making, teaching, and/or studying animation. Sources of information for current articles on animation can also be found in *The Readers Guide to Periodical Literature* and film periodical indexes (*Film Literature Index, New Film Index, International Index to Film Periodicals,* and *The Critical Index*). It's also nice to know that sometimes there are student rates and other discounts offered on subscriptions. And often, articles and supplements to these magazines can be found on the Internet.

This section is divided into Animation Magazines, which is dedicated to those magazines that discuss animation in general; Computer Magazines, which lists those that offer a great deal of information on the newest technologies available to animators and others; and E-zines and the Internet, which lists additional animation resources found exclusively on the Web.

ANIMATION MAGAZINES

Animation Journal. AJ Press; 2011 Kingsboro Circle; Tustin, CA 92780-6733, http://www.chapman.edu/animation/. About $20 for 2 issues/yr. and index to back issues. Peer-reviewed scholarly journal devoted to animation history and theory from Chapman University School of Film and TV in Orange, CA.

Animation Magazine. Thoren Publications; 30101 Agoura Court, #110; Agoura Hills, CA 91301-4301. (818) 991-2884. Monthly, $36/yr. Created as a forum of the newest techniques and classic styles of the animation process for animators, afficionados, and general audiences to enjoy. Promotes the art and business of animation and gives recognition to those animators and technicians who make the world of animation what it is today.

Animation Report. P.O. Box 2215, Canoga Park, CA 91396 (818) 346-2782. 10 issues/yr. $250. Newsletter with current events, commentary, and analysis of the animation industry. Includes detailed reports on studios, articles on the economy of animation, and interviews with producers.

ANiMATO! 92 Thayer Road, Monson, MA 01057-9445. Quarterly, $18/yr. http://members.aol.com/thelmascum/animato.htm. Loaded with articles written by fans and industry people for animation fans; some articles are well-written, others not. Highly recommended.

Cinefex: The Journal of Cinematic Illusions. P.O. Box 20027, Riverside, CA 92516-0027. (909) 788-9828. Quarterly magazine devoted to cinematic special effects, which increasingly means computer animation. Provides insight into all motion picture illusions, from computer-enhanced effects to the trickery of the makeup department. Interviews with special effects creators are also included.

Daily Variety. Cahners Publishing Company, Inc.; 5700 Wilshire Blvd., #120, Los Angeles, CA 90036-3659. (213) 857-6600. Daily Mon-Fri, $187/yr. With special animation issues about every three months. Created for professionals and observers of the global entertainment business, such as film, television, video, cable, music, theater, etc. Includes coverage of financial, regulatory, and legal matters pertaining to entertainment.

Digital Video Magazine. Miller Freeman, Inc.; Harrison St., San Francisco, CA 94107. (415) 905-2200; or subscriptions: (888) 776-7002. $20/yr. http://www.livedv.com. Monthly magazine focusing on the needs of digital video and digital media creators, including animators.

Film & Video, The Production Magazine. Phillips Publishing International; 8455 Beverly Blvd., #508, Los Angeles, CA 90048-3416. (213) 653-8053. Monthly, free to industry members—request free copy to see criteria, $55. Written for producers of motion pictures, television programming, commercials, music videos, and multimedia. Designed to educate and inform readers of the changes, technologies, and emerging trends in the film, television, and interactive industries.

Hollywood Creative Directory (Film and TV Production Companies and Executives), ***Hollywood Distributor Directory*** (Distribution and Syndication Companies) and ***Hollywood Interactive Directory*** (Title Developers and Publishers). 3000 W. Olympic Blvd., Ste. 2525, Santa Monica, CA 90404. (310) 315-4815; (800) 815-0503; www.HollyVision.com. Semiannually, $50. Great resources for the most up-to-date info.

The Hollywood Reporter. Billboard Publications, Inc.; 5055 Wilshire Blvd., 6th floor, Los Angeles, CA 90036-4396. (213) 525-2000; http://www.hollywoodreporter.com. Daily Mon–Fri, $189/yr. Written for members of each segment of the international entertainment industry, from movies and television to legitimate theater and music. Focuses on emerging new media and corporate finance. Animation special issues throughout the year. Also publishes the ***Hollywood Reporter's Blu-Book Directory*** annually in January, which has listings for production companies and a multitude of resources, including those for animation.

The Independent Film and Video Monthly. Foundation for Independent Video and Film; 304 Hudson St., New York, NY 10013-1015. (212) 807-1400. http://www.virtualfilm.com/aivf. Monthly, $45/yr. ($25/yr. for students.) The official publication of FIVF. Covers new film and video releases from the smaller, sometimes lesser-known producers. Also includes profiles of independent film and video makers, festival listings, information on new media technologies, and legal and business news.

Millimeter. Intertec Publishing; 5 Penn Plaza, 13th Floor, New York, NY 10001. (212) 613-9700. Annual animation issue; information on new techniques and equipment.

Producer's Masterguide. 60 E. 8th St., 34th Fl., New York, NY 10003. (212) 777-4002. http://www.producers.masterguide.com. Annually, $125. Comprehensive reference for film, TV, and media, national and international production. Web site has table of contents, sample pages, and ordering information.

COMPUTER MAGAZINES

3D Artist is a printed how-to magazine for desktop 3-D graphics (PC, Mac, Amiga, Windows NT). The Web site supplements the news, resources, and articles that appear in the magazine. Most articles are written by the readers themselves. The target readership is freelance artists who use 3-D tools, and people whose goal is to become professional 3-D users. *3D Artist* appears at intervals of approximately six to eight weeks at $37/yr.

3D Artist Magazine. Columbine, Inc.; P.O. Box 4787; Santa Fe, NM 87502. (505) 982-3532; http://www.3dartist.com/

3D Design. P.O. Box 420432, Palm Coast, FL 32142-0432. (800) 829-2505. Monthly issues for $30/yr.

Computer Graphics World. 10 Tara Boulevard, 5th Floor, Nashua, NH 03062-280. (603) 891-0123; fax: 603-891-0539; www.cgw.com; dru@pennwell.com. 3D, CGIs, Design Products, Information and Reviews. A graphics industry magazine. Product reviews, industry news. Annual subscription of 12 issues, $40.

Digital Magic. Monthly. P.O. Box 122; Tulsa, OK 74101. (918) 832-9257; fax: (918) 831-9497. Info on digital technology, reviews, product information and reviews, 3-D effects animation and more.

MacUser. Monthly. P.O. Box 56986; Boulder, CO 80322-6986. (415) 547-8600. Macintosh products, review and help.

MacWorld. Monthly. P.O. Box 54529; Boulder, CO 80328-3429. (800) 288-6848. Macintosh products and reviews.

PC Computing. P.O. Box 58229, Boulder, CO 80322-8229. (800) 676-4722; www.pccomputing.com. Monthly, $25/yr. Another source for info on PC hardware and software.

PC Magazine. P.O. Box 54093, Boulder, CO 80322-4093. (303) 665-8930; www.pcmag.com; $50/yr. (22 issues). Great resources for the latest in PC hardware and software.

Video and Multimedia Producer. 701 Westchester Ave., White Plains, NY 10604. Monthly; $53/yr. Info on digital technology, video, film, broadcast and game production, 3-D animation, reviews, Internet, new products.

The Visual Computer, Computer Graphics Society (CGS); http://www-ci.u-aizu.ac.jp/VisualComputer/k-myszk@u-aizu.ac.jp. A good graphics journal.

E-ZINES AND INTERNET SITES

3D Site
www.3dsite.com
Dedicated to 3-D computer graphics.

Animation and Cartoon Heaven
http://www.geocities.com/Hollywood/Hills/1684/
Popular Web site includes news, features, and reviews of interest to the animation community and fans.

Animation World Magazine
http://www.awn.com/mag/index.phtml
6525 Sunset Blvd., Garden Suite 10, Hollywood, CA 90028.
(213) 468-2554.
Animation World Network's monthly Internet magazine of animation has current news, features, profiles, software, reviews, festivals, conferences, and employment opportunities.

Animejin
http://www.animejin.demon.co.uk/
The UK's on-line anime fanzine. Back issues were printed—now available only on web.

Entertainment Directory
http://www.edweb.com
Contains Yellow Pages. Excellent for searching for animation resources all over the world.

NYU Animation Station
http://www.nyu.edu/tisch/filmtv/animation/station.html
All kinds of information for animators: books, films, schools, organizations, festivals, and more!

SIGGRAPH
http://www.siggraph.org
International ACM Conference on Computer Graphics and Interactive Techniques. Contains informative articles from past SIGGRAPH conferences.

Visual Magic Magazine
http://www.visualmagic.awn.com
Devoted to visual affects and 3-D animation, including news and information on the tools and techniques of visual imagery.

Warner Brothers Animation
http://www.wbanimation.com
Offers a step-by-step look at the Warner Brothers process of creating animated series, from beginning to end.

ORGANIZATIONS

The following are national and international organizations that support the art of animation. Some of these are devoted to animated filmmaking and scholarship and some maintain a strong interest in animation within a broader context of support for all forms of the moving image media.

Do not overlook organizations at the local level that can provide important resources for people interested in animation. Schools and colleges with film-production programs are usually a center for animation interest. Museums, libraries, and special film screening centers exist in many cities. State arts councils are important supporters of the film arts as well as other art forms, and strong regional film study and film production centers are developing throughout the United States.

AMERICAN FILM INSTITUTE (AFI)

AFI is an independent, nonprofit organization established in 1967 by the National Endowment for the Arts. AFI preserves films, operates an advanced conservatory for filmmakers, gives assistance to new American filmmakers through grants and internships, provides guidance to film teachers and educators, publishes film books, periodicals, and reference works, supports basic research, and operates a national film repertory exhibition program.

Main Office:
2021 N. Western Ave.
Los Angeles, CA 90027
(213) 856-7600
http://www.afionline.org

INTERNATIONAL ASSOCIATION OF ANIMATED FILMMAKERS (ASIFA)

ASIFA is an international animated film association composed of the leading animation artists from more than fifty countries. The organization sponsors annual festivals of international animation and has participated in the publication of a series of books on animation techniques. Membership/subscription is $18 a year or $40 for both local and international membership.

ASIFA International
c/o Borivoj Dovnikovic
Secretary General, ASIFA
Hrvatskog Proljessa 36
41040 Zagreb, Croatia

ASIFA Canada
Case postale 5226
St.-Laurent, Quebec
Canada H4L 4Z8
(514) 842-9763

ASIFA Canada/Vancouver
c/o Leslie Bishko
Graphics and Multi-Media Research Lab
Centre for Systems Science
Simon Fraser University
Burnaby, B.C.
Canada V5A 1S6
(604) 291-3610
http://fas.sfu.ca/cs/people/GradStudents/bishko/personal/anim/asifa.html

ASIFA Central
Deanna Morse—Membership Secretary
School of Communications
Lake Superior Hall
Grand Valley State University
Allendale, MI 49402
(616) 895-3101
http://www.swcp.com/~asifa or
http://www.asifa.org/~asifa/

ASIFA East
c/o Michael Sporn Animation
Jim Petropoulos
632 Broadway, 4th Floor
New York, NY 10012
(212) 966-8887 x110
http://www.yrd.com/ASIFA/

ASIFA Hollywood
725 S. Victory Blvd.
Burbank, CA 91502
(818) 842-8330
(818) 842-5645 fax
http://www.awn.com/asifa_hollywood/index.html

ASIFA San Francisco
P.O. Box 14516
San Francisco, CA 94114
http://www.awn.com/asifa-sf/

Association of Independent Video and Film Makers (AIVF)
$45 for membership, which includes *The Independent* monthly

304 Hudson Street, 6th Floor
New York, NY 10013
(212) 807-1400
(212) 463-8519 fax
http://www.virtualfilm.com/AIVF

Educational Film Library Association, Inc. (EFLA)
Nadine Covert (212) 246-4533
A membership service that includes individuals and institutions other than libraries. Provides information services, film evaluations, library. Annually coordinates the American Film Festival.

International Coordinating Bureau of the Institutes of Animation (BILIFA)
Robert J. Edmonds (312) 663-1600
An organization of U.S. and foreign animation schools. Encourages an exchange of curricula, student films, and general information. High schools, colleges, and universities are eligible for membership.

Motion Picture Screen Cartoonists, Local 839 IATSE
4729 Lankershim Blvd.
North Hollywood, CA 91602-1864
(818) 766-7151
http://www.awn.com/MPSC839/839INDEX.HTM

National Film Board of Canada
1251 Avenue of the Americas
New York, NY 10020
(212) 596-1770
www.nfb.ca
Search engine for their titles.

Main Office:
Operational Headquarters
Norman McLaren Bldg.
3155 Còte de Liesse Road
St. Laurent, Quebec H4N 2N4
(514) 283-9000 or (800) 267-7710

Quickdraw Animation Society (QAS)
A nonprofit, artist-run film production co-op dedicated to the production of independent animation and the appreciation of all types of animation. To this end, they offer Free Film Nights, Animation Workshops, and courses in Calgary.

#300, 209-8 Ave. SW
Calgary, Alberta, Canada T2P 1B8
(403) 261-5767
http://www.awn.com/qas

Society for Animation Studies
Supports animation scholarship in a variety of ways, including an annual conference, special exhibits, and a newsletter published five times a year informing members of animation news and printing reviews of current books in the field. Fosters research and writing focused on animation topics, with several competitions.

Richard J. Leskosky, Pres.
University of Illinois
2117 Foreign Language Bldg.
707 S. Matthews Ave.
Urbana, Illinois 61801
http://www.awn.com/sas

ANIMATION SCHOOLS

Programs that teach animation will—and should—be constantly revising their approaches and personnel. And new programs are popping up all the time. The listing here will provide a starting point, but for more current information, check out the Worldwide Directory of Film/Video/Communications Schools and Programs, which is available at the UFVA Web site at http://www.temple.edu/ufva/index.html, the Animation World Network's Animation School Directory at http://www.awn.com, and ASIFA's List of Animation Schools available on-line at http://asifa.hivolda.no or via mail from Gunnar Strom at More & Romsdaal College, P.O. Box 188, N-6101, Volda, Norway. Also check out *The Complete Guide to Animation and Computer Graphics* and *The Complete Guide to American Film Schools,* both by Ernest Pintoff.

American Animation Institute
4729 Lankershim Blvd.
N. Hollywood, CA 91602
(818) 766-0521

California Institute of the Arts
24700 McBean Parkway
Valencia, CA 91355
(800) 545-ARTS (Nationwide)
(800) 292-ARTS (California)

Columbia College Chicago
600 S. Michigan Ave.
Chicago, IL 60605
(312) 663-1600 Ext. 367

Computer Arts Institute
310 Townsend St.
Suite 230
San Fransisco, CA 94107-1607
(415) 546-5242

MIT Media Laboratory
20 Ames Street
Cambridge, MA 02139
(617) 253-5114

New York University
Institute of Film and TV Animation Area
721 Broadway, 8th Floor
New York City, NY 10003
(212) 998-1781

Northwestern University
1905 N. Sheridan Road
Evanston, IL 60208
(312) 489-4191

Pratt Institute
200 Willoughby Avenue
Brooklyn, NY 11205
(718) 636-3633

Rhode Island School of Design
Film/Animation/Video Dept.
2 College Street
Providence, RI 02903
(401) 454-6300

Ringling School of Art & Design
2700 N. Tamiami Trail
Sarasota, FL 34234
(941) 351-5100

Rochester Institute of Technology
One Lomb Memorial Drive
Rochester, NY 14623-0887
(716) 475-2411

Rocky Mountain College of Art and Design
6875 E. Evans Avenue
Denver, CO 80224-2359
(303) 753-6046

San Francisco Art Institute
800 Chestnut
San Francisco, CA 94133
(510) 771-7020

San Francisco State University
Dept. of Cinema
1600 Holloway Avenue
San Francisco, CA 94132
(415) 338-1649

Savannah College of the Arts
548 E. Broughton St.
Savannah, GA 31410
(912) 238-2400

School of the Museum of Fine Arts
230 The Fenway
Boston, MA 02115
(617) 267-1218

School of Visual Arts
209 E. 23rd Street
New York, NY 10010-3994
(212) 592-2000

Sheridan College
Visual Arts Dept./Classical Animation
1430 Trafalgar Road
Oakville, Ontario
L6H 2L1, Canada
(416) 845-9430

UCLA Animation Workshop
Dept. of Film and TV
405 Hilgard Avenue
Los Angeles, CA 90024
(310) 825-5829

University of the Arts
320 S. Broad St.
Philadelphia, PA 19102
(215) 875-1020

USC School of Cinema
Television Animation Dept.
University Park
Los Angeles, CA 90089
(213) 743-3205

SUPPLIERS

Animation Suppliers

Below are a few suppliers of animation tools that will ship throughout the United States and Canada. It is best to try obtaining your supplies through local resources, so check out your Yellow Pages and contact art and animation schools or organizations in your area. A local dealer with expertise is often your best resource, since they are convenient, and usually eager to help someone who may become a loyal customer.

Alan Gordon Enterprises, Inc.
1430 N. Cahuenga Blvd.
Hollywood, CA 90028
(213) 466-3561
(213) 871-2193 fax
www.A-G-E.com
—registration devices, production equipment (lights, cameras, audio, etc.), drawing tables, etc.

Cartoon Colour Co. Inc.
9024 Lindblade St.
Culver City, CA 90203-2584
(310) 838-8467; or (800) 523-3665
(310) 838-2531 fax
—animation paints and art supplies, registration devices, books on animation, etc.

Norris Film Products
1014 Green Lane
La Canada, CA 91011
(818) 790-1907
(818) 790-1920 fax

Oxberry
180 Broad St.
Carlstadt, NJ 07072
(201) 935-3000
(201) 935-0104 fax
—cameras, animation stands, supplies

Software and Hardware Suppliers

The following is a list of software publishers for programs related to animation, most of which are mentioned in this volume. The list for companies producing hardware is so vast that is better to start your research by browsing dedicated computer magazines for the most current product and pricing information. It is often cheapest and most convenient to purchase computer hardware and software products through mail-order companies like PC and Mac Connection (800-800-0005) and MacWarehouse (800-255-6227), especially if you know exactly what you want.

Adobe
P.O. Box 1034, Buffalo, NY 14240-1034
Customer Support: (800) 833-6687
http://www.adobe.com
Products: Photoshop, Premiere, Illustrator, After Effects

Allegiant Technologies, Inc.
9740 Scranton Road, Suite 300, San Diego, CA 92121
(619) 587-0500
http://www.allegiant.com
Products: SuperCard

Apple Computer, Inc.
1 Infinite Loop, Cupertino, CA 95014
(408) 996-1010
http://www.apple.com
Products: HyperCard, Quick Time

Claris Corporation (FileMaker, Inc.)
General Info: (800) 544-8554
http://www.claris.com; www.filemaker.com
Products: FileMaker Pro

Corel
1600 Carling Avenue, Ottawa, Ontario, K1Z 8R7
(800) 772-6735
http://www.corel.com
Products: Lumiere

Deneba
7400 SW 87th Ave., Miami, Florida 33173
(305) 596-5644
http://www.deneba.com
Products: Canvas

FormZ
(Division of Autodesk)
(614) 488-8838
http://www.formz.com

Fractal Design
Orders: (800) 846-0111
http://www.fractal.com
Products: Painter, Ray Dream Studio, Ray Dream Designer

Hash Inc.
2800 East Evergreen Blvd., Vancouver, WA 98661
(360) 750-0042
http://www.hash.com
Products: Animation Master

Kinetix
(Division of Autodesk)
642 Harrison St., San Francisco, CA 94107
(415) 547-2000
http://www.ktx.com
Products: 3D Studio MAX

Linker Systems
13612 Onkayha Circle, Irvine, CA 92620
(714) 552-1904
http://www.linker.com
Products: Animation Stand

Macromedia
600 Townsend Street, San Francisco, CA 94103
(415) 252-2000
Sales: (800) 457-1774
http://www.macromedia.com
Products: SoundEdit 16, Director, Extreme 3D, Flash, Free-Hand, SmartSketch

Opcode
3950 Fabian Way, Suite 100, Palo Alto, CA 94303
(650) 856-3333
http://www.opcode.com
Products: Audioshop, Studio Vision Pro

Softimage
3510 St. Laurent, Suite 400, Montreal, Quebec
(514) 845-8252
http://www.softimage.com
Products: Softimage 3D, Softimage 3D Extreme

Specular
7 Pomeroy Lane, Amherst, MA 01002
(413) 253-3100
Orders and Info: (800) 433-SPEC
http://www.specular.com
Products: Infini-D

Strata
2nd West St. George Blvd., St. George, Utah 84770
(801) 628-5218
http://www.strata3d.com
Products: Studio Pro

Toon Boom Technologies Inc.
3601 W. Alameda Ave., #201, Burbank, CA 91505
(818) 954-8666
http://www.toonboom.com
Products: TicTacToon, USAnimation

25.3 Cartoon virtuosity: A master animator's art is not simply to make a series of drawings move but also to make them move with recognizable elements of character, emotion, and life. This series of key drawings is reproduced from *Animation—How to Draw Animated Cartoons* by Preston Blair.

• DISTRIBUTION •

Written with Jan Cox

Now that you've made a film, what do you do with it? And where do you go to see other independently made animation? Let's tackle this by defining the categories of distribution. First you must realize that there are several types of rights associated with your film. In general, they are theatrical, television, and video rights.

THEATRICAL DISTRIBUTION

A big screen in a dark theater with an excellent sound system will best highlight your efforts, or magnify a film's flaws. For this reason, one hopes you have spent some time thinking about the production quality of your film. There are two paths to follow to theatrical distribution, and you should pursue both of them.

The Festival Circuit. Film festivals are basically contests, which offer awards and prizes, and have rules of entry. Typically, a jury of animation experts will judge the film. The rules of the festival may limit the entries to student works or first films, films made since a particular date, or films with a theme. There are many, many festivals all over the world, so you should pick and choose which to be involved in. The amount and/or type of prize may help you decide. Usually, a monetary prize is awarded, but sometimes, the prize is simply the honor of having won. In some cases, just your ticket to attend the festival will be gratis, and in others, your travel and lodging expenses will be covered. Contact ASIFA for their list of sanctioned festivals, and search the Internet and entertainment industry magazines for announcements. New festivals pop up like weeds, so you may want to check out the awards offered before you leap. These festivals will have entry fees, and you must supply the print or video of your film and pay for shipping. This can be an expensive hobby if your film isn't in the running for an award. However, the exposure gained may help your film progress to the next type of distribution: licensing.

Theatrical Licensing. There are entertainment companies that may be interested in acquiring the rights to your film for the purpose of distribution in theaters. Rather than pay to have your film shown, you will get paid. This may sound great, but there are pitfalls to watch out for. You should ask about the ability of the entertainment company to distribute your film. How many theaters will it go to, in how many cities? How will the company advertise your film? Does the company have a track record of successful theatrical distribution? If so, have you personally seen any films they have distributed? These are all questions you should ask them. You should always retain permission to circulate your film in competitive film festivals for awards. And if the company buys the rights to your film, and neglects to use them within a short time, you should get the rights back. Otherwise, your hard work is legally tied up, and no one gets the opportunity to see it. The bottom line is, go for the widest possible distribution you can find. To find these companies, look in industry directories and film festival guidebooks, and ask other animators.

TELEVISION DISTRIBUTION

With the advent of cable and satellite television, the outlets for animation have widened considerably. Consult a library for *Bacon's Media Guide—Televi-*

sion. Depending on the theme of your film and the intended audience, you could match your film to many types of programming, especially if it is a short film. Another option is to sell the television rights to an entertainment company, who then sells your film and pays you a royalty or a fee each time it is sold. As before, check the track record before you assign your television rights. Be aware, you are NOT giving up your character rights, just the rights to screen one particular film. In addition to the actual film, the character you have invented is a commodity. You may sell the character rights separately, or you may want to develop more stories involving the same character.

VIDEO DISTRIBUTION

This area of distribution is where clout and connections are most important. It does you no good to have a video produced if the public can't find it, so make sure that whomever you assign the video rights to has adequate distribution. That should include shelf space in major video chains. And that old bug again: What is their track record? Don't worry that it will be your only chance. Once you give up the rights, you can't go back when a better offer comes along. In the meantime, you could try to sell a video of your film or compilation all by yourself. Contact a duplication company (your cost will be about $4 per tape for a small quantity) and spread the word among friends and family. Place ads in local 'zines and see if your local video store will carry some copies. In the meantime, there are many companies who distribute videos, and eventually you'll find one who wants to add yours to their product line. One of the most comprehensive animation catalogues is *The Whole Toon Catalog* (800-331-6197), which also lists some of the production companies of the many videos they sell. Another type of research you should do is to simply walk into your local video store (sales, not rental) and see who has space on the shelf in the animation department.

WORDS OF WISDOM

There are animators all over the world, doing a wide variety of projects for different purposes. Explore via trade magazines and the Internet. A great place to start is at www.awn.com (Animation World Network), which is a complete source for anything to do with animation. Attend film festivals and meet the people who are showing or producing films. Join an ASIFA chapter and/or Women in Animation (818-759-9596). Write letters or call people you'd like to meet. Keep taking classes. It helps you grow and meet people AND it helps the educational programs to stay alive. Most people in animation are incredibly nice, and networking is crucial to your artistic development and to learn about career opportunities. Another resource is MPSC, the Motion Picture Screen Cartoonists (818-766-7151). And if you get to the point where you need an agent, find one who specializes in animation. One place to start is Animanagement (805-260-2950). Ask other animators for references. The bottom line is, get connected!

DISTRIBUTORS OF ANIMATION

The following is a selected list of distributors of animation. For specific titles and current prices, request the distributor's catalogue. For the most recent listings and information, contact the resources mentioned earlier.

Alice Entertainment
1539 Sawtelle Blvd., Suite 4
Los Angeles, CA 90025
(310) 231-1050

American Federation of Arts
41 East 65th Street
New York, NY 10021
(212) 988-7700

Canadian Filmmakers Distribution Centre
37 Hanna Ave., Suite 220
Toronto, Ontario
Canada M6K 1 W8
(416) 588-0725
http://www.cfmdc.org

Carousel Films, Inc.
260 Fifth Ave., Suite 405
New York, NY 10001
(800) 683-1660
specialization: multicultural, social issues

Enold Films, USA, Inc.
16501 Ventura Blvd., Suite 306
Encino, CA 91436
(818) 907-6503

Expanded Entertainment
30101 Agoura Court, Suite 110
Agoura Hills, CA 91301
(818) 991-2884

The Filmmaker's Cooperative
175 Lexington Ave.
New York, NY 10016
(212) 889-3820

Film Roman, Inc.
12020 Chandler Blvd., Suite 200
North Hollywood, CA 91607
(818) 761-2544

FilmWright
231 State Street
San Francisco, CA 94114
(415) 864-5779

INI Entertainment Group
11845 Olympic Blvd.
Suite 1145W
Los Angeles, CA 90064
(310) 479-6755

International Film Bureau
332 South Michigan Ave.
Chicago, IL 60604
(312) 427-4545

Lee Mendelson Film Prods.
330 Primrose Road., Suite 310
Burlingame, CA 94010
(415) 342-8284

Link Television Entertainment, Inc.
14333 Addison St., Suite 311
Sherman Oaks, CA 91423
(818) 981-8980

Manga Entertainment, Inc.
964 5th Ave., Suite 330
San Diego, CA 92101
(619) 531-1695
http://www.manga.com

Mellow Manor (aka Spike & Mike)
7488 La Jolla Blvd.
La Jolla, CA 92037
(619) 459-8707
http://www.spikeandmike.com

Moon Mesa Media
P.O. Box 7848
Northridge, CA 91327
(818) 360-6224

M3D Studios, Inc.
15820 Arminta St.
Van Nuys, CA 91406
(818) 785-6662

Nelvana Communications
4500 Wilshire Blvd., 1st Floor
Los Angeles, CA 90010
(416) 588-5571

Nest Entertainment
6100 Colwell Blvd.
Irving, TX 75039
(212) 402-7100

New Yorker Films
16 West 61st Street
New York, NY 10023
(212) 247-6100

Pulse Entertainment
2444 Wilshire Blvd., Suite 303
Santa Monica, CA 90403
(310) 264-8320

Pyramid Film and Video
2801 Colorado Ave.
Santa Monica, CA 90404
(310) 828-7577; (800) 421-2304
http://www.pyramidmedia.com

Rembrandt Films
Ballyhack Rd.
Brewster, NY 10509
(914) 279-4158

Silverline Pictures
11846 Ventura Blvd., Suite 100
Studio City, CA 91604
(818) 752-3730

Whamo Entertainment
1850 S. Sepulveda Blvd., Suite 201
Los Angeles, CA 90025
(310) 477-0338

Yellow Ball Workshop
62 Tarbell Ave.
Lexington, MA 02173
(617) 862-4283

FESTIVALS

The following is a select list of U.S. and foreign festivals either entirely devoted to animated films or with special categories for animation. Unless otherwise noted, these are annual events. Contact the individual festival you are considering for entry rules and firm deadlines. They often require the entries to be in a particular format (16mm, 35mm, video, etc.) and of a particular length. Also check out the UFVA Web site at http://www.temple.edu/ufva/ for the UFVA International Festival Directory for Students, and the Directory of On-Line Festivals at http://www.awn.com/awneng/fest-av.html

U.S. Festivals

Aspen Filmfest
July deadline; September awards; $30–$40 entry fee
Tel: (970) 925-6882
http://www.aspen.com/aspenonline/directory/mac/sponsors/artscouncil/sponsors/filmfest/index.html
Formats: 16mm, 35mm

Chicago International Film Festival
July deadline; October awards; $35–$250 fee.
Tel: (793) 644-3400
http://www.chicago.ddbn.com/filmfest
Formats: 16mm, 35mm, 3/4", VHS

Chicago Underground Film Fest
May deadline; August awards; $25–$35 entry fee
Tel: (312) 866-8660
http://www.cuff.org
Formats: Film, Video

Cine Golden Eagle Film & Video Comp.
February & August deadlines (semiannual); $100
Tel: (202) 785-1136

D.FILM Digital Film Festival
Touring festival of short films using technology in new ways, including digital/desktop video, nonlinear editing, 2-D and 3-D Animation, digital cameras, etc.
The entry deadline is ongoing and there is no entry fee.
Tel: (415) 541-5683
http://www.dfilm.com

Honolulu Underground Film Festival
September deadline; November awards; $20 entry fee
Tel: (808) 735-2242
http://www.lava.net/huff

Los Angeles International Animation Competition
November deadline; March awards; $75 entry fee
Tel: (818) 991-2884

New York Underground Film and Video Fest
January deadline; March awards; $30 entry fee
Tel: (212) 925-3440
http://www.nyuff.com

Palm Springs International Shortfilm Fest
June/July deadline; July/August awards; $25 entry fee
Tel: (619) 322-2930

San Francisco Int'l Film Fest
April awards
Tel: (415) 929-5016

Sinking Creek Film Fest
June deadline; November awards; $30–$60 entry fee
Tel: (615) 322-4234
Formats: 35mm, 16mm, 3/4", Beta, VHS

Slam Dance International Film Fest
October deadline; January awards
Tel: (310) 204-7977

South By Southwest
December deadline; March awards; $20 entry fee
Tel: (512) 467-7979
http://www.sxsw.com
Formats: VHS only

Spike and Mike's Festival of Animation
The Festival plays in over thirty cities across the U.S. and Canada, along with the "Sick and Twisted" for more mature audiences.
No official deadline
Tel: (619) 459-8707
http://www.spikeandmike.com/
Formats: Pencil test or demo reel on VHS

The UFVA Student Film and Festival
National Tour of selected work
Tel: (800) UFVA or (215) 923-3532
http://thunder.ocis.temple.edu/~ddoyon

World Animation Celebration
November deadline, February awards
Tel: (818) 991-2884
http://www.wacfest.com

Overseas Festivals

Annecy International Film Festival and Market, France
January deadline; May/June awards
6 avenue des Iles—BP 399
74013 Annecy Cedex—France
Tel: 011-33-04-50-57-41-72
Fax: 011-33-04-50-67-81-95
Format: 16mm, 35mm

Bilbao Int'l Fest of Doc. and Short Film, Spain
September deadline; November awards; no entry fee
Tel: 011-34-4-4245507
Fax: 011-34-4-424-5624

Bradford Animation Festival
May deadline; June awards
Tel: 011-44-1274-725347

Cardiff International Animation Festival
June awards
Tel: 011-44-171-494-0506
Fax: 011-44-171-494-0807
http://www.vital-animation.org

Drambuie Edinburgh Film Festival
May deadline; August awards; no entry fee
Tel: 011-44-312284051
Fax: 011-44-312295501
88 Lothian Road
Edinburgh, EH3 9BZ, Scotland
(131)2284051 (Fax) (131)2295501
http://www.edfilmfest.org.uk/
Formats: 16mm, 35mm

East Kilbride Animation Fest, Scotland
January deadline; March awards; no entry fee
Tel: 011-353-1-679-2937
Fax: 011-353-1-679-2939

Fest Int'l de Cinema Fantastic Sitges
July deadline; October awards
Tel: 011-343-415-3938
Fax: 011-343-237-6521

Festival du Dessin Anime, Belgium
November deadline; February awards; no entry fee
Tel: 011-32-2-534-4125
Fax: 011-32-2-534-2279

Hiroshima International Animation Festival, Japan
March deadline; August awards; no entry fee
Tel: 011-81-82-245-0245

Holland Animation Film Festival
October deadline; November awards
Tel: 011-31-0302331733
Fax: 011-31-030-2331079

Internationales Trick-Film Stuttgart
December deadline; March awards; no entry fee
Tel: 011-49-711-262-2699
Fax: 011-49-711-2624980
http://www.itfs.de

London Effects and Animation Festival, UK
November awards
Digital Media International
10 Barley mow Passage, Chiswick, London W4 4PH
Tel: 011-44-0-181-995-3632
Fax: 011-44-0-181-995-3633
http://www.digmedia.co.uk

Montreal International Short Film Festival
December deadline; April awards; $30 entry fee
Tel: (514) 285-4515
Fax: (514) 285-2886
4205 Rue St-Denis, Bureau 326
Montreal, Quebec, H2J 2K9, Canada
Formats: 35mm, 16mm, Super 8

Ottawa International Animated Film Festival
Canadian Film Institute
2 Daly Ave., Ottawa, Ontario, Canada K1N 6E2
Summer deadline; fall awards; no entry fee
Tel: (613) 232-8769
Fax: (613) 232-6315
http://www.awn.com/oiaf

ZAGREB, Croatia
February deadline; June awards; no entry fee
Koncertna Direkcija Zagreb
Kneza Mislava 18 10000 Zagreb, Croatia, Europe
Tel: (385 1) 46 11 808/ 46 11 709
Fax: (385 1) 46 11 808/ 46 11 807

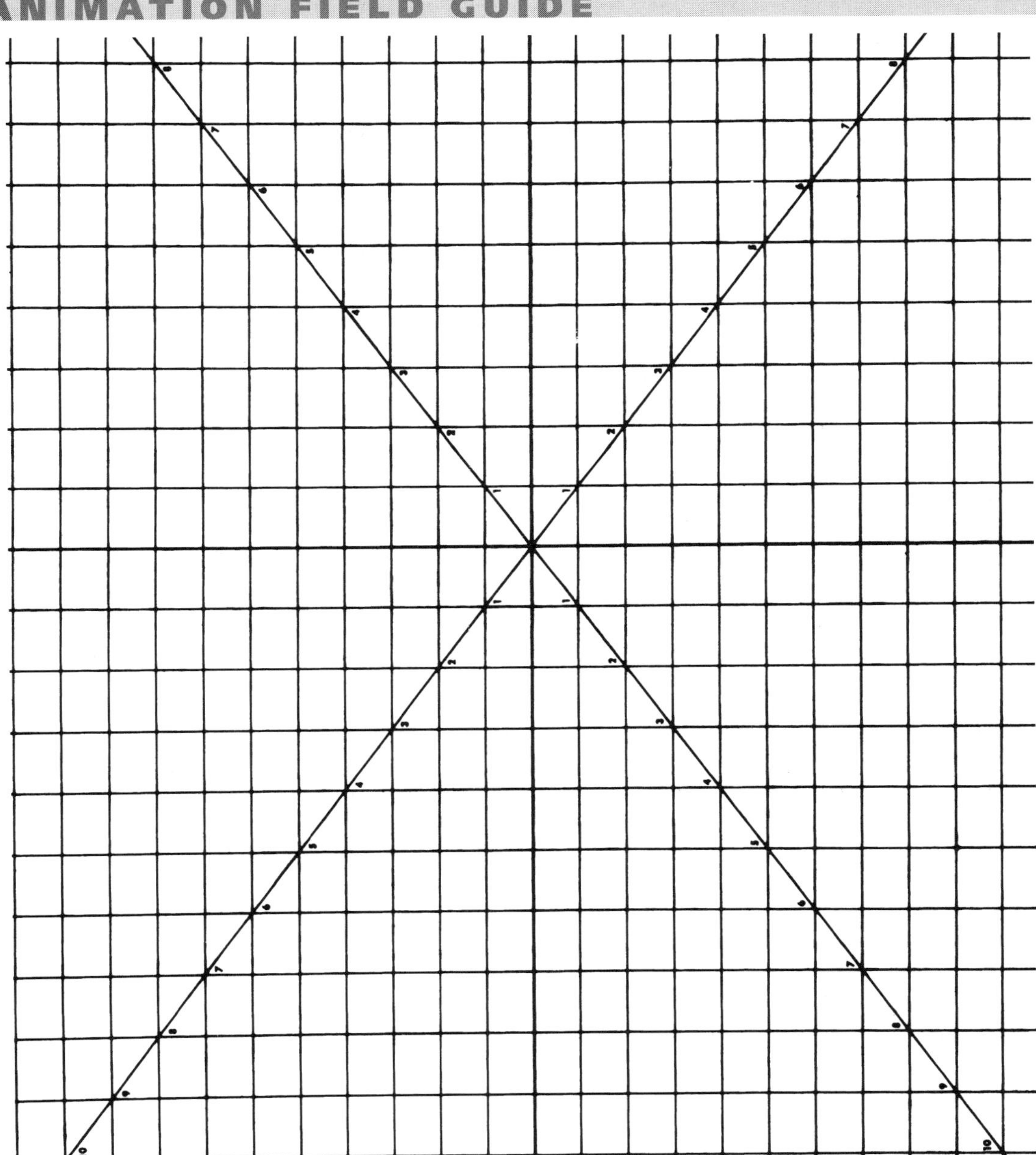

25.4 16MM Animation field guide: Reproduced in full scale, this is a portion of the standard field guide. Full-sized guides are printed on thick acetate, cover a 12 field (12 inches wide), and come with professional peg holes cut into them (either Oxberry or Acme registration systems are available). This guide can be modified for your own use if you punch it to fit the pegging system that you use. Note that a test must be done to make certain that the center of the field guide corresponds to the center of the camera's frame.

EXPOSURE SHEET

SCENE	TITLE	ANIMATOR	FOOTAGE	SHEET NO.

ACTION	DIAL	4	3	2	1	BKG	DIAL	CAMERA INSTRUCTIONS
	1						1	
	2						2	
	3						3	
	4						4	
	5						5	
	6						6	
	7						7	
	8						8	
	9						9	
	0						0	
	1						1	
	2						2	
	3						3	
	4						4	
	5						5	
	6						6	
	7						7	
	8						8	
	9						9	
	0						0	
	1						1	
	2						2	
	3						3	
	4						4	
	5						5	
	6						6	
	7						7	
	8						8	
	9						9	
	0						0	
	1						1	
	2						2	
	3						3	
	4						4	
	5						5	
	6						6	
	7						7	
	8						8	
	9						9	
	0						0	
	1						1	
	2						2	
	3						3	
	4						4	
	5						5	
	6						6	
	7						7	
	8						8	
	9						9	
	0						0	

25.5 Exposure sheet: This sheet can be duplicated so that you will have enough copies to record every frame in an animated film. Regardless of technique, an exposure sheet should be employed whenever one is working with a pre-recorded sound track.

A

B

25.6 Kinda-mation™: George Evelyn is a great wit as well as a wonderful character animator from (Colossal) Pictures. Along with the accompanying pictures, George offers these . . .

Semi-Scholarly Remarks about Kinda-Mation™

"Adapting one's vision to a budget, schedule, or production reality is a time-honored human endeavor, long employed in the practical realm of graphic designers, illustrators, engineers, architects, and so on. Kinda-Mation™ means applying this point-of-view to animation, which, by it's very nature (frame-by-frame generation of a motion picture), always has the potential of becoming grueling and complicated.

"In addition to the traditional parameters of time and money, some animators have new technical aspects to consider: things like the narrow bandwidth of the Internet and available memory (on CD-ROMs, hard drives, etc.) that affect design and frame rate. At the same time, with the advent of desktop cinema, lots of new production tools are available to the animation artist. Each new program or bit of machinery offers both opportunities and idiosyncratic limitations as far as the "look and feel" of the finished work is concerned.

"The practitioners of Kinda-Mation™ might be considered to be part of the older tradition of independent artists. These dedicated souls work alone to produce personal animated films for an audience far smaller than that of the "industrial strength" studio animation one sees in feature films or TV series. In so doing, their work helps broaden the public's perception of what animation is really about. This makes possible an ever more sophisticated audience who can learn to appreciate the wider spectrum of stories and worldviews, in a process which has been described as *the democratization of animation.*"

And now to the pix. *(A)* samples George Evelyn's rollicking but minimalist style as used in a music video produced by Walt Disney Records for the feature *George of the Jungle* (which was inspired by Jay Ward's original Saturday morning cartoon). *Courtesy Jay Ward Productions, Inc.* The frame from *Dr. Zum, (B),* was also designed by George Evelyn for MTV. *Courtesy MTV's* Liquid Television. Both *(A)* and *(B)* were animated by Vargo Studios in Budapest, Hungary. It is rumored that this new branch of animation is deeply suited to the Eastern Bloc aesthetics.

If you are feeling overwhelmed by the realities of animation—limits of budget and technology and time and clients—try adopting the semi-Zen, semi-fatalistic, semi-bemused perspective that comes from working within the constraints. Don't freak out. Forget animation and, like George, surrender yourself to Kinda-Mation.™ Incidentally, the phrase itself is a trademarked term devised, with tongue in cheek, by George Evelyn and Drew Takahashi at (Colossal) Pictures. Some years back, they did the same thing when describing the stylistic range of Colossal's animation as Art-That-Moves.™

Index

ABOUT THE AUTHOR

Kit Laybourne is a highly successful independent producer who has worked on such television series as MTV's Emmy-winning animation series *Liquid Television* and Nickelodeon's *Eureeka's Castle* and *Gullah Gullah Island.* During the 1995–96 academic year, he taught the Advanced Animation Seminar at NYU's Tisch School of the Arts. His full-time gig is Chief Creative Director at Oxygen, a new media company developing interactive content for the convergence of TV and computers. He is also Executive Producer of *Hank the Cowdog*, an animated feature film and TV series being developed by Nickelodeon.